The
CREATION
of
SCIENTIFIC
EFFECTS

JED Z. BUCHWALD

∴

The
CREATION of
SCIENTIFIC
EFFECTS

Heinrich Hertz and Electric Waves

The University of
Chicago Press
Chicago and London

Jed Z. Buchwald is the Bern Dibner Professor of the History of Science at MIT and director of the Dibner Institute for the History of Science and Technology, which is based at
MIT. He is the author of two books, both published by the University of Chicago Press:
*From Maxwell to Microphysics: Aspects of Electromagnetic Theory in the Last Quarter of the Nine-
teenth Century* (1985) and *The Rise of the Wave Theory of Light: Aspects of Optical Theory and Experiment in the First Third of the Nineteenth Century* (1989).

The University of Chicago Press, Chicago 60637
The University of Chicago Press, Ltd., London
© 1994 by The University of Chicago
All rights reserved. Published 1994
Printed and bound by CPI Group (UK) Ltd, Croydon, CR0 4YY

03 02 01 00 99 98 97 96 95 94   1 2 3 4 5

ISBN: 978 0 226 07887 8 (cloth)
       978 0 226 07888 5 (paper)

Library of Congress Cataloging-in-Publication Data
Buchwald, Jed Z.
    The creation of scientific effects : Heinrich Hertz and electric waves / Jed Z. Buchwald.
        p.   cm.
    Includes bibliographical references and index.
    1. Electric waves.   2. Hertz, Heinrich, 1857–1894.   3. Physicists—Germany.   I. Title.
    QC661.B85   1994
    537—dc20   93-41783
                                                                                      CIP

CONTENTS

FIGURES

TABLES

PREFACE

In the late spring of 1988 I decided to write a brief article on a peculiar early deduction of "Maxwell's equations" that was produced in 1884 by Heinrich Hertz, later famed for his production of electric waves. I was at the time lecturing at the University of Copenhagen, where I had been invited by their historian of mathematics, Jesper Lützen. His perceptively puzzled questions in response to my attempted explanations of Hertz's odd deductions eventually forced me to realize that something quite deep underlay them, something that spoke to Hertz's own puzzlement concerning central, but essentially unstated, points in the physics of his day. This book resulted from that realization, and it is, in part, an extended attempt to work through what I came to see as themes that guided Hertz's work throughout his tragically short career.

Although my previous two books, especially the second, had paid some attention to experiment, I had there used the laboratory mostly to explore conceptual structures that were bound to the work that was done within it. So, for example, in discussing Fresnel's diffraction experiments in my second book, or the discovery of the Hall effect in my first, I had been concerned rather with what they had to say about Fresnel's developing understanding of waves or about conceptions of fields than with the experimental work per se. The story I tell here is considerably different, because it concentrates directly upon Hertz's wide-ranging experimental work for its own sake and not for its exemplification of something else.

"Something else" is nevertheless quite markedly present in Hertz's laboratory, but it is not precisely the same kind of thing that informed Fresnel's or even Hall's experimenting. Both of them set out to find something that they had reasonably precise theoretical grounds for believing in or at least for suspecting. Hertz's experimenting was quite different in character, because it was designed either to show that something does not occur or else to find something new that was not required by the kind of physical scheme that Hertz deployed. Hertz's experimental work, unlike theirs, had little to do with theory. This is, however, not at all to say that Hertz worked in his laboratory without a program. On the contrary, a central contention here will be that Hertz's work was bound tightly to a certain physical scheme, one that had powerful implications for doing as well as for thinking. Throughout much of this book we will

be exploring through particular examples the differences between highly articulated knowledge, or theory, and this essentially unarticulated kind of knowledge that may undergird both theory and experiment, thereby uniting the two
in a way that places neither one in a subordinate relation to the other.

This book should, however, not be read as a case study written to exemplify
something about the nature of scientific practice. It is meant to stand as history,
not as methodology. I did not set out to prove something about science when I
began to work on Hertz; I set out to understand what he said and what he did.
Naturally my own views concerning the nature of science shape the kind of
story that I tell, but it is the story, and not the views, on which I hope the
reader will concentrate. My intention is to take the reader into Hertz's world,
eventually into his laboratory, in a way that does not permit one to stand apart
from it in understanding. I will not offer an Archimedean platform from which,
for example, to grasp what we might nowadays say took place in Hertz's evaporation apparatus, in his cathode-ray tubes, or even (perhaps I should say especially) in his electric-wave devices.

Many people have given me advice and help over the last six years. I have
already mentioned Lützen, who listens and comments better than almost anyone I know. So many others have come to my assistance that to name them all
would be to exhaust the field of potential reviewers. I must, however, thank
those who provided concrete aid. Manuel Doncel and Gerhard Hertz gener-
ously put at my disposal their work on, and translation of, Heinrich Hertz's
laboratory notebook. Doncel's comments at a late stage in my writing also
helped to clarify a number of significant points. My Toronto colleague Trevor
Levere sought out Hertz correspondence at the Deutsches Museum in Munich.
Geoffrey Cantor in England helped me acquire copies of the Hertz manuscripts
at the London Science Museum. Christoph Scriba in Hamburg obtained material concerning Hertz's youth. Lew Pyenson in Montreal gave me the benefit of
his extensive knowledge on, among other things, German education. David
Cahan in Nebraska provided much material on, and discussion about, Helmholtz. Naum Kipnis at the Bakken Museum in Minneapolis collaborated with
me in reproducing Hertz's early experiments.

This project (and its continuation in a subsequent volume) was made pos-
sible by grants from the Social Sciences and Humanities Research Council of
Canada and by two years of leave on a Killam Fellowship from the Canada
Council.

O N E

Introduction:
Heinrich Hertz, Maker of Effects

Toward the end of December 1887 an ambitious young German physicist named Heinrich Hertz decided that he had fabricated a new phenomenon in a Karlsruhe lecture hall. Soon physicists throughout Europe built devices based on Hertz's description that, most agreed, produced this new effect. Within a decade and a half greatly mutated descendants of Hertz's original device, known as the dipole oscillator and the resonator, had become technological artifacts of little interest to most research physicists. These events might be said to constitute the first artificial production and exploitation of electromagnetic radiation. They might also be said to constitute the creation of a technical archetype, of a device that was modeled and remodeled in novel environments.

These two statements are not equivalent to one another. Neither are they mutually exclusive. According to the first there always was something in the world—electromagnetic radiation—that required only an appropriate device to manifest its presence. According to the second an instrument was created that behaved in certain ways. This instrument was then used as a model to produce further devices for entirely novel ends. Realists will prefer the first statement; agnostics will prefer the second. Historians have different tastes in such matters, but they must at least act as professional agnostics because one of their tasks is to uncover piecemeal how the fabricator of a fresh device produced it without relying unduly on statements about what must, or must not, have been going on.

Heinrich Hertz's apparatus was more widely copied and argued about in a shorter period of time than most laboratory gadgets before it. Precisely because of this speedy dissemination and mutation, Hertz's device has in retrospect been divorced from what it was taken to have produced, *electromagnetic radiation.* It is time to put the machine back into the effect, not least because Hertz's device never did become entirely unproblematic, even in research laboratories. On the contrary, it was troubled by difficult pathologies that frustrated its acquisition of the kind of status that, for example, low-frequency induction coils had long ago acquired. In fact, Hertz's device never became a research instrument at all. Devices that were intended to be instances of it were produced, and they did fabricate effects that appeared to be similar to (but that were far from identical with) the ones that Hertz produced. But the devices were always worried, and scarcely ever in similar ways. Like most novel objects in experimental science Hertz's device had to be crafted and used in certain ways to produce the sought-after effects.

The goal of this book, which is the first of two volumes, is to understand the origins and character of Hertz's craft in a way that casts strong light upon his production, use, and interpretation of the dipole oscillator and resonator. That craft was deeply embedded in Hertz's homelife, his upbringing in 1860s Hamburg, his punctuated education, his highly strong sensibility and intense competitiveness, and his complete absorption of the peculiar ethos that characterized the Helmholtz-centered physics community of the 1870s and 1880s. We shall look at Hertz's early environment and career, at the pressures upon him, and at his response to those pressures. We shall examine as well the kind of physics that Hertz's master, Hermann Helmholtz, constructed in Berlin during the 1870s. Here, in the environment established by a man whose complex influence on Hertz ran so deep that it can scarcely be exaggerated, we will find the inmost sources of Hertz's approach to physics and of the motivations that drove him so intensely throughout the 1880s to seek novelty, to try repeatedly to fabricate new effects in the laboratory. His eventual production of the oscillator and the resonator was only the last, and most successful, of many such attempts. Yet the device was not originally built to produce electromagnetic radiation; the route from the apparatus to this novel effect was not at all straight. nor was it at first altogether clear to Hertz just what he had produced.

Although the German physics community was a constant, ghostly presence lurking behind Hertz's workbench and haunting his notebook, for he was intensely driven by competitive urges, he worked in substantial isolation from his colleagues and without students between the time he left Berlin and his call to Bonn scarcely a half-decade later. When Hertz did work with, or at least in association with, someone else in his years as an assistant at Berlin, he subtly but unmistakably challenged his colleague's expertise and authority. At Karlsruhe he had a laboratory assistant, and he had objects made locally; he also fabricated many things himself. The young, driven Hertz of Berlin, Kiel, and Karlsruhe had little directly to do with industry (with one partial exception), was uninterested in metrology (though not in units), was impatient with mathematical rigor or undue abstraction, and strongly disliked testing things thought up by other people unless there was the chance of finding novelty. From the moment he left Helmholtz's physical presence, Hertz moved from topic to topic, from technique to technique, seeking always to make something new come into being.

This first volume ends with Hertz's reproduction of the device on paper, with his attempt to turn it into a piece of canonical theory. The paper dipole was just as difficult and problematic a fabrication as the physical device was, and the relations between the two were far from unambiguous. Indeed, there is a very strong sense in which Hertz's paper dipole, which was itself replicated many times by physicists throughout Europe and Britain during the ten years following Hertz's creation of it, stood for a rather mysterious, and absent, object, something that may have strongly furthered its usefulness in other paper creations.[1] We end, then, near the time when Hertz's device left his exclusive control, and when its reproduction in other places and by other physicists and engineers changed both the apparatus and Hertz's social position in the German physical community. His work changed as well after this, shortly becoming much more abstract than it had been before. Our story concerns the period before this major change occurred, the years of the ambitious, high-strung, stubborn, laboratory-focused Heinrich Hertz of Berlin, Kiel, and Karlsruhe, not the famous, preoccupied, and increasingly ill heir-apparent to Hermann von Helmholtz of the early to mid-1890s. This is at once the story of a device, of a kind of physics, of a career, and of the environment that birthed them through the agency of a brilliant, passionate, and complicated product of the era, Heinrich Hertz.

The focus here is accordingly on Hertz himself—on his thoughts, his practice, and his career before he achieved fame. A second volume will turn to the intricate story of his last years, and later, which will take us far beyond Hertz and the Helmholtz-centered context in which he lived for so long. By the fall of 1888 Hertz was becoming a very important public figure in German physics in his own right, and he rapidly became an international one as well. Reproductions of his device, as well as replications of some of his experiments, occurred at many sites. At several of these sites, in particular at ones where replications rather than reproductions took place, problems arose that undermined Hertz's original claims, and indeed that undermined the character of the device itself as a stable effect-generator. Unpublished correspondence between Hertz and the interrogators of wave-producing apparatus will offer the opportunity, in the second volume, to discuss the transformation of a device, the production of a new physical regime (the electromagnetic spectrum), and the construction of novel relations between physical and mathematical argument.

In Helmholtz's Laboratory

TWO

Forms of Electrodynamics

2.1. THINGS AND STATES

Imagine oneself in a laboratory circa 1870 about to measure charge or current. Such things must in the first place be produced by a device that will generate enough of an effect for some other device to detect it. Charge would be produced by means of an electrostatic machine, which operates by friction; current would usually be produced by means of a battery or an electromagnetic generator.[1] A device that detects charge—an electrometer—works by measuring in some manner the deflection produced in one charged object by another one against the action of gravity, spring tension, or some such force. Similarly, current detectors—galvanometers—work by measuring the deflection of a magnet or of another current-bearing body by a given object. In both cases the goal is to measure a force acting upon a material object in the requisite state of "chargedness" or "current-bearingness."

Although this no doubt sounds like an elementary introduction to a positivist view of the connection between theory and measurement, nevertheless its implications must be thoroughly assimilated in order to grasp the differences between contemporary forms of electrodynamics in Germany and Britain, since they differ most powerfully over the meanings of "to be charged" and "to carry a current." Let us begin by specifying the central tenets of the two major forms that were prevalent in Germany by the time Heinrich Hertz came to Berlin in 1878 to study under the direction of a master of the subject, Hermann Helmholtz, putting the British temporarily to the side. This will eventually permit us to return to our positivist-like account of charge or current measurement in a way that will bring out its significance for both the German and the British context.

One of the two German forms based electromagnetism on six distinct principles that derive preeminently from the work of Franz Neumann and Helmholtz himself:

Helmholtz's Principles for Electrodynamics

1. Electrically charged bodies interact through the Coulomb force, which can be obtained through the gradient of a potential U_s that de-

pends solely upon the distance between the charges and their magnitudes.

2. Current-bearing circuits interact with one another by means of a mechanical, or material, force[2], which can be obtained through the gradient of a potential function U_d that depends upon the distances and the orientations of the circuit elements and upon the intensities of the currents.

3. Current-bearing circuits also interact with one another through an electromotive force that is given exactly by the time derivative of the same U_d that governs the mechanical force.[3]

4. The current in a circuit is proportional to the net force that drives it.

5. The charge density ρ at a point will change only if the current C there is inhomogeneous according to the equation of continuity $\partial \rho / \partial t + \nabla \cdot C = 0$.

6. All interactions in electrodynamics require the existence of a corresponding *system energy.*[4]

Wilhelm Weber developed the second form of electrodynamics in the 1840s on the basis of a physical hypothesis for the electric current due to his colleague at Leipzig, Gustav Fechner. We shall accordingly denote with "Fechner–Weber" whatever satisfies the four principles of this group. They are

Fechner–Weber

1. Charge consists of groups of two kinds of electric particles, or *atoms of electricity.*

2. The electric current is the equal and opposite flow of these particles.

3. The particles exert central forces on one another that depend upon their distances, as well as upon the first and second time derivatives of these distances. Consequently all electromagnetic forces—mechanical and electromotive—derive from a single, fundamental action.

4. The current in a circuit is proportional to the net force that drives the electric particles.

These principles differ in a fundamental way from Helmholtz's six, for they are based directly upon the physical structure of the current, whereas the Helmholtz principles in no way at all specify what either charge or the current might be. Consider, for example, the respective ways in which Fechner–Weber and Helmholtz yield the Ampère force between the elements of current-bearing circuits. In Helmholtz's scheme the Ampère force emerges directly out of a mathematical process that employs only the objects that already occur in the interaction energy (currents and circuit elements) and that are considered to be theoretical primitives. Fechner–Weber, however, must first calculate the forces between particles (which are its primitives) and then divide circuits into elements that each contain these interacting objects. The result reduces to an apparent interaction directly between the elements, but it lacks deeper physical significance. Similarly, electromagnetic induction can be represented in Fechner–Weber by a function that has precisely the same form as the potential that had been developed by Franz Neumann, but this again lacks any deeper significance, whereas in Helmholtz's system the Neumann potential (or its generalization by Helmholtz in 1870) has a natural energetic interpretation.

Indeed, the longest-lasting and deepest assumption of Helmholtz's work in electrodynamics from 1847 until the late 1880s, and even after, was precisely that Neumann's potential U_d (or something very like it) is a fundamental quantity. There are two general (and connected) reasons for his conviction in the potential's importance. First, and of overriding importance, it has immediate energetic significance. Second, an electrodynamics based upon U_d, unlike Fechner–Weber, remains independent of assumptions concerning electric atoms and their concomitant forces; it acts as a constraint upon any model purporting to explain charge and current. It is, therefore, a much more general theory than Fechner–Weber, and, like energy conservation itself, it can survive extreme changes in the higher-order principles of electrodynamics.

In the very piece that set out energy conservation in 1847 Helmholtz had insisted that actions which depend upon speeds and accelerations, or that are not central, must involve infinite force (Helmholtz 1847, p. 27). Weber's electrodynamics, Helmholtz initially (and incorrectly) argued, therefore violated energy conservation.[5] This provided the technical foundation for a controversy between the two masters of German electrodynamics that erupted during the 1870s, several years before Hertz came to Berlin. Helmholtz eventually ceded the fullness of his original claim, but in the early 1870s he turned to special cases of it, one of which he eventually had Hertz examine in the laboratory. In going to Helmholtz Hertz was (perhaps unwittingly) taking sides in a great argument of the day that divided German physicists into antagonistic camps.

Between 1853 and 1869 Helmholtz published only two articles on electrodynamics, neither of which touches fundamental issues of this kind. But between 1869 and 1875 he published ten articles, every one of which does so. Most of them contain extensive discussions, often in reply to criticism, of Fechner–Weber as well as detailed presentations of Helmholtz's electrodynamics. In 1882 the first volume of his *Wissenschaftliche Abhandlungen* was printed, and it collected in one place all of these recent articles as well as everything else that Helmholtz had written on electrodynamics, including his latest understanding of the constraints imposed by energy conservation. The protracted controversy in the early 1870s over the fit between energy conservation and Weber's force law had at least two significant effects. First, it led Helmholtz to probe in greater detail than he might otherwise have done the structure of his electrodynamics in order to clarify the differences between it and Fechner–Weber; second, it focused his protégé Hertz's theoretical eye on energy and the potential and his experimental eye on distinguishing Helmholtz's electrodynamics from Fechner–Weber.

Hertz's first publication, in 1880, contains only one citation—to a major article that Helmholtz published in 1870 in which he attempted to subsume all possible forms of electrodynamics, including Weber's, under a single, general potential, in which he attacked Weber's theory directly, and in which he developed a highly influential understanding of Maxwell's work. One can scarcely exaggerate the influence of this and others of Helmholtz's articles on Hertz.[6] But Helmholtz's influence on him, though profound, was not entirely straightforward. Within five years, during a period when he turned to abstract, rather than directly experimental, work, Hertz was reading Helmholtz in ways that the master had not himself intended as he sought to grasp the inner meaning of Helmholtz's effort to force Maxwell's theory into a particular mold.

2.2. Objects and Interactions

Return now to the positivist-seeming account of measurement with which we began, and let us consider the process from the point of view of Fechner–Weber electrodynamics. This theory postulates a direct transference between the laboratory measures and the physical interactions that produce them. The charged or current-bearing object disappears from Fechner–Weber, to be replaced by a region in which electric particles move (or do not move) about within a matrix of material particles. The electric particles (which are mere points) exert specific electrodynamic forces on one another, and each responds to a given force with a fixed acceleration in the force's direction. Through some unspecified mechanism the forces between these electric point atoms are transferred directly to the particles that form the bodies in which they occur, so that the laboratory indicator actually measures the *net force* between the electric atoms themselves. The bodies in which they exist are merely containers for them that are, as it were, carried along by an interaction between the electric atoms and the particles that comprise the laboratory objects. In Fechner–Weber the electric atoms therefore act directly as *sources* for the actions whose overall effects are detected. The interactions consequently do not occur between the laboratory objects themselves, but they do occur between entities that subsist in these objects and that act upon their constituents.

In Britain, where the Faraday–Maxwell form of electrodynamics was rapidly becoming predominant during the 1870s, laboratory measurements were thought about in a very different way from this, for Faraday–Maxwell does not postulate the direct transference that Fechner–Weber requires. Instead of envisioning entities that are different from, but that subsist in and act on, the material objects, Faraday–Maxwell dispenses altogether with the objects as electrodynamic entities in their own right and instead introduces something different from them. This entity, or field, subsists in the space occupied by the objects as well as in the space between them. When an object is in a "charged" or "current-bearing" condition, it does not, properly speaking, interact with the field; rather, its condition is a shorthand way of referring to the *local state* of the field, a state that depends upon the local presence of matter. The object's electromagnetic condition reflects, and is reflected by, the energetic structure of the ambient field, which determines the tendency of the object to move.[7] Since the local change in the field's state affects states throughout the field, objects may rather loosely be said to "interact" with one another. Here the laboratory measure does not emerge from a forcelike interaction between entities that compose the objects but from a connection of an utterly different character between the states of the objects and the state of an entity that comprises nothing but itself.

Yet we see that, despite these profound differences between them, in neither Fechner–Weber nor in Faraday–Maxwell do the laboratory objects interact directly with one another. In Fechner–Weber they interact with electric particles, which in turn interact with one another. In Faraday–Maxwell they are linked at any given moment only to the local state of the field, and the existence of other laboratory objects at the moment of object–field interaction is irrelevant to the process: only the field state at the object's locus activates the detectors. In 1870 Helmholtz created a theory that differs radically from both of these in refusing to abstract from the laboratory objects in the fashion of Fechner–Weber and yet also in refusing to introduce something entirely different in nature from them in the manner of Faraday–Maxwell. He substituted instead what one might call a *taxonomy of interactions* for the unitary forces of Fechner–Weber and for Faraday–Maxwell's duality between field and object.

Seen through Helmholtz's eyes the objects in the laboratory remain entities in their own right—they are not to be reduced to, say, groups of particles. The question for Helmholtz's physics is what conditions objects may have—what states they may be in—such that these states specify an interaction energy between the objects. A "charged" object accordingly differs from an "uncharged" one in its acquisition of a condition that it did not previously have in relation to other objects that are also charged (i.e., that have a state such that two objects, each in this state, determine an interaction energy). Similarly, current-bearing objects have no mutual interactions before they acquire the current-bearing state; afterward they do.

Contrast, for example, the three ways in which electrostatic effects could be viewed circa 1870. According to Fechner–Weber a charged body is in itself no different from an uncharged one; it merely has more of one of the two kinds of electric atoms than the other kind. Consequently, in Fechner–Weber electrostatic interactions are in fact *always* present between any two conductors, but net forces do not always result, and in any case the interactions are between electric atoms. According to Faraday–Maxwell also, a body remains essentially unchanged by its charged state. However, the *field* at its surface, or within it, has its state changed, with an accompanying alteration in the field's energy gradient at the body's surface. This translates into a force upon the object due to whatever the unknown connections are that link the body's state to the state of the field. Here, as in Fechner–Weber, the object does not in itself determine the actions. Moreover, in Faraday–Maxwell it is also indifferent to the simultaneous presence of other bodies, although the state of the field at any given moment and place depends upon the previous positions of other objects at other places.[8] But now consider how this appears from Helmholtz's perspective. There are no electric atoms to transfer force to the laboratory objects; there is no local field for the bodies to interact with. Rather, the *states* of the bodies at any given instant directly determine their mutual interaction. A charged body acts instantly and directly upon another charged body, and vice versa, but it is incorrect to say that the charge of the one body acts upon the charge of the other: the bodies act, and "charge" is just a short way of specifying the nature of the interaction.

We can now grasp what is meant by a taxonomy of interactions. In the broadest sense electromagnetism for Helmholtz is merely one among many different classes of possible interactions between laboratory objects. As a member of a given class an object can have one or more *states*. The critical point to grasp about Helmholtz's electrodynamics—the point that makes it radically different from Fechner–Weber and from Faraday–Maxwell—is that every distinct kind of interaction requires a unique specification of the states of the interacting objects. At the simplest level there are two states and two interactions. An object may be charged (the first state, call it q), and it may also carry a current (the second state, call it c). Charged objects can affect one another, and so can current-carrying objects. Consequently, at least two kinds of interactions exist, namely qq and cc. And every type of interaction requires, we shall see, a specific energy determined by the pair of objects in the specific states. This has radical implications indeed, because it means, for example, that charged objects and current-carrying objects have no necessary interactions with one another *whatsoever* in Helmholtzian theory. If they do interact, then a new entry must be made in the taxonomy, and a new form of interaction energy invented.

There are other striking differences. Nearly every physical theory embraces a certain number of canonical problems, ones that it seems designed to explicate. Field theory, in Faraday's form, seems at once to encompass the generation of electromotive force by motion through a magnetic field, although even here there are subtleties and obscurities.[9] Conversely, certain problems may escape the theory's easy grasp. So, for example, it is hardly obvious from the principles of Faraday–Maxwell that a magnetic field will exert a force on a moving, charged body, although it does follow from them after a rather lengthy analysis. Moreover, the problems that are easy in one theory may be just the ones that are difficult in another theory, and vice versa. To continue with these two examples, Fechner–Weber requires a fairly elaborate computation of the forces between moving particles to specify what occurs when the conductors are themselves in anything but the simplest relative motion. But if the motion is simple, as when a charged conductor moves slowly relative to a current-bearing one, then the forces are easy to compute, and it is almost obvious that the charged conductor will be affected by a net force. So in Fechner–Weber the generation of electromotive force by motion can be a complicated affair, whereas the deflection of a moving charged object by currents is simple to understand. In Faraday–Maxwell the situation is reversed.

In Helmholtz's theory canonical problems of this kind do not exist, although certain types of problems do require more intricate analyses than other types. This unusual, and disturbing, characteristic derives from the theory's fundamental structure, from its insistence that electromagnetic processes must be construed as *states* of bodies, and electromagnetic interactions as unmediated relationships between bodies in these states. The theory does not introduce, as both Fechner–Weber and Faraday–Maxwell do, a third entity that intervenes between bodies (electric particles for Fechner–Weber and the field for Faraday–Maxwell). Such a thing must be present in a theory in order for canonical problems to occur in it because the canon generally reflects the elementary properties of the tertium quid. In Fechner–Weber problems in which the interactions between the electric particles seem obvious are canonical; in Faraday–Maxwell problems in which the field's behavior within and at the surfaces of bodies is simple to specify are canonical. In both cases the canon resides in a third entity whose relationship to bodies can be, and often is, problematic in complicated situations. But in Helmholtz's theory there is no third entity. There are instead a potentially very large (even, we shall see, infinite) number of different electromagnetic states that bodies may possess. None of these states has in itself a privileged position, so that no special kinds of problems seem to be more closely tied to the deeper recesses of the theory than any other kinds. Helmholtz's theory has no elementary problems of the usual sort.

But it does have a spectrum of interaction energies. Although Helmholtz never thoroughly laid out the theory's structure, nevertheless the many articles he wrote to explicate it, articles that Hertz carefully studied in the late 1870s and early 1880s, together reveal it.[10] The electromagnetic world according to Helmholtz's theory consists of bodies in various states. At its most simple the theory takes account only of conductors and the states that they may have. Suppose, for purposes of argument, that n distinct states exist, some of which are mutually exclusive. Represent a conductor C_a in the ith state by $(C_a)^i$. According to Helmholtz's theory, if two conductors C_a and C_b interact with one another in the respective states j and k, then there must exist an *interaction energy* (though Helmholtz never called it that), which depends upon $(C_a)^k \times (C_b)^j$ where the sign $\times$ indicates that the interaction depends upon the two states. The essence of Helmholtz's method involves taking this energy and determining how it varies when the bodies are moved or the magnitudes of their states change.

Again a simple example helps to clarify the situation. Consider two conductors, each of which carries a current or, better put, each of which is in a *current-carrying state*. Then construct (from "experiment") an interaction energy between them. If the conductors change their distances from one another, then the interaction energy may also change. By "energy conservation" such a change will translate into a mechanical force on each of them.[11] The essential point to grasp however is that there is no third entity that intervenes between the bodies, no *object* that must be thought of as having its own, *independent* existence which can act upon the bodies.[12] The interaction energy is not itself such a thing since it does not itself have states; neither is it determined by anything other than the conditions and the mutual distance of the interacting objects, nor (in Helmholtzian conception) does it move independently about.

In appearance this theory is an apotheosis of instrumentalism because it seems not to go beyond the laboratory objects and their unmediated interactions with one another. Even force is absent from the theory as an entity in its own right because it emerges only as a result of an energy calculation, as an epiphenomenon of the interaction energy.[13] Helmholtz did not, for example, consider interactions to involve the exertion of a force by one body on another, because forces do not inhere in bodies properly speaking: they are functions rather of bodily states than of bodily existence or even of the spatial and motional relationships between bodies. And indeed Helmholtz's theory never does move far from the laboratory. But there is a heavy price to pay for instrumentalism of this kind because it requires in effect for many situations an a posteriori specification of energies that in the other two theories can be derived a priori.

Although the structure of the new electrodynamics first appeared nearly full-blown in 1870, during the next six or so years Helmholtz became increasingly concerned to probe it in the laboratory, to distinguish it from Fechner–Weber in particular and to cross-link it to Faraday–Maxwell. Indeed, he devoted himself with growing avidity to his new theory and then to related aspects, such as Faraday's electrochemistry, which led him to deploy the very useful, and also instrumentalist, concept of the *ion*.[14] The amount of work that he devoted to electrodynamics and its offshoots during this decade, in particular to its defense and elaboration against attacks from such Webereans as Hermann Herwig and Friedrich Zöllner, bespeaks its importance to him.[15]

2.3. The Fine Structure of Helmholtz's Electrodynamics

Eddington once remarked of the British physicist Joseph Larmor that nothing was clear to him until it could be formulated in terms of the principle of least action. It could equally well be said of Helmholtz that after the early 1870s nothing was clear to him until it could be formulated in terms of interaction energies. Helmholtz first clearly brought together, generalized, and explored the difficult principles of his energy-based, relational, and deeply empirical electrodynamics in the major 1870 article that was so profoundly to influence Hertz and almost every other Continental physicist during the 1880s and 1890s. In later years, after Hertz's discovery of electric waves, physicists looked to this work for a way to understand field theory. However, when Helmholtz wrote the article, his primary aim was certainly not to obtain a variant of field theory but rather to probe the structure of his electrodynamics and to contrast it with Fechner–Weber: only seventeen of the article's eighty-three pages in the *Wissenschaftliche Abhandlungen* concern the kinds of effects that are uniquely important for propagation. The rest concentrates on characteristics linked directly to the potential function. This lack of emphasis on field theory is what one would expect because, in 1870, it had neither Maxwell's *Treatise* to explain it nor Hertz's experiments to publicize it. Indeed, Maxwellian field theory as yet lacked many of the conceptions that, three years later (when the *Treatise* reached print), forged a unified system out of what were until then disparate and occasionally undeveloped parts.[16]

Helmholtz began his 1870 analysis by noting that the only empirical limitation on the form of the interaction energy Φ_0 between elements carrying currents is that it must yield the Ampère force between closed (or what I shall hereafter call reentrant) currents.[17] Given this condition, he demonstrated, the most general form of Φ_0 consists of the old (Neumann) expression

$$-A^2 \int_r \boldsymbol{C} \cdot \int_{r'} \frac{\boldsymbol{C'}}{r_d} d^3r' d^3r$$

supplemented by the scalar product of the current with a certain gradient:[18]

$$\Phi_0 = -A^2 \int_r \boldsymbol{C} \cdot \int_{r'} \left[\frac{\boldsymbol{C'}}{r_d} + \frac{1}{2}(1 - k)\nabla_r(\boldsymbol{C'} \cdot \nabla_{r'})r_d \right] d^3r' d^3r$$

Rewrite this by introducing Helmholtz's vector potential function, U, together with his auxiliary scalar, Ψ, such that

$$\Phi_0 = -A^2 \int U \cdot \boldsymbol{C} d^3r$$

$$U(r) \equiv \int \frac{\boldsymbol{C'}}{r_d} d^3r' + \frac{1}{2}(1-k)(\nabla_r \Psi)$$

$$\Psi(r) \equiv \int (\boldsymbol{C} \cdot \nabla_{r'} r_d) d^3r'$$

The continuity equation links these potentials to the charge density ρ and therefore to the scalar potential ϕ_f:

$$\nabla \cdot C + \frac{\partial \rho}{\partial t} = 0$$

$$\nabla^2 \phi_f = -4\pi\rho$$

From the last three equations and "partial integrations" (Buchwald 1985, app. 10) follow

$$\nabla^2 \Psi = 2 \frac{\partial \phi_f}{\partial t}$$

$$\nabla^2 U = (1 - k)\nabla \frac{\partial \phi_f}{\partial t} - 4\pi C$$

$$\nabla \cdot U = -k \frac{\partial \phi_f}{\partial t}$$

These three equations constitute Helmholtz's basic analytical structure. He remarks that setting k equal, respectively, to 0, -1, or 1 gives the expressions for Φ_0 to which Maxwell's, Weber's, and Neumann's theories lead. However, for Maxwell and Weber Φ_0 must be deduced; Neumann gives it at once in this form.[19]

Helmholtz's procedure for obtaining the electromotive and the mechanical effects from Φ_0 was to vary the energy of the system while holding fixed either the spatial coordinates or the time: variation with fixed spatial coordinates yields an electromotive effect, and variation with the time fixed yields a mechanical effect (see appendix 2 for details of the latter derivation for extended conductors).[20] In the end the "forces" that involve a current C can be expressed in terms of the potential function U. In order, then, to deduce the action between two objects, one must begin with the energy of the system that they mutually determine. This novel character of Helmholtz's electrodynamics produced a great deal of confusion among his contemporaries. The point is worth pursuing through a laboratory example because it illustrates just how odd the scheme seemed to those trained in Weberean conceptions.

2.4 WEBEREAN INCOMPREHENSION

In 1874 the Weberean Hermann Herwig built a device in which an electrodynamic force arises even though the configuration of the apparatus remains completely symmetrical—thereby implying, it would appear, that the electrodynamic potential cannot change, and so, in seemingly direct Helmholtzian consequence, entailing that no force at all should arise. In Herwig's somewhat misleading diagram (fig. 1) a magnet stands vertically. A brass hanger *abc* pivots at *a* at the magnet's top end, with its side *bc* also vertical. In familiar demonstration experiments (though not in this one, as we shall see in a moment) such a hanger usually terminated at *c* in a vat of mercury, and a circuit was completed through the mercury to the hanger's upper end. It follows easily

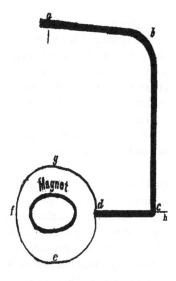

FIGURE 1   Herwig's apparatus (Herwig 1874)

from the Ampère law as applied to circuit elements that the hanger will rotate about the magnet's axis as a result of forces that are exerted primarily on its vertical portion *bc*.

Helmholtz's potential law, one might think, cannot accommodate even this rotation because the current-bearing hanger seemingly remains symmetric with respect to the magnet during its rotation, so that the interaction potential cannot change, and so forces cannot occur. However, Helmholtz pointed out in some detail that as the hanger swings round, the circuital path formed through the mercury changes continuously. This alters the parts of the mercury that are inolved in the interaction, and a force thereby comes into being.[21] Since this occurs within the mercury, where the hanger makes sliding contact with the liquid, it follows that the action that moves the hanger must lie there as well.[22] This contrasts directly with the Ampère law, which locates the action on the vertical portion *bc* of the hanger. The difference between the two analyses clearly cannot be tested with this sort of apparatus (and indeed Helmholtz had already argued that it can never be examined as long as all of the current-carrying circuits are closed).

Herwig reasoned that the difference between Ampère and Helmholtz could be probed by removing the mercury and connecting the lower end of the hanger *c* rigidly to the end of a fixed wire *cd* placed to one side of the vertical magnet. To replace Helmholtz's changing paths through the circuit-completing mercury, Herwig formed a thin metal wire *def* of *fixed* length, which he folded into a semicircle whose center coincided with the magnet's axis, and which therefore lay in the horizontal plane (i.e., in Herwig's diagram *def*, as well as *dgf*, is actually perpendicular to the vertical portion *bc* of the wire).

With this configuration the hanger cannot rotate freely, but it can be twisted, and Herwig set out to examine the effect. He found that a twisting torque does exist, despite the fact that Helmholtz's changing fluid paths had been removed. Herwig was, however, well aware that the fine wire *def* will necessarily break its symmetrical position with respect to the magnet's axis because one end of it is tied to the end of the hanger, and the hanger must move somewhat in response to the torque. This naturally entails a change in the circuit's configuration, and so just the kind of alteration in potential that Helmholtz's account required. Herwig accordingly tried to find some way of demonstrating that the torque on the hanger cannot possibly derive from an effect that is localized in the connecting wire *def.*

Therefore, he attached another wire *ch* in such a fashion that it lay along the extension of the lower, horizontal part *cd* of the hanger. In a first experiment, Herwig insulated *h*, so that the entire hanger bore the current. Here the torque occurred in full measure. He then insulated point *a* of the hanger and passed the current through *h*, so that only the part *cd* of the hanger also carried a current. Here he found that the torque dropped to a very small fraction of its previous value. From this he concluded that the rotational action must occur between the magnet and the vertical part *bc* of the hanger (as the Ampère law, but not Helmholtz's potential, required) since it occurs only when *bc* carries a current.

Helmholtz (1874c, pp. 767–71) replied almost immediately. Herwig had wrongly assumed, Helmholtz remarked, that the added wire *ch* is electrodynamically ineffective. Certainly the potential does require that the hanger's vertical part cannot be the locus of any action that occurs, and yet the action vanishes when only *dch* carries a current. But, Helmholtz argued, the action does not vanish because the current in *bc* has been removed, as the Ampère law would have it. The situation is, he continued, rather more complicated than Herwig had assumed, for the second experiment is no more equivalent to the first one from the standpoint of the potential law than it is from that of the Ampère law. With a current flowing from *h* through *c* and on through *def,* Helmholtz noted, a slight displacement of the lower end of the hanger will change the potential of *both def* and *ch* with respect to the magnet. Furthermore, and of critical import, the change for the one is opposite to the change for the other. As a result they are impelled in opposite directions to one another, which in the end produces a vanishing net result. Far from demonstrating that *def* cannot be the source of the action, Helmholtz concluded, Herwig's experiment is fully consistent with that claim.

Herwig had missed this point because, as a Weberean, he tended to think of pieces of wire as objects that constrain interacting particles. In order to analyze the device he mentally broke it up and estimated the mutual actions of the particles in the pieces on one another. The order and placement of the device's parts consequently acted as *essential and unalterable constraints* for his Weberean analysis: to understand what happened, the regions to which the particles were confined must be givens of the problem. Herwig did not think to pay any attention to the small deformation that the wire *ch* must necessarily experience when it is torqued, especially since, undeformed, the wire has no potential at all with respect to the magnet. Such a thing would simply make the problem much too complicated to solve, for if the particles' loci could not be delimited, then their actions could not be estimated.

Helmholtz's view could not have been more different from this one. He saw malleable volume elements that determined the device's system energies as a function of the distances between the elements. He, like the Weberean, also broke the device up into pieces—albeit into volume elements—but he remained with the pieces and estimated their interaction energies, not the emergent forces between them. For him the order and placement of the parts were merely the starting point of the analysis; they were not essential constraints. Far from fixing the parts, he had actually to deform them in order to calculate the virtual changes in the energies that translate into bodily forces. The Helmholtzian therefore tended naturally to think about deforming or mutating a system, whereas the Weberean rarely, if ever, did. Here, then, we again see how the primitive differences between the schemes could come to life in the laboratory.

By the early 1870s Helmholtz himself certainly found it very hard to think in any other way about electrodynamic interactions. He remarked in concluding his analysis of Herwig's mistakes (as well as similar ones generated by Friedrich Zöllner):

> The potential law requires one and the same simple, proportionate mathematical expression to encompass the entire, experimentally known realm of electrodynamics, ponderomotive and electromotive effects, and to the area of the ponderomotive effects it brings the same great simplification and lucidity that the introduction of the idea of the potential brought to the study of electrostatics and magnetism. I myself can bear witness to this, because for thirty years I have applied no other fundamental principle but the potential law and have needed nothing else to find my way through the rather labyrinthine problems of electrodynamics. (Helmholtz 1874c, p. 772)[23]

The result was to make the Helmholtzian laboratory a much more flexible, manipulative place than a Weberean laboratory could possibly be. In the latter, one would concentrate for the most part on measuring constants using devices with given, unaltered structures. These pieces of *measuring equipment* would rarely, if ever, be used in conjunction with other devices whose behavior could not be thoroughly calculated *from theory* beforehand. Whereas Helmholtzian laboratories were places for seeking out unknown phenomena, Weberean laboratories were places for measuring unknown constants. One might rather crudely put the difference this way: in the Weberean laboratory one builds and, having built, analyzes; in the Helmholtzian, one builds, probes, and mutates until the device behaves as a system in a satisfactory manner. This was the special *craft,* the unique experimental *culture,* that Hertz at least imbibed in Helmholtz's laboratory.

2.5. A Taxonomy of Interactions

Although Helmholtz certainly never made the point explicit, his general scheme for electrodynamics requires a careful listing of the several actions that can affect the current. He allowed three: electrostatic, the action of other currents (electromotive and electrodynamic), and the various chemical, thermal, etc., effects that can produce currents. These actions are necessarily unrelated to one another. Each of them involves a distinct interaction energy that is immediately determined by the physical characters of the sources: for every pair α, β of distinct kinds of sources that interact with one another there must exist a unique energy function $U_{\alpha\beta}$. So, for example, the energy function for a charge–charge interaction, call it U_{qq}, is distinguished from the energy function U_{cc} for current interactions by the sources involved. The physical effects of one source on the other emerge from variations. Varying the spatial coordinates of the charge–charge energy yields the mechanical electric force that moves the charged body. Doing the same with the current–current energy also yields a mechanical force, this time electrodynamic, that moves the current-bearing body. However, there is a fundamental difference between charges and currents that is represented by the continuity equation and that permits the existence of a kind of current–current interaction for which there is no parallel action between charges.

Integrate the continuity equation over all space under the assumption that the charge and current densities vanish at infinity. Then, by Green's theorem, the term in $\nabla \cdot C$ integrates to zero, which means that the total quantity of charge in the universe is conserved, whereas current quantity is not. Unlike charge, current is an intrinsically variable quantity. In direct consequence, a force can in principle exist that acts directly to alter current quantity, whereas no such force can exist for charge, whose value at a point can be altered only as a result of inhomogeneities in the current there. In Helmholtz's electrodynamics such a force is determined by the time derivative of the very same function (U) whose curl generates the electrodynamic or mechanical force between current-bearing bodies. Taking the time derivative of the scalar potential (ϕ_f) yields in itself no further actions between charges.

The result of this scheme, a result that both intrigued and troubled Hertz in the mid-1880s, was to enforce an absolute separation between the forces that can affect a given object but that derive from different kinds of sources. For example, the fact—should it be one—that a charged body can be moved physically by a changing current as well as by another charged body does not suggest any relationship at all between the forces that act in the two cases. Two completely unrelated interaction energies must be involved. Both actions move charge and so might be termed "electric" in a rather vague sense, but that is the only sense in which they are similar to one another. Consequently, many things that in field theory (or in Weber's electrodynamics) are not issues, or are at best subsidiary ones, raise fundamental questions for Helmholtz. These questions can best be appreciated by first recognizing that only three broad kinds of things can happen, whatever interactions may be occurring: bodies may be physically moved; conductors may become charged; and currents may be produced. Contemporary terminology tended to call actions that move charged bodies "electric"; actions that produce currents were "electromotive"; and actions that move current-bearing bodies were "electrodynamic" or "magnetic" (though sometimes "electrodynamic" can refer to the electromotive force that engenders currents). Here we can begin to see how utterly different the requirements of Helmholtz's conceptions were from those of his contemporaries, whether Webereans or partisans of field theory. To make the point completely explicit, we can specify, in a way that Helmholtz never did, the possible kinds of interactions that his scheme required.

Suppose we assume for a moment that charged bodies act electromotively on conductors with the same force that they would exert electrically if there were a unit charge at the point where the action occurs, but we do not further assume that generators of electromotive force also act electrically (i.e., to move charged bodies). Then table 1 summarizes the kinds of actions that must occur between bodies according to the implicit demands of Helmholtz's scheme. The horizontal labels are the objects that act upon the ones that are listed vertically, and the entries in the table are the kinds of actions involved. In each case the only action listed is the one that is due to the particular phenomenon that is specified in the top row (e.g., the action of a body in which the current is changing on a conductor bearing a steady current is listed as electromotive only, even though there is also an electrodynamic force that depends upon the value of the changing current at a given moment—this force appears under the entry for the action of a body with a steady current). Table 1 distinguishes between actions that neither field theory nor Weberean electrodynamics distinguish from one another. For example, the interaction between a charged conductor and an uncharged conductor is listed as "electromotive," which is distinguished as a separate kind of interaction from "electric": the former generates currents; the latter moves charged bodies. In field theory both derive from the electric field; in Weberean electrodynamics all forces are due to interparticle actions so that this kind of distinction lacks meaning. But in Helmholtz's scheme one must above all else specify the sources to know the actions. Difficult enough to understand (as we shall see) where conductors alone are concerned, this fundamental requirement easily leads to a rococo maze when (as in the table) dielectrics intrude (on which see appendix 18).

TABLE 1   Chart of Interactions

	CU	CC	DUP	DSP	CSC	CVC	DCPS	DCPV
CU	—	emo	—	emo	—	emo	—	emo
CC	—	es = emo	—	es = emo	—	emo	—	emo
DUP	—	emu	—	emu	—	emu	—	emu
DSP	—	es = emu	—	—	emu	—	emu	es = emu
CSC	—	emo	—	emo	ed	emo	ed	emo
CVC	—	emo	—	emo	ed	emo	ed	emo
DCPS	—	emu	—	emu	ed	emu	ed	emu
DCPV	—	emo	—	emu	ed	emu	ed	emu

Objects:
CU = conductor, uncharged
CC = conductor, charged
DUP = dielectric, unpolarized
DSP = dielectric, constant polarization
CSC = conductor bearing a steady current
CVC = conductor bearing a changing current
DCPS = dielectric with polarization changing at a constant rate
DCPV = dielectric with polarization changing at a variable rate
Actions:
es = electric
emo = electromotive with Ohm's law
emu = electromotive without Ohm's law (or assuming infinite conductivity)
ed = electrodynamic

2.6. BEYOND ELECTRODYNAMICS

The foundation of Helmholtz's scheme in system energies more than permitted, it actually impelled, its practitioners not only to abjure a priori models but also to think carefully about connections between disparate areas. The essence of the method, recall, is to establish a system energy for every set of states that a duo of *objects* can have. Models have no function here because the *objects* are treated as having *states,* not as being composed of other things that do not have, but that in groups effect, states.[24] Now with such a scheme it will often be the case that two physical systems will behave in very similar, perhaps even in identical, ways—however different they may be as physical systems—because their energies have the same mathematical dependence on their respective states. The analogical relationship between different phenomena consequently reflects the analytical identity of their energetic structures, and, in consequence, solving one system must necessarily solve the other—*because there is nothing more to be said or done.*

Perhaps the clearest way to understand this unique approach is to draw a contrast with similar British methods, as illustrated, for example, in J. J. Thomson's 1888 *Applications of Dynamics to Physics and Chemistry.* Thomson's *Ap-*

plications is in places extremely close in character to Helmholtzian relational physics. Like it, Thomson's method relies upon system energies and generates processes through variational technique. However, there is an essential difference between the British and the German methods. For Thomson the system energy always stands as the analytical representative of a process that, in his words, "may be either that of parts of the system, or the surrounding ether, or both; in many cases we should expect it to be mainly the ether" (Thomson 1968 [1888], p. 15). And if—as Thomson believes must in almost every case ultimately prove to be the case—it *is* mainly the ether, then the energy function does not characterize the relationship between the *objects*, in Helmholtz's sense, that constitute the system but rather represents the effect of their existence on the local state of the ether. Moreover, the difference between their approaches reveals itself even where the ether is not directly involved, as in their development of the same mechanical analogy for the second law of thermodynamics.

Both J. J. Thomson (1968 [1888], secs. 46–49) and Helmholtz (1884) develop essentially the same way of obtaining the second law of thermodynamics from mechanics.[25] Each of them distinguishes certain inner coordinates of a system that cannot be individually controlled (as Thomson put it) or that do not appear in the system's kinetic energy (as Helmholtz put it) from those that can be controlled or that do appear in the energy. Each of them then combines Lagrange's equations with energy conservation, to obtain the result that only part of the energy communicated to the system via the uncontrollable or cyclic coordinates can be converted into work done through the controllable or acyclic coordinates.

Though they reach the same result using the same fundamental procedures (Lagrange's equations and energy conservation), Thomson stops with the isolated system. Helmholtz presses further to consider two "coupled" systems that together determine a "total energy" and a "total entropy." If the kinetic energy corresponding to the cyclic coordinates is the same in each of them, then they must behave precisely like two systems that are in thermal equilibrium. The two systems are, as it were, linked indissolubly together through their "*Gesamtenergie.*" Thomson, like other Maxwellians, was never concerned to exhibit a connection between systems but to derive novel properties of a single system from its energy function. Where, for Thomson, systems are either singular or else are linked to the ether—but never directly to one another—for Helmholtz, the connection is always bipartite.

One consequence of this difference was that Helmholtz's followers were forced to consider the connection between ether and matter within precisely the same scheme they employed to consider the links between material systems: the only difference is that one of the two systems is ether rather than matter. Helmholtz's objects always interact without any mediation whatsoever: the ether merely adds another interacting object to the scheme; it does not essentially change it. Maxwellians were able almost completely to ignore the ether-matter link because they regarded it as the *only* link that exists, since the ether is (in principle) the unique depository of all energies between interacting objects. For them the link between ether and matter is mysterious because it is univocal. For Helmholtz the link must not be mysterious, because it is *not* univocal. Indeed, it must be essentially the same as the link to any other dielectric, since the ether is simply a member of the general class "dielectric," whereas to Maxwellians *only* the ether was, properly speaking, a dielectric. As a result Helmholtz's physics faced a spectrum of difficulties that had no exact counterpart in Maxwellianism.

Those who came to work in Helmholtz's laboratory during the 1870s probably absorbed a great deal of the spirit that animated his relational understanding of physics. Yet Helmholtz himself was perhaps not the most avid or radical pursuer of every aspect of his own electrodynamics, just as Maxwell was not, in several respects, the most ardent Maxwellian. Like Maxwell, Helmholtz had struggled to realize a new and powerful way of doing physics, but, again like Maxwell, he did not himself actively extend its grasp, particularly after he failed to elicit specifically Helmholtzian behavior in the laboratory. That attempt preoccupied Helmholtz during the 1870s and forms the immediate context that Hertz encountered when he arrived in Berlin.

$$\cdot\,\cdot$$

THREE

Realizing Potentials in the Laboratory

3.1. AN EXOTIC TOOL

Several years before Hertz arrived, Helmholtz had a series of experiments performed in his laboratory to elicit specific novel features of his electrodynamics,
in particular to distinguish it from Weber's.[1] These experiments relied on the
implications that Helmholtz drew from his general potential that inhomogeneous currents should exert three novel actions, one electromotive and two mechanical.

To find these actions we divide with Helmholtz the electrodynamic potential
Φ into two parts, Φ_1 and Φ_2:

$$\Phi = \Phi_1 + \Phi_2$$

$$\Phi_1 = -A^2 ij \iint \frac{ds \cdot d\sigma}{r}$$

$$\Phi_2 = -A^2 \frac{1-k}{2} ij \iint \frac{\partial^2 r}{\partial s \partial \sigma} ds d\sigma$$

Here ds and $d\sigma$ are circuit elements, and i,j are the current-strengths in their
respective circuits. This division isolates the effect of the disposable constant
k in a single term, Φ_2. That term always vanishes, whatever the value of k may
be, if at least one of the two circuits is closed.[2] Since this is always the case in
Helmholtz's experiments, k had little experimental importance—it was primarily a device to achieve formal agreement between certain aspects of the several
theories for electrodynamics. Helmholtz did derive the novel electrodynamic
actions that the k term involves, but his focus in the laboratory was always on
the k-independent potential, Φ_1.

The potential Φ_1 has precisely the same form as Franz Neumann's potential
function—and yet Helmholtz will obtain entirely novel actions from it. There
are two connected reasons for this. First, Helmholtz's potential function represents a *system energy;* second, the energy represents the interaction of *circuit
elements.* Neumann's original function had no comparable significance and was
therefore never treated as Helmholtz would now treat his interaction energy.[3]
The core of Helmholtz's novel scheme lies precisely in his insistence that the

electrodynamic potential specifies an element–element interaction—that, in other words, circuit elements have intrinsic physical significance. This assumption reflects the underlying tenet of Helmholtz's physics that any object, whatever its dimensions, can in principle interact with any other object if an interaction energy exists between them. Helmholtz's novel actions, which we will now follow him in deducing, accordingly exhibit mathematically the very deepest convictions of his new creed.

Assuming, then, that circuit elements are electrodynamically active, meaning that they specify interaction energies in and of themselves, Helmholtz could apply the d'Alembert principle (by this time the principle of virtual work) to them. In its integral form, which he first writes down, the principle requires that the work done on the system during a virtual displacement of its parts must be compensated by a change in the system's energy, that is, by a corresponding variation in the interaction energy:

$$\iint F \cdot \delta r_s + \delta \Phi = 0$$

Here δr_s represents the virtual displacement of an element of the s circuit, while the σ circuit, with which s interacts, remains undeformed.

To facilitate the calculation Helmholtz introduced a pair of scalar parameters, p and ω, such that the path lengths s and σ are specified respectively by them. This enabled him to rewrite $\delta \Phi_1$ in the following way:

$$\delta \Phi_1 = -A^2 ij \iint_{\omega p} \left(\frac{dr_s}{dp} \cdot \frac{dr_\sigma}{d\omega} \right) \left(\nabla \frac{1}{s r} \right) \cdot \delta r_s dp d\omega - A^2 ij \iint_{\omega p} \frac{1}{r} \left(\frac{dr_\sigma}{d\omega} \cdot \frac{d\delta r_s}{dp} \right) dp d\omega$$

where the gradient is taken with respect to the deformed circuit s, and r is the magnitude of the distance $r_s - r_\sigma$ between the elements ds and $d\sigma$. Helmholtz's next step is the critical one, and it bears some resemblance to one taken a decade later by Maxwellians. He assumes, without discussion, that it is legitimate to split the integral into parts, each of which can be taken on its own terms. That is to say, he considers that each of these distinct parts can be *partially integrated* independently of every other part, which means that contributions to the complete integral from the termini of every element may now come into play, as follows.

Taking the second part of the expression just obtained for the variation in the potential, Helmholtz in effect[4] rewrote it through a partial integration (in which d/dp is treated as the operator) over the element ds, on which the force that is being calculated acts (not, at this point, over the element $d\sigma$ as well):

$$\iint_{\omega p} \frac{1}{r} \left(\frac{dr_\sigma}{d\omega} \cdot \frac{d\delta r_s}{dp} \right) dp d\omega = \left(\delta r_s \cdot \int_\omega \frac{dr_\sigma}{d\omega} \frac{1}{r} d\omega \right)_{r_a}^{r_b} - \int \delta r_s \cdot \frac{d1/r dr_\sigma}{dp \, d\omega} d\omega$$

The complete expression for $\delta\Phi_1$ has now become

$$A^2ij\left(\delta r_s \cdot \int_\omega \frac{dr_\sigma}{d\omega}\frac{1}{r}d\omega\right)_{r_a}^{r_b} - A^2ij\int_\omega \delta r_s \cdot \frac{d1/r}{dp}\frac{dr_\sigma}{d\omega}d\omega - A^2ij\int_p\int_\omega \nabla\left(\frac{1}{r}\right) \cdot$$

$$\delta r_s\left(\frac{dr_s}{dp}\cdot\frac{dr_\sigma}{d\omega}\right)d\omega dp$$

The variation must be equal to the force that acts over ds. It divides into two parts: one part, due to the first of the three contributions, acts on the ends r_a, r_b of ds and results in a torque; the other part acts on its center. Since the virtual displacements are arbitrary, the second and third terms may be combined by eliminating the parameters p and ω. Putting the terms together, Helmholtz now had the several forces (including the ones that result from $\Phi_2{}^5$) as:[6]

Source	Object	Force
Center of $d\sigma$: Φ_1	Center of ds	$F_{ds} = A^2ij\dfrac{ds}{ds}\times\displaystyle\int_\sigma d\sigma\times\nabla\dfrac{1}{r}$
Center of $d\sigma$: Φ_1	Ends of ds	$F_{\text{center on ends}} = A^2ji_{end}\displaystyle\int_\sigma\dfrac{1}{r}d\sigma$
Ends of $d\sigma$: Φ_2	Ends of ds	$F_{\text{ends on ends}} = \dfrac{1-k}{2}A^2i_{end}\displaystyle\sum_\sigma j_{end}\nabla r$

Here i_{end}, j_{end} are respectively the currents at the termini of the element ds, on which the force in question acts, and at the termini of the element $d\sigma$, with which the interaction takes place. The sum is taken over all such elements $d\sigma$.[7]

Instead of writing i_{end}, j_{end}, Helmholtz, in a move that was certain to cause confusion among those who did not thoroughly grasp his analysis, used the continuity equation to replace the linear currents with the changing charges (respectively, de/dt, $d\varepsilon/dt$) in which they terminate. Unless the analysis is followed through, this replacement can easily lead to the conclusion that the interactions occur between current elements and changing charges or between pairs of changing charges. On the contrary, the charges appear solely because of the continuity equation, not because of any interactions that directly involve them as states. The electrodynamic potential itself has nothing at all to say about charges, and these forces emerge directly from it. Why then did Helmholtz here introduce charges at all? He did so in order to specify the circumstances under which the novel forces that he had obtained will have an effect.

A circuit can be divided in an indefinite number of ways into elements. According to Helmholtz's scheme the novel forces act at the termini of every element in any of these divisions. However, unless an element has a terminus that bounds the circuit, it will contact two other elements, one at either end. Now in order properly to compute the novel forces on the ends, care must be taken to give the correct sign to the currents that pass through the ends. In particular, if some point bounds two elements, then a positive current for one element constitutes a negative current for the other element because it leaves the one and enters the other. Since the new forces are linear functions of the current, it follows at once that their net result vanishes for points that lie within the circuit proper. The circuit's boundaries—where (and only where) the continuity equation entails that changing charge must occur—are not compensated in this way,[8] and so here, but only here, the new forces can be detected. Helmholtz substituted changing charge for current to emphasize this point. We shall, however, see below that confusion persisted concerning the actions that take place on internal elements, eventually requiring Helmholtz (and Poincaré) to provide an explicit, careful discussion of the point and to generalize his analysis by applying it to the realistic case of extended conductors.[9]

Helmholtz also deduced a new electromotive force, according to which the open end of a circuit will itself act to alter currents. The new term is a function of the changing charge density at the circuit terminus, but again, solely because of the continuity equation. The actions are in fact due only to currents in the usual sense. This effect must be extremely small because it depends on the second derivative of the charge density with respect to time. The other two, mechanical, effects might conceivably be detected given contemporary instrumentation. The first experiments undertaken in Helmholtz's laboratory in 1873 were accordingly designed to detect the new mechanical action between a circuit terminus and a closed circuit.

Since Helmholtz felt that his potential function could capture every major contemporary form of electrodynamics, these early experiments were not at all done with an eye to choosing between theories. They were done rather to elicit hidden novelties that every one of these theories could in principle accommodate, but that none had ever identified. Here we have nothing to do with a choice of any kind between theories but rather with a deliberate, careful attempt to bypass theory altogether. To do this required Helmholtz and those who worked under his direction to build new devices and to use familiar ones in unfamiliar ways. Perhaps the best way to understand these early experiments is to think of them as attempts to activate subtle effects—not as attempts to distinguish between theories or to confirm some particular theory. Of course the potential is itself a theoretical construct, but according to Helmholtz it is something that gathers within itself every theory and so stands beyond all of them.[10]

Until Helmholtz's new way of thinking made them interesting objects, open circuits had been applied in the electrodynamic laboratory solely as devices (actually as parts of devices) to produce something else that was useful. By

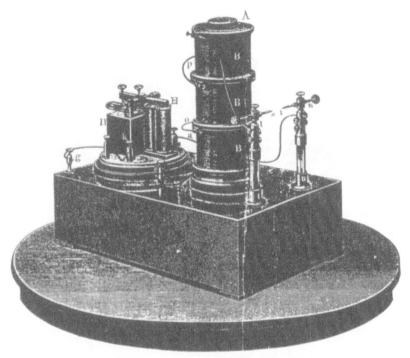

FIGURE 2   Vertical coil (G. H. Wiedemann 1885)

1870 high-voltage currents had long been generated through various kinds of induction coils (figs. 2 and 3). Although coils came in many forms, they consisted essentially of two distinct but structurally identical components, one of which was used to produce the desired effect in the other. Both parts were composed of wire coils. The activating coil was connected across a powerful battery (either a Daniell's or a Grove's cell) and contained a switch (or "hammer") to open its circuit (figs. 4 and 5). The activated coil surrounded the first one, and it was connected across whatever it was that needed high-voltage current. Once the battery had established the comparatively low voltage current in the first coil, the switch was thrown open, and as the current rapidly waned, a high-voltage current was produced by electromagnetic induction in the second coil. To produce a reasonably steady action the switch often formed part of an "interrupter" that broke the first circuit and reestablished the current in it many times a second. In 1870 the standard for such devices was the reliable and powerful Ruhmkorff coil.

The point to notice here is that the device used open circuits only to generate something else. In itself the open circuit was completely uninteresting, and in any case the apparatus as a whole served as a source of high-tension current— as a producer of an important laboratory tool—rather than as an object of in-

FIGURE 3   Horizontal coil (G. H. Wiedemann 1885)

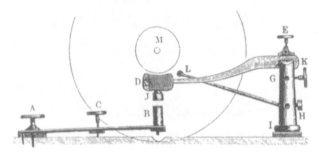

FIGURE 4   Wagner hammer (G. H. Wiedemann 1885)

quiry in itself. Consequently, the open circuit remained thoroughly unfamiliar as a generator of effects. To suggest (as Helmholtz now did) that it may do strange things troubled the electrodynamic experimenter's sense of instrumental stability just as much as it roiled theoretical waters. And the tool that Helmholtz's laboratory used to explore this alien domain of open circuits, the oscillating current, was itself an unaccustomed laboratory presence.

3.2. SEEKING NEW EFFECTS

Helmholtz probably had the first experiments performed sometime during the winter or spring of 1873; in the absence of laboratory notebooks, we cannot be sure precisely when he began work. Nor can we know in detail the strategies that he and his assistants developed to deal with recalcitrant devices, or the instrumental adjustments and changes that they made. Most of these results were published serially over several years in articles that are hardly polished pieces of work. They are rather attempts to contain damage, to reply to criticisms, and to reach further, all mixed with ongoing reports from the laboratory. Consequently, if read with careful attention they can provide unmistakable

FIGURE 5   Helmholtz interrupter
(G. H. Wiedemann 1885)

hints (though certainly not unambiguous evidence) about what went on in the laboratory. Unlike final reports that present results which the experimenter thinks have been successful—and that, accordingly, are intended to close the point at issue—these reports from Helmholtz's laboratory leave nearly everything open until quite late, when Helmholtz bound them to conceptions that, until then, he had kept far from the laboratory.[11]

One of the early experiments sought a *positive* result: something that should happen only according to Helmholtz's system. Another sought a *negative* result: something that should not happen according to Helmholtz but that should happen according to the usual interpretation of the Ampère law.[12] In the first experiment a steel ring coiled about with wire hung vertically; immediately beneath it a circular capacitor hung from a torsion cord (fig. 6). The coiled wire surrounding the ring magnet was taken off at two points and attached to the centers of the capacitor's plates. A current supplied to the coiled wire produced a magnetic field within the ring, and it also produced radial currents in the capacitor plates that terminated at their circumferences, where the novel action accordingly arose.

Helmholtz noted that under these circumstances the charge on the capacitor will "oscillate," which, though Helmholtz said nothing about this, will occur in such a way as to have a major effect only if the circuit forms the secondary in an induction coil.[13] Here the secondary has itself been broken open and made thereby the object of inquiry instead of the source of power. In common electrodynamic usage electric oscillations were solely tools for generating secondary currents; here they were made to serve a novel function, although in themselves they still remained tools rather than objects of intrinsic interest.

According to the "Ampère law" the ring magnet should not affect the capaci-

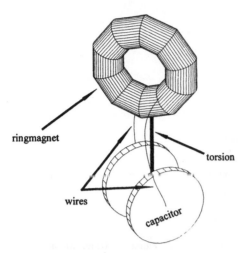

ringmagnet

torsion

wires

capacitor

FIGURE 6   Ring magnet and circular
capacitor

tor, because the magnetic field is confined to the ring.[14] But according to Helmholtz there will be an action between the closed magnetizing current around
the ring and the diverging current at the circumference of the capacitor that
must twist the capacitor about the cord. Helmholtz did not provide the analysis,
but one can qualitatively understand how the action arises by considering
Helmholtz's force $A^2j(\partial\rho/\partial t)r \cdot ds/r^2$ between the current that terminates at the
capacitor's circumference and the elements of the wire wound about the ring.
Since the distances of the wire elements from any given point on the capacitor's
circumference vary, every winding on the ring will exert a net force on the
point. For the same reason, that force differs for every point of the capacitor's
circumference. Symmetry considerations indicate that for a ring magnet the
net result will be to twist the capacitor about the axis that joins its center with
that of the ring magnet. To detect the effect, which will be very small, requires
setting the capacitor into torsional oscillation, and to do that the current that
feeds the device must itself oscillate.

In the second (negative-seeking) experiment the circular capacitor sat about
an axis that passed through the center of a cylindrical electromagnet placed
beneath it (fig. 7). The wire that spiraled round the cylinder terminated at the
capacitor's plates. As before, radial currents occurred on the plates and terminated at the peripheries. The Ampère element–element law requires that the
capacitor twist about its axis because the radial currents in it occur within a
sensibly uniform magnetic field that is perpendicular to the plates. But—and
this is a critical feature of Helmholtz's system—nothing whatsoever distinguishes one part of a capacitor plate from any other part in *this* experiment;
consequently, all parts of a plate must have precisely the same electrodynamic
potential, and so the magnet cannot exert any force on it. Nothing should
happen.

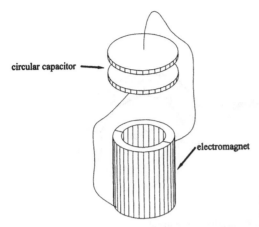

circular capacitor

electromagnet

FIGURE 7   Cylindrical electromagnet and circular capacitor

When Helmholtz attempted these experiments, they failed repeatedly to produce what he hoped for (though no details of the tests apparently remain). The failures must have been quite unambiguous, and Helmholtz immediately sought for an explanation in hidden pathologies of the device. The "least asymmetry" in the device, he remarked, permitted electrostatic forces to swamp the electrodynamic ones he was looking for (Helmholtz 1873c, p. 701, footnote). In these first, failed experiments Helmholtz had tried to find a torsional motion (or lack of it) on the object that bears the changing charge. But if the objects were not highly symmetric, then the electrostatic actions between the rapidly changing charges upon them and any neighboring conductors would also twist them. It was an obvious step, which was soon taken in Helmholtz's laboratory, to modify the experiment by keeping the object that bears the changing charge stationary and then looking for a torsion of the magnet. This would completely eliminate the electrostatic pathologies.

In the summer of 1874—about a year after the early experiments—a Russian scientist named Nikolaj Schiller visited Helmholtz's institute, where the new experiment was performed.[15] Schiller's experiment was essentially an inversion of the one that Helmholtz had performed the year before with a ring magnet. But now, instead of observing the action of the magnet on the charged body, a rapidly growing charge was used to test for a torsion of the ring magnet (fig. 8). To make the action as large as possible, Schiller, perhaps at Helmholtz's suggestion, made a critical change in the basic structure of these experiments. Instead of using induction coils to feed a capacitor, he used a powerful electrostatic induction apparatus—a "Holtz machine"—to discharge into a wire that terminated in a point placed near the ring magnet.

The device, first described several years before (1865), had been invented to produce very high charge densities by electrostatic means—so high that a spark can be permanently maintained across the gap between the two small metal spheres shown in fig. 9.[16] When operated without producing a spark (by

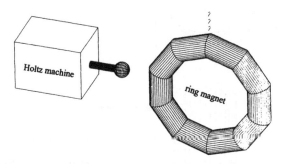

FIGURE 8   Schiller's experiment

increasing the gap between the metal spheres), Holtz's machine has the obvious advantage for Helmholtz's new experiments of permitting the magnet's deflection to be detected over time (as the charge density on the device increases) without forcing it to oscillate. This use for the apparatus symbolized instrumentally the disturbing new associations that Helmholtz's system suggested in the laboratory: the Holtz machine was an electrostatic device being used to produce an electrodynamic effect that had nothing to do with the currents, in the usual sense, that effect the changing charge densities which it generates. On the contrary, Helmholtz was looking specifically for something that no one had previously suspected that the device would produce, namely an electro-dynamic effect of terminating currents. In suggesting that such an effect might exist, Helmholtz at once rendered problematic in electrodynamic application a device (the Holtz generator) that had already been thoroughly assimilated into contemporary German practice.[17]

Of course the extraordinarily large charge densities produced by the Holtz apparatus—and which were themselves its usual purpose—had to be prevented from affecting the measuring device. To ensure this, Schiller encased the ring magnet in a conducting enclosure and went so far as to crisscross with wires the glass window through which the magnet was observed. Under these conditions, then, the electrodynamic force that should be exerted by the changing charge at the end of the point will twist the ring magnet about an axis that is perpendicular to the line joining the point with the magnet's center. Since the magnet hung down, and the point was placed on a line with its center, the ring should have twisted about its central, vertical axis.[18]

But there was no deflection at all, or at least nothing like what it should have been. Schiller repeated the experiment under much more precise conditions when he returned to Moscow and determined conclusively that he would have

FIGURE 9 A Holtz electrostatic generator (G. H. Wiedemann 1885)

detected the action had it existed since the magnet should have been deflected by an easily observable twenty-seven divisions on his scale. This must have been an astonishing and disheartening result for Helmholtz, though we do not know what his immediate reactions were. He apparently waited for the results of Schiller's more precise measurements before undertaking new ones of his own to see where the problems lay.

In Schiller's experiment the very high potentials that are generated by the Holtz machine led to disruptive discharge at the point placed near the ring magnet. In Helmholtz's unadorned scheme the rapidly moving, charged air particles that flee the point should have nothing but an electrostatic effect, and so should in no way influence Schiller's apparatus. They are moving, charged bodies, not current-bearing conductors, and there is no interaction energy between such things and currents. Helmholtz concluded that perhaps the air particles that convect electricity do somehow act electrodynamically, in which case Schiller's results must be interpreted with this in mind.[19] In Helmholtz's words:

> It is accordingly to be concluded that either the effects of current ends indicated by the potential laws do not exist, or else, in addition to the electrodynamic effects indicated by these laws, yet others due to the convectively

> driven electricity exist, that the potential law is therefore incomplete if, in it, one only takes account of the distance actions of the electricity that flows in the conductors. (Helmholtz 1875, p. 781)

But this alone hardly explained why Schiller's experiment failed: even if convected charge does in some way act electrodynamically, why was there no action at all on Schiller's ring magnet? To answer that question in a way that actually supported his fundamental outlook, Helmholtz discussed an experiment that he had carried out himself, and whose implications he developed in the same article that first reported Schiller's year-old results.

All three experiments that had been performed thus far sought to measure the mechanical force that must, according to Helmholtz, be exerted by the open end of a current-bearing conductor. None of the three had in any way at all been concerned with electromotive actions—for the reason that the new electromotive force required by Helmholtz's potential would be extremely difficult, and perhaps technically impossible, to measure under these circumstances. Although we cannot know this with certainty, it seems very likely that Helmholtz set about designing a new experiment that would show effects when he was first faced with the failure of Schiller's to show electrodynamic actions, but before Schiller had demonstrated, in Moscow, to Helmholtz's satisfaction that there is no detectable action. That is, Helmholtz reacted to Schiller's results in Berlin by deciding to put aside the question of electrodynamic actions and to examine electromotive ones instead, since as he remarked in his published account a year later: "Connected with this difference [between his theory and others] in the determination of the ponderomotive force there is another such in the determination of the induced electromotive force in unclosed conductors" (1875, p. 781). But since the new electromotive forces required by Helmholtz were too small to measure directly given contemporary technique, a reasonable course would be to perform an experiment in which something electromotive should not happen according to Helmholtz but which should happen according to Fechner–Weber. This line of reasoning probably led Helmholtz to perform the experiment whose results, together with the highly convincing ones that were obtained by Schiller when he returned to Moscow, generated strong but unavoidable stresses that Hertz brooded over nearly a decade later. This experiment for the first time introduced something beyond Helmholtz's basic system into the laboratory. It introduced *theory*—before this entirely absent—as well.

3.3. THEORY INVADES HELMHOLTZ'S LABORATORY

The principle of Helmholtz's experiment exploited a peculiar, and obviously confusing to his contemporaries (he had several times to explain the point), aspect of his system concerning how electromotive forces are brought about by the motion of a conductor in a magnetic field. Consider what might seem

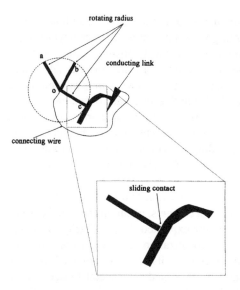

rotating radius

a

b

conducting link

o

c

connecting wire

sliding contact

FIGURE 10   Helmholtzian induction by *Gleitstelle*

to be a simple situation as Helmholtz described it (fig. 10). A conducting radial arm rotates about a fixed point in a homogeneous magnetic field that is perpendicular to its plane of rotation. At some point in its swing the arm comes into contact with a conducting link, thereby producing a complete circuit by means of a wire that connects the link to the arm's pivot. While the arm is in contact with the link, a current flows in the circuit. This is simple to understand in Fechner–Weber or, say, in terms of the cutting of magnetic lines of force by the rotating arm. Both of these theories, however, assert not only that a current flows through the circuit when the arm contacts the link but that there must be an electromotive force that is directed along the arm whether or not it touches the link. That electromotive force, which is always present, generates the current in the circuit. When it cannot generate current, the radial arm must be charged at its ends.

This kind of situation, which almost prototypically represents electromagnetic induction, is comparatively difficult to understand in Helmholtz's system. Indeed, one's first thought is that there should be no current at all: since the magnetic field is homogeneous, the rotating arm is at every moment in precisely the same circumstances, and so it should have always the same interaction energy associated with the external sources. In which case no actions of any kind can occur, because they arise only from changes in the electrodynamic potential. However, the premise of this argument is in fact wrong: the rotating arm is *not* in the same circumstances at every moment, even though the field is homogeneous. When the arm first comes into contact with the conducting link, it begins to slide along it over some finite distance. During this time a circuit exists whose length changes as the arm slides along the link. At other times *no* circuit exists. Consequently, at the locus of the sliding connection—or *"Gleitstelle"*—a conducting region acquires an electrodynamic potential as part of a circuit, which it did not previously possess before the circuit formed. This constitutes a change in potential over time, one that takes place in the direction of the sliding contact, whence an electromotive force arises there, but only there: there is no electromotive force at all generated in the arm itself.[20] According to Helmholtz, therefore, the isolated arm should not be charged at its ends during the rotation, but a current should exist in the circuit during sliding contact, whereas according to Fechner–Weber or an interpretation in terms of lines of force, the arm should also be charged. Helmholtz's experiment probed the difference.[21]

He designed an apparatus that could be used in two kinds of experiments. In one, only electromotive forces, but not currents, occur. In the other, currents also occur. To accomplish this Helmholtz exploited the peculiarity of his system that we just examined: namely, its localization of the electromotive force in the region where the moving circuit changes its extension. In his device (fig. 11) two cylindrical metal sections (*b*) are joined by a metal bridge (*a*) and fastened to a central axis about which the object can rotate. Two other cylindrical sections (*c*) rest in fixed positions concentric with the movable ones and diametrically opposite one another. When the inner sections are opposite the outer ones, the latter are grounded (*A*); when they have rotated another quarter-revolution, the outer sections are placed in contact with a Kohlrausch condenser, whose charge is measured with a Thomson quadrant electrometer. The whole is placed within a homogeneous, axially symmetric magnetic field.

According to the common understanding, the inner sections will be charged oppositely from one another as they rotate through the field. Consequently, when opposite *c*, they will induce a charge on these outer plates of equal magnitude but opposite signs to the ones that they carry, because the *c* are grounded. When (*B*) the inner sections have passed by, the *c*, now in contact with the condenser, transfer their charge to it, and this is measured by the electrometer. Helmholtz's description is apposite:

> If the plates *bb* are charged positively by magnetic induction, then the plates *cc* in position *A* will be negative with respect to ground, and indeed the apparatus thereby works as a condenser, so that a moderate electromotive force piles up a proportionately significant quantum of electricity. If the plates move to position *B*, then the entire negative electricity, whose potential is considerably enhanced by the distance of the positive plates *bb*, is conveyed into Kohlrausch's condenser and piles up in the latter until its insulating plate itself acquires the potential of plates *cc* in position *B*. (Helmholtz 1875, p. 784)

But according to Helmholtz's system the condenser should never become charged, because the inner cylinders have no electromotive force generated in them. On the other hand, if the rotating sections make sliding contact with the

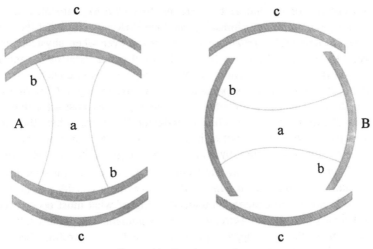

FIGURE 11 Rotating capacitor

outer sections, then according to Helmholtz's as well as the usual understanding, a current will flow that terminates on the *c*, producing a charge that will be measured in the electrometer. Helmholtz certainly found the condenser to be charged when *Gleitstellen* were present. But he also found that "the plates of the rotating condenser will be charged by means of an induced electromotive force even when slip-joints are absent" (Helmholtz 1875, p. 785). This result for electromotive action, when combined with the news of Schiller's continued failure in his more accurate experiments to find the mechanical force, left no doubt in Helmholtz's mind that his original scheme had to be married to the ether, that it could not stand alone, without the support of theory.

Both experiments had in the end failed to show the results that Helmholtz's electrodynamics apparently required. And neither Helmholtz nor anyone in his laboratory questioned this failure in detection. They did not, that is, question that their devices were capable of revealing the appropriate actions if they existed, *and if there were no compensating effects.* The instrumental stability of the experiments was in their view simply too powerful to permit arguments based on poor accuracy or inadequate execution. Helmholtz was, however, well aware that in one of them—the Schiller experiment—powerful convection currents probably occurred. If they occurred also in his own experiment, then that link between them might hold the clue to the failure. However, in Helmholtz's experiment there is no disruptive discharge and so, it seems, no comparable convection current. The experiments are not at all similar to one another in this respect, so that convection per se does not provide an obvious solution.[22] Although in the absence of documentary evidence one can only conjecture the course of Helmholtz's reasoning, nevertheless his ultimate resolution of the quandary provides hints of what probably happened.

Concentrate for the moment on Helmholtz's own experiment, and think about it as Helmholtz must himself have done.[23] Here we find that the device behaves as though *Gleitstellen* are always present even when physical contact between the inner and outer cylinders does not occur: *Gleitstellen* must be there even if they do not seem to be. When they pass by one another without touching, the cylinders are separated by air. Consequently, *Gleitstellen must exist in the air*. That is, the air between the passing plates must act with respect to external currents or magnets like a conductor. It is perhaps here that Helmholtz realized a solution lay, not with convection, properly speaking, but rather with the continuous creation by motion of new regions that interact with the external currents. One may say that Helmholtz was forced by his own devices, which had been constructed *without theory*—without considering at all the polarizable ether—to change his understanding of what must occur even in prototypically Helmholtzian circumstances. Argument, he evidently felt, could not destabilize these unhappy experiments that had so carefully been performed under his auspices.

The essence of Helmholtz's inspiration was determined by necessity: whatever physical process goes on in the *Gleitstelle* must be able to produce charge accumulation on the outer cylinders just as though conducting contact had been maintained and a current had terminated there. But, of course, the physical conductor has been replaced by air. Suppose, however, that the air can be polarized by electromotive force. (This means that there must be an interaction energy between currents and polarization, so that the total system energy will change as the cylinders slide past one another and alter the size of the polarized region.) With metallic contact the outer cylinder becomes charged as a result of the continuity equation, since a current terminates upon it. But with air replacing metal the outer conductor will charge as a result of electric force, the air's polarization inducing the charge by static means. With metal contact the charging process involves a single interaction energy between current and current, as well as the continuity equation. With polarized air the process involves two interaction energies—first, between the air's polarization and the external currents and, second, between the air's polarization and the charge of the outer conductor—whereas the continuity equation plays no significant role at all.

Helmholtz had found, to a high degree of accuracy, that air and metal *Gleitstellen* have the same relation to electromotive force. The only possible conclusion, if the result is due to polarization, is that the surface charge at the boundaries between the separating air and the bracketing cylinders is to a very high degree of accuracy the same in magnitude as that which exists on the outer cylinder when metallic contact occurs. To explain the effect we can retain the original requirement that pure motion does not generate an electromotive force and, as we just saw, treat the separating air as a nonconducting *Gleitstelle* at which polarization is generated by the change in potential that results from the slide. Or we can assume that pure motion does generate an electromotive force, in which case conduction charge will be produced on the moving arm by the motion via the continuity equation. These two alternatives, only the first of which is based on the notion that the air can be polarized by a *Gleitstelle* effect, yield different values for the conduction charge on the condenser. But the empirical result can be obtained, whichever alternative holds, only if the susceptibility of the air is extremely large. (Appendix 4 discusses Helmholtz's analysis in detail and probes its difference from the Maxwellian one for this experiment.) However, a major virtue of Helmholtz's conclusion that the susceptibility of the air is very large is that it can also be used to explain Schiller's results.

The absence of mechanical force in the Schiller experiment must mean (continuing to think within Helmholtz's framework) that the force due to the end of the conductor is completely canceled by some other force. The air will be polarized, under Helmholtz's assumption, by the terminating charge on the wire in Schiller's apparatus (which appears there as a result of the continuity equation). As the charge on the wire's end changes, so does the surface polarization charge of the air that immediately embraces it. That charge is opposite in sign to the conduction charge and, if the air's susceptibility is extremely large, nearly equal to it in magnitude. If we assume (as, we have seen, was entirely natural in all theories) that the changing polarization acts electrodynamically, then this polarization current will completely cancel the mechanical action of the changing conduction charge. In other words, the action is not missing because the corresponding force does not exist but because it is *canceled* by another force of the same kind.[24]

Taken together, then, Schiller's, Helmholtz's, and Rowland's experiments, in the context of Helmholtzian theory, required that the air's susceptibility must be very large indeed (all three), that changing polarization must act electrodynamically in the same fashion as a changing charge density (Schiller and Rowland), and that electromotive force must generate polarization (Helmholtz). As we have seen, the electrodynamics of polarization current raised few fundamental issues, whereas Helmholtz's requirement undoubtedly requires a new and mysterious interaction energy. In conductors there is no need for a direct connection between charge and electromotive force. In dielectrics there must be. The implication of Helmholtz's experiment, therefore, was that his basic scheme by itself failed to represent the full spectrum of interaction energies through the static and vector potentials.[25] Air polarization must be taken directly into account.

This was a stunning shift in the original thrust of the experiments. What had begun as an attempt to elicit novel effects that every contemporary theory (Helmholtz thought) could encompass ended in the embrace of a particular form of electrodynamics—Helmholtz's version of Maxwellianism—since the unacceptable alternative was to reject the possibility of founding electrodynamics on a potential.[26] This had the effect of constraining subsequent Helmholtzian laboratory work in a much stronger fashion than the potential itself did, because now one had to build devices in ways that paid close attention to the intricate, confusing theoretical details of changing polarization. The failure of the early experiments to reveal novelties, one might say, transformed a laboratory where work was molded about new ways of doing things rather than about high theory into a place where theory greatly restricted the freedom to manipulate and to create devices, and where attention turned rather to discriminating between theories than to finding new things that could be common to them all. So it is hardly surprising that the Berlin Academy offered a prize in 1879 for an experiment—the first one that the newly arrived Hertz undertook—to test a theory (Weber's): because Helmholtz's system had by then been bound to a particular form, rather than to a multitude of possibilities, the question of which possibility was empirically tenable, which had not previously been pressing, was now unavoidable.

PART TWO

Information Direct from Nature

FOUR

A Budding Career

4.1. THE MAGDALENENSTRASSE

A carefully posted photograph taken in 1857 displays the infant Heins on his seated mother's lap. To Heins's left, standing but with left elbow casually leaning on the back of a heavy carved chair, Gustav Hertz, gaunt and broodingly handsome, stares intently, his right arm placed protectively behind wife and son. The mother, Anna Elisabeth, is plain and rather thin. Her long nose and broad cheeks are offset by lustrous dark hair. Heins's eyes stare quizzically from a round face. Portrait of a young bourgeois family in Hamburg during the 1850s in the midst of a sustained period of economic expansion and political retrenchment.

Gustav Hertz, an attorney, had married well and fittingly. Anna Elisabeth, eight years his junior, was the daughter of a Frankfurt physician, Dr. Johann Pfefferkorn (fig. 12).[1] Over time Gustav's career prospered, and the family moved from an apartment in the crowded Poststrasse to a solid three-story building on the wider Magdalenenstrasse, half-covered in ivy and with an impressive drive, where Heins grew up. Gustav's father was Jewish but his mother was not, and he was raised a Lutheran, though neither he nor Anna Elisabeth were concerned to raise *their* children in a religious atmosphere. Formalities had nevertheless to be observed, and propriety respected, so Heins was confirmed, but the family engaged itself rather with the children's intellectual and moral character than with the well-being of their souls.[2]

Or, rather, Heins's mother concerned herself with such things. His father was for many years only an evening presence whose direct influence upon Heins and the younger children can perhaps be gauged from Frau Hertz's recollections:

> The greatest influence of all on the boys' development was their father. Not only could they feel his most tender and affectionate care at all times, but because of his extensive knowledge and very good pedagogic disposition he became their teacher as well. At meals, *the only time during the week when my husband saw the children,* it was their interests that occupied us and our intimate chats made the hour dear to us. (J. Hertz 1977, p. 7; emphasis added)[3]

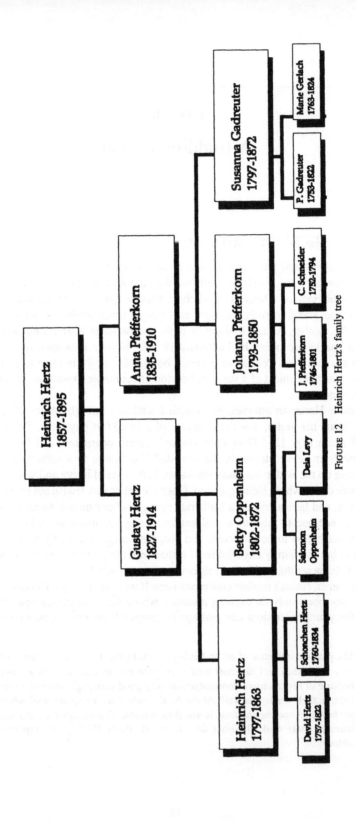

FIGURE 12 Heinrich Hertz's family tree

After he reached adolescence Heins, or Heinrich by then, became more interesting to his father, but until then he was in the hands of his mother, "the servants," and, at the age of six, the schoolmaster. Anna Elisabeth Hertz did not,
however, spend a great deal of time with the young Heins; an hour or so reading
a book to him after dinner apparently sufficed.

Heins's early years were spent little differently from those of many other
children of the north German, Protestant bourgeoisie at midcentury. Home,
play, servants, a concentrated moment of familial bustle at dinner, then a quiet
hour with his mother over a picture book. Later there was music training, an
essential for the properly cultured, and lessons in handicrafts (to elicit manual
dexterity and to inculcate patience). Heins had no ear for music, but he had an
excellent hand for building things, and he possessed what his mother termed
"perfect memory." He was also enthusiastic and perhaps rather nervous, evidently requiring continual stimulation and absorption in activity.

At age six Heins was sent to Friedrich Wichard Lange's school, where he
remained for nine years. Lange was a student of the school reformer Diesterweg, a follower of the Swiss pedagogue Pestalozzi.[4] Lange, true to his master's
convictions, had insisted that Heins not be prepared in any way, preferring no
doubt to put his own impress on the unformed clay. That impress must have
been rather forceful; at seventeen a worldly-wise Heinrich remarked:

> All of us, or at least the better scholars among us, were unusually fond of that
> school, despite the hard work and the great strictness. For we were ruled
> strictly; detentions, impositions, bad marks in neatness and behaviour rained
> down upon us; but what particularly sweetened the strictness and profusion
> of work for us was, in my opinion, the lively spirit of competition that was
> kept alert in us and the conscientiousness of the teachers who never let merit
> go unrewarded nor error unpunished. (P. 15)

Lange and his teachers were not alone in impressing discipline and a "spirit of
competition" on Heins; his mother avidly participated. Anna Elisabeth, by then
the mother of three boys, read and criticized everything, at least during the first
year and a half. She was herself a strict and impatient taskmaster with a "quick
temper," waiting anxiously each Saturday for the weekly school reports. On
the whole she gave the boys a conspicuous amount of freedom but insisted on
punctilious obedience.

Young Heins was at first rarely at the front of his class, according to Lange
because of "slight fluctuations" in "diligence and neatness." But over time
Lange and his mother damped Heins's fluctuations sufficiently that he became
the acknowledged head of the class. This would not have endeared him to his
fellows in a contemporary English public school, but the north German bourgeois environment put a considerable premium on demonstrated and superior
accomplishment. It was not enough to be adequate or even good; it was essential to be outstanding, to stand out from the crowd by virtue of dedicated appli-

cation and natural ability. Still, the hot-house environment of a boys' school, even in midcentury Germany, tends not to reward classroom accomplishment with social esteem, something Heins no doubt realized:

> "Well," [Heins] said, "the teacher also asked who was the brightest and most ingenious among us [students], and then they all pointed to me." "Oh, that must have made you happy?" "Yes," replied Heins, "but at that moment I like to have crawled *under* my desk." (P. 7)

When he was about eleven Heins began also to attend the local *Gewerbeschule,* or industrial high school, on Sundays, where he learned drafting. From his earliest years Heins was absorbed to an unusual degree by modeling, drawing, constructing, in short by the manipulation, depiction, and making of objects: "I often found a piece of his work," wrote Anna Elisabeth, "a mill that actually turned, a forge, country houses, and so on, and sometimes a pretty drawing." His parents encouraged Heins's infatuation with construction and depiction, to the extent that "on the advice of experts" they did not force him to follow the classical curriculum when the class at Lange's school divided. Little wonder that "experts" had to be consulted in such a momentous decision. The class that he joined emphasized "arithmetic and the natural sciences"; the other emphasized Latin. Only the latter led to university.

Although the secondary-school system varied somewhat from principality to principality in preunification Germany, nevertheless since the early nineteenth century the division between practical and higher studies had been strongly maintained. During the 1810s Wilhelm von Humboldt, among others, had begun to forge a school system in Prussia that was based on the sociocultural concept of *Bildung.* This powerful, yet slippery, notion originally had a strong democratic component, in that it was to guarantee individual freedom by infusing in students a desire for (and selecting among them on the basis of) intellectual and aesthetic learning in its widest possible sense—but learning that determinedly excluded practical training.[5] Over time this neohumanistic *Bildung* became equated with training in ancient languages, and in short order a social divide of nearly unbridgeable proportions developed between those trained in the classical *Gymnasien* and those trained in the practical *Realschulen* and the technical *Gewerbeschulen.*[6] By the late nineteenth century, particularly after a period during which the boundaries between the schools were redrawn, the antagonism had become sufficiently intense that contemporary memoirs refer to insults and fistfights between students from the different kinds of schools (Albisetti 1983, p. 31). "The antipathy," notes one historian, "of the neohumanists for anything even remotely tainted with practical training meant that the *Gymnasium* neglected the contribution that manual skills might make to *Bildung;* even drawing played at best a tertiary role in a *Gymnasium* education." The *Gymnasium* had a monopoly (through its *Abitur,* or leaving, examination) on training for university education, which meant in particular on the training of all higher civil servants.

Hamburg was a commercial city, and despite his training as a lawyer there can be little doubt that Heinrich's father, Gustav, must have had considerable ongoing contact with that very practically oriented world, which may have influenced the decision that Heinrich need not follow the usual classical path. This was not by that time quite so limiting a choice as it had been a decade earlier (or in Gustav's own youth), because after 1870 graduates of first-class *Realschulen,* which had greatly expanded following new regulations in 1859, could matriculate in university philosophical faculties and could teach mathematics, the sciences, and modern languages at *Realschulen* (but not at *Gymnasien*).[7] There was great uproar over this, but it perhaps eased the way for the Hertzs to permit Heins not to immediately follow a classical curriculum.

Heins, however, soon came to feel that the classics students were learning more than he was, perhaps because they were learning things that he had never been introduced to formally, whereas they at least had learned something of "arithmetic and the natural sciences" before the class had been divided. There is moreover little doubt that he encountered some of the disdain that incipient *Gebildete* directed at the "practical hack" (Albisetti 1983, p. 31). It was in any case hardly likely that Hertz's parents did not intend him eventually to take the *Abitur,* even if Heins did become an engineer, as he intended, since with it his military service would be reduced to one year in the officer corps. Gustav was by this time beginning to take a deeper interest in Heins and arranged for additional, private classes with a Dr. Köstlin, with the result that Heins would have the advantage of both kinds of training from his twelfth through his fifteenth year.

The next two years (ages fifteen through seventeen) were critical ones for Heins's development. He did not attend the *Gymnasium* but continued his studies with Köstlin and began to take mathematics with a Herr F. Schlottke, in addition to persevering at the *Gewerbeschule* and attending a daily exercise class. Most of the day he studied at home. His mother paints a vivid portrait of this period:

> When he sat with his books nothing could disturb him nor draw him away from them. His desk stood in a room through which I often had to pass, but I always saw him bent over his books in the same way, deep in his work. We never exchanged a word. At 12:30 we both had lunch with our little Otto [his youngest brother]. Half an hour's play with the little boy was his only relaxation. Then he studied afresh, mostly until dinner time at 5 o'clock. (Pp. 15 and 17)

During these years Heinrich's father displaced his mother as the major influence on him:

[H]e found full support and inspiration in his father, who followed Heinrich's studies with great interest. Their lively conversations gave me great pleasure. Unfortunately I understood very little myself. I remember one evening he wanted to tell me about his mathematical studies and I exclaimed, "Oh Heins, I am too stupid for that." He put his arms around me tenderly and said with deepest feeling. "Poor Mama, that you have to miss this pleasure!" (P. 17)

In the evenings after dinner he worked at a turner's lathe, forgoing all study. As usual he "received instruction," this time from a "master turner" named Schultz. And he soon began to produce "various types of physical apparatus, cutting each brass screw, pouring the little weights, and making all the essential parts by himself, with incredible patience."

The separation we see here between mathematical study and physical manipulation endured for over a decade. It strongly distinguished Hertz from other German physicists, who insisted on seating physics in elaborate mathematical abstraction, and throughout his life powerfully molded his approach to physical theory. Very much unlike his contemporaries in the classical *Gymnasien*, Hertz passionately embraced a universe of devices. His mother recalled a significant incident:

He tried to make a spectroscope. His father had promised him the prisms for it and had written about it to Herr Schroeder, who was famous for his optical glass. The reply was that Heinrich should come to see him on Sunday, he would be in his office until 12 o'clock and would fix the prisms for him. Heinrich's joys and hopes were great. He worked on his apparatus with fiery zeal, got up at 5 o'clock on Sunday, hardly permitted himself time for breakfast in order to be ready, and then set out with his father. But because they had apparently miscalculated the time required, they did not get there until five minutes past 12 o'clock and found the place closed and a sign saying that Herr Schroeder would be away for several weeks. Heins came home inconsolable, and it cut me to the quick to see him crying in his quiet way, one big tear after another rolling down his cheeks. (P. 17)

For many years Hertz continued to separate his passion from his interest, mathematics.[8] This is quite apparent in his earliest publications, where the mathematical structure remains entirely subordinate to the demands of experiment and is indeed chosen precisely to fit the necessities of the laboratory. Had Hertz been able—as he was not—to remain enmeshed in the laboratory after his training under Helmholtz, then he might never have merged passion with interest, in which case the interpenetration of theory with laboratory practice, which more than any other single thing characterizes his mature work, would never have evolved. Conversely, the early separation between the universes of mathematics and of devices was itself essential for Hertz to achieve his rooted understanding of laboratory electrodynamics untroubled by the deeper problems of contemporary theory.

When he was seventeen Heinrich returned to the *Gymnasium* for one year (in order to take the *Abitur*) and excelled in the classics curriculum, which he had striven privately for two years to master, at the three-centuries-old Johanneum. The school was difficult because the new director, Richard Hoche, seemingly determined to make his mark, failed nearly half the class.[9] Hertz must have found the Johanneum under this stern Prussian something of a contrast to his memories of Lange's school, particularly since Hoche was strongly antagonistic to the comparatively liberal Lange. One doubts that Lange (though himself a Prussian), the apostle of Diesterweg, would have led the school in joyous celebration of the Kaiser's birthday. Hoche was thoroughly different; he sought to turn his students into good Germans.[10]

Heinrich did well, but (despite his mastery and even enthusiasm for languages)[11] he did not go on to university. Instead he decided at first to pursue his original goal—a career in engineering—by apprenticing to a Prussian architect in Frankfurt am Main. Having satisfied his need to prove himself the master of classics as well as mathematics and science, Heinrich remained enthralled by the concrete. The tension we see here between, as it were, the practical (engineering) and the abstract (mathematics) was particularly sharp among the midcentury German bourgeoisie. On the one hand, there was the continuing pull of the traditional classical curriculum, of the lofty old cultural concept of *Bildung*, which emphasized the superiority of unsullied contemplation (Jungnickel and McCormmach 1986). But by midcentury this was rather heavily counterbalanced by a parctical, though still highly formal, alternative. The resulting stress between *Bildung* and the concrete betrayed itself in the *Gymnasium* and in the university. Hertz's father was a successful attorney (and later a senator), and he was no doubt educated in a thoroughly traditional manner. However, he evidently regretted the rigorous discipline of his youth,[12] which he perhaps associated with the old educational system, and indeed Gustav encouraged Heinrich's practical bent.

A career in science would certainly have been more prestigious socially than one in engineering at this time, because the natural scientist was still a member of a philosophical faculty at a university and admirably *Gebildete*, but it was also more risky because the chances of success (i.e., of becoming an *Ordinarius*, a full professor, at a prestigious university) were not large.[13] Accordingly, Heinrich, having demonstrated that he could master the classics (and, more important, having obtained the requisite credentials), remained committed to a practical career and decided to apprentice himself for a time to architecture. "Only," he wrote at the time, "if I were to prove unsuited for the profession or if my interest in the natural sciences were to increase further, would I devote myself to pure science."

4.2. CIVIL ARCHITECTURE, THE MILITARY, AND MUNICH

The Young Hertz's Wanderjahre

Hamburg
Home and hearth
1856–1875

Frankfurt am Main
Civil architecture and boredom
Spring 1875–spring 1876

Dresden
The Polytechnic: engineering mathematics, Darwin, and Kant
Spring 1876–fall 1876

Imperial Berlin
The military: discipline and more boredom
September 1876–October 1877

Munich
The Polytechnic
October 1877–November 1877
The university and physics
November 1877–August 1878

Helmholtz's Berlin
September 1878

Heinrich (fig. 13) arrived in Frankfurt am Main in the spring of 1875 and went immediately to work for Baurat Behnke on redrafting designs for public-works buildings. Since his mother came from the town, and her brother-in-law (Emil von Oven) was well known locally, Heinrich had scarcely left the nest altogether. From the outset, however, he felt restless and occasionally irritable. The office work neither challenged nor interested him, and in any case filled at most six hours a day: from nine in the morning until noon, and then from three until six. He tried to occupy his luncheon time with reading but found it hard to find an appropriate place to do so. "How I could make some good use of my day," he wrote home, "even with only 6 working hours, I do not know." His parents perhaps sensed the homesick boy's ill ease and visited him in May, which revived his spirits. He plunged into a campaign to master whole new areas of learning and craftsmanship, and to pursue old ones further. Sculpture, Greek literature and philosophy, and architectural history were gobbled in turn. But nothing satisfied him. "If only I were educationally or usefully occupied here I would just as soon stay," he complained in September.

But then he rediscovered the allure of physics. In his increasingly desperate search for something interesting and satisfying to do, Heinrich tried to join the

FIGURE 13   Heinrich Hertz. By permission of the Burndy Library, Dibner Institute, MIT.

local Physics Club, which was run by Professor Rudolph Boettger, an able producer of demonstration experiments (Jungnickel and McCormmach 1986, vol. 1:126). At first he had trouble contacting Boetgger but nevertheless plunged ahead with his usual enthusiasm, reading Wüllner's *Physics* (a widely used elementary text of the day filled with diagrams of intricate instruments) and the German translation of Tyndall's *Heat as a Mode of Motion*[14] "I have regained a strong inclination for natural science through reading Wüllner," he wrote in late October, "but I cannot convince myself to give up what I have set as the most desirable goal for myself." Boettger's public lectures, with elaborate demonstrations, enthused him: "My outlook, my thoughts of the future, change with every day. Since I read a great deal in Wüllner's *Physics* I am again turning very much towards natural science." The signs must have been unmistakable to Gustav and Anna Elisabeth over the Christmas holidays, but Heinrich did not in any case have long to spend in Frankfurt. In the early spring of 1876 he left for Dresden to study engineering at the local polytechnic. Here at least Heinrich could grapple directly with science, but he remained lonely and homesick. For the most part, and apparently for the first time, mathematics captivated him—"sometimes marvellous things come in view that make one's head swim"—and his expertise in it grew rapidly. He also read Kant's *Critique of Pure Reason* and studied "Darwinism." Then, in September, Heinrich began his compulsory year in Berlin serving in the military of Bismarck's five-year-old German empire.

Hertz found military discipline a good antidote to laziness, but he was soon bored: "each day and each hour is like every other, and every day that has passed is regarded as a day conquered." By the end of February he was brooding over his future, gloomily writing home that "day by day I grow more aware of how useless I remain in this world." His mood fluctuated, but he was in good form by summertime. In October he left Berlin for Munich, where he intended to study at the polytechnic to further the formal engineering background that he had begun in Dresden. His determination did not last long. By the end of the month he had decided that engineering was not for him, that he was gripped by the natural sciences, and therefore that he had to enroll instead at the university. Such a major decision naturally required his parents' consent, and on November 1 he wrote them a long and touching letter:

> I cannot understand why I did not realize it before now, for even in coming here it was with the best intention of studying mathematics and the natural sciences and with no thought at all about surveying, building construction, builders' materials, etc., which were supposed to be my main subjects. I would rather be an important scientist than an important engineer, but rather an unimportant engineer than an unimportant scientist; yet now, as I stand on the brink, I think that what Schiller said is also true: "And if you don't dare to stake your life, you can never hope to win the strife," and that too much caution would be folly. (P. 63)

His father's permission arrived on November 7, and Hertz decided to "stake his life" on physics.

He at once contacted the Munich professor of physics, Philipp von Jolly, a well-known experimentalist who urged mathematics, mechanics, and home study on him. Hertz took his advice to heart, including Jolly's recommendation to concentrate on the "old sources," and he spent the next four months intensely engaged with Lagrange ("or sometimes another mathematics text, for Lagrange is dreadfully abstract") and with Montucla's ancient and lengthy history of mathematics. He found little useful in the "new mathematics" (including non-Euclidean geometry) for the physicist "for I find it so abstract, at least in parts, that it no longer has anything in common with reality." Already he itched for the laboratory though he was not even studying experimental physics: "ideas for a hundred experiments I should like to undertake now occur to me, and I already look forward with pleasure to the distant homecoming in the long vacation, when the turner's lathe will really turn again." His "main wish" was for a copy of Wüllner's *Physics,* the text he had first examined in Dresden. Wüllner's work was filled with diagrams of intricate experimental devices and careful, though not deeply theoretical, explanations of how they worked. But Hertz immersed himself for the present in Lagrange and Montucla, occasionally turning for relief to "a significantly easier mechanics text by Poisson." He read slowly and carefully, trying to reason out for himself the implicit meanings of physical and mathematical concepts and their relationships with one another: "I need much time to ponder over matters myself, and particularly the principles of mechanics (as the very words: force, time, motion indicate) can occupy one sufficiently; likewise, in mathematics, the meaning of imaginary quantities, of the infinitesimally small and infinitely large and similar matters."

In the next semester, at the beginning of May, laboratory courses began. Hertz joined the group that "had never done it before" and had to learn at an elementary level how to work a balance and so on—he had to learn what sort of a place the laboratory was, and how people generated meaningful knowledge in it. He progressed fast and immersed himself in it, working in Jolly's laboratory at the university for six hours a week and in Beetz's at the polytechnic for eight hours. "I start at 7 o'clock in the morning," he wrote, "and when I return from work at 6, I am quite tired, especially my eyes, so that I cannot do very much more, and almost the whole afternoon is spent." Professor Bezold (director of the university's Physics Institute) tried to restrain Hertz's obvious propensity to bury himself in the laboratory, urging him "not to turn to physics too early."

By the end of July Hertz had decided not to stay at Munich. He does not tell us why, but he probably felt that the courses in experimental physics available there were not sufficiently powerful.[15] He had a long conference with Beetz at the polytechnic about where to go, and he must have emphasized his overwhelming desire to find laboratory work since Beetz (exaggerating the reality) told him that he "would find a laboratory and a chance to work in it anywhere" (Cahan 1985, p. 4). Beetz offered introductions, and Hertz eventually decided on the Berlin Academy, or, more precisely, on Helmholtz's renowned laboratory.

4.3. Berlin versus Göttingen

> Of the many students now scattered over the earth there is not one who will not to-day think of his master with love as well as admiration. (Hertz on Helmholtz in 1891)

In the summer of 1874 Arthur Schuster, having just obtained his Ph.D. in Kirchhoff's laboratory at Heidelberg, visited Göttingen for a few months, where the seventy-year-old Weber had just handed over his laboratory to Eduard Riecke. "There was," Schuster recalled, "only one student at work beside myself; for the purpose of his doctor dissertation he was magnetising ellipsoids and testing magnetic formulae in the orthodox fashion." The laboratory that Riecke had inherited from Weber was for measuring constants and testing properties. At the end of the summer Schuster traveled on to Berlin, where he found "a laboratory of very different character and ambitions," a place where "promising students from all parts of Germany," crammed into a few small rooms, "were preparing their doctor dissertation on some subject arising directly out of Helmholtz's work." Here, Schuster felt, exciting new work was being done, work that went far beyond mere measurement and testing. In Helmholtz's laboratory "no efforts were made to push numerical measurement beyond its legitimate limits, and though most of the work done was quantitative in character, qualitative experiments were not discouraged" (Schuster 1911, pp. 15–16).

By the early 1870s laboratories concerned with physical research were quite common throughout Germany, but very few among them were either well housed or well funded, and physics as a discipline still had to make claims on resources by emphasizing its usefulness to students in such areas as medicine, pharmacy, and the natural sciences. Despite the poverty of material foundations, and the ever-present demands of often elementary teaching, the "research ethos" was widespread by this time, particularly in laboratories whose directors had the energy and forcefulness to fight for resources and the craftiness to play one university off against another.[16] Schuster's vivid contrast of Göttingen with Berlin reveals that considerably different kinds of work went on within the overall ambit of German laboratory research.

In Weber's and Kohlrausch's laboratory at Göttingen, as well as in Kohlrausch's influential later establishments, *measurement* was the goal, the elemental purpose of research.[17] Even when Kohlrausch did formulate a new hypothesis concerning electrolytic conductivity, it was precisely tailored to the demands of measurement (see Cahan 1989a, for an example). He scarcely used the laboratory as an engine for discovery; he used it as a generator for constants. Helmholtz's "two or three rooms" in early 1870s Berlin had a different ethos. There measurement per se engaged little interest.[18] One doubts, for example, that Helmholtz would have looked with great interest on a project to pursue deviations in the accuracy of Ohm's law, though Schuster had done just that at Göttingen, where Weber had "entered with great spirit into the question."

The division between Helmholtz-trained or associated physicists and those who took their guide from the Weberean ethos became acute by the early 1880s. Max Planck, for example, who took over from Hertz at Kiel in 1885, and who had been much impressed by Helmholtz and Kirchhoff during the year he spent as a student in Berlin [19] was bruised by this rupture in 1887, though he himself was not an experimental physicist. Planck, like Hertz, strongly felt the competitive pressures of the discipline, but because he had turned exclusively to theoretical physics, the difficulty of establishing a reputation, and of getting a position, was particularly sharp. "All the more compelling," he later wrote, "grew in me the desire to win, somehow, a reputation in the field of science" (Planck 1949, p. 20). This led him to compete for an 1887 prize offered by the Philosophical Faculty of Göttingen on "The Nature of Energy." Though he won second prize, no first was given, and the reason he had failed of complete success, Planck noted, was to be found in a single sentence of reservation in the judges' decision, to wit that "the Faculty must withhold its approval from the remarks in which the author tries to appraise Weber's Law." Planck continues:

> Now, the story behind these remarks was: W. Weber was the Professor of Physics in Göttingen, between whom and Helmholtz there existed at the time a vigorous scientific controversy, in which I had expressly sided with the latter. I think that I make no mistake in considering this circumstance to have been the main reason for the decision of the Faculty of Göttingen to withhold the first prize from me. *But while with my attitude I had incurred the displeasure of the scholars at Göttingen, it gained me the benevolent attention of those of Berlin, the result of which I was soon to feel.* (1949, pp. 22–23; emphasis added)

The "result" Planck referred to was quite concrete indeed: in 1889 he was invited to replace the recently deceased Kirchhoff at Berlin as an *Extraordinarius,* or associate professor, and in only three years he became an *Ordinarius,* or full professor. According to his own testimony Planck's early work on the foundations of thermodynamics before the late 1880s had aroused little interest,[20] so that his call to Berlin almost certainly reflected Helmholtz's conviction that Planck was a powerful ally in an important battle. This is not however to say that Planck's major body of work did not also have a place in the Berlin scheme. The Berlin Philosophical Faculty's report on the candidate praised

Planck "for carrying through 'the strong consequences of thermodynamics without interference from other hypotheses'" (Jungnickel and McCormmach 1986, vol. 2:52). They praised him, that is, *not* for his work on the foundations of thermodynamics, but for *applying* thermodynamics to such areas as physical chemistry, whereby he could obtain results without—and this would have been a critical point of approbation in Berlin—"other hypotheses," such, for example, as atoms. This view of Planck's work became, if anything, even more firmly entrenched during the next few years.[21]

F I V E

Devices for Induction

5.1. At Helmholtz's Suggestion

The time Hertz spent in "the reading rooms" learning enough to do the experiments necessary for a prize competition set by Helmholtz must have been well spent, though Hertz was a quick study, for by November 1878 he was busy arranging things in the laboratory. The prize question concerned a subject of great interest to Helmholtz—namely, whether or not the electric current possesses inertia—for in his understanding it should not, whereas in Weber's electrodynamics it must.[1] Hertz was already well aware of the highly competitive environment he had joined, and he was at first reluctant to commit himself fully to the project "since I may fail." Helmholtz encouraged him: "I reported to Prof. Helmholtz yesterday that I had thought over the matter up to a point and would like to begin. He went with me to see the demonstrator and was kind enough to spend another 20 minutes in discussion as to how best to begin and what instruments I should need."[2] On that day, November 5, Helmholtz began to mold Hertz, a process that developed extremely rapidly as Hertz tackled the prize problem and that was, if not complete, then thoroughly formed by early February, when Hertz decided that he had finished.

During the fall Hertz settled into a routine: "an interesting lecture every morning; then I go to the laboratory, where I stay until 4, with a short break; afterwards I work at home or in the reading room; till now I have been kept busy gathering material on extra currents." As well as ingesting how to do laboratory work Hertz also began to understand that he had entered an extremely competitive profession, which both stimulated and frightened him. He wrote that

> when I went to sign up with Prof. Borchardt [for his lectures on analytical dynamics], I took the opportunity of looking into the several registration books of the other students and I was really shocked by their zeal and stamina. Most of them had signed up for 2 hours Weierstrass, elliptical functions, 2 hours Kummer, number theory, etc., one thing after the other, as if it were nothing, so that I was properly ashamed of myself with my two courses [the other being Kirchhoff's on electricity and magnetism]. Then when I took my seat, I could see my neighbor's hand and he too had signed up for the most difficult branches of mathematics every day from 8–12 and in the afternoon

as well. When Prof. Borchardt opened his lectures with the words, "Mathematics has been called, more in jest than in truth, the science of what is self-evident," a pen started writing rapidly near me, and as I turn around I see that he takes down every word in shorthand and is already on the second sentence. To my horror, he went on like this the whole two hours. He looked quite healthy. Even though this zeal is, perhaps, just a little exaggerated, it is true that people work tremendously hard here; and that it is more or less contagious and that I already have some qualms about whether I should not have taken more courses. Anyway this spirit pleases me very much, when it is not exaggerated; it is a comfort that there is no need to apologize when one has some work to do. (J. Hertz 1977, p. 97)

During these early months Hertz, who already had a good measure of competitiveness (not, as we see from his remarks, unusual for German students of the time), learned that he would have to work very hard indeed to make a career in physics. The rapid series of publications that he produced during the next five years show how thoroughly he assimilated this intense research ethos.

Hertz was awarded the prize in early August 1879, and this event marked his proper entry into the profession:

The evaluations will be printed and you will be able to read them. Another paper was entered on the same subject, but it had no luck and received such a devastating judgment that it was greeted with general merriment. The author must have been possessed of great innocence when he sent it in. (J. Hertz 1977, p. 113)

He had lost his own "innocence" sometime the previous fall, and the prize award stamped him in his and his colleagues' eyes as a promising new force in the profession. Although Hertz had for some time been imbued with the sentiments of the German university student, his early success powerfully embedded the profession's elite, competitive ethos. That, we shall see, undoubtedly stimulated him to pursue laboratory discovery, but it also exacerbated his tendency to self-doubt and melancholia, rather common traits among contemporary German professionals.

5.2. How to Detect Electric Mass

The experiment that won Hertz the prize and that first brought him to the attention of the wider physics community in Germany was expressly designed to examine a major point of contention between Helmholtz and Weber: the question of whether or not the current has mass. Or, it would be more accurate to say, the purpose of the experiment was solely to pick holes in the Weberean conception of the current, since Helmholtz's electrodynamics had nothing at all to say about the current and was in principle compatible with the existence of current mass. Helmholtz's argument against Weber was based on a remarkable conclusion that he (1870b, p. 589) had drawn from the Weberan assumption that the current carriers possess mass: namely, that currents on a conducting sphere whose radius is less than a certain quantity would increase over time without limit (see appendix 6). The Weberean reply simply asserted that the electric mass must be so small that the instabilities never occur. No practicable experiment could possibly answer that claim, but experiment could raise doubts about the Weberean position among physicists who were not prepared to accept this sort of a priori invulnerability.[3] If a delicate experiment could reduce the mass's upper limit impressively far, then the unconvinced in the physics community might very well be willing to take that limit as a reasonable value for the mass and thereby permit an empirically based computation of the instability.[4]

Hertz's results are not in themselves of great contemporary significance. They could hardly change Weberean opinion. Yet the experiment was quite important for Hertz's career and for the development of his characteristic laboratory techniques in electrodynamics. He built a reputation as a meticulous experimenter and developed a particular kind of electrodynamic skill in constructing, analyzing, changing, and reanalyzing his devices. The article—his first—that Hertz wrote for the *Annalen der Physik* bears many traces of the working and reworking that he had had to go through in order to reach what he considered to be a satisfactory conclusion. The apparatus and its mutations are rather confusingly discussed in comparison with the limpid descriptions in his later experiments on electric oscillations. But we have here the opportunity to catch Hertz in the process of learning how to deploy Helmholtz's conceptions in just the right way to bring out what interests him.

Hertz began on November 6, 1878. He described his first two days in the laboratory in detail to his parents:

> Yesterday and today I began to set up my equipment. I have a little room all to myself, as large as our morning room [i.e., large], but 1 1/2 to 2 times as high; so I can come and go as I please. You can thus perceive that there is elbow room enough and to spare. But everything else is also capitally arranged; there are tools, gas, running water, a glassblowing bench at hand. A special advantage is that only very few people seem to work there, I have only seen 4 or 5, and, assuming that the rest of the building is similarly populated, there can be hardly 10–15 in it. Therefore there is little bother and everything can be used well. By now I have met two of the assistants, very nice (young) people. It is to be hoped that the results I will obtain will come up to the environment; things could not be more convenient for me than they are right now. Meanwhile I am busy with preparatory work, but even now I take the greatest pleasure in utilizing so many resources. My galvanometer, for which the safest place at home was on the lathe, now stands on an iron bracket let into the wall (Helmholtz said, "I regret you have no massive stone pillar"); if it nevertheless jitters whenever a dray truck passes by on the Dorotheenstrasse, that is just too bad. The telescope can be adjusted in all directions by screws, which is somewhat more convenient than propping it up on

books. I can build a battery of any strength for myself in a special room (to prevent the acid fumes from causing trouble); the wire goes through the wall. As I said, I can only hope the results will satisfy me, but this pleasure will soon turn into habit, and will then be no more. (J. Hertz 1977, pp. 95–97)

Here we see nicely combined Hertz's nearly tactile love for laboratory equipment and his moody realization that "pleasure will soon turn into habit, and will then be no more." Hertz did characteristically become bored with the same equipment after a while, and this was probably a powerful factor in stimulating him to alter old devices and to build new ones.

The instability of the galvanometer's mounting soon proved to be very troublesome, and he was given a different room, which he had to share, but which had a "stone pillar in the wall." Helmholtz came in every day "for a few minutes." By the end of November Hertz was in the laboratory most of the day, usually engaged in routine activities ("cutting corks, filing wires, etc."). But he found the laboratory deeply satisfying, and he was himself rather stunned at how rapidly he had moved to the frontiers of research:

> I should not like to forego these pursuits; I cannot tell you how much more satisfaction it gives me to gain knowledge for myself and for others directly from nature, rather than to be merely learning from others and for myself alone. As long as I work only from books I cannot rid myself of my feeling that I am a wholly useless member of society. It is also rather strange that I am now occupied with quite specialized subjects in the study of electricity, whereas a good six months ago I hardly knew more about it than what I had managed not to forget from Dr. Lange's time. I hope this will not harm my work. At the moment, it looks promising; the difficulties Helmholtz listed as principal ones at the beginning are already behind me, and if all goes well, I shall have finished a preliminary solution in two weeks and I shall then have time to work it up properly. (J. Hertz 1977, p. 99)

Following Helmholtz's initial suggestions, Hertz had within a few weeks reached a point where he thought that he could quickly obtain a "preliminary solution," though we shall soon see that it was no easy matter deciding what counted as a "solution," since if (as hoped by Helmholtz), the experiment failed to yield a positive result, then it could only provide an upper bound. The work went "steadily, unfortunately not quite directly towards the goal, but deviating sometimes to the right and sometimes to the left." A major setback in early December "threw [him] back almost to the beginning" (J. Hertz 1977, p. 101). That setback was the occasion for Hertz's first attempt to manipulate the objects of investigation in a way that deliberately exacerbates the kind of coupling that he was looking for.

The problem posed by the prize competition was a difficult one indeed. To obtain a convincing limit for electric mass required extreme accuracy in measuring currents. The way to detect the mass is to look for effects that occur during acceleration, during transient changes in current magnitude. That much is obvious, and the Philosophical Faculty of the university, who set the problem as a prize (undoubtedly at Helmholtz's instigation), suggested using "extra-currents flowing in opposite directions through the wires of a double spiral" (Hertz 1880a p. 3).[5] Helmholtz's laboratory (of course) had a pair of such spirals, and the essential idea was to see whether inductive coupling could be used to detect electric mass. This was, to say the least, a proposal that taxed Hertz's ingenuity.

By the late 1870s highly accurate galvanometers were readily available. To measure small, transient currents the ballistic galvanometer had been in use for some time and was commercially available (Hertz used a "Meyerstein"). In this type of device (which was still being used productively as late as the 1970s) the magnetic indicator has a fairly large moment of inertia. A transient current will vanish before the indicator moves very much, and so the torque that the current exerts upon it remains essentially constant during the current's existence. Consequently, the indicator acquires an angular momentum that is directly proportional to the integral of the current over time, or to what Hertz termed the current's "integral flow," which connects directly to the circuit's coefficient of induction and therefore to whatever mass the current may have. To measure the integral flow consequently requires watching the indicator's swing for its maximum excursion from equilibrium. This device, together with a standard galvanometer to measure the intensity of steady currents, remained completely unproblematic throughout Hertz's experiment. Whether he had succeeded in detecting the current's putative mass or not, the galvanometric measures would never have raised any questions other than (very important) ones involving accuracy—these devices remained outside the orbit of theory in this experiment, and indeed in all of Hertz's experiments. They functioned, that is, purely as pieces of measuring equipment. Other pieces of equipment were designed to exacerbate the effect of mass, should it exist, and these devices, unlike the measuring equipment proper, were ineluctably bound to theory.

The foundation of Helmholtz's electrodynamics in energy systematics provided Hertz with a simple way to express the connection between a putative current mass and "integral flow," $\int i dt$ (hereafter denoted by J). In the absence of mass the work done on a current i by an electromotive force A during a time dt must equal the Joule heat that accumulates along with the increase in the circuit's potential energy during that same time:

$$Ai dt = i^2 r dt + (1/2) d(Pi^2)$$

where r is the circuit's resistance, and P is the coefficient of induction. If however the current has mass, then kinetic energy must be included in the balance, and this requires adding a term to the right-hand side of the equation. Such a term will be proportional to the square of the current since, under the present assumption, the current represents the flow rate of electric substance. Its contribution to the differential work $Aidt$ may accordingly be written as $(1/2)d(mi^2)$, where m is proportional to the mass. Simple manipulation then shows how m affects the law of induction:

$$ir = A - \frac{Pdi}{dt} - \frac{mdi}{dt} = A - \frac{(P + m)di}{dt}$$

The new term increases J, the integral flow: if we remove the external electromotive force A and assume that mass does not exist, then J will be simply $(i/r)P$; if mass does exist, then m must be added to P, and J increases by (im/r). Here i is the initial current. Hertz had to construct a circuit that could detect very small additions to the swing of the ballistic needle.

Since J can only be measured ballistically, it cannot be determined using a single circuit, for the initial, steady current produces a large deflection from equilibrium. The procedure for measuring J (as when determining the strength of a magnetic field) was to use a branched system, in particular, a Wheatstone bridge. Here an electromotive force applied across AA' generates a current in the two parallel branches $A'BA$, $A'B'A$, which are bridged across BB', where the ballistic galvanometer (G) sits (fig. 14). The resistances in the various legs are adjusted so that there is no voltage across BB' in the steady state, in which case G will not be deflected. The inductors in the diagonally opposite arms BA' and BA will in general produce a voltage across the bridge when the circuit is opened. Consequently, the galvanometer's deflection can measure the flow J that occurs for this set of resistances and inductances when the current collapses. This flow represents the combined actions of the inductors in the opposite branches.[6] If one of the inductors remains permanently fixed, whereas the other can be altered, then the measured flows can be used to determine the latter's inductive effect. The fixed inductor, in other words, can be treated as a given for purposes of flow measurement and can in consequence be ignored in the computations.

This would, however, require rather precise measurements of resistances (as well as of the initial currents) because J is given by $i(P + m)/r$: since m is no doubt very small, having to determine resistances as well as inductances would vitiate the experiment's accuracy. It occurred to Hertz that simply making certain that the circuit's resistance never changes from experiment to experiment would bypass this problem, for it is comparatively easy, and highly accurate, to adjust the resistance of the bridge in the steady state to ensure that it remains constant.[7] Since the resistance is always the same, then the quotient $i(P + m)/J$ remains constant, and so the value of m can be determined from a pair of experiments that yield the sets (i, J, P) and (i', J', P'):

$$m = \left(P\frac{J'}{i'} - P'\frac{J}{i}\right) \Big/ \left(\frac{J}{i} - \frac{J'}{i'}\right)$$

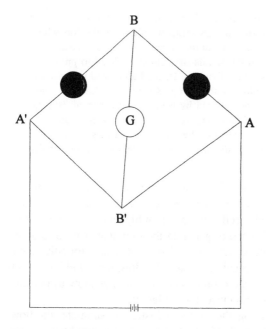

FIGURE 14 Hertz's spirals and bridge

The expression depends critically upon the magnitudes of the initial currents, so Hertz required that they, like the resistances, must also be the same in every experiment. This can be ensured through calculation rather than manipulation since the integral flow varies directly with the initial current.[8] To simplify even further Hertz required the inductance (P') in one group of experiments to be so large that m can be ignored, with the inductance (P) in the other experiments being small enough to bring out the effect of the mass, should it exist. Consequently m becomes

$$m = P\left(\frac{JP'}{J'P} - 1\right)$$

5.3. COMPOUNDING MARGINAL ACTIONS

And now Hertz faced the daunting task of generating meaningful results from devices that he would have to push to the very limits of contemporary technique. He was at first stumped:

> The chief difficulty in these measurements was to be met in the smallness of the observed extra-currents. It is true that by merely increasing the strength of the inducing current the extra-currents could be made as strong as desired, but the difficulties in exactly adjusting the bridge increased very much more quickly than the intensities thus obtained. With the greatest strength, which

> still permitted permanently of such an adjustment, a single extra-current only moved the galvanometer needle through a fraction of a scale division, whilst the mere approach of the hand to one of the mercury cups or the radiation of a distant gas flame falling on the spirals [inductors], sufficed to produce a deflection of more than 100 scale divisions. Hence I attempted to make use of very strong currents by allowing them to pass for a very short time only through the bridge, which was adjusted by using a weak current. But the electromotive forces generated momentarily in the bridge by the heating effects of the current were found to be of the same order of magnitude as the extra-currents to be observed, so that it was impossible to get results of any value. (Hertz 1880a, p. 6)

Hertz found himself in a quandary. The bridge could not be reliably adjusted if the current was stronger than a certain magnitude (because the electromotive force of the battery—a Daniell's cell—fluctuates at higher currents), but if the same magnitude of current is used to generate the flow that is used to adjust the bridge, the effect is too weak. Yet if a weak current is used for adjustment, and a strong current, briefly applied, generates the flow, then (Peltier) effects due to heating swamp the indicator altogether. He had somehow to multiply the flow without using a powerful current to initiate it.

Hertz conceived of adding together in the ballistic galvanometer the flows of a series of distinct currents each of which is too weak individually to produce much of an effect. To do so he built an ingenious commutator that reverses the battery connections to the bridge after a flow takes place; at the same time it also reverses the connections to the ballistic galvanometer. In this way the second (reversed) flow will travel through the galvanometer in the opposite direction to its predecessor, so that the indicator will receive a series of kicks that are all in the same direction (twenty times in two seconds in the actual device). The reversal takes place when one of two copper prongs at the ends of wooden shafts attached to a disk slips from one mercury trough, through which the current flows, to the next (figs. 15–16).

> A circular disc revolving about a vertical axis has attached radially to its edge twenty amalgamated copper hooks which just dip into the mercury contained in the vessels B and C. They are alternately nearer to and farther from the axis, so that the inside ones, and the outside ends of the nearer ones, lie on the same circle about the axis. They reverse the current in passing over the vessel B, and the galvanometer in passing over C. The vessel B is not exactly opposite to C, but is displaced relatively to it through half the distance between successive hooks, so that a reversal of the galvanometer occurs between every two reversals of the current. (Hertz 1880a, pp. 13–14)[9]

With a steady but weak current passing through the bridge, Hertz turned the device around once, its rotation being stopped by a peg. He watched as the indicator swung out and then back; when it returned to the initial point, he again rotated the device, producing twenty new kicks to further amplify the

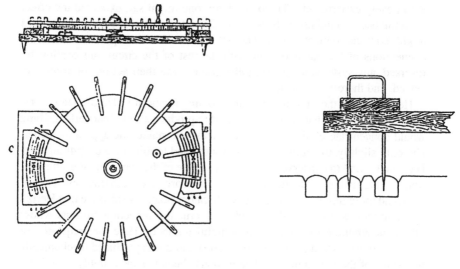

FIGURE 15   Hertz's commutator (Hertz 1880a)

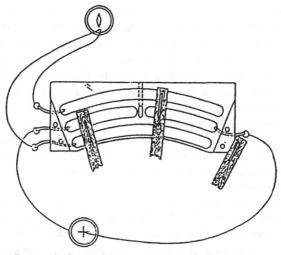

FIGURE 16   Detail of the commutator switch (Hertz 1880a)

swing. He could do this about seven to nine times before damping effects (which had to be carefully measured for each particular circuit arrangement) became too irregular to compensate through calculation.

Hertz's mutation, the commutator, consequently had two significant effects. First of all, it multiplied the action that he was interested in to a point where it could be repeatedly distinguished from the random fluctuations that are inherent in every experiment. The commutator, one might say, extracted the effect from the background noise. Second, it eliminated whatever permanent actions might exist that deflect the galvanometer from its neutral position: since the connections of the galvanometer with the rest of the circuit are continually reversed, any systematic electromotive forces have their actions on it also reversed, and therefore canceled.[10]

Here we have the first instance of a common pattern in Hertz's later work. Faced with devices that could not be used to produce a significant effect, Hertz would concentrate on the problematic areas and mutate the apparatus, generally only slightly, to amplify what he wanted and to minimize everything else. He usually tried to conserve as much as possible of the apparatus at each mutation. At least two other instances of the pattern are evident in this same experiment and more might be if we had his laboratory notes. We also see that Hertz's first mutation was designed to produce alternating induction currents whose effects accumulate to generate meaningful numbers. Here the frequency of alternation was extremely slow (about 10 Hz) and was produced mechanically by means of the commutator.[11] The response becomes measurably significant because the very weak and short-lived induction effects that are produced by the alternation combine over a comparatively long period of time. The measuring equipment itself (the ballistic galvanometer) accordingly constrained the mutations that Hertz could make since the long time during which the commutator produces its effect was required by the structure of Hertz's indicator. The experiment's design treated the galvanometer's mechanical behavior in relation to the magnitude of the electric flow as an unproblematic given. The remaining two major components of the device were not givens, but they nevertheless differed in character from each other. One of them—the commutator—accumulated the problematic effects that the other—the inductor—generated. Both commutator and inductor (unlike the measuring apparatus proper) could be fruitfully mutated, in Hertz's view, but changes to the commutator would serve primarily to amplify given effects, whereas changes to the inductor could alter the effects themselves.

Living theory accordingly played no active role at all in the process of measurement itself since the principles that governed the detector—the galvanometer—were effectively set in concrete. Moreover, neither here nor in his later work did Hertz ever attempt to perfect the devices that produce the numbers that are recorded in laboratory notebooks. He either took such things as objects that must be used essentially as they stand and that consequently constrain the interesting parts of the experiment, or else he turned away from them altogether, as in his later experiments with electric waves, where Hertz needed no *measuring* device beyond a simple ruler.

Following the suggestions of the prize's proposers (which certainly means at least Helmholtz), as well as using what he had immediately to hand, Hertz first tried a pair of spirally wound wire spools (fig. 17). Each spool had two

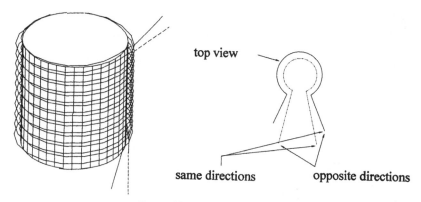

same directions   opposite directions

FIGURE 17   Double-wound spiral

independent strands wound about it, each strand forming several layers (hence "double-wound" spirals). Inserted into a circuit, such a thing will have a reasonably *large* inductance if one of the two independent strands in one of the two spirals is switched into a separate circuit (case P') carrying the initial current: for then it will couple inductively to its brother strand, which still forms part of the Wheatstone bridge, when the current in it collapses. A vastly smaller inductance results if both strands are switched into the bridge in such a fashion that the currents in them flow in opposite directions (case P). The inductance can be calculated for this second case, albeit only through a tedious series of approximations.

This set the pattern for Hertz's first experiments (fig. 18). Perform two sets of experiments: in one set use the powerful inductance P'; in the second set take care to keep the resistances the same as before (by using an adjustable slide) but use the very weak inductance P. In the P' experiment (fig. 19) a current first flows steadily in one of the two spirals through its strand that is switched out of the bridge, and this current is measured with the galvanometer G'. The remaining strand of the spiral remains in circuit with the bridge, as does the other spiral, whose strands are so connected that a current will flow through them in opposite directions. When the current collapses, the independently circuited strands in P' couple inductively with one another, generating a current in the bridge. In the P experiment (fig. 20) a current is first established throughout the bridge, and both strands of both spirals remain in circuit, connected in the appropriate manner. Here Hertz used his commutator to produce measurable flows, which was not necessary in the P' experiment because of the large coupling.

The two sets of experiments differ in another essential respect as well, since only in the second did Hertz actually calculate the inductance; in the first he measured the inductance P' in an independent (and so third type of) experiment in which the great magnitude of P' permitted him to ignore the effect of

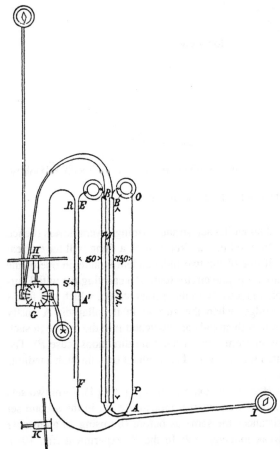

FIGURE 18   Hertz's experiment
diagram (Hertz 1880a)

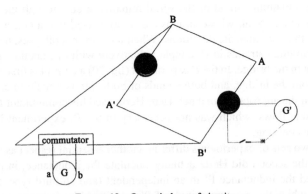

FIGURE 19   One spiral out of circuit

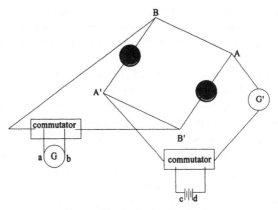

FIGURE 20   Both spirals in circuit

electric mass. Here, he found, P' could be measured to an accuracy of about 1 part in 40. The small inductance P was punctiliously calculated, with a final accuracy of about 1 part in 60. Consequently, induction experiments provided the flows J' and J, as well as a value for P'; computation and experiment provided the value for P. If the mass function m is zero then $J'P$ should be equal to JP' within the limits of the experiment. But limits are precisely what this experiment was about, and, in Hertz's words, he found that the "difference between the two amounts to little more than 1/100 of the total value, while the errors of experiment and of calculation at most may amount to 1/30" (Hertz 1880a, p. 25).

Hertz's extreme care in estimating experimental error, which is a hallmark of his later work, has its origin here, in the demanding (indeed, nearly impossible) task Helmholtz had set him of significantly delimiting an effect that Helmholtz believed not to exist. One major limiting factor in the experiment derived from Hertz's method of multiplying the very small flows. In order to translate the large mechanical excursions of the ballistic needle into an electric flow, Hertz had to assume that the small, component flows in the P experiment that conspire to produce the measurable effect are equal to one another. Whatever inequality does exist limits accuracy, and so Hertz sought to determine the size of the deviations by computation. To do this he assumed that the individual deviations, which must be the results of irregularities during the circuit's opening and closing, "may be regarded as [produced by] an impact in a direction opposed to the [needle's] motion which is proportional to the duration of the closing of the circuit and to the velocity of the needle" (Hertz 1880a, pp. 8–9). Hertz could then calculate a relationship between successive swings that involves the irregular intervening impact. From the observed excursions he could then compute the impacts. He found that "the greatest difference [between impacts] amounts to less than 1/30 of the whole" (Hertz 1880a, p. 11), which translates directly into the maximal error due to this cause in the measured flow, though the actual error is usually closer to 1/50 and can be found empirically by comparing the flows that result from several trials with different primary currents.

There were other errors, most especially the ones involved in calculating the small inductance P and in measuring the large one P', which was too complicated to compute. Because of the error involved in calculating P and the practical impossibility of computing P', Hertz concluded that it would be useless "to make further experiments with spirals." This was most likely the "setback" that he wrote to his parents about early in December. If the error in determining the inductances (through calculation and experiment) is of the same order as the experimental errors, then no possible improvement in the latter could pull the accuracy into the neighborhood of the differences that he was actually detecting, since these were much smaller than even the experimental errors.[12] But, one might ask, so what? Why proceed further? One could simply stop here and use the 1/30 error to compute an upper limit for the current's kinetic energy since the experimental values fall well within the anticipated errors. One could say, as Hertz wrote, that "at most 1/20 to 1/30 of the very small extra-current from doubly-wound spirals can owe its existence to a possible mass of the moving electricity" (Hertz 1880a, p. 25). But that result would hardly have been very satisfying for an experimental investigation, because this limit resulted as much from difficulties in computation as from inherent observational errors.

What Hertz needed was a new arrangement that permitted the inductances, which were the center of the experiment, to be calculated accurately, thereby reducing the errors due to calculation well below those due to observation. Then the upper limit on the current's mass would be determined entirely by the results of measurement. To that end he conceived of using two wires so connected that the currents can flow either in the same or in opposite directions in them, and so arranged as to form a pair of nested rectangles, the whole being placed on the floor beneath the bridge (fig. 21). Although the flow errors remained essentially what they had been (as fractions of the total flow), the inductances could now be very accurately calculated for both configurations ("their errors can hardly amount to 1/100"), and this greatly lowered the upper bound on the current's mass, which was now determined almost entirely by inaccuracies in measuring the flows. And here a very interesting thing occurred, for the new arrangement provided values that, taken as they stand, seemed to indicate the presence of a mass effect.[13]

Hertz tabulated eighteen values for the second of his two series of experiments with rectangular wires. First he calculated what the flow ratio should be in the absence of mass. Then from the eighteen experiments for the flow produced when the currents are in the same direction in the wires (and so for P', the large inductance) Hertz obtained a mean value for J'. From this and the

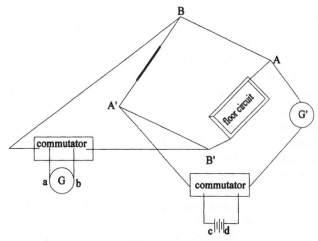

FIGURE 21   Spiral replaced with rectilinear circuit

calculated ratio he then found a value for J. But this computed value differed
by about 7% from the observed mean, whereas the new experiments were more
accurate than the previous ones and so could not support more than about 3%
of error.[14] Hertz immediately sought an explanation for this result in hidden
experimental pathologies, which he teased out of the data in the following way.
This series of experiments involved successively larger initiating currents, and
the earlier ones in the series had correspondingly small flows. After dividing
the data into two blocks—the first seven trials and the last eleven—Hertz in
effect remarked that the second block shows a deviation from its mean of only
2.7%. So, he argued, the untoward deviation can be ignored because "it only
occurs in the case of those observations which were already uncertain because
of the smallness of the effects" (Hertz 1880a, p. 29), that is, in experiments for
which the fractional errors in the flows are quite large.

From these results Hertz could put a much lower upper bound on the mass
effect, though not without a caveat.[15] In his words: "The kinetic energy of the
electric flow in one cubic millimetre of a copper wire, which is traversed by a
current of density equal to 1 electromagnetic unit, amounts to less than 0.008
milligramme-millimetre" (Hertz 1880a, p. 32). So much is unproblematic. But
then Hertz goes on to insert this upper limit into Helmholtz's formula for instability in Fechner–Weber (see appendix 6) to conclude that "in a [conducting]
sphere of 1 cm. radius the first 90 component currents, almost the whole current, might increase indefinitely." This is a most peculiar remark. According
to Webereans the mass is de facto so small that it does not give rise to actual
instabilities; Hertz's experiment merely confirms this assumption. Hertz has,
however, taken his upper bound and used it to compute the Helmholtz instability as though the mass, if it existed, would actually have this maximum value.

But why should it? Hertz's ingenuous remark no doubt reflects his understanding of Helmholtz's concern with the issue. He felt called to make some comment about his experiment's theoretical significance, and so he naturally sought to infuse his mentor's mathematics with empirical meaning.

Careful estimation of error, concentration on inductive actions aided by alternating currents, and mutation of the effect-generating parts of the device to bring out the latter by reducing the former characterize not only this first of Hertz's experiments in electrodynamics but also his later ones. In this first experiment Hertz had transformed his initial device in a very specific way, stripping it of anything that obscured the inductive links between the objects composing it. Hertz envisioned the device in its two major forms as a set of inductive couplings. In the device's initial configuration the couplings occur between the separately wound strands in each of the two spirals. In the experiment that probes small inductances the strands couple to one another in the same way in both spirals; in the experiment that probes a very large inductance, the coupling remains the same in one spiral but is multiplied over 200 times in the other spiral. When the weakly coupled system proved, in experiment, to yield results that were vitiated by the difficulty of calculating the smaller, and measuring the larger, inductance, Hertz created new systems with a pair of rectilinear circuits whose coupling could always be very accurately computed, whether large (for currents in opposite directions in the circuits) or small (for currents in the same directions in them). Yet in apparent paradox the rectilinear arrangement actually *reduces* the larger inductance and leaves the smaller one essentially the same in magnitude. Its advantage is not grossly to magnify a small effect to bring it within the ambit of the fixed detectors but rather to *purify* the inductive couplings in a way that makes them accurately computable. By concentrating on the objects that together form the system, Hertz was able rather simply and directly to extract the inductive links.

SIX

Hertz's Early Exploration of Helmholtz's Concepts

6.1. LEARNING HOW TO COMPETE

With success in the prize competition came pressure. Helmholtz's eye was focused very closely on Hertz. The prize competition had tested Hertz's competence in the laboratory, as well as his ability to deploy Helmholtz's methods and concepts. Having brilliantly demonstrated his abilities, Hertz could now be trusted to do much more, to probe issues that had a direct bearing on Helmholtz's own electrodynamics. Early in July the Berlin Academy proposed a new prize for exactly this kind of work. Helmholtz pressed Hertz to compete, only this time the project would take quite a bit longer:

> [W]hen I had waited for Prof. Helmholtz in his laboratory last week to thank him [for his help in the first competition], he was exceptionally friendly and congratulated me repeatedly and suggested a new project to me, one that would however occupy me completely for two or three years, and that I would not undertake if it were not for the particular honour of Helmholtz's invitation and the way he offered it, and by which he promised me his continued support and interest. (J. Hertz 1977, pp. 114–15)

Hertz had in the meantime written down his ideas for experiments on the new project and given them to Helmholtz to read. But the master apparently had his own experiments in mind and was not interested in the neophyte's suggestions. Hertz became distressed and a bit testy, for he did not in any case wish to devote such a long time at such an early stage in his career to a project with uncertain outcome. If the results, like his first ones, were negative, his reputation would hardly be advanced. It is worthwhile quoting in full his remarks at the time, for they date the beginning of Hertz's attempt to distance himself a small amount from the master who was seeking to bind him even closer:

> I am working in the laboratory again now [November 4], but not on the [new] prize problem of the academy. When I first went to see Helmholtz, I showed him my paper [Hertz's proposals for experiments to address the 1879 prize questions] and asked him to look it over, which he promised to do; but when I came back a few days later, he said that he had not got around to it yet. Since he has not returned it to me, nor otherwise referred to it, he must simply

have put it aside. So I could have spared myself the labour of copying it. Otherwise he was very kind and pointed out which experiments he thought practicable [for the new prize questions]. However, since these experiments [the ones Helmholtz himself proposed] required too much equipment, and since I had no chance to talk with him about simpler experiments that I had in mind, I said that I should at least like to think the matter over first and would rather undertake something else. I could have wished that he had taken a closer look at my paper, if only because of the calculations that I had carried out in addition and that take up most of the space.[1] As far as the immediate purpose of this work is concerned, as preparation for experiments which I have postponed, I am very glad of the way it all turned out; for I was properly afraid that Helmholtz would approve the experiments and would thus force me to go on with them. The thought of working in *secret* for three years was a nightmare. If these experiments can be performed easily, then they will not require three years, and there will still be time to turn to them. (J. Hertz 1977, pp. 115–17)

6.2. A Proposal for Helmholtz's Eyes

When Helmholtz suggested to Hertz that the July 1879 prize questions would be a fitting topic for research, Hertz had at once started to work. Although we will see that Hertz did uncover ways to approach the questions in the laboratory, he did not go ahead with the project, primarily because Helmholtz apparently had his own ideas about what to do. These, Hertz felt, would require much more equipment than the ones that he himself had in mind. But even his own experiments would have taken a long time, and Hertz did not relish the "nightmare," as he put it of "working in *secret* for three years." For Helmholtz three years of a promising assistant's time devoted to a critical problem no doubt seemed well spent. To the young and ambitious Hertz it looked like an eternity in solitary confinement. Helmholtz nevertheless did not bring too much pressure to bear. Indeed, there is no evidence that Helmholtz ever *forced* a student to do anything, though he seems to have been quite good at exerting less-overt pressure. He did find a way to keep Hertz temporarily in Berlin, and he may have hoped that Hertz would eventually come round to the problems.

Despite the fact that Hertz never did perform the experiments that he drew up for Helmholtz and that no publications of any kind followed directly from this work, these *undone experiments* reveal particularly well how Hertz had assimilated Helmholtz's laboratory-based electrodynamics. The proposal that he produced for Helmholtz maps the trajectory of his subsequent work because it contains in one form or another the ancestral patterns for many of the problems with which he later grappled, including those for his dissertation. Hertz wrote down his thoughts in a long manuscript intended for Helmholtz's eyes and entitled "Demonstration of Electric Effects in Dielectrics," dated in his hand August–October 1879.[2] We have here the opportunity to witness a quick-witted neophyte in the throes of learning how to mesh a complicated structure with laboratory practice *before* actual laboratory manipulation took place, yielding insight that is not obscured by the inevitable and intricate rethinking and reworking that always accompanies genuine practice. The prize question asked for "decisive experimental proof":

> for or against the existence of electrodynamic effect of forming or disappearing dielectric polarization in the intensity as assumed by Maxwell . . . [or] for or against the excitation of dielectric polarization in insulating media by magnetically or electrodynamically induced electromotive forces. (Bryant 1988, p. 7)

Hertz, who almost certainly knew nearly nothing at all of Maxwell at this point, ignored the last phrase in the first question and concentrated on envisioning experiments that would produce unambiguous results—results that were not likely to be questioned. This was an extraordinarily difficult task and required Hertz to probe inductive relationships in ways that took him beyond contemporary laboratory experience.

The introduction to his discussion—which, it is critical to recall, was written expressly for Helmholtz to read—distinguished three major effects that Hertz felt could reasonably be sought. These distinctions were themselves hardly obvious and certainly took him some time to reach.[3] He differentiated between three kinds of experiments:

1. Examination of a dielectric's *electromotive* effect on a neighboring conductor.
2. Examination of a dielectric's *electrodynamic* action on a current-bearing conductor.
3. Examination of dielectric polarization generated by *electromagnetic induction* (SM 245).

The first and second kind of experiment fit the first of the Berlin Academy questions, though we shall see in a moment that the second involves something else as well. The third type of experiment in the list fits the second question, and it is particularly important because Hertz's analysis here led directly to his doctoral dissertation.[4]

To uncover the structure of Hertz's discussion it is instructive to reformulate these experiments in the following way:

1 and 2. Both experiments examine the coupling between conductors in current-bearing states and dielectrics in states of changing polarization. The first experiment examines the effect of a temporal variation on the coupling; the second examines the effect of a space variation.

3. This experiment seeks a coupling between a dielectric in a state of unchanging polarization and a current-carrying conductor when the two are in relative motion.[5]

The special importance of considering Hertz's proposals in this way derives from their intimate connection to Helmholtz's electrodynamics, for every one of them directly employed equations and conceptions that were unique to Helmholtz's work and of which Hertz was thoroughly aware, as, we shall see below, his manuscript considerations strongly suggest. Indeed, a Maxwellian scanning Hertz's proposals would rapidly have decided that his first experiment at least would show nothing at all and that Hertz's analysis for the second was thoroughly flawed.

6.3. STIMULATOR AND INTERACTOR

The overall structure of Hertz's experiments 1 and 2 was probably suggested to him by the sequence of researches in Berlin that sought effects specific to Helmholtz's potential,[6] as well as by the experience with inductive actions that he had gained during the prize competition. In essence he strove first of all to create situations for *conduction currents* that involve effects peculiar to the potential. Then he substituted a dielectric for the conductor, with the substance of the analysis remaining otherwise unaltered. Helmholtz's potential was accordingly hardly a side issue in the investigation; it lay at the very center. Successful completion of any one of these experiments would, outside Berlin, certainly have raised more issues than those that were explicitly addressed in the prize questions.

Since the problem revolved about current interactions of one sort or another, Hertz naturally thought to use inductively coupled circuits, just like he had for the earlier prize competition. There was, however, an important novelty. In the earlier experiments Hertz had taken the measuring devices (galvanometers of various kinds) to be unproblematic; on the other hand, the interacting circuits had to be treated as problematic because the point at issue concerned the nature of their coupling, viz., whether it was modified by a mass factor. In the new experiments the coupled circuits that produce the induction current are themselves *not* at issue; they must be thoroughly analyzed, but the structure of the analysis must raise no more questions than, say, the behavior of a galvanometer does. Only the effects that these parts of the apparatus engender in a distinct device raise questions.

Hertz's thoughts here are deeply buried in a confusing morass of detail. They must have become clear to him only over time, and (thankfully for the historian) he never recast his initial proposal for Helmholtz. Precisely because he never did carry out these experiments he left behind a pure record of his thoughts, one that might otherwise have vanished or at least have been overwritten and ripped apart in subsequent working. If we tease out his understanding from this pristine record of something that he never did, then we may see how closely it connects to his earlier laboratory experience—how he built carefully on his previous experience to form his new experiments. He began with a pair of coupled induction coils—indeed, the ones he had in mind were the very same double-wound spirals that he had used for the previous prize competition. They were now to function as *stimulators* for another piece of apparatus by feeding inductively generated current into it. We shall see in a moment that unless this apparatus had a very large capacitance, its interesting behavior would depend critically upon the actual oscillations that the stimulator underwent.

Figure 22 represents Hertz's coupled inductors. The primary (on the left) connects to a battery through an interruptor; the secondary bridges the experimental device, through which flows current generated in the secondary by the coupling. Figure 23 is a modern rendering of the circuits. From it we see that each circuit contains resistance, capacitance, and inductance, and that they couple to one another through the mutual inductance Π. The system is thus governed by the following pair of equations:[7]

$$\left(p\frac{d}{dt} + w \right) I + \Pi \frac{di}{dt} + \int \frac{I}{c} \, dt = 0$$

$$\left(P\frac{d}{dt} + W \right) i + \Pi \frac{dI}{dt} + \int \frac{i}{C} \, dt = 0$$

A nearly complete set of equations for interacting coils had already been provided and analyzed by Schiller while he was working in Helmholtz's laboratory in 1874 (and he had even used coils, albeit unsuccessfully, to seek the electromagnetic effects of dielectrics) (Schiller 1874).[8] Schiller, whose purpose was to "prove the theoretical laws of alternating currents," had undertaken the work at Helmholtz's suggestion, and so we know that already in 1874 Helmholtz had had it in mind to use alternating currents to investigate the behavior of dielectrics.

In Hertz's experiments 1 and 2 the device couples either electromotively (1) or electrodynamically (2) to the stimulator, which accordingly serves two distinct functions. First, it generates in the secondary circuit a current that activates the experimental device; second, the effect in the device is actually measured through its interaction with the currents in both the primary and the secondary circuits. Consider first the electrodynamic coupling, which caused Hertz a great deal of difficulty in calculation. Here, as in his experiments for the prize competition, Hertz intended to measure in the end an impulsive action. However, the action in this new situation does not come from a current impulsively setting an unproblematic measuring device (the ballistic galvanometer) into motion. Instead, it results from the actions of both the primary and the secondary currents on a very problematic device, which is connected to the secondary circuit.

Let us assume that at any given moment the current in the device is directly proportional to the secondary current *i*, which somehow engenders it. The measurable action generated by the effect results from an interaction between

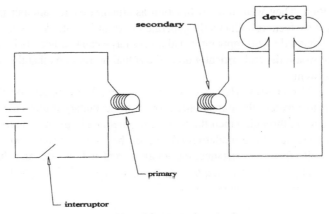

FIGURE 22   Hertz's inductive stimulator

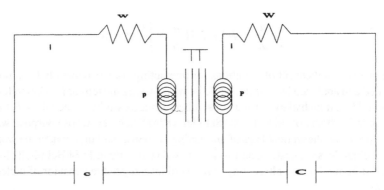

FIGURE 23   Couplings in the stimulator

this current and the ones in the primary and secondary circuits, that is, between something proportional conjointly to i and, respectively, to the currents I, i. The resulting impulse will therefore have the form $\int(i^2 + Ii)dt$, integrating from zero to infinity. This is what Hertz had to calculate—and Schiller had already done it for him (Schiller 1874, p. 538).[9] Hertz distinguished three cases: first, oscillations may occur in both primary and secondary circuits, that is, both have finite capacities; second, the primary has effectively infinite capacity; third, both primary and secondary have infinite capacities. Hertz felt that he could use these cases to produce an upper and a lower limit for the impulse. The first, upper limit derives from case 3 and, under the further simplification that the two circuits have the same characteristics, requires the following relations:

$$\int i^2 dt = \frac{I_0^2 P}{4W} \quad \text{and} \quad \int Ii dt = \frac{I_0^2 P}{2W}$$

The second, lower limit takes case 2 and requires that

$$\int Iidt = 0 \quad \text{and} \quad \int i^2 dt \text{ is bounded by } 0 \text{ and } \frac{I_0^2 P}{4W}$$

These values bound the impulses that the apparatus can generate. "Experiment," Hertz wrote, must decide which is the better fit, but he felt that a good guess was that the total impulse would be near $(I_0^2/2)$ P/W. This expression, then, must be used to estimate whether or not an effect can actually be measured.

6.4. COMING TO GRIPS WITH HELMHOLTZ'S POTENTIAL

Hertz discussed two kinds of experiments that might detect the electrodynamic action of a polarization current. Both experiments deploy as an unproblematic tool of investigation the very action that Helmholtz and his assistants had striven so hard to elicit during the previous decade—the force that arises as a result of end-effects. In both of them the object of investigation is a physically isolated element that has a current generated in it by electrostatic action. The element may be either a conductor or a dielectric. Hertz, we shall see, treats the two cases in precisely the same way.

In figures 24 and 25 a pair of conducting plates are connected across the secondary in Hertz's stimulator. Between the two plates sit another, smaller pair, which are directly connected to one another but which are separated on either side from the larger plates by a small distance c. These constitute Hertz's element, which may be thought of as an improvement on Helmholtz's terminated-circuit devices of the 1870s. In those experiments a closed circuit interacted with another circuit, which was broken open at an appropriate point. The novel effect that Helmholtz sought arose at that terminus, which defined the physical boundaries of the open circuit and therefore determined what was to appear as the appropriate entity in the interaction energy. Hertz's proposed experiment improves considerably on this, because he has created a physical circuit element by breaking the object away from the stimulator. Whereas in Helmholtz's experiment the open circuit itself contains the source of electromotive force that drives it, in Hertz's experiment the electromotive force arises outside in a way that permits the object of inquiry to be examined in ways that Helmholtz's arrangement did not. The most astonishing (in retrospect) characteristic of Hertz's proposal is that he gives no indication *at all* that his experiment might be used to investigate the unique characteristics of the potential itself. On the contrary, Hertz uses these properties as an unproblematic tool for investigating something else that is problematic—namely, the behavior of a dielectric current element. Helmholtz's potential has, as it were, been transformed by Hertz into a device that is as neutral as the behavior of closed circuits themselves. He will use it, not query it.

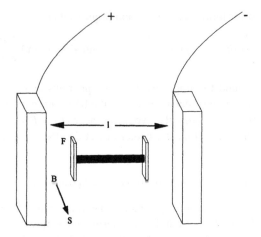

FIGURE 24   Action on a linear element

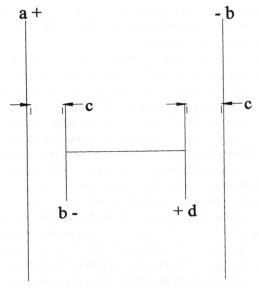

FIGURE 25   Setup in fig. 24 seen from above

The device operates in the following way. When the stimulator activates the large, outer plates they will charge the inner device electrostatically. As time goes on, the current from the stimulator changes and therefore so do the induced charges on the device. According to the continuity equation a current flows across it. In a second inspiration Hertz eliminated an external, separate circuit for the device to interact with: the apparatus as a whole sits within the stimulator in such a fashion that the stimulator's magnetic field is itself parallel to the plates and normal to the inner element. According to both the potential law and the usual Ampère law the element should be deflected in the direction BS. Substituting a dielectric for the conducting element does not change the situation in any significant way, although it does considerably alter the nature of the current state that arises. If the inner element is a conductor, the current results from the combination of Ohm's law with the continuity equation. If it is a dielectric, however, Ohm's law is not involved. Instead each point of the dielectric is the locus of a polarization current that is determined solely by the rate of change of the conduction charge on the outer plates.[10] Since the polarization charge that is induced on the dielectric will be less than the charge induced on a conducting element, the polarization current will also be proportionately smaller. The experiment's goal would be to measure the element's deflection.

Without further discussion Hertz abandoned this type of experiment for a related one in which a rotation, rather than a linear deflection, would be measured. This has the advantage, first, of amplifying the effect because it increases the size of the element and, second, of being much simpler to observe. A rotational experiment nicely brings out the special features of Helmholtz's potential. Hertz concentrated his attention exclusively on it. In figure 26 the circular outer plates are connected to the stimulator's secondary; the inner device is either a conducting or a dielectric cylindrical shell. The apparatus as a whole sits within the stimulator in such a fashion that the magnetic field is parallel to the cylindrical axis. It is also essential, for reasons that will be clear presently, that the inner cylinder *not* be concentric with the stimulating plates; it must have its axis displaced somewhat from theirs. As the stimulator's secondary sends current to the plates, they charge electrostatically and so charge the inner device by induction.

To understand what should take place, consider, with Hertz, a general situation that encompasses both linear and rotational experiments. In the broadest terms both of them are designed to detect the motion of an isolated current-bearing object that is surrounded by a coil, the element lying in a plane at right angles to the coil's axis. The coil's magnetic field, near the locus of the element, is uniform, parallel to the coil's axis, and normal to the element—though it is best here not to think in terms of magnetic fields but rather, as Hertz did, in terms of interactions between the element and the surrounding coils.[11]

Consider the effect of one of the coils (there are two) on the element. Suppose that the element carries a current α and that the coil carries a current I. If, Hertz wrote, the coil has q windings, and its outer and inner radii are, respectively, R and r, then according to the Ampère element–element law the element will in general undergo the following actions, depending upon whether it is located outside or inside the coil:

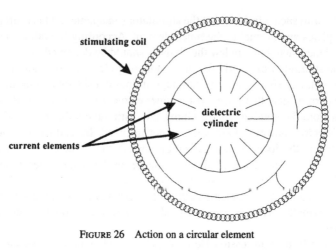

FIGURE 26   Action on a circular element

Action according to the Ampère Law

1. Element outside the coil: no force at all
2. Element inside the coil: element experiences a force that is normal to it and that pushes it toward the surrounding windings

The differences between Helmholtz's potential and contemporary alternatives are worth reemphasizing here. All forms of electrodynamics yield precisely the same electrodynamic force between *closed* circuits. They differ tremendously concerning the interactions between circuit elements. In Hertz's proposed experiments a closed circuit—the coils—interact with a circuit element, and under these circumstances the Helmholtz potential yields forces on the element that are unique to it. Hertz had envisioned a way to make use of these forces— not to query their existence. Not only, asserts the Helmholtz potential, should a force on the element exist *outside* the coil, but its magnitude in all cases depends upon the angle ω between the element and a line drawn from the element's center to the coil's axis. If the distance between the element's center and the axis is ρ, then, Hertz calculated:[12]

The Potential between Element and Coil

1. Outside the coil: potential $= (\tfrac{2}{3})q^2\pi(R^3 - r^3) \sin \omega \, I\alpha/\rho$
2. Inside the coil: potential $= q^2\pi(R - r)\rho \sin \omega \, I\alpha$

It is at once obvious from this that linear forces and torques must both occur because distance and angles can be independently varied:

The General Actions according to Helmholtz's Potential

1. A torque that tends to align the element at right angles to the line joining its center to the coil's axis
2. A linear force toward the coil's windings

Comparing the Amperean results with Helmholtz's, then, Hertz had determined that *outside the coil* the Ampère law gives no force at all, whereas the potential law gives both a torque on the element and a force toward the coil's axis. *Within the coil* the potential law continues to yield a torque but produces no linear force on the element at the coil's axis proper (since ρ is zero there); the Ampère law gives no torque anywhere, but it requires that the element experience a linear force even at the coil's axis. Hertz emphasized the difference in order to point out that his experiment takes advantage of Helmholtz's unique forces as tools to be used for investigation.

Hertz now proposed to use precisely this difference to examine the effects of dielectrics. He proposed, that is, to subsume Helmholtz's novel forces into his experiment as givens and to examine whether dielectric elements behave in essentially the same manner that conducting elements do.[13] If the experiment had been performed successfully to his and his contemporaries' satisfaction, it would have been a tremendously stabilizing achievement, though it would also have placed the difficult issues presented by polarization states directly in the center of attention.

He envisioned an experiment along the following lines (fig. 27). Place the clyindrical shell, whether a conductor or a dielectric, between the plates of the capacitor. If it is a conductor, currents will arise over its surface according to Ohm's law and the continuity equation as a result of electrostatic induction. If it is a dielectric, polarization currents arise at each point of the surface. In either case a net torque will arise on the cylinder as a whole provided that its axis is not concentric with that of the surrounding coil (since otherwise the torques at each point of the surface will balance overall).[14]

The torque can easily be estimated from Hertz's previous computation of the interaction, which was based on the assumption that the induced currents are proportional at any moment to the current i that emerges from the secondary.[15] However, in order to do so Hertz had to calculate the capacities, as well as the inductive properties, of his double-wound spirals. He calculated that the "integral rotational moment," or impulsive torque, will be $2.8\,I_0^2$, where I_0 is the initial current in the primary strand of the spiral. Using Grove's cells,[16] Hertz calculated that this would be easily detectable for a conducting cylinder, "a very measurable quantity," as he put it. If, instead, a dielectric doughnut is used, then the effect will still be at least half the magnitude.[17]

Here, then, was a detectable electrodynamic effect that depended directly on Helmholtz's potential. What now of electromotive interactions? Unlike their spatially dependent brothers, they could not be detected by directly observing

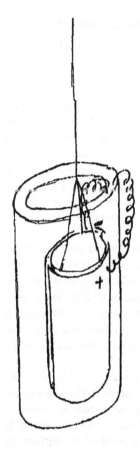

FIGURE 27   Hertz's apparatus for detecting rotational
effects. Reproduced from the Communications collections of
the Science Museum, London.

the experimental device (i.e., the inner apparatus). Whatever actions of this
kind may exist will affect the rate at which the current develops in the device
and the rate at which it changes in the stimulator, for they are coupled together.
There was no simple way to measure the *currents* on the inner device, so whatever novelties occur had to be observed in the stimulator proper (e.g., in, say,
the rate at which the current in it decays). The question was, then, whether
such an effect could be measured.

Suppose a thin dielectric shell with radius R_1 is placed within the stimulator,
whose radius is R_2. If the shell has polarization ξ, then such a thing will, Hertz
easily calculated, exert an electromotive force equal to $-A^2 2\pi \partial^2 \xi/\partial t^2 (R_1^2/R_2)$
on the surrounding coil, where A is the usual electromagnetic constant. The
coil, carrying a current i, itself exerts an electromotive force equal to
$-A^2 2\pi R_1 di/dt$. "Accordingly," Hertz wrote, "the equations of motion" of the
system are

$$\kappa_i = -A^2 \left(2\pi R_2 \frac{di}{dt} + 2\pi \frac{R_1^2 d^2\xi}{R_2 dt^2} \right)$$

$$\frac{\xi}{\varepsilon b} = -A^2 \left(2\pi R_1 \frac{di}{dt} + 2\pi R_1 \frac{d^2\xi}{dt^2} \right)$$

where ε is the dielectric constant, and b the number of windings in the surrounding coil. The problem now was to solve this system for its measurable consequences. This was, to say the least, difficult.

Hertz's discussion of his results is extremely brief, but we can understand what he found in the following way. If the dielectric were not present, the stimulating current i would decay exponentially with a damping factor equal to $\kappa/2\pi A^2 R_2$. In the presence of the dielectric, Hertz showed, the solution contains three distinct additive terms. The first is the same as in the dielectric's absence. The second and third terms both have essentially the same form, and for them the damping factor and angular frequency are

$$\text{damping factor} = \frac{\kappa R_1^2}{4\pi A^2 R_2 (R_2^2 - R_1^2)}$$

$$\text{angular frequency} = \frac{R_2^2}{2\pi \varepsilon b A^2 R_1 (R_2^2 - R_1^2)}$$

We may accordingly compare the rate at which these extra solutions damp to the rate for the first solution (which is the same as in the dielectric's absence) to find a ratio between them equal to $R_1^2/2(R_2^2 - R_1^2)$. In order for the solutions to decay at the same rates the dielectric cylinder must therefore nearly fill the coil. In any event the new terms damp at more or less the same rate as the first one, so that if the latter were detectable, then, all other things being equal, the new ones would be as well. However, the frequency of the extra terms is another matter.

In the time interval Δt during which the first solution will drop to $1/e$ of its initial value, the new ones will oscillate a number of times equal to $R_2^3/[2\kappa \varepsilon b R_1 (R_2^2 - R_1^2)]$. In the system Hertz used, κ is on the order of 10^{-17}, so that during the life of the main current the extra terms oscillate many millions of times. Whatever their actual magnitude may be, averaged over a complete cycle they must vanish, and so it follows at once that they cannot possibly have an observable effect on the main coil current.[18]

What, though, about the polarization on the dielectric cylinder itself? The solution for it, like that for the current in the coil, also has three components. Consequently, the first of the three, which is purely damped and corresponds to the one for the coil in the dielectric's absence, might possibly have observable effects.[19] Such effects could not be detected galvanometrically, but they might perhaps be observed electroscopically. If that were to be possible then the integral of the induced polarization over the life of the stimulating current would have to lie within a detectable range. It does not: Hertz calculated that the effect would be about a millionth of the magnitude of the stimulating current, which is "not observable."

For the coupling between conduction and polarization currents, then, Hertz had drawn two conclusions: first, that the effect of the coupling on the conduction current involves an additional oscillation at very high frequencies; second, that the total electrostatic action of the polarization over its lifetime cannot be detected electroscopically. The first effect evidently piqued Hertz's interest because it seemed to be so different from what would occur, one might think, for a similar coupling between conduction currents. Hertz accordingly decided to probe the latter case, but he did not do so in the same way. In dealing with polarization currents Hertz used no theoretical tools beyond the polarization and current proper, calculating for each of them the electromotive force generated by the other. But here, in dealing with coupled conduction currents, he turned instead to more advanced theory, in fact to differential equations satisfied by the vector potential for closed currents.[20] This is the first time that Hertz had probed so deeply into the more elevated reaches of Helmholtz's electrodynamics. Though of no great contemporary significance, Hertz's work on this point began to familiarize him with the subtle kinds of effects that can occur in Helmholtz's electrodynamics.[21] (For details see appendix 7.)

So, in Hertz's view, electromotive experiments stood no chance of success at the frequencies he could produce in the laboratory, whereas electrodynamic ones might very well work. What is particularly striking about this result is that when, seven years later, Hertz did detect dielectric action he found *electromotive,* not electrodynamic, evidence for it. Such a drastic reversal of his earlier convictions directly reflects an equally drastic change in his understanding of what to use as a detector and as a stimulator. In these early years Hertz used as detectors what everyone else, whether in Berlin or elsewhere, had long employed, namely, galvanometers and electroscopes of various kinds. These kinds of devices were carefully calibrated objects that were designed and used for purposes of exact measurement.

The numbers produced by these devices, and others like them, measured the physical deflection of an object: of a galvanometer's needle or of an electroscope's indicator. To obtain results with these kinds of instruments it was accordingly essential to measure some object's motion. In his prize competition experiment, for example, Hertz had followed the usual path; he generated numbers by reading off the deflections of his ballistic galvanometer. Here, in his proposal for Helmholtz, he proposed to do more or less the same thing. He proposed actually to measure the effect. That was his intention and his hope. It was also his bête noire, because exact measurement ill fitted an enterprise whose purpose was to find out whether or not something occurs at all.

Consider again the prize competition. Here, we saw, Hertz's results could easily be dismissed by Webereans because they merely provided a numerical value, an upper bound for electric mass. Webereans had only to say that the mass was lower than that. If we invert historical reality for a moment, then we can uncover something quite interesting about this. Suppose that a contra-Hertz working for the aged Weber in, say, the late 1870s had decided to *measure* electric mass. The contra-Hertz does not question the existence of the mass, or at least of something that behaves like mass; he wishes, however, to know its magnitude. So he performs an experiment like Hertz's (unlikely though this would have been, albeit for reasons having to do with experimental technique), and he obtains some upper bound. "Excellent," he concludes, "this is a useful, though hardly definitive, number to use for electric mass, at least until experimental delicacy improves sufficiently to directly detect its effect." Within a community that takes electric mass for granted, a Hertz-like experiment would be the beginning of a program of measurement, not the end of a research tradition. For this community numbers are the very essence, the goal, of experiment.

For Hertz as a disciple of Helmholtz numbers were not in and of themselves very interesting. Certainly measurement was important in order to determine the magnitude of something known to occur. But it was more important to show *that* something occurs than to measure it precisely. It was however extremely difficult to disentangle the goal of Helmholtzian experiment—to prove existence—from the typical end product of the contemporary electrodynamic laboratory in Germany, which was a number. The influence and prestige of measuring physics were immense and tended to impel the generation of laboratory numbers. As long as Hertz continued to think that he had to produce such things, then the kinds of experiments that he described for Helmholtz in his prospectus for research, ones in which subtle, hitherto undetected effects were being sought, posed nearly insuperable problems of design and computation. If numbers had to be produced, then an entire class of effects—electromotive ones—could never, Hertz realized, be detected. Electrodynamic ones could possibly be found, but to distinguish the physical deflection that would occur in Hertz's design from systematic effects required computing their magnitude, and this itself posed very great difficulties. Here (had Hertz carried out the experiments and reported success) major issues of his judgment in calculating approximations would surely have arisen. Indeed, they troubled him already in his manuscript, where much effort was devoted to such things. Hertz was still gripped by the conviction that the purpose of an experiment was to measure something as exactly as possible. Paradoxically, in order to make progress in the laboratory Hertz eventually decided that exact measurement was not very important.

6.5. THE EFFECTS OF MOTION

Much of Hertz's attention in the manuscript was devoted to a considerably different kind of situation, namely, to analyzing the interaction between current and polarization charge that occurs when a dielectric moves in a magnetic field. Unlike the first two experiments that Hertz proposed, this one could raise problematic issues for field theory, but it did not do so here. It is important to understand why not, because one of the difficulties that Hertz had to face in later years was to create a theory for the electrodynamics of moving bodies where, strictly speaking, one had not previously been necessary.

In field theory moving bodies could raise difficult issues because decisions had to be made about the relationship between the motion of the body and the state of the ether at and about its locus. This could be avoided only in situations where the body, as it were, disappears in itself from the analysis, as it did in the typical Maxwellian approach to motion. The Maxwellian body was usually ignored qua object; it was instead treated as defining a region within which the field has some particular state. As the body moves, the affected region of the field "moves" with it. Motion for Maxwellians was therefore a case of the transfer of a field state from point to point. This amounted to the tacit assumption that bodies do not interact at all with the ether except by changing its electromagnetic condition—that, to put it somewhat crudely, they do not drag ether. However, Maxwellians certainly did not carry this position to an extreme; for them it was merely an approximation designed to facilitate computation, and the actual relationship between moving matter and ether remained outside the ken of field theory (appearing only when questions at the boundaries of the discipline were infrequently broached).[22]

Nevertheless, because field theory considered electromagnetic states to be ether states it had perforce to deal, even if tacitly, with how a body affects the ether through which it moves. To ignore the effect of motion was implicitly to assume that the ether itself remains stationary as bodies swim through it. Motion posed no problems at all for Helmholtz's electrodynamics in the absence of an ether, although the solution of a given problem that involved motion might be rather difficult. The very structure of the scheme, its foundation in variational methods, already carried the solution to motional problems. Motion, after all, is merely a continuous change in coordinates over time. Helmholtz's standard technique for solving problems involved varying coordinates. Consequently, a situation involving motion is in effect equivalent to one in which coordinates vary continuously rather than differentially. In practical terms this meant that Helmholtz's electrodynamics of moving bodies merely required calculating the rate at which the interaction energy of a pair of objects changed with their motion. Since a basic assumption was that the energy at a given instant must depend on the mutual distance of the objects at that same instant, the subject was completely unproblematic, and Hertz himself thought so at this time. Indeed, as far as Hertz was concerned, Helmholtz had already produced general equations that were sufficient to deal with it.

Dielectrics raised no new issues of any consequence here, Hertz apparently believed, because they are merely another kind of object whose interactions with other things obey the same basic requirement. Solve the problem for interactions between conductors and then simply put in a dielectric for one of the conductors, making appropriate adjustments in the analysis for the replacement of Ohm's law by the proportionality of polarization to electromotive force. Since Hertz had not as yet thought very deeply about the furthest reaches of Helmholtz's scheme (since he was still absorbing it), his analysis remained superficial, concerned, that is, with applying an accepted structure rather than with probing for structural weaknesses. His proposal for a third experiment shows this nicely.

Hertz's study would have been even simpler than it was had it not been for the fact that Helmholtz's equations were not the only ones available. As long ago as 1863 Emil Jochmann, working loosely within a Weberean context, had obtained equations for precisely the situation that Hertz intended to treat, namely, a sphere rotating under magnetic influence. Despite the Weberean context even a quick perusal of Jochmann's work indicates that it is in fact independent of the overt differences between Weberean and Helmholtzian mathematics, though Jochmann's motional equations are in appearance not at all the same as Helmholtz's. Hertz had accordingly to uncover the connection between the equations, which he did with considerable skill. He found that Jochmann's term was included in Helmholtz's more general expression, but that only Helmholtz's extra term could be used in an experiment with dielectrics, since it implied that spinning objects in a magnetic field acquire a static potential, whereas Jochmann's term only generates currents where Ohm's law holds. In-

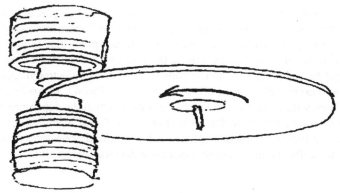

FIGURE 28   Rotating dielectric disk. Reproduced from the Communications collections of the Science Museum, London.

deed, Hertz's success here indicates that by this time his ability to deploy mathematics appropriate to electrodynamics was nicely refined (for details see appendix 8).

Hertz concluded that a sphere with a radius of 100 mm rotating about the local vertical at the equator ten times per second would acquire by induction (from the earth's magnetic field) an electric potential of about 1/100,000 of a Daniell's cell (i.e., about 1/100,000 of a volt).[23] Since it seemed that the earth's field could not produce detectable results, Hertz suggested instead a variant of the antique Arago disk, replacing the latter's metal plate with a dielectric (fig. 28). Using a sulfur disk 10 cm in radius rotating fifteen times per second in a reasonably strong laboratory-produced magnetic field should, Hertz calculated, produce a charge about the same as that generated by ten Daniell's cells in an air capacitor whose plates are spaced a millimeter apart. This could be detected by placing a metal sheath *very* near the rotating disk. Hertz was not sanguine about the likelihood of success even here, because it would be hard to place the sheath close enough to the disk without any contact at all, and even the smallest direct contact would produce electrification by friction, utterly swamping the induced effect.[24]

Hertz in the end concluded that the only practical experiments involved the electrodynamics of oscillating currents:

> I have been able to find only two sorts of experiments which hold out some prospect of success: the experiments with oscillating induction-spirals and those by means of continuous rotation. Success is doubtful with both; from the beginning, the first has had more to recommend it
> 1) because it demands simple means,
> 2) because it gives an interestingly complete test in case of success, and
> 3) because even in case of failure it would not be entirely useless.
> The second kind of experiment demands very great means and could easily fail. (SM 245, pp. 45–46)

These were not encouraging conclusions, which may in part explain Helmholtz's silence since he no doubt hoped for something more promising. Though Hertz was quite relieved that Helmholtz did not press him to carry out the experiments, the exercise of writing the proposal had been far from useless. He had first of all learned to use parts of Helmholtz's electrodynamics which he had known before, if at all, only abstractly. Second, in working through problems with dielectrics he touched points at which the coherence of Helmholtz's scheme tended to fragment. Finally, he found a limited but solvable topic—induction in rotating spheres—that he could use impressively to display his skills, a topic perfectly fitted for a doctoral dissertation.

PART THREE

Berlin's Golden Boy

Rotating Spheres

Having succeeded in deflecting Helmholtz's pressure Hertz turned away from the laboratory, perhaps in reaction, to work on a "rather theoretical subject" for his doctoral dissertation: the currents induced in conducting spheres that rotate under magnetic influence. Although the subject was an intricate and difficult one, requiring much clever approximation (at which Hertz was becoming expert, having honed his talents on the first prize competition), he took care to connect his results very firmly to laboratory measures. Since the topic had been dealt with before according to Weberean principles by Emil Jochmann, and since Hertz made use of Jochmann's work, the subject was eminently suitable for his major debut in the profession: he could reasonably hope both to avoid controversy and to make a favorable impression. The techniques he learned here were turned years later to good measure in his computation of the field of a radiating dipole. Of more immediate significance, and despite Hertz's probable intention to avoid controversy, the investigation suggested questions with important meaning for Helmholtz's electrodynamics.

Hertz sought to solve completely the induction of currents on rotating solid and hollow spheres. He began formally, pointing out what had been done before and what he intended to do that was novel. Jochmann had solved the problem for small velocities, neglecting "self-induction." Maxwell, in the *Treatise,* had done so for an infinitely extended, thin plate. Using Jochmann's equations together with Maxwell's analytical technique, Hertz tackled this intricate, but straightforward, problem for spheres.

Jochmann's equations were less general than Helmholtz's, which latter were:

$$E_{\text{HELM}} = A \left[-v \times (\nabla \times A) - (v \cdot A) \right]$$

The term in the curl of the vector potential was Jochmann's. For the purposes of his dissertation, which concerned currents, Hertz could ignore the extra, Helmholtz term (and thereby avoid controversy).[1] The paper was very carefully constructed and clearly written, more so than Hertz's earlier work on the current's kinetic energy. He proceeded almost magisterially, section by section, starting first with a careful definition of symbols and units. He then analyzed what occurs when self-induction is neglected; next, he included it, but only for infinitely thin spherical shells; finally, he removed the limitation on thickness.

In the next three sections he considered first the forces exerted by the induced currents and then turned to rotating *magnetizable* spheres, culminating in a series of related problems. The last section examines "special cases and applications." The work's limpid, logical structure gives it a seemingly unproblematic aura, as Hertz certainly intended, and he was not again so thoroughly to achieve such lucidity in a written work until a decade later.

Hertz's fundamental equations had the following general form, with ω representing rotational velocity (i.e., the linear velocity v is $\omega \times r$)

Everywhere: $\kappa C = -\nabla\phi + (\omega \times r) \times (\nabla \times A)$ and $\nabla \cdot A = 0$

Within the metal: $\nabla \cdot C = 0$

At the surface: $r \cdot C = 0$

Hertz went beyond Jochmann primarily by including self-induction in computing the vector potential A. What he meant by this, however, had nothing to do with the electromotive force generated by a changing current on itself, because he considered only objects rotating uniformly about their axes of symmetry—Hertz's currents are all steady. Instead, by "self-induction" he referred to a complicated hierarchy of interactions that depend upon the following assumption: that an object in a given current state will have that state modified by the object's own motion just as though the object were moving in relation to a distant, spatially fixed system of currents. Or, put in a different way, the object is considered to carry currents that are engendered by its motion through its own magnetic field. This assumption (to which Webereans would also assent) leads at once to an infinite series of correction terms that must be included in the vector potential because each such current state in turn induces another one on the moving object. In Hertz's words:

> In accordance with usual views[2] we regard the total induction as compounded of an infinite series of separate inductions; the current induced by the external magnets induces a second system of currents, this a third, and so on *ad infinitum*. We calculate all these currents and add them together to form a series which, so long as its sum converges to a finite limit, certainly represents the current actually produced. (Hertz 1880b, p. 50)

Here we find the methodological germs of Hertz's later (1884) attempt to construct field equations by considering an infinite series of interactions, each one generating the next in the process.

Hertz first expanded the vector potential of the external magnetic source in spherical harmonics and proceeded to solve the problem for a finite spherical shell. In a next step he demonstrated that each term in the series that represents the magnetic potential of the induced current (neglecting self-induction) involves an infinite series of further terms. Under appropriate assumptions[3] this series converges in such a way that the conclusions he had drawn for the approximate case could be carried over in form to the general situation. The principal difference between the approximate and the general case was a rotation of the current flow lines from their positions in the approximate solution.

Hertz produced graphical representations of the lines of current flow on his conducting spheres (fig. 29). These diagrams were carefully calculated for quite specific situations,[4] and Hertz offered practical advice for carrying out experiments:

> If measurements are to be made in experiments on the rotary phenomenon, very thin spherical shells should be used; for in their case the calculation can be easily and exactly performed. The simplest form of experiment would be one in which such a spherical shell is made to rotate under the influence of a constant force. The rotation of the current planes might be demonstrated either by the effect of the currents on a very small magnet, or better by a galvanometric method. (Hertz 1880b, p. 120)

He then applied his results to various known experiments, such as a conducting sphere rotating between the poles of a suddenly excited electromagnet.

Formal and thorough, Hertz's dissertation was an exercise in how to generate solutions to a complicated general problem by a method of approximation that was adequate for practical purposes. Though more sophisticated and carefully wrought than the previous-summer's report for the prize competition, his dissertation shared with it a fine sense of how to manipulate general relationships to obtain results that would be meaningful in the contemporary laboratory. There was no new theory—the equations Hertz used had originally been developed by Jochmann on the basis of Weberean relationships and subsequently rederived (and supplemented) by Helmholtz on the basis of the electrodynamic potential. But even here Hertz was able to bring to bear a sense of how to tackle the problem in a way that was strongly tied to the laboratory, since his infinite sequence of inductions was useful because in most *practical* situations it converged to a finite result. Precisely because Hertz connected abstract analysis to actual, physical configurations he was able to work through to novel, useful solutions.

7.1. THE EXAMINATION AND ITS AFTERMATH

> I have to boil and simmer many substances, sulfur, sealing wax, pitch benzene, and the like, which is quite a trial in this heat. (Hertz to his parents, July 1, 1880; J. Hertz 1977, p. 125)

Hertz's doctoral research progressed with extraordinary rapidity—so much so that he had to request special permission to be examined, which took place on the evening of February 5, 1880. The examiners were Kirchhoff, Zeller (in philosophy), Kummer (in mathematics), and Helmholtz. Hertz visited them ahead of time. "I spent only a moment with Helmholtz," he wrote home, "he frightened me a little by receiving me with the words, Now then, Herr Hertz,

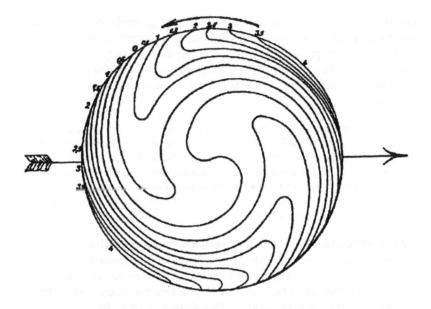

Figure 29   Representation of current flow lines from Hertz's doctoral
dissertation (Hertz 1880b)

are you ready so that we can proceed with the examination on Thursday?" The
examination lasted from six until eight in the evening. "As soon as it started,"
Hertz wrote,

> I saw that I should not fail. I think I could have answered at random as long
> as I did not show the most blatant ignorance. However, sometimes the ques-
> tions did go somewhat deeper, but as soon as I began to show some uncer-
> tainty the examiners hurried to encourage me by the easiest questions, and
> by changing the subject. (J. Hertz 1977, p. 121)

This experience, like the prize judging the previous August, again brought out
the competitive instincts that were so common in the profession: "doctorates
in my class [magna cum laude] from this university," he wrote home the next
day, "are very few in number, especially Helmholtz and Kirchhoff are said not
to have awarded many." He lamented that "the fear of a bad outcome takes
away the pleasure of a good one." His intense immersion in "theoretical work"
sated his appetite for it, and by the end of February he was back in the laboratory "from 9 in the morning until 9 in the evening"—but not working on *Helmholtz's* problems. At this time he received the galleys for his dissertation, and,
he wrote with self-conscious solemnity, "what was previously so simple and
easy to change now stands unalterably before me, so that the responsibility for
the complete accuracy of what I have written begins to weigh upon me."

During the spring Hertz sensed that he was receiving special approval by the great ones of Berlin science.[5] He rummaged around for a project, and though he began one by midsummer Hertz was feeling ill at ease. "Staying on here is becoming very disagreeable," he wrote home. "I regret to say I am suffering from weariness and distaste for work in general and my specialized work in particular, which would disconcert me if I had not the experience of last year and two years ago, when I felt the same at this season but got everything going again in the autumn." Hertz was always subject to these fits of depression, and his state of mind was no doubt not improved by the fact that he did not have a position of any kind. This changed when, on August 8, 1880, Helmholtz offered him a two-year, renewable assistantship at his institute, one that had recently been vacated by Kayser. When he accepted, Hertz truly became a fledgling physicist, a member, though at the lowest level, of the intensely competitive, research-oriented Berlin community.

By early October Hertz had not yet had time to begin his own laboratory work, but he was being pulled ever more tightly into Helmholtz's social and professional circle, dining with the family and attending Physical Society meetings. "I grow increasingly aware, and in more ways than expected, that I am at the centre of my own field; and whether it be folly or wisdom, it is a very pleasant feeling." But he still kept a wary eye on Helmholtz, whose slow speech now irritated him, and whose opinion Hertz "did not care to put forth" his own against. Though he wanted very much to please Helmholtz, he wanted to compete with him as well, to have "confidence in [his own] accomplishments vis-à-vis" the master.

During the early months of his tenure in the Berlin laboratory Hertz became increasingly familiar with apparatus. He also became fascinated with theory for a time. He wrote home about his current research that "all these papers are theoretical ones; I find it almost impossible to do experimental research when I have something else in mind, and I have resigned myself to regarding the beautiful laboratory as a luxury for the moment." During this period Hertz deepened his theoretical insight in two areas.[6] First, he delved into Helmholtz's electrodynamics itself, whose subtleties he had now begun to grasp (Hertz 1881a). Second, he struck off in a new direction by applying elasticity theory (in the form he had learned from Kirchhoff) to the compression by contact of two isotropic bodies (Hertz 1881c, 1882a).

7.2. ELECTRIC REDISTRIBUTIONS

This first work of Hertz's after his prize competition essay and inaugural dissertation has the innocuous-sounding title "On the Distribution of Electricity over the Surface of Moving Conductors" (1881a). Here nevertheless Hertz for the first time came to grips with the effects of Helmholtz's insistence that there is no such *thing* as electricity—indeed, that all physical interactions involve special states of known objects. The problem he sought to analyze is deceptively simple to set out. Suppose a group of charged conductors that are initially in electric equilibrium are set into relative motion. Then, Hertz remarks, as a result of their relative shift in position "the distribution of free electricity at the surface varies from instant to instant." And, he continues, "This change produces currents inside the conductors which, on their part again, presuppose differences of potential." Note Hertz's wording. He does not say that the change in relative position of the conductors engenders currents that alter the electric distribution. Not at all. He begins instead with an unproblematic fact—something no one would deny—namely, that the motion will alter the electric distribution. Such a change "produces currents inside the conductors": currents do not engender the new electric distribution; they are produced *by* it. And if currents exist, there must be "differences of potential." Or, one might say, alterations in states of charge instantaneously "produce" states of current via the continuity equation.

This is not circumlocution. In fact, it makes the cause of the unproblematic *fact* of redistribution highly problematic. The situation Hertz considered had nothing to do with electromagnetic induction and therefore did not at all involve interactions between states of current. Rather, one began with an interaction between states of charge and then altered its terms by changing the distance. It is *this* alteration, a change in the charge interaction, that "produces" something different, namely, the state of current. Precisely how that happens is, we have already seen, perhaps the central obscurity of Helmholtz's electrodynamics, because it requires postulating an interaction energy between state-of-charge and state-of-current. Hertz skirted this intensely destabilizing issue by referring only to the necessary existence of (scalar) potential differences that are associated with the currents. But this is not all. Hertz continued:

> In forming the differential equations we assume that the only possible state of motion of electricity in a conductor is the electric current. Hence if a quantity of electricity disappears at a place A and appears again at a different place B, we postulate a system of currents between A and B, *not a motion of the free electricity from A to B.* The explicit mention of this assumption is not superfluous, because it contradicts another, not unreasonable, assumption. When an electric pole moves about at a constant distance above a plane plate the induced charge follows it, and the most obvious and perhaps usual assumption is that it is the electricity considered as a substance which follows the pole; but this assumption we reject in favour of the one above mentioned. (Hertz 1881, p. 128; emphasis added)

Begin with the assumption that Hertz rejected, which is of course the Weberean one. According to it the electric redistributions that arise when charged conductors are mutually displaced result from motions within the conductors themselves of a distinct stuff, the stuff of electricity. In Hertz's evocative words, an electric pole in motion over a plate will be "followed," in this way of thinking, by the induced charge on the plate. This, Hertz wrote, he entirely rejects.

He did not believe that electricity moves at all in the sense of the displacement of an object from one locus to another. There is no motion of electricity, but there is a *state* associated with the electric redistribution that indubitably does take place, namely, the electric current. And the current is itself "a state of motion of electricity," which is to say, as Hertz used the phrase, that a current is associated with the motion of an electric state; it is not a displacement of electricity.

What is extraordinary about Hertz's point of view is that it seems to make no difference at all. Why should he so strongly insist, we might wonder, that his assumption is not the "usual" one when whichever is chosen makes, one might think, no difference whatsoever to the subsequent analysis and examples, which occupy almost the entire article? Does it make the slightest difference whether electricity moves bodily through a conductor or whether the electric state is transferred via currents? Is it not the case that the moving electric objects constitute currents and that, in direct consequence, the analytical structure of the problem is independent of Hertz's "assumption"?

In Hertz's day, and indeed for several decades before, it would not have been at all apparent that what Hertz intended by "currents," both conceptually and pragmatically, were analytically equivalent to what Webereans meant by currents. Weber's currents were not merely flows of electric particles; they were *Fechner* flows—equal but opposite motions of two electricities. For Webereans the continuity equation was therefore not entirely transparent. If net electric charge accumulated somewhere, then this had to mean that the Fechnerean flows into the region were *unbalanced,* that more of one kind of electric particle flowed in per unit time than did the other kind. Each flow might individually be inhomogeneous, but only if the inhomogeneities were different from one another could net charge appear.

Consequently a situation of the kind that Hertz envisioned could not possibly be dealt with by a Weberean solely through the continuity equation, which was, per contra, the only tool that Helmholtz's electrodynamics permitted.[7] That equation hid a welter of complicated physical changes, changes that might be very important to take into account in a given situation. This was such a situation, because a Weberean would attack Hertz's problem by considering directly the several interactions between the physically moved electric particles in the conductors.[8] Indeed, more likely than not, he would refuse to attack the problem at all because, to him, it would appear to be much too complicated to solve. But for Hertz the problem was comparatively simple precisely because there were no such things as moving electric particles to worry about.

As always for Hertz abstract considerations had little meaning in themselves. They had to be tied to realizable situations. He accordingly invented one, and here we find him beginning to think about how to do experiments that could reveal something without requiring much in the way of accurate measurement or elaborate calculations. Both of these had troubled his work

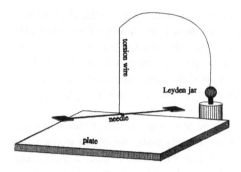

FIGURE 30   Helmholtzian interactions realized

for the prize competition the year before; and both—but particularly the for-mer—had discouraged him from pursuing Helmholtz's new prize questions. Here, accordingly, he sought for something that would elicit very noticeable—not subtle—effects without hiding them beneath complicated (and so argu-able) computations. "It remains to inquire," he wrote, "in what practically rea-lisable case the effects discussed could become appreciable." He felt that he had succeeded—succeeded where, recall, Helmholtz and his previous assis-tants had failed.

Take some "mirror glass" (which has a small but quite measurable conduc-tivity) and above it suspend from a wire a "needle" that terminates on both ends in brass plates, which always remain parallel to the plate (fig. 30). Charge the plates up. Then, Hertz wrote:

> the bound electricity [induced on the glass by the plates] was compelled to follow the motion of the needle, and ought, according to the preceding, to damp the vibration of the needle. *Now such a damping actually showed itself.* The needle was connected with a Leyden jar, of which the sparking distance was 0.5mm., whilst [the distance of the needle above the glass] was 2mm. The needle was found to return to its position of rest without further oscilla-tion, though previously [before connection with the Leyden jar] it had vi-brated freely; even when the [distance] was increased to 35mm., the increase of the damping at the instant of charging was perceptible to the naked eye. And when I charged the needle by a battery of only 50 Daniell cells, while [the distance] was 2mm., I obtained an increase of damping which could be easily perceived by mirror and scale. It was impossible to submit the experi-ment to an exact computation, but by making some simplifying assumptions I was able to convince myself that theory led to a value of the logarithmic decrement of the order of magnitude observed. (Hertz 1881a, p. 135; empha-sis added)

The idea was simple: if Hertz was correct, then as the charged needle oscil-lated, thereby altering the locus of the induced *free* (i.e., conduction) charge on the glass, currents would flow in the latter, dissipating energy in Joule heat. By energy conservation it followed at once that the needle's motion had to be more strongly damped when the plates on its ends were charged than otherwise. It was.

Hertz provided no computations; neither did he provide any experimental data. He wrote simply that he saw something happen that accorded with his qualitative expectations, and that "simplifying assumptions" convinced him that the "order of magnitude" of the effect was also what he expected. Nor did Hertz write anything concerning the ability of Weberean electrodynamics to deal with the effect. Here for the first time Hertz had found something *positive* to report, which contrasts markedly with the purely *negative* quality of his prize competition paper. And here also we find a noteworthy change in how Hertz discussed his results. The prize competition paper was filled with calculations, many tabular data, and error estimates. This one has few calculations and no error estimates or data at all. Showing that something does not produce an effect required the full panoply of exact measurement. But showing that something does produce an effect required only a reportorial description. Devices that do not work must be thoroughly dissected; devices that do work need not be taken thoroughly apart. Or, one might say, exact measurement, which was above all emblematic of Weberean investigations, was for Hertz associated primarily with the *failure* to generate something. The appropriate language with which to describe persuasive research at the frontier emphasized the qualitative over the quantitative. Hertz replicated this pattern in his later work, and it applies particularly to his fabrication of electric-wave devices.

Hertz had apparently completed this journey by the early winter, since he probably read his work to the Berlin Physical Society in early December (1880).[9] He regarded this investigation, as well as a very different one he began at just this time, as "theoretical": "all these papers" that he was reading to the society or working on, he wrote, "are theoretical ones." By this Hertz did not mean that they were not applied to practical cases. On the contrary, he always gave numerical examples. He meant that they did not take their shape and content from the laboratory but were built about abstraction. His prize competition work, and the unfulfilled proposals for investigating Helmholtz's new questions, had been embedded first of all in the laboratory; they were *about* experimental work. The inaugural dissertation, this paper on electrified, moving conductors, and his latest work on elasticity were *about* theory, even though the work on conductors had led to an experiment. Hertz was ambivalent about this: "I find it almost impossible to do experimental research when I have something else in mind, and I have resigned myself to regarding the beautiful laboratory as a luxury for the moment" (J. Hertz 1977, p. 141). However, his novel work in elasticity had the advantage of taking him beyond the physical research community proper; it brought him into contact with scientifically literate industry.

EIGHT

Elastic Interactions

In pursuing elasticity Hertz came into close contact with Kirchhoff. Despite the favorable opinion that he held of this star pupil of his close colleague Helmholtz, Kirchhoff was not undividedly impressed when the young man trod into Kirchhoff's own fields. Hertz had first sent his elasticity paper to Kronecker for inclusion in Borchardt's *Journal für die Reine und Angewandte Mathematik,* where it eventually did appear—reflecting the paper's "theoretical" character. This had been in late January or early February of 1881; when he did not receive an answer by late April, he went to see Kronecker, which must have seemed rather forward to the distinguished mathematician. Hertz described what happened:

> [H]e said the delay arose because the paper had been sent to Prof. Kirchhoff for review; the latter was very interested in it, and they would be glad to take it, but Prof. Kirchhoff had some criticisms of the form of the paper, and since it would be several months before it was set up in print, he would like to let me have the paper back for now, so that I could improve it in the meantime. He then showed me how Kirchhoff had thoroughly annotated the paper and had rewritten three or four pages in another form in the margins. At first I was surprised and even flattered that Kirchhoff had gone over it so thoroughly, but apart from a wrong sign that I had indeed overlooked, his comments seemed only to say the same thing (and by no means better) that was in the paper. In part the points were expressed in a manner peculiar to Kirchhoff which I do not like at all, and which I should be very unwilling to have imposed on me. (J. Hertz 1977, p. 147)

Having bothered Kronecker, Hertz decided to confront Kirchhoff himself. Kirchhoff, he wrote home, was very friendly but persisted in thinking that the paper contained critical errors "and seemed to believe that [Hertz] had reached the right results so to speak by accident." Hertz dug in his feet, and within a few weeks he had apparently forced Kirchhoff to back down:

> [I]n looking over my paper, I found that Prof. Kirchhoff himself had made the main error with which he had reproached me (what I had written was merely not quite clear) *and I have demonstrated it to him.* (J. Hertz 1977, p. 149; emphasis added)

Having already deflected Helmholtz's pressure to perform secretive experiments, Hertz had seemingly also forced Kirchhoff, the master of mechanics, to confess error—little wonder that he flourished in the highly competitive atmosphere of the physics community, rejoicing in "testing his strength" against that of his colleagues.

The problem that captured Hertz's attention was not a traditional one in this highly mathematical subject. Hertz wanted to find out what happens when two bodies are pressed together either by an external force or by the forces of impact. Traditional problems in elasticity involved only one deformable body; what the body touched was considered to be immutable or simply to be given (e.g., a wall with a flexible beam embedded in one end, or a sphere whose surface is subject to a given stress). Hertz had to invent some way to retrieve a simulacrum of the usual situation from his own in order to deploy the mathematics of stress that he had learned from Kirchhoff. This was not simple to do; it apparently produced Kirchhoff's first objection.[1]

Hertz had first to create a system of coordinates that, on the one hand, would express the contact of the two bodies as they press together and, on the other hand, could also be used to calculate stresses and deformations. These two desiderata do not mesh nicely. The former demands a system that depends upon the relation between the two bodies; the latter demands a fixed surface for boundary conditions. These requirements are in apparent conflict because the one seems to demand a movable system of coordinates, whereas the other seems to require a fixed one, and this was perhaps one reason for the comparative neglect of the problem among traditional elasticians.

Hertz's clever solution, which would not (and in fact did not) appeal to rigorous elasticians, was this: he made the system of coordinates *itself* approximate and mutable; he made it something that depended upon the physical character of the interaction. This was not an easy solution, as is amply evident from Kirchhoff's having required Hertz extensively to rewrite his initial explanation of it, and from the fact that, for example, the British elastician A. E. H. Love used Hertz's new coordinates inconsistently in discussing the theory (Love 1944, sec. 137). The essential idea was this. When two elastic bodies press together, Hertz argued, the resulting deformation will usually be limited to a small region near their surface of contact. Far away from that region they will remain undeformed, but not unmoved: these far regions will move close together very nearly *as rigid bodies*. So, Hertz decided, the appropriate thing to do is to introduce two systems of coordinates. Each system is rigidly connected to the undeformed region of one of the two bodies and moves with it. As the bodies press together, then, their respective systems of coordinates also move together through some distance α. One goal of the theory was to find α.

Since the "compressed area," or region of contact, is common to the two bodies, it can serve, as it were, as the mediator between the demands of the physical interaction and the demands of mathematical elasticity, in the following way. In order to apply traditional methods Hertz had to refer the displacement of a bodily element to a fixed system, and such a system is provided for each body by the axes that are fixed to its undeformed parts. Certainly these axes move through space with the undeformed parts as the compression occurs near the region of contact, but elasticity theory concerns itself only with relative displacements. Consequently, Hertz's two axial systems can serve, each for its proper body, as appropriate coordinates for elastic computations. Having represented the deformations in this way for each body, Hertz could then use the region of contact to mediate between them, because it is common to both bodies. To do this he distinguished between the systems with his z axes: z_1 points away from the region of contact and *into* body 1; z_2 points away from the compressed area and *into* body 2, so that the z_1, z_2 axes are antiparallel.

Assume with Hertz that when the bodies simply touch one another—when they are in what he termed "geometric contact"—the distance between a point on body 1 and a point on body 2 that have the same x, y coordinates is $Ax^2 + By^2$, where A and B are constants.[2] After the compression these same two points have undergone displacements w_1, w_2 with respect to their axes z_1, z_2. In addition, the origins of the two coordinate systems have been moved together through a distance α. Consequently, the distance between the points is increased by $w_1 - w_2$ (recalling that the z axes are antiparallel) and decreased by α, yielding $Ax^2 + By^2 + w_1 - w_2 - \alpha$ as the new distance between them. Within the region of contact itself the points are coincident, producing the following requirement for displacements at the common boundary:

$$w_1 - w_2 = \alpha - Ax^2 - By^2$$

In the amended version that reached print, Hertz introduced his coordinates in the following words:

> Further, we imagine in each of the two bodies a rectangular system of axes, rigidly connected at infinity with the corresponding body, which system of axes coincides with the previously chosen system of xyz during the mathematical contact of the two surfaces. When a pressure acts on the bodies these systems of coordinates will be shifted parallel to the axis of z relatively to one another; and their relative motion will be the same in amount as the distance by which those parts of the bodies approach each other which are at an infinite distance from the point of contact. The plane $z = 0$ in each of these systems is infinitely near to the part of the surface of the corresponding body which is at a finite distance, and therefore may itself be considered as the surface, and the direction of the z-axis as the direction of the normal to this surface. (Hertz 1881c, p. 148)

Much of the corresponding passage in the manuscript (SM 250) was crossed out by Kirchhoff and then entirely rewritten by Hertz. The original paragraph had treated the new coordinate systems as conveniences for calculation and explained very little about them. The new one, as we see, carefully explained them. Kirchhoff no doubt insisted on a much more careful explanation of what were, after all, entirely novel and admittedly approximate coordinate systems. It was not that Hertz introduced an approximation after having laid out the exact conditions of the problem, which was the traditional procedure. Rather, he *began* with an approximation for the problem's elementary mathematical structure. This was evidently sufficiently novel to be disturbing to a rigorous elastician like Kirchhoff, who insisted on Hertz's crossing all of his *t*'s and dotting all of his *i*'s.[3]

Kirchhoff's emendations did not cease here. There was one other paragraph of this kind—of changes to an explanation—which this time Hertz did not accept. More important, Kirchhoff insisted on very large changes indeed to Hertz's mathematics, or rather he insisted on Hertz's putting in mathematics that was missing, that Hertz had jumped over via qualitative argument. The nub of Kirchhoff's objection was, it seems, this. Aside from the complexity introduced by his novel coordinate systems, Hertz's analysis had the following structure. He first provided a differential equation and a set of boundary conditions. He then (following what was by now becoming his common practice) introduced a potential function of a mathematically suspect kind and went from it without direct analytical proof to expressions for the displacements, arguing along the way that the relevant conditions of the problem were thereby satisfied. Having done this he then introduced a particular expression for the potential, and, again qualitatively, argued that it satisfied all conditions and so represented the unique solution to the problem. Kirchhoff rejected almost all of Hertz's original argument here and himself wrote out several pages of direct analytical demonstration, reaching in the end the very same results that Hertz had. Kirchhoff's emendations were used by Hertz verbatim in the printed version, without any mention that they differed considerably from what he had originally submitted (fig. 31–32). (For further details, see appendix 10.)

Despite his ultimate acquiescence, Hertz strongly felt that Kirchhoff had not done anything new, that he had merely reformulated the results in a way that Hertz strongly disliked. He was angry that Kirchhoff thought that he, Hertz, "had reached the right results so to speak by accident (he continually spoke about his calculations and mine as though they were different, whereas they are quite the same)." Hertz in the end dealt with the problem "by substituting his formulation for mine" in nearly every instance. A not inconsiderable portion of the paper as printed is Kirchhoff's wording and mathematics, not Hertz's.

Hertz had not done any experiments for the elasticity paper, though he had calculated something that had not been calculated before: namely, the duration of contact between colliding elastic bodies.[4] Kirchhoff's critique would soon alert Hertz to the less-than-favorable reaction his novel approach would receive among elasticians. They, or rather the wider German physics community, were Hertz's desired audience; they were the ones who could transform his work

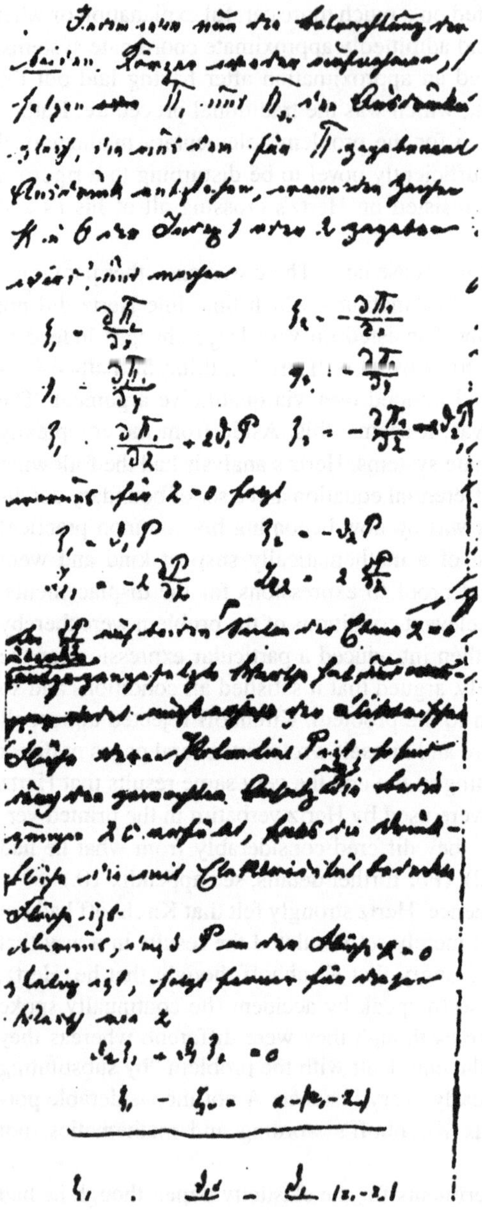

FIGURE 31 A sample of Kirchhoff's corrections. Reproduced from the Communications collections of the Science Museum, London.

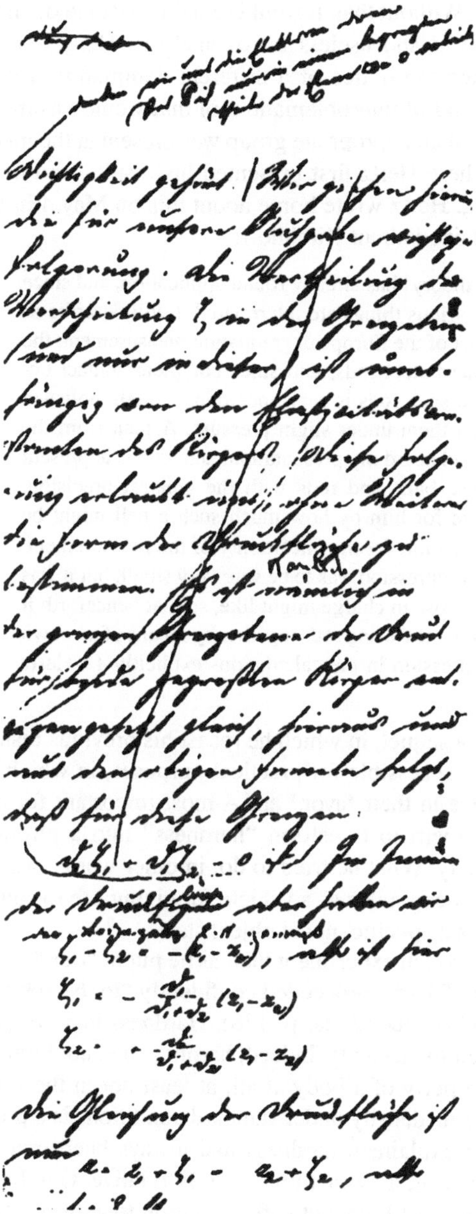

FIGURE 32 Hertz's original. Reproduced from the Communications collections of the Science Museum, London.

into a part of accepted physics. Without that, it would remain an isolated, undeveloped, and probably (to physicists) useless bit of analysis. Hertz's work, however, serendipitously gained the notice of a different community, which would turn his work into the kind of unproblematic tool that would ensure its perpetuation. A representative of an appropriate group was present at the meeting of the Physical Society where Hertz first presented his results, and he at least found them quite exciting. Hertz wrote home about this on May 4 in the same letter in which he complained about Kirchhoff:

> Incidentally it [the elasticity theory] has already found application, and since that shows very nicely how various things are interrelated, I will describe it. In a triangulation which is part of the European graduation measurement, the base is measured with standard steel rods, and since complete contact between the end surfaces of two such rods is very uncertain, a small, polished glass ball is inserted between them under slight pressure. A man from the standards commission, who had heard my presentation, and who is at present charged with measuring these balls and rods with the greatest precision, asked me if I would calculate for him by how much such a ball might be compressed, in order to have a limit for the uncertainty of the measurement. I did that and found that the compression was to be sure very small, but it was still presumably greater than those in charge might like, so that henceforth it might be preferable to press the measuring rods together by means of a known weight and include the compression in the calculations explicitly. (J. Hertz 1977, p. 149)

This stimulated Hertz to write a sequel, in which he recast his previous results in a way that made them more accessible for practical purposes, in which he provided experimental evidence in their favor,[5] and—most important for our purposes—in which he used them to transform "hardness" into a physical property similar to, say, elasticity. What he tried to do, in other words, was to use the community of industrial engineers as a vehicle for giving life to something that had been designed as a specific, novel object state.

Having convinced himself through experiment that, as he put it, "our formulae are in no sense speculations," Hertz proceeded confidently "to the application now to be made of them" (Hertz 1882a, p. 178). Hardness had not previously been directly connected to elasticity. Indeed, hardness had not hitherto been treated as an inherent property of a body at all, at least not in the same sense that a body's elasticity or its density under standard conditions is a property. More precisely, as Hertz explained, hardness had always been treated, according to mineralogical practice, as something entirely relative. One body might be hard in relation to a second body but soft in relation to a third. This, Hertz argued, raises a great difficulty because the mineralogists rely on operational criteria that are not manifestly connected to one another.

> The hardness of a body is usually defined as the resistance it opposes to the penetration of points and edges into it. Mineralogists are satisfied in recognis-

ing in it a merely comparative property; they call one body harder than another when it scratches the other. The condition that a series of bodies may be arranged in order of hardness according to this definition is that, if A scratches B, and B scratches C, then A should scratch C and not *vice versa;* further, if a point of A scratches a plane plate of B, then a point of B should not penetrate into a plane of A. The necessity of the concurrence of these presuppositions is not directly manifest. (Hertz 1882a, p. 178)

Hertz was convinced that hardness could only have true physical meaning if it did not depend upon a particular method of measurement for its very definition. More to the point, he wanted hardness to be "a property of the bodies in their original state" and not something that might be an artifact of the measurement process.[6] Instead of looking for the forces that permanently alter bodies, Hertz insisted, one should instead find the maximum forces that can be applied to them without inducing permanent change. Such forces would depend only on the object proper rather than upon the measurement technique; they would accordingly define a specific, unique object state. As Hertz put it, "Since the substance after the action and removal of such forces returns to its original state, the strength thus defined is a quantity really relating to the original substance, which we cannot say is true for any other definition." And his analysis of interacting elastic objects provides a method for determining hardness according to this definition.

To produce an absolute scale for hardness, Hertz argued, one should first determine the "principal stresses" (viz., the diagonal elements of the stress tensor) that an element of an object can sustain before the limits of elasticity are exceeded. Then create a sort of hodograph—a map in stress-space—in which the components of the radius vector are these stresses. This produces for the particular material a surface (the yield surface in modern parlance)[7] such that states of stress that lie within the surface do not permanently deform the body; any stress that lies outside it produces a permanent deformation (called "set"). Hertz assumed that he could specify the "hardness" of a material by means of its yield surface, and this required specifying a standard method for constructing the latter. Hertz proposed to do so in the following way. First, he specified, put two bodies into contact in such a fashion that the resulting deformations (before set) involve a circular surface of pressure. Using Hertz's theory one could calculate the corresponding pressure. Then, he continued, increase the normal pressure at the center of the circular region of contact until "in some point of the body [whose hardness is being measured] the stress may just reach the limit consistent with perfect elasticity." Since the two bodies in contact can very well be formed from the same material, Hertz continued, "we therefore do not require a second material at all to determine the hardness of a given one." And, therefore, the procedure produces an "absolute measurement," that is, one that depends solely upon the material in question.[8]

Supposing that the yield surfaces exist, Hertz's measuring procedure will certainly enable materials to be compared with one another with respect to yield. But does it do so as well for hardness, or, put a different way, does the scale that results for yield correspond to the scale that traditional methods produce for hardness? Hertz thought that it did, and he offered the following argument in support of his belief:

> [S]uppose two bodies of different materials pressed together; let the surface of pressure be circular; let the hardness, defined as above, be for one body H, for the second softer one h. If now we increase the pressure between them until the normal pressure at the origin just exceeds h, the body of hardness h will experience a permanent indentation, whilst the other one is nowhere strained beyond its elastic limit; by moving one body over the other with a suitable pressure we can in the former produce a series of permanent indentations, whilst the latter remains intact. If the latter body have a sharp point we can describe the process as a scratching of the softer by the harder body, and thus our scale of hardness agrees with the mineralogical one. (Hertz 1882a, p. 181)

Hertz's claim depends upon a major assumption, one that was never widely adopted by any community: namely, that rupture (here, scratching) can be assimilated physically to set (permanent deformation).[9]

In the paper that he wrote explicitly for engineers Hertz sought to transform an ill-defined, operational concept (hardness) into a physical property that would be no different as such from elasticity proper. Hardness was now to be a true property of a body *considered in itself;* it was not to be comparative by definition. To say that body 1 is harder than body 2 now meant that an interaction between 1 and 2 in which the objective hardness states are brought to bear produces a specific change in the state of one of them (to wit, denting). All other aspects of hardness must be reduced to this one, including the traditional scratch and the more recent puncture tests, which Hertz reduced to cases of denting. Where, for mineralogists, bodies were only hard with respect to one another, for Hertz bodies had hardness absolutely. For him physical processes could only be predicated of interactions between bodies in objective physical states.

N I N E

Specific Powers in the Laboratory

In June of 1882 Hertz spoke to the Berlin Physical Society about his recent work on evaporation. An article in the *Annalen der Physik und Chemie* resulted from it that provides us with an extraordinary opportunity, outside electrodynamics, to see Hertz the disciple of Helmholtz at work in the laboratory. The manuscript for the article exists (SM 248). It bears witness to radical changes that Hertz introduced as his laboratory investigations, which seemed at first to elicit the effect that he was trying to produce, turned difficult.

Hertz perhaps did not always write down measurements, comments, and guesses as he went along in a well-organized laboratory notebook. Nothing of the kind apparently exists before 1887. Or, better put, except for a few scraps of calculations, nothing distinct from manuscripts in nearly final form exists until his fabrication of electric waves. These manuscripts (as we have already seen in the case of elasticity) may differ from the printed versions, but sometimes the difference is altogether negligible. Nevertheless, both the manuscripts and the printed articles contain many clues to the course of his work, because Hertz treated them rather like a laboratory notebook. He may not always or even usually have begun writing during the actual course of an investigation, but he permitted the final product to reflect both his earlier conceptions as well as his final thoughts. The published articles, in other words, are not always carefully reworked to hide the many byways that Hertz traveled. We will see, for example, that the manuscript on evaporation (which is, like most of the others, quite close to the published paper) bears conceptual witness to changes in Hertz's investigation as he tried to clarify a new effect he thought he had found. Instead of entirely rewriting the article in the light of his final results, Hertz retained in it substantial vestiges of a now-lost original draft that he probably drew up when he thought that his experiments were providing good evidence for a new effect. Here, and in Hertz's other papers as well, we find that the published artifact need not be treated as a highly worked product designed to erase the sequence of guesses and tries that preceded its production. On the contrary, from it we can gain a very great deal of understanding about the problems that Hertz faced and about how he attempted to overcome them.[1]

Evaporation became interesting to Hertz as a by-product of a brief investigation he did early in 1882 on hygrometry (1882b). He had developed a new method for measuring humidity by weighing a hygroscopic substance.[2] Such a thing absorbs water from the air until the saturated vapor pressure above it (and produced by it) becomes equal to the pressure of the unsaturated vapor "actually present in the air." Shortly afterward Hertz became interested in vapor saturation as a property.

What happens, Hertz queried, when a liquid evaporates "in a space which contains nothing but the liquid and its vapour" (1882c), when, that is, no pressure acts on the liquid's surface other than that of its own vapor?[3] The kind of evaporation, or condensation, that occurs in Hertz's hygrometer takes place under conditions in which the atmospheric pressure vastly exceeds the vapor pressure and, further, in conditions of near-equilibrium. In a space filled solely with vapor far from equilibrium with the liquid, Hertz argued, two things could happen. Either (1) the liquid will continue to evaporate at a rate limited solely by the speed with which heat can be conveyed to its surface, or else (2) every liquid has an intrinsic *maximum* to its rate of evaporation, no matter how rapidly heat can flow through it. If, as Hertz believed, the second possibility holds true, then "the rate of evaporation will depend upon a number of circumstances, but chiefly upon the nature of the liquid, *so that there will be for every liquid a specific evaporative power*" (1882c, p. 187; emphasis added). Every liquid will, that is, have a property or *quality* that determines its evaporative character. One might then conclude that two liquids would interact with one another in ways that depend upon their respective "evaporative powers," upon qualities of the objects per se. If Hertz was thinking about evaporation in this way, we would expect to find that he designed experiments with at least two liquid surfaces evaporating into (and so interacting with one another through) the same space. That is precisely what we do find.

Suppose with Hertz that several liquid surfaces evaporate into a given enclosed space. If, Hertz claimed, the evaporation rate were unlimited, then, supposing that heat can be supplied rapidly enough, "all liquid surfaces in the same space must assume the same temperature; and this temperature as well as the amounts of liquid which evaporate are determined by the relation between the possible supply of heat and the different areas." Or if, as Hertz believed, the rates are intrinsically limited, then

> there may be surfaces at different temperatures in the same space, and the pressure and density of the vapour arising must differ by a finite amount from the pressure and density of the saturated vapour of at least one of these surfaces: the rate of evaporation will depend upon a number of circumstances, but chiefly upon the nature of the liquid; so that there will be for every liquid a specific evaporative power. (Hertz 1882c, p. 187)

Hertz's cryptic remarks require considerable explanation in order for one to grasp why an unlimited rate requires the vapor to be saturated and all evaporat-

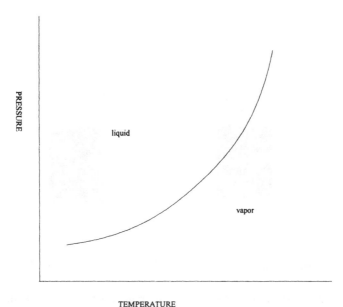

TEMPERATURE
FIGURE 33   Vapor-pressure curve

ing surfaces to have the same temperature, whereas limited rates do not. These consequences hardly leap unaided to mind.

Hertz's conclusions depend upon the properties of a saturated vapor and upon what might happen in the absence of equilibrium between a vapor and its liquid. When, he clearly knew, a liquid and its vapor coexist in equilibrium in some region, then the state of the vapor must lie somewhere along the vapor-pressure curve that divides the pressure–temperature space into liquid and gaseous regions (fig. 33). Raising the temperature causes evaporation and increased pressure; lowering the temperature causes condensation and decreased pressure. But in quasi-static conditions the system always sits on the vapor-pressure curve. Under these circumstances the vapor is said to be *saturated*. If the pressure at a given temperature were, say, suddenly increased—swiftly enough to preclude a rapid return to equilibrium—the vapor would become *supersaturated*, and much of it would condense. A rapid increase in temperature produces *superheated* vapor; evaporation then occurs until the system returns to the vapor-pressure curve. At equilibrium the system sits on the vapor-pressure curve, the liquid and vapor have the same temperature, and the pressure on the liquid surface and due to the vapor equals the vapor pressure.

To uncover what Hertz had in mind, consider figure 34. Here we have two heat sources: one at the high temperature T_2; the other at the lower temperature T_1. During evaporation heat flows through each source to its surface, which is separately represented in the figure and which may have its own temperature

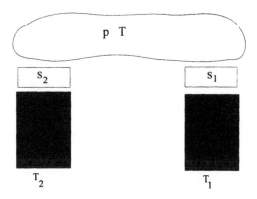

FIGURE 34   Hertz's evaporation system: vapor (pT), surfaces (S), and sources (T).

(T_S^1 and T_S^2). The vapor in the surrounding region has some temperature T and pressure p. Suppose that the high-temperature source is hotter than its surface, which is hotter than the surface of the low-temperature source, this last source being coolest of all ($T_2 > T_S^2 > T_S^1 > T_1$). Heat accordingly flows *from* source 2 to its surface, where vapor forms. At the low-temperature surface heat flows *to* the source from condensing vapor and from conduction. For the system to reach equilibrium the rate of evaporation from 2 must equal the rate of condensation on 1, which requires that there be no net heat flow into or out of the vapor between the surfaces.

   Saturated vapor can form in such a system provided that the evaporation rate has no limit. If there is no limit, then *all* of the heat that flows to surface 2 from its source can be used to produce vapor; none remains to accumulate at the surface, increasing its temperature. However, the actual rate at which vapor formation absorbs heat must not be greater than the rate at which the heat formed by condensation at the low-temperature surface can be removed by conduction to its reservoir. Assuming that both surfaces have the same area, then equilibrium can occur, and saturated vapor can form only if they both also have the same temperature—whatever that might be.[4] If, per contra, the evaporation rate were limited, the common surface temperature necessary to form saturated vapor might never be reached.[5]

   Although Hertz did not do so, we can elucidate his reasoning with a bit of simple algebra. Consider the heat fluxes through the two surfaces on the supposition that the surface temperatures are intermediate between T_1 and T_2:

Flux across S_2: $+F_2^F$ from T_2
$-F_2^e$ due to evaporation
$+F_2^c$ due to condensation

Flux across S_1: $-F_1^F$ from T_1
$-F_1^e$ due to evaporation
$+F_1^c$ due to condensation

We can use this to establish equilibrium conditions, that is, conditions under which the net flux through each surface separately vanishes and, therefore, such that the surfaces have the same temperature. They would, in essence, form two parts of the same surface:

Conditions for Equilibrium

(1) $\qquad\qquad F_2^F - F_2^e + F_2^c = 0$ and so $F_2^e = F_2^F + F_2^c$

(2) $\qquad\qquad - F_1^F - F_1^e + F_1^c = 0$ and so $F_1^e = - F_1^F + F_1^c$

We see that for equilibrium to subsist the rates of heat flux due to evaporation for each surface (which measure the rates of evaporation) must equal the sum of the corresponding rates due to flow and condensation. This at once implies that the system cannot reach equilibrium unless the evaporative rates are essentially unlimited, since the reservoir temperatures (and so the heat fluxes from them) can be made arbitrarily large or small. But if the rates are intrinsically unlimited, the surfaces must have the same temperature, and the vapor must therefore be saturated. If the pace of evaporation cannot exceed a certain maximum, then for some reservoir temperature the heat flux to or from the surface will exceed the rate at which the heat can be turned into, or condensed from, vapor.

Hertz accordingly had to find some way to determine, for a given liquid, the temperature t at its *surface*, the pressure P upon it, and the height h of the liquid layer that evaporates in unit time. He evidently planned originally to measure at the high-temperature surface only, seeking to find whether its temperature corresponds to a saturation pressure that differs from the pressure of the surrounding vapor (thereby implying a limit to the rate of evaporation). He thought, we shall presently see, that he could obtain the vapor pressure without directly measuring it. The temperature raised problems, because the very large temperature gradient from the surface to its reservoir meant that "if we dip a thermometer the least bit into the liquid it does not show the true surface temperature," but Hertz was convinced that he could overcome these problems.

He built the apparatus, using mercury as the liquid. Figure 35 reproduces his illustration of the device.

> Into the retort A, placed inside a heating vessel, was fused a glass tube open above and closed below; inside this and just inside the surface of the mercury

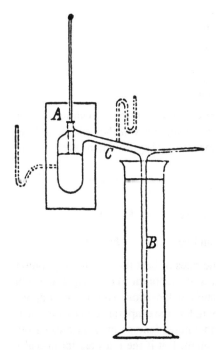

FIGURE 35   Hertz's first evaporation device
(Hertz 1882c)

was the thermometer which indicated the temperature. To the neck of the
retort was attached the vertical tube *B*, which was immersed in a fairly large
cooling vessel, and could be maintained at 0° or any other temperature. By
brisk boiling and simultaneous use of a mercury pump all perceptible traces
of air were removed from the apparatus. The rate of evaporation was now
measured by the rate at which the mercury rose in the tube *B*. (Hertz 1882c,
p. 189)

Uninterested in an absolute measure for evaporation, Hertz intended to operate entirely with relative values, measuring solely the liquid's height in his
apparatus under different conditions. Similarly, he did not care to know the
pressure *P* in absolute measure; he needed only its relative value. Indeed, he
at first thought that he could avoid measuring the pressure altogether, on the
following grounds: whether the mercury surfaces in *A* and *B* differ in temperature or not, the pressure on either surface "could not exceed the pressure of
the saturated vapour at the lower temperature," viz., the saturation pressure
corresponding to 0°.[6] This meant that he would not have to measure *P* because
he could control it directly by heating the low-temperature source.

Hertz rapidly uncovered an oddity:

[W]hen the temperature [of the high source] began to exceed 100°, and the
evaporation became fairly rapid, the vapour did not condense in the cold tube

B, but in the neck or connecting tube at *C.* This became so hot that one could not touch it; its temperature was at least 60° to 80°. This cannot be explained on the assumption that the vapour inside has the exceedingly low pressure corresponding to 0° [i.e., *B*'s temperature]; for in that case it could only be superheated by a contact with a surface at 60°, and could not possibly suffer condensation. (Hertz 1882c, p. 189)

If the vapor pressure was not governed by the low-temperature source (*B*), then perhaps it was governed by the higher temperature of *A.* He attached a manometer at the neck, *C,* and measured the pressure at different rates of evaporation—that is, at different source temperatures of *A.* "But," he reported, "this did not show any change from its initial position when the rate of evaporation was increased." This truly startled him, because it seemed to mean that the vapor pressure somehow did not depend solely on the states of either of the active surfaces.

"I began to doubt," he wrote, "not whether these magnitudes [surface temperature and pressure] were necessary conditions, but whether they were sufficient conditions for determining the amount of liquid which evaporates." As he thought the problem through it occurred to him that the vapor pressure should be distinguished from the actual pressure on the evaporating surface: since the vapor is ejected with a certain speed, it must exert a reaction force on the surface, and this must be added to the intrinsic vapor pressure, whatever the latter might be. Nor would such a reaction be at all negligible: the vapor moved so fast at higher source temperatures that "when the drops of mercury on the glass attained a certain size they did not fall downwards from their own weight, but were carried along [by the vapor] nearly parallel to the direction of the tube." The vapor's kinetic energy might itself produce a great deal of heat, and this could account for the otherwise surprising high temperature of the neck, with the temperature of the vapor still being determined by the cooler reservoir. Far from being super*heated,* the vapor is, as it were, super*saturated* by the pressure kick it gets from the ejecting surface. Contact with the neck then raises the neck's temperature toward the point on the vapor-pressure curve that corresponds to saturation at this higher pressure, but droplets form as the vapor resaturates at its own temperature. The neck produces condensation by *slowing* the supersaturated vapor, not by cooling it through conduction.

Was this reaction pressure, however, sufficient to produce the necessary kinetic energy? To find out Hertz attached a manometer directly to reservoir *A* and measured the pressure at several reservoir temperatures (and so rates of evaporation). "It turned out that there was a very perceptible pressure," amounting to "2 to 3mm. when the thermometer stood at 160° to 170°." Or, it would be more complete to say, the pressure that Hertz had measured would be substantial *if* the vapor were so cold that its intrinsic (vapor) pressure could be ignored, leaving only the reaction force to press the surface. He did not at this time wonder whether the *vapor pressure* itself at these temperatures might have produced the reading, because the invariance with evaporation rate of the pressure measured in the neck had already convinced him that the vapor state was not governed by the high-temperature source.[7]

The original apparatus had failed to work properly, Hertz now felt, because the rapidly streaming vapor made it impossible to control the vapor's native pressure, though Hertz believed it to be that of saturated vapor at the temperature of the lower source. He decided to alter the apparatus in order to slow the stream, which required keeping it under pressure. That way the measured pressure on the surface would also be the native pressure of the gas. The goal of the experiment would then be to see whether that pressure corresponds to saturated vapor at the temperature of the surface. If not, then the evaporation rate had to be limited.

His redesigned apparatus appears in figure 36. The two most striking differences from the first experiment are the connection of an extension tube (C) to reservoir A and the removal of the cooling bath from the condensing tube B. The vapor still condenses in B, but since Hertz would no longer directly control the low-temperature source, it made no difference what the temperature there might be. His attention concentrated entirely on reservoir A. The extension tube permitted measuring the pressure on A by reading the difference in height of the liquid in A and C, while the absolute height of the liquid in A over time measured the rate of evaporation. Because mercury's density varies markedly

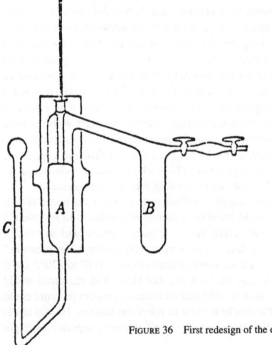

FIGURE 36   First redesign of the evaporation device (Hertz 1882c)

with temperature, and tube *C* was not itself heated to the temperature of *A*, Hertz had to correct his pressure measures. Some of these corrections "were much larger than the quantity whose value was sought." Nevertheless, Hertz asserted, the pressure measure could be relied on to about 0.1 mm, and the evaporation measure to about 0.02 mm.

The surface temperature was "the most uncertain element."

> I thought it was safe to assume that the true mean temperature of the surface could not differ by more than a few degrees from the temperature indicated by the thermometer when the upper end of its bulb (about 18mm. long) was just level with the surface; and it seemed probable that of the two the true temperature would be the higher. For the bulk of the heat was conveyed by the rapid convection currents; these seemed first to rise upwards from the heated walls of the vessel, then to pass along the surface, and finally, after cooling, down along the thermometer tube. If this correctly describes the process, the bulb of the thermometer was at the coolest place in the liquid. (Hertz 1882c, p. 191)

The bulb of the thermometer was, as it were, immersed in the same stream of hot liquid that, beginning deep within the fluid mass, creeps up the tube's walls and bathes the surface itself. Confident that his device correctly measured the surface temperatures, Hertz "carried out a large number of experiments at temperatures between 100° and 200°, and at nine different pressures." He altered the pressure by admitting air into the device through valves located near the condensing tube *B*. The results were precisely what he had hoped for: "The observed pressure P was always smaller than the pressure P_t of the saturated vapor corresponding to the temperature *t* of the surface," and by a considerable amount indeed—at circa 180° the difference amounted to over 7 mm of mercury, which was vastly larger than any possible error in measuring the pressure. But what about the temperature? To lower the pressure of the corresponding saturated vapor by 7 mm requires a temperature 30° lower than the one Hertz had measured. No measuring error in pressure could have occurred; "nor do I believe," he wrote, "that the second [temperature error] could."

When Hertz first wrote these last words he was apparently convinced that he had been able to elicit a reliable limit to the evaporative rate of mercury. The words remain in both the manuscript and in the printed article. Indeed, it seems quite likely that Hertz originally intended to close the experimental part of his article at this point, concluding with a "theoretical" discussion of how to calculate the limits given certain data. These theoretical remarks do indeed appear at the end of his paper and indicate that a new effect should exist, but between the report on his first two sets of experiments and the concluding, theoretical remarks, he inserted a discussion of yet another series of experiments, ones that he had certainly intended solely to provide further evidence for his claim to have fabricated something novel (and which he would presumably not have described in the detail that he did if they had merely corroborated his previous results). These experiments manifestly contradict Hertz's claim to discovery, and they forced him to redraft thoroughly what he had planned (and probably begun) to write when the second set of experiments had worked to give him something new.

Impelled by native inclination, by Helmholtz's conceptions, and by the competitive drive of the German physics profession to fabricate novelty, Hertz might have stopped with his initially successful results. Thoroughly convinced that his temperature measurements simply could not be in error by nearly 20%, he could have gone into print with these results. Why didn't he? One thing pressed him on. Not doubt that he had produced novelty—of that he was certain. Rather, he felt that he needed to provide more secure quantitative measures in order to be able to specify accurately the limiting conditions on the evaporative rate. Contemporary demands, particularly in the German physics community, for exact measurement would have made that seem to be essential. His second set of experiments could be improved upon to that end in two respects. First of all, they required very large corrections to be made for mercury's expansion in determining the pressure. These corrections had required "a careful application of theory and . . . special experiments," which, Hertz was convinced, made the result reliable to 0.1 mm. In his third set of experiments Hertz sought to manipulate the device into a form that did not place elaborate corrections between measure and pressure. Second, and much more important, even though Hertz was certain that his temperature measures could not have been off by 20% (i.e., by 30°), nevertheless, they might admittedly be inaccurate by "a few degrees."

In the second mutation of his original device, Hertz produced the apparatus of figure 37. The first obvious change is the absence of tube C, which had required measuring corrections because it lay outside the heated vessel. Instead the two arms now both lie within the vessel immersed in a paraffin bath. Pressure measurements could be read off directly from the height difference between the mercury in the arms. But the temperature required an entirely separate device based on equating the heat that flows to the surface in unit time to the heat absorbed during that time in evaporation.

> This [surface] temperature is equal to that of the bath [reservoir], less a correction which for a given apparatus is a function of the convection current only which supplies heat to the surface. The known rate of evaporation gives us the required supply of heat; from this again we can deduce the difference of temperature when the above-mentioned function has been determined. (Hertz 1882c, pp. 192–193)

Hertz had observed large-scale convection currents in his earlier apparatus and had concluded that most of the heat conveyed to the surface comes to it in this way. He needed a device untroubled by evaporation to pin down the heat flow. The right-hand diagram in figure 37 illustrates it. In Hertz's words:

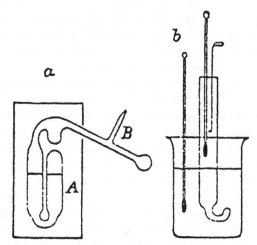

FIGURE 37   The second mutation of the evaporation device (Hertz 1882c)

A piece of the same tube from which the manometer was made, was bent at its lower end into the shape of the manometer limb. This was filled with mercury to the same depth as the manometer tube; above the mercury was a layer of water about 10cm. deep, and in this a thermometer and stirrer were placed. This tube was immersed up to the level of the mercury in a warm linseed-oil bath, the temperature of which was indicated by a second thermometer. A steady flow of heat soon set in from the bath through the mercury to the water. The difference between the two thermometers gave the difference between the temperatures of the bath and of the mercury surface; the increase of the temperature gave the corresponding flow of heat. (Hertz 1882c, p. 193)

Hertz's clever device measures the mercury's surface temperature by putting it in contact with another liquid that, when stirred, would be in thermal equilibrium with the surface. Instead of immersing a second thermometer in the mercury proper, Hertz placed it in the surrounding heat reservoir (here of linseed oil), thereby obtaining the temperature at the source of the convection current that feeds the surface.[8] In this way he found that a layer of water 117 mm high would be heated at a rate of 0.48° per minute by a 10° difference between bath and surface temperature.

The apparatus in figure 37 yielded an evaporation rate of 0.057 mm of mercury per minute and a pressure difference between the two arms of the manometer of 0.26 mm for a temperature of 118° in the paraffin bath. Given from elsewhere how much heat (in unit weights of water) is necessary to vaporize (a unit weight of) mercury at 118° (and given as well the ratio of the specific gravity of mercury to that of water), Hertz concluded that the amount involved here would heat 117 mm of water (on the same base) at a rate of 0.48° per minute, which is why he had sought for this specific difference in his temperature-measuring apparatus. Consequently, the evaporating surface (in the left-most tube in fig. 37) must have been 10° colder than the paraffin bath in which the manometer sat, making its temperature 108°.

The mercury surface in the rightmost limb of the manometer, however, bounds an enclosed region and thus produces saturated vapor at the temperature of the encompassing bath, that is, at 118°. Now if the evaporative rate were unlimited, each surface would be in contact with saturated vapor at its respective temperature of 118° on the right and 108° on the left. If this were the case, then, Hertz noted (using values for the vapor pressure of mercury that he himself generated in a separate series of experiments), the pressure difference across them would be 0.27 mm. This was in fact only 0.01 mm more than he had actually measured—leaving only an experimentally insignificant amount to represent the effect of a limited evaporative rate. Hertz felt that only one conclusion was possible given the great care that he had taken in temperature and pressure measurement: "the positive results obtained by the earlier method had their origin partly, if not entirely, in the errors made in measuring the temperature."

Hertz's confession that he had *failed* to produce something new contrasts remarkably with his assertion, following his description of the second experiments, that the *positive* results he had obtained relied on pressure and temperature measurements that were not subject to substantial doubt. In his words, again, "The first-mentioned error [pressure] could not have occurred; *nor do I believe that the second* [temperature] *could*" (emphasis added). Both statements—that the temperature measures in the second set of experiments are reliable, and that they must not have been—appear in the same published article, as well as in the manuscript (which scarcely differs from the printed version). Both cannot be correct, nor would Hertz likely have acknowledged both consciously and simultaneously. We have evidence here for one aspect of the neophyte Hertz's laboratory work habits.

The article as printed divides into (but was not so divided by Hertz) five distinct parts: first, an introduction that raises the issue of a new effect and elliptically discusses the results of the experiments; second, the initial experiments, which immediately revealed flaws in his assumptions; third, the experiments that seemed to reveal the new effect; fourth, the more elaborate experiments that contradicted the results of the third part; finally, "theoretical" considerations, which have the effect of justifying the ultimately negative results without giving up belief in the effect's existence (see appendix 11). The manuscript for the printed article has no physical marks of discontinuity between these several parts (with, we shall momentarily see, one exception). And yet we have just seen that statements in parts three and four concerning temperature measurement manifestly contradict one another. It seems reasonable to conclude that the article as it was finally written contains in its second and third parts substantial vestiges of a *first* article, now lost, in which Hertz had announced success. After, or perhaps even while, he was writing this lost account, he decided that the force of the paper, the impression that it would make on the German physics community, would be considerably strengthened if he could provide something more than evidence that the new effect simply exists—if he could, that is, pin it down quantitatively to something better than "a few degrees": "But I could not conceal from myself," he wrote in recollection of what had prompted him to undertake the new experiments, "that the results, from the quantitative point of view, were very uncertain." He accordingly built the apparatus of part four, and he then discovered, undoubtedly to his great surprise and consternation, that the two sets of experiments did not tally at all with one another. We can be nearly certain that he must have checked his results very carefully indeed, but that he could see no way out: his experiments had *failed* to bring the effect to laboratory life.

Some physical evidence remains in the manuscript that may at least indicate the repercussions of these disconcerting events. The second paragraph of the manuscript (and of the printed article) tells the reader that Hertz had undertaken experiments "which have only partly achieved their aim," the aim being to arrive "at an experimental decision" concerning a limit to evaporative rates. At that precise point in the manuscript—beginning at the start of the paragraph and ending with "have only partly achieved their aim"—Hertz pasted over whatever he had first written (fig. 38). Evidently he had difficulty deciding how to phrase his results, given their essentially negative character, and he tried with considerable effort to put things in as positive a way as possible. He might, for example, have written that his experiments provide no evidence whatsoever for the existence of an evaporative limit. Instead, he wrote here (in part one) that they "have partly achieved their aim"; in part four (deep within the article) he wrote somewhat less positively that "the net result of the experiments is a very modest one." The partial success or modest result amounted to this: that within the temperature and pressure ranges Hertz had examined an effect that no one had previously thought to exist in fact did not exist. Little wonder that Hertz had trouble sounding positive, that he had to write the important introductory remarks on what he had achieved in the laboratory, and that he tried to keep as many of his earlier, positive remarks as possible, producing in the end literal incoherence.

These events reveal both Hertz's deep-seated adherence to Helmholtz's implicit insistence on interaction and his absorption of what one might call the Berlin edict: "Go forth and discover." To make a place for himself within the highly critical and competitive Berlin physics community, and within the wider German environment, Hertz felt in his marrow that he had to produce something positive, something new, something that went far beyond an improvement in accurate measurement. He needed to find a new *effect,* and he kept looking

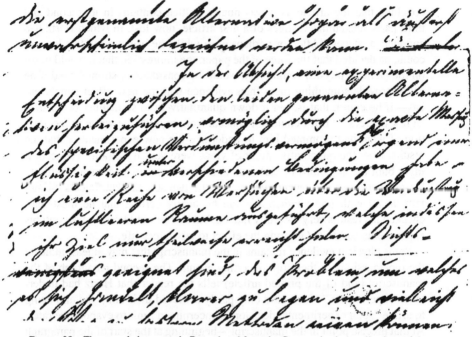

FIGURE 38   The emended paragraph. Reproduced from the Communications collections of the Science Museum, London.

for one. His work on elasticity had been a creative application of *Helmholtzian* belief, but it was (in Hertz's own words) "theoretical"—it did not bring to experimental life a novel, previously unsuspected effect. One could (and Hertz did) certainly calculate things from it that could be examined in the laboratory, but the basis of the calculation did not itself emerge *out* of laboratory work.

The evaporation investigation paralleled Hertz's elasticity work in its manifest insistence on interaction but differed from it in being seated from the outset in experiment. Hertz's initial evaporation devices, we have repeatedly seen, differ from apparatus designed to measure the vapor-pressure curve for mercury in two very important respects: first, there are always *two* liquid surfaces involved, whereas vapor-pressure experiments (such as the ones that Hertz himself performed to provide data for evaporation) involve a *single* surface; second, the surfaces in the evaporation apparatus interact dynamically with one another, whereas in the vapor-pressure device the surface interacts statically with its own vapor. The evaporation device acts like an engine: one surface constantly evaporates, decreasing the volume of liquid below it, while the other receives the evaporate and increases its corresponding volume. The vapor-pressure device sits unchanging as long as the temperature remains constant. Just as, when he thought about elasticity, Hertz had naturally turned to the *interaction* of elastic objects at their surfaces, so here too he turned to the *interaction* of evaporating liquids at their surfaces—at the surfaces because the kinds of connection involved required direct contact between the two objects (elasticity) or between one object and material that formerly belonged to the other object (evaporation). Both investigations exemplify this characteristic emphasis.

However, the evaporation experiments were the first ones that Hertz undertook in order actually to fabricate a novel interaction, to bring it reliably to life in experiments. Object–object interactionism constituted the intellectual bedrock of Helmholtz's physics. Producing new effects constituted its professional imperative. Hertz's reputation, and indeed his prospects for satisfying employment, depended critically upon achieving novelty, and he indubitably knew it, as we can see from comments in his letters home. In early November of 1881 he wrote home disparagingly about significant changes he had undertaken to his experiments on the upper limit to a current's kinetic energy (of which more below), experiments that continued to yield no evidence for such a thing: "it is a very unimportant little thing and the result is negative; *of course, if it had been positive, it would have given me quite a reputation*" (J. Hertz 1977, p. 153; emphasis added here and below). He continued, referring to what were almost certainly the evaporation experiments: "The work that I have in hand at present (or rather that I am at the moment prevented from completing) will give me great pleasure *if it fulfils the hopes I have for it*." Negative experiments did not improve professional standing; positive ones did. And the competition was intense: Hertz remarked, about a talk that he gave to the Physical Society on his hygrometer (a novel *device,* no doubt, but hardly a novel *effect*), that "in the Physical Society it did not receive acclaim, because it was already quite late, and *in general it is accomplishment enough if one is not torn to pieces after a lecture*" (J. Hertz 1977, p. 155). By February of 1882 Hertz was still working hard at the evaporation experiments. He wrote a letter home on the twenty-third that typifies the irritation and depression that, we shall several times see, he usually felt when his career seemed to be drifting and his experiments had not as yet proved out:

> It is very sad that one grows older so quickly and accomplishes so little in the short time, but considering its inevitability, one has to accept fate. Looking back on what was accomplished, I am not very well pleased with the past year and I can only hope for something different from the future; however, in general I was quite content and happy during this year, and that is already more than [miserable mortals] can expect. I hope that the work I am now doing will take shape nicely and will yet prove worth the great amount of time I have spent on it, which is to say a large part of the past year, since it deals with an apparently obvious and yet completely unexplored subject *and could indeed lead to important discoveries if I am lucky;* but in the meantime I can neither arrive at a lucid theory nor do my experiments have the precision

that I had hoped for, with all the trouble I took. Now I have to hurry to complete the experiments before the vacation, for then follows an interruption of
several weeks and then the new semester, at the beginning of which I shall
not have a great deal of time. Thus I cannot wonder that progress is so slow,
but it irks me. (J. Hertz 1977, pp. 158–59)

About February 23, then, seems to be when Hertz decided to improve his previ-
sion—to begin the experiments that became the fourth part of his account.
Within about two weeks he knew that he faced failure:

> As far as my present progress is concerned, I have come nowhere near a
> satisfactory result; on the contrary, some new experiments showed me, one
> day, that I have spent a large part or most of my efforts on barren ground, that
> there are sources of error of a magnitude that could hardly have been ex-
> pected, *and that the beautiful, positive result I had thought safe in my hands
> was changed into a completely negative one. At first I felt completely beaten,
> but now I have somewhat consoled myself and have more courage than be-
> fore; I am only sorry about the beautiful time that is lost for good.* (J. Hertz
> 1977, p. 159)

Hertz was beaten by the transformation of a "beautiful, positive result" into a
"completely negative one." Success, and reputation, had as it were been
snatched from his hands, after a year of effort, by *his own attempts* to push
further the precision of his experiments. Little wonder that Hertz turned melan-
cholic for a time. At the beginning of May he wrote home a powerful, plaintive
cri de coeur while on military exercises in the reserves. His mother had found
his recent letters cheerless. "If you are expecting enthusiastic reports from me,
I must beg to be excused, for you could get them from me only with great
difficulty." He complained of the officers' "carefree life" in words that say
much about his unhappiness with his career and perhaps even with his future
prospects:

> [F]or me it is more like a spectacle that I am watching, but in which I do not
> take part as an actor, so that I cannot tell whether the outcome is good for me
> or not. Dear Mama, you should always judge the good or bad spirit of my
> letters accordingly; it takes a great deal for me to declare myself content, for
> I have the somewhat superstitious feeling that fate does not allow anyone a
> hairsbreadth more than that with which he declares himself content. Further-
> more, *the lofty ideals I have formed for myself are terribly fixed and inflexible;
> and if the world does not fit them in ever so many points, I can manage to
> keep silent and avoid preaching sermons in the wilderness, but it is impossible
> for me to change them in order to conform them to reality or to suppress the
> consciousness of the conflict;* as long as I am alone this conflict naturally does
> not arise, but it does so the more when I am suddenly thrust into a larger
> company. If that is wrong, dear Mama, then I am sorry for it, but you are
> more responsible for my nature than I am myself. (J. Hertz 1977, p. 161)

Hertz was writing here about the conversation of his fellow officers, but he might as well have been writing about the recalcitrance of his apparatus. He had molded it to design, to elicit from it an effect that he had at first only hoped for. The device that he used operated strangely, in a way that, Hertz eventually decided, might be due to something associated with the very effect that he was trying to produce. He mutated the apparatus to take advantage of this unexpected action. Success seemed to be at hand: the apparatus worked. Then unanticipated failure. Hertz seems almost to have felt as though the (natural) world had "failed" him, that it had somehow refused to fit "the lofty ideals I have formed for myself," just as his officer companions failed to do. Hertz wrote that the conflict between ideal and reality that he felt did not occur "as long as I am alone." On the contrary, it seems that it was precisely then, when he *was* alone in his laboratory with apparatus of his own creation, that he felt the conflict between ideal (how things should go) and reality (what he could make the apparatus do) most intensely. For the young Hertz experimental devices were not tools to investigate a fixed, static nature. They were things to make nature do what Hertz wanted it to do, and when they did not work as he had hoped they would, he took it very nearly personally; they, and by extension the world, had failed him.

By early May Hertz was in better spirits. His formal paper on elastic contact (1881c) had by then elicited interest from Förster, the director of the Standards Commission, and as a result Hertz decided, as we have seen, to produce a different version, to which he added his account of hardness, for a "technical journal" (in the event, the *Verhandlungen des Vereins zür Beförderung des Gewerbefleisses*). "I am only sorry," he wrote home on May 1, "that it [the original paper] has no sequel yet, for although I have every hope of finishing, in the next few weeks, with the project that occupied me all last year [on evaporation], it will not be anywhere near what I had hoped it would be" (J. Hertz 1977, p. 163). Determination to succeed returned, and Hertz set himself a daily schedule of tasks:

> My time is divided as follows: Every morning from 6:30 to 10:30 I work undisturbed for myself, then until 3:30 in the laboratory also—disturbed— for myself; then I go out for a meal and take a longish walk until about 7 or 7:30, then I read or work a little more and go to bed about 11:30. Thus I have regular working hours while I am still fresh every day. (J. Hertz 1977, p. 163)

Still, he remained unhappy with the outcome of his work on evaporation, though he had substantially completed the paper by June 13.[9] He began again to cast around for something promising to look into. But now his search changed slightly in character. Hertz's previous laboratory work, indeed his entire approach to experiment, had certainly evinced Helmholtz's implicit emphasis on object–object interactions. However, these early efforts also bear the unmistakable mark of the exact-measurement credo: in every case Hertz strove not only to bring about something but also to corral it with numbers. Indeed, he had brought out the inadequacy of his first experiments on evaporation by seeking more accurate measurements for it—not to prove its existence, which he did not initially doubt (hence his subsequent great surprise and disappointment), but rather to accommodate the widespread emphasis on exactitude. His failure with evaporation not only depressed him for a time, it also began to push him away from experiments that, he felt, contemporaries would demand exact measures for. He turned instead to the wellsprings of Helmholtz's physics and began looking for something entirely novel that could be brought to life without worrying much about exactness. He had found it by late June in an area of special interest in Berlin, one that his fellow assistant and friend Eugen Goldstein specialized in: the luminous glows produced by electric discharge in Geissler tubes. *These* investigations proved fruitful in German context, and for the first time Hertz's career moved decisively forward.

T E N

The Cathode Ray as a Vehicle for Success

10.1. THE "WITCH'S KITCHEN"

Hertz wearily wrote up a paper on his evaporation work in early June 1882. Frustrated by his failure to elicit a novel effect with his evaporation experiments and fed up with excruciatingly exact measurements that in the end led him nowhere, he looked for something else to investigate, something that held out the hope of yielding a prestigious, *positive* result. Until then he had worked areas that no one else in the Berlin laboratory had claimed, but he had not succeeded in eliciting the type of result on which the Helmholtz environment put the highest premium: a new kind of object or a new state of a known object. This time he decided to make a space for himself in an area that had for nearly a decade been a major Berlin preoccupation: the beautiful, puzzling luminosity of the Geissler tube, a field of inquiry that his colleague Eugen Goldstein had long ago appropriated. Here was a doubly attractive subject: it was replete with strange effects, and it had nothing at all to do with exact measurement. It had the disadvantage of belonging to Goldstein, which meant that Hertz had to find some way of working productively without violating the laboratory's norms of behavior, which evidently put a very high premium on a worker's property rights in the area that he had specially cultivated. Hertz solved this social difficulty by consulting Goldstein at every step, asking his advice, and following up leads that Goldstein gave him. He told his parents about his new work at the end of June:

> I keep busy from morning to night with luminous phenomena in rarefied gases, the so-called Geissler tubes, only you must not take them for the kind that are usually exhibited publicly. I feel like changing for once to a broader experimental field, leaving exact measurements aside; moreover, the aforesaid field is very obscure and unexplored, and its exploration is probably of great theoretical interest. So I am looking into it for material for a new project, but so far I only keep rushing about without any fixed plan, familiarizing myself with what is already known, repeating the experiments, and setting up other experiments as they occur to me; this is great fun, for the phenomena are mostly very beautiful and infinitely varied. Much depends here on the glass blowing; I am much too impatient to order a tube from the glassblower

today and only get it several days later; so I prefer to limit myself to what my own small skill can produce. (J. Hertz 1977, p. 165)

Hertz buried himself in the laboratory, playing with tubes, blowing glass for new ones, working the air pump, watching and manipulating the strangely attractive, changing patterns (fig. 39). "I have seen most beautiful and interesting new things," he wrote, physically exhausted, two weeks later, giving a vivid description of the ambitious young experimenter surrounded—indeed, captured—by his devices:

> [I]t is no fun to sit in a completely darkened room, with the sun streaming down on the shutters and making the room hot, and with all sorts of wires strung about that can give quite appreciable electrical shocks when touched carelessly. However there can be no progress any other way. In the middle of it all, to my consolation, stands the luminous belljar in which the gases perform the maddest antics under the influence of discharges and produce the strangest, most varied, and most colourful phenomena. My place now really looks very much like a witch's kitchen. (J. Hertz 1977, p. 267)

During July he made his first discovery. Half a year later he wrote a formal report about it for publication:

> [Figure 39] represents a discharging apparatus, which consists of a glass tube, not too finely drawn out, and of two electrodes, one inside the tube, the other attached to it outside near the opening. When this apparatus is placed under

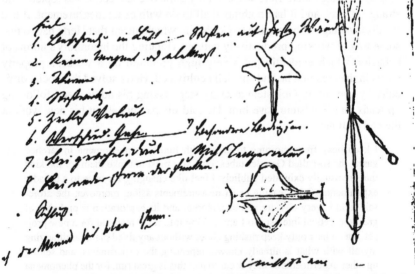

FIGURE 39   Excerpt from Hertz's manuscript (SM247) on Geissler-tube discharge. Reproduced from the Communications collections of the Science Museum, London.

the receiver of an air pump, the receiver filled with well-dried air and exhausted down to 30 to 50mm. pressure, and the discharge from the induction
coil then sent through, the following phenomenon is observed: Near the cathode is the blue glow; it is succeeded towards the anode by the dark space,
one or more millimetres wide, and from its end to the anode the path of the
current is marked by a red band 1 to 2mm. in diameter . . . in addition I
observed a jet, brownish-yellow in colour, and sharply defined, which projected in a straight line from the mouth of the tube; it was some 4cm. long,
and its form was like that shown in the drawing. The greater portion of the
jet appears to be at rest, and only at the tip does it split into a few flickering
tongues. The jet does not change its shape appreciably when the current is
reversed. But when a Leyden jar is joined up, an important change occurs:
the jet becomes brighter, and is straight for a distance of only 1 to 2cm.; then
it splits up into a brush of many branches, which are violently agitated and
separate in all directions. (Hertz 1883b, pp. 216–17)

Perhaps Goldstein directed Hertz's attention to, or helped him to concentrate
on, the effect, which would for Goldstein certainly have been a comparatively
unimportant side issue in the context of the many new processes that he had
brought about during the years he had worked with the tubes.[1] For Hertz, on
the other hand, the jets were his ticket to discovery, not only for themselves
but because he could use them to begin formulating an argument that reached
fruition months later in a different experiment.

To understand why Hertz's jets were sufficiently novel to be worth a month's
work and the writing of a paper, begin with the processes that had been reproduced and refined for many years, and, particularly in the 1870s, by Goldstein
himself. Figure 40 shows a glowing tube. The tube's ends each carry metal
electrodes, which can be attached to any of several different kinds of machines
that generate electromotive force in the region between the electrodes (the kind
of machine used can, we shall see, make a substantial difference for some
processes). The tube is filled with a rarefied gas, evacuated typically in the
1870s and 1880s to tenths or hundredths of a millimeter by means of the mercury air pump, designed originally by Geissler in 1855. The negative electrode,
or cathode, on the top is surrounded by a dark space whose size increases with
evacuation. To that succeeds a smoothly lit region (the "negative glow"), then
another dark space, and finally a region lit with striations (the "positive glow").
Some sort of ray emanates from the cathode and causes suitable objects to
phosphoresce when it strikes them.

Quite a few effects of these rays, several of them first produced by Goldstein,
had been successfully reproduced in many British and European laboratories,
and there was a broad range of agreement about them, for example, that the
"cathode rays" are normal to the plane of the cathode, that magnetic action
alters the paths of the cathode rays in much the same way that it alters the path
of an inelastic, flexible wire carrying a current, and that decreasing density

FIGURE 40 The discharge tube as reproduced by J. J. Thomson
from an article by E. Wiedemann (Thomson 1893)

expands the negative dark space. It was also agreed that current processes of some sort must be going on in the tube. Glowing Geissler tubes were pursued most avidly in Germany, where they were the province of experimental craft, not of high-level theory, during the 1870s and 1880s. As such, they were perfect vehicles for discovery.

To say that high-level theory (though not lacking) had little to do with this laboratory work does not at all imply that the experimenters operated in a conceptual vacuum. On the contrary, Goldstein and Hertz deployed a common understanding of objects and states: the insistence on manipulative perturbation. Neither of them would, in the mid-1880s, have insisted that cathode rays should be mapped to other, previously known or described object states, but they both did insist or imply that the rays are sui generis. Indeed, the purpose of Hertz's tilling of Goldstein's fields was to construct an experimental argument for that precise point, one that Goldstein had himself tried to develop throughout his own researches.

10.2. GOLDSTEIN'S RAYS

Although the details of Goldstein's investigations (begun in 1874) are rather intricate, the important point to remark about them for our purposes is that, according to Goldstein, the glow in a Geissler tube does not directly involve either conduction or electrolysis. It does not, that is, represent the effects of the transfer of electric charge from point to point throughout the evacuated tube. Instead, the discharge involves a series of successive, localized "ray" generations: where this takes place, a particle becomes the source of new rays, in much the same way that the rays originally emanate from the cathode. These new rays in turn stimulate other particles of the residual gas to produce rays.

> [E]ach secondary negative bundle [of light produced by processes that begin at the cathode] represents a *motion which, excited at the point of origin of the bundle, is transferred to the surrounding medium;* hence each particle[2] affected, as far as the excitation is propagated, assumes the characteristic form of motion which is produced at the *point of origin* of the rays; whilst a comparison of the discharge at any point with conduction in metals and electrolysis can afford a guide only for the relationships at the point itself. (Goldstein 1880, p. 185; emphasis added)

For Goldstein a "motion" (state) begins at the cathode, which elicits an interaction between the cathode and the "surrounding medium" that generates "rays" in the latter. These rays bear the stamp of the interaction, and when they strike another particle, an interaction between the medium and the particle occurs that endows the latter with the same *state* (the particle "assumes the characteristic form") that the cathode itself possessed in its ray-generating interaction with the medium. The process then repeats, coupling particle to medium and medium to particle throughout the path marked by the secondary light. The nature of the process, Goldstein insisted, must be distinguished from conduction. For here, among the particles in the glowing tube of rarefied gas, the last interaction in the sequence bears the stamp of the very first one.[3] Goldstein's rays accordingly represented whatever state of the medium interacted with the corresponding state of the cathode or the other objects (particles) that were embedded in it.

But what are these rays that are produced by the interaction between the medium and the particles or cathode? In 1883, after Hertz claimed to have definitively proved that the cathode rays cannot be electric currents, Helmholtz wrote from Berlin to congratulate him (Hertz had by then accepted a position at Kiel):

> For some time I have been wondering whether the cathode rays might not be a mode of propagation of a sudden impact on the Maxwellian electromagnetic ether, in which the electrode surface forms the first wave front. For as far as I can see such a wave should be propagated exactly as these rays are. That would then also make deflection of these waves by magnetization of the medium possible. Longitudinal waves could be more easily postulated and could exist if the constant k in my electrom. theories were not zero. But transverse waves might also be generated. You seem to have similar thoughts in your own mind. (J. Hertz 1977, p. 349)

This requires unpacking to grasp. One of Goldstein's major discoveries was that the rays are not emitted from the electrode in the same fashion that light is emitted from a luminous object, because they stream away normally to the surface instead of proceeding in all directions from every point of it. Perhaps, he had early reasoned, this is because the rays are waves produced by a "sudden impact"—presumably so sharp and rapid a blow that the entire electrode actually constitutes the initial wave surface proper. Now it would be difficult, though not impossible, to envision an impact on the ether so sharp and so rapid as to produce a *transverse* effect with this property, especially since (presuming Maxwell's identification of light with transverse electric waves) these kinds of waves do not in other circumstances merge their initial front with the emitting surface. But *longitudinal* waves, Helmholtz here remarked, are much better candidates because in them the disturbance, and so the impact, parallels the direction of propagation. If, moreover, the longitudinal wave propagates extremely rapidly, then the disturbance may move sufficiently far away before a second impact occurs to perturb its conformity with the emitting surface.

By 1879 Goldstein had developed a rather intricate way of understanding cathode rays that linked a propagatory effect to a precipitate disruption in the equilibrium of the ether that takes place when the potential over the cathode's surface equalizes. In Goldstein's understanding, as the potential rapidly equalizes across the cathode, an unspecified kind of "motion" propagates away from the equalizing region at high speed through the ether. When this motion strikes the "atoms" that remain in the tube, it sets them vibrating, and they in turn generate ordinary light, viz., transverse ether waves, which are observed and which therefore mark the path of the unspecified motion. Goldstein in fact thought of the motion as a species of open current, that is, he conflated longitudinal waves in the ether with unterminated currents, conceiving that the ether might possess conductivity.[4] For him cathode rays were simultaneously a kind of current and a kind of wave.

Despite his 1883 letter to Hertz, Helmholtz had years earlier urged Goldstein to consider the rays to mark the paths of moving electric particles—not of an ether motion—and he had apparently even found Goldstein's early suggestions that something different was involved "distasteful." Arthur Schuster much later recalled that

> Goldstein [in the summer of 1874] was working at his electric discharges in high vacua, and trying to explain effects he discovered near the kathode by a theory which I know to have been distasteful to Helmholtz; yet no word did he ever say to discourage the purely experimental side of these experiments. (Schuster 1911, p. 17)

And what was the distasteful theory? Schuster continues some pages later:

> Goldstein's experiments, in which a second kathode placed parallel to the first was observed to repel the radiation coming from the first, seemed absolutely conclusive in favour of some theory of projected particles, but Goldstein himself took a different view, *in spite of the fact that Helmholtz, in whose laboratory these experiments were made, urged him—as he subsequently assured me—to adopt the corpuscular hypothesis.*[5] (Schuster 1911, p. 55; emphasis added)

So *Goldstein*, not Helmholtz, first insisted on considering the rays to involve something very different from particle motion.

10.3. HERTZ'S JETS

By the time of Hertz's arrival in Berlin, then, Goldstein had for some time been insisting that the glows, positive as well as negative, must be treated as sets of localized processes, and that each such process marks an *open* current: "From the kathode as from a number of points lying between the two electrodes, which correspond to the limits of the positive layer towards the kathode, issue a number of open currents, which render the gas incandescent in their path, and reach so much the further the greater the exhaustion is" (Goldstein 1880, p. 189–90). These open currents, which begin and end seriatim between the electrodes, are the cathode rays. This must be taken quite literally. According to Goldstein there is no transfer of electricity from the end of one such ray to the beginning of the next:

> *The kathode-light, each bundle of secondary negative light, as well as each layer of positive light, represent each a separate current by itself, which be-*

gins at the part of each structure turned towards the kathode, and ends at the end of the negative rays or of the stratified structure, *without the current flowing in one structure propagating itself into the next,* without the electricity which flows through one also traversing the rest in order. (Goldstein 1880, p. 183; emphasis in the original)

Granting that the rays must be currentlike in this sense (viz, open) solves the problem that considering them to be closed currents otherwise raises, namely, that there is no evidence at all for a return current (for a current that curves back to the origin, thereby closing the path):

There is, however, no *action* of this hypothetical return-current to be observed. The magnet diverts the electric rays only in the manner required by the current flowing from the kathode towards the ends of the rays. The hypothetical return-current does not cause the least appearance of light, although it is present in the same medium, and certainly not of greater section than the direct current which fills the whole width of the tube. Any manifestation of light due to it must become visible when the direct current is diverted by the action of the magnet to one side of the tube. In the space thus rendered free, any possible luminous effect of the return stream would show itself. Experiment proves, however, that this space is *dark.* (Goldstein 1880, p. 180; emphasis in the original)

Furthermore, Goldstein continued, in tubes where the electrodes do not lie along a line the cathode rays nevertheless stream straight out from the cathode; they do not bend toward the anode, as they might be expected to do if they traced a current path between the electrodes (see fig. 41). Goldstein accordingly developed the following understanding. First, and most essential for him, all rays originate in local processes, in local states of some sort. Second, these states correspond in some way to open currents: that is, there are multiple regions in the gas that sustain current processes without the current in one region passing into the current in the other region. Something must accordingly be going on at the boundaries between these regions, and we have already seen that he assimilated *that* boundary process to a propagatory motion. According to this way of thinking, the cathode rays take on a striking sort of regenerative individuality. Goldstein's language nicely reflects this sense that he was manipulating coherent entities when he applied a magnet to the tube:

[I]t is *not the absolute position and expansion of the rays which determines the position they take when magnetized,* but the intimate relationship which exists between all the points of a ray and its *point of origin,* in consequence of which *each luminous body springing from a given point appears as a single coherent whole.* (Goldstein 1880, p. 187; emphasis in the original)

Given that the rays were currents, Goldstein also felt it necessary to require the discharge itself to be intermittent in order to effect the seriatim regeneration of the rays down the tube.[6]

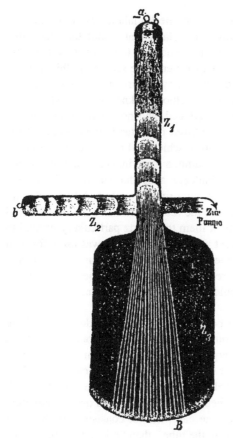

FIGURE 41   Goldstein's illustration
showing cathode rays diverging from the
path between electrodes (G. H.
Wiedemann 1885)

As he played with Geissler tubes, spoke with Goldstein, and read the literature, Hertz began to see how to open a space for himself within this complicated, fertile area. He apparently soon realized that Goldstein understood his
glows to constitute local states sui generis. Playing with Geissler-tube glows
in Berlin therefore amounted to the manipulation of a new kind of effect.
Goldstein had partially separated the new effect from an old, familiar one—
the electric current—by insisting that it had at most to be a special kind of
current, an open one. That claim, Hertz rapidly understood as he probed
Goldstein for knowledge and craft, was an entirely subsidiary one in that
Goldstein certainly did not regard it as constitutive of his central body of work.
Indeed, assimilating the rays even to open currents to some extent reduced
their novelty as new states.

Hertz made his move precisely here. He could dispute the claim that the rays
had anything at all to do with currents. This apparently runs counter to one of
Goldstein's claims, but it actually fortifies the publicly available arguments for

Goldstein's core belief. The coherence of the Berlin laboratory would thereby be strengthened, not weakened, through a claim that would look like, that would certainly *be,* an insistence that prior belief had to be changed. Not only would Hertz's reputation as a fruitful experimenter be secured in this fashion, but it would be done in precisely the right way to further the wider claim of the Berlin laboratory that it had a unique method for knowledge production.

What did Hertz think that he had produced, and how had he done it? To answer the last question first, Hertz deployed four standard apparatus of the German electromagnetics laboratory, all of which had probably been used by Goldstein in his own work: a mercury pump, a large induction coil (probably driven by an alternator rather than a battery), a Leyden jar, and a Holtz machine. Sometimes he used just the coil; other times he inserted the Leyden jar in the circuit. The Holtz machine drove the tube by itself. The new jets could always be produced, but the aspects of them that Hertz particularly wanted to emphasize appeared to greatest effect when the Leyden jar was used, for then "the top [of the jet] consisted of flames, which shot violently apart in all directions." This violent jet exerts pressure, heats any obstacle it encounters, and in general behaves "as liquid jets would do if they emerged from the mouth of the tube."

But one thing it does not do is respond to the magnet, or indeed to anything electric: "A magnet has no action on the jet. Neither have conductors, when brought near, even when they are charged." Observing the jet through a rotating disk, Hertz found that "in reality it consists of a luminous cloud, which is emitted from the tube with a finite velocity." He could produce the effect with several gases, though with some of them it could be detected only by the pressure it exerted. If, as seemed to be the case, the jet consists entirely of "a luminous portion of gas escaping from the tube," then "it is natural to assume that the projective impulse [that expels the jet] is the force of expansion occasioned by the rise in the temperature of the gaseous content." In that case the jet might not be particularly interesting, since its condition might be due purely to gaseous thermal effects. If, however, the production of the jet did involve something beyond that, then its behavior might reflect conditions within the tube itself, though Hertz did not explicitly draw that conclusion.

Hertz accordingly made the jet more interesting by demonstrating that it could not have been expelled thermally. To do so he perturbed the apparatus:

> But if we place the electrode, which previously was outside the tube, close to the mouth of the tube inside it, or if we allow sparks to pass inside a glass tube sealed at both ends and possessing a lateral opening, in these cases also jets escape from the mouth of the tube; but they are much weaker than those which would be produced if the spark also passed through the opening. If rise of temperature were the cause of the emission, such a difference could not exist. (Hertz 1883b, p. 221)

Hertz's jets could, then, be signs of the special processes that go on in the tube. Yet they had no electric properties at all: Hertz had produced an object that exhibited some glow-tube properties but that had nothing to do with currents. He now pushed further, this time by seeking completely to divorce cathode rays from the tube current, thereby granting the rays their own, independent status.

10.4. THE MAKING OF A NEW STATE

Old experiments that have made it into the physicist's book of lore tend to be so encrusted with the detritus of later work that they can hardly be separated from what happened years after they had been performed. Heinrich Hertz's trials with cathode rays constitute an exemplary case of encrustation, because they are invariably thought about in terms of J. J. Thomson's laboratory identification of cathode rays with moving electrons more than a decade later. Because Hertz had shown that cathode rays did not evince electric or magnetic properties, in the years following Thomson's contrary claim physicists often turned back to Hertz's work to see what he had done wrong. It is important to understand these reinterpretations because they help to clarify J. J. Thomson's own work and its reproduction. But attempting to follow reinterpreters in extracting mistakes from the past does not usually illuminate the original workers' labor.

After-the-fact explanations are almost always possibility-stories, since the now-problematic device (which has long since been dismantled or changed into something else) may in fact be neutral in respect to them. The device might have given incorrect results because, says new physics, too many complicated things went on within it to permit the original investigator's understanding of its operation. In that sense the device becomes in hindsight an improperly built piece of equipment rather than an incorrectly worked one. The unacceptable claims of the earlier worker are then assigned to his having ignored this or that effect which, it is said, turns out to have vitiated the device's usefulness as a detector. But, one may ask, how can we know? If one could take the original equipment and make it do something that it should do, according to the new physics, but that it should not according to the old, then one might say that the device could have revealed the inadequacy of the old interpretation. It might never have occurred to anyone at the time to manipulate it in such a manner, but the new physicist could nevertheless say that the device was inherently capable of discriminating between old and new claims.

This is a rare event, because old devices usually cannot be manipulated in this fashion. They have to be turned into something else, into new apparatus with a very different structure, to do what the new physics wants. This new device may (or may not) reveal what the new physics wants it to, but it may also have very little to say about how the old device worked. In which case the new physicist must deal with the old device by explaining why, according to his understanding, it just could not detect anything. But this hardly touches the context in which the original device was built and successfully used. Instead of looking for what Hertz did wrong, we shall participate with him—as his colleague and expert Geissler-tube investigator, Eugen Goldstein, his master, Hermann von Helmholtz, and many others did—in what they all understood as the unproblematic probing of an entirely new physical state, the cathode ray, a state that bore no direct connection to anything that had previously been incarnated in an electric or magnetic device. In this context *Hertz was right:* whatever cathode rays may be, they are very different from electric currents.[7]

In September 1883, having played since the previous June with Geissler tubes, Hertz wanted to find a way to continue with this line of work. He decided in particular to query the claim, which was widely accepted among Hertz's immediate colleagues, that the discharge process must be intermittent. Eugen Goldstein, concentrating on the striated character of the tube's glow, had argued in a rather elaborate fashion for the seriatim creation of *open currents* down the length of the tube. These open currents constituted the cathode rays proper, and they did not transfer charge from the cathode to the anode (hence their openness). This seriatim process required time, and Goldstein held to the view that this is due to intermittent discharge: the stimulating action builds up and then rapidly collapses, producing a snaplike disturbance within the tube that propagates away from the cathode, giving rise at each visibly glowing striation to a new ray (*open current*).[8] For Goldstein intermittence enabled the discharge process to generate electric objects that propagate but that do not transfer charge. Hertz saw an opening here and decided to build a device that would show that the tube current is not sporadic and that punctuated discharge could be replaced by another conception.

10.4.1. *Intermittence Becomes an Artifact*

The claim that the cathode discharge is intermittent originated with Gassiot, who, using a battery and a rotating mirror, had found that the discharge (in Hertz's words) "could be decomposed into a number of partial discharges following each other very rapidly." This, or rather its status as a general fact, had recently (1879) been challenged by Hittorf, who was able to produce situations in which the discharge could not be visually decomposed. The next year (1880) Eilhard Wiedemann calculated that a discharge rate of 100,000 or higher per second would escape detection by a rotating mirror. Goldstein and others did not care what the rate was, as long as it existed. "The point," Hertz concluded, "may therefore be regarded as still an open one," which Hertz tried now to close. To do so he set about successfully producing conditions under which intermittence cannot be detected by the rotating mirror even if it is present, conditions in which "none of the ordinary symptoms of intermittence" were apparent.

THE CATHODE RAY AS A VEHICLE FOR SUCCESS 143

Hertz's attempt to probe invisible intermittence was not, he knew, without precedent. Several years, before, Warren de la Rue and Hugo Müller in England claimed to have detected punctuated discharge using electromagnetic induction in circumstances which do not show its "ordinary symptoms." They had initially prosecuted their researches "without the knowledge of much that has been done by other workers in the same field" (de la Rue and Müller 1879). Concentrating particularly on the conditions that determine the behavior of the glowing strata, they had produced a device to detect intermittence even when a rotating mirror showed no evidence of it. Their device differed considerably from the usual induction-based apparatus for stimulating tubes. The commonest such device included a capacitor, an induction coil, a battery, and a switch, all in series circuit; the tube connected across the coil's secondary. De la Rue and Müller modified the common arrangement in the following ways. First, they inserted the tube directly in circuit with the battery, the capacitor, and the primary of the induction coil. Second, the capacitor was placed in parallel with the battery and a resistance. Two switches (K and K') operated the device; K switched the battery in and out; K' effectively switched the resistance in and out. Finally, the induction coil's secondary was connected across a galvanometer instead of across the tube.

Here is how they described the working of the device (fig. 42–43):

> The apparatus being so arranged, if K is pressed down the condenser charges up (more rapidly when K' is also pressed down than when the current is allowed to pass through the resistance FR''), and when it has reached a certain potential the current commences to pass steadily though the tube and the primary of the induction coil 819, and will continue to do so after the connection with the battery is interrupted by raising the key K.
>
> The condenser discharges itself more or less rapidly through the tube according to its resistance, and the external resistance introduced into the circuit at FR, FR', and the megohm [a large-resistance apparatus]. The fluid resistance FR'' has no effect upon the time occupied by the charge running out of the condenser, as the current does not pass through it when the key K is allowed to rise so as to disconnect the battery. . . . The only use of FR'' is to regulate the rate of *inflow* of the charge, and it may be so adjusted as to feed the condenser exactly as fast as it loses its charge. By pressing down the keys K' and K the condenser is maintained at the highest potential possible with a given number of elements, a given tube, and a certain external resistance. (de la Rue and Müller 1879, p. 228)

De la Rue and Müller's arrangement serves effectively to remove the usual kinds of oscillations from the experiment, because in it the condenser discharges directly into the tube's high resistance. The coil accordingly has an altogether different function from its normal one. Instead of being used to produce oscillations, it was now used to detect *punctuated* (intermittent) currents.

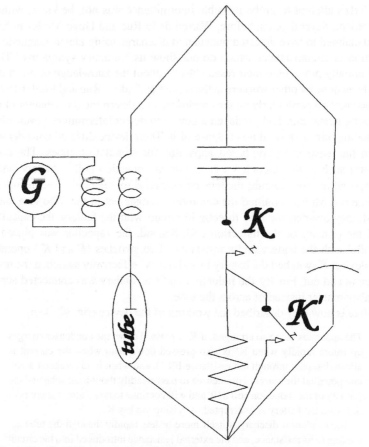

Figure 42   Circuit diagram for de la Rue and Müller's intermittence detector
(de la Rue and Müller 1879)

It is quite evident that even if pulsations do take place in the current through
the tube, no effect would be produced on the galvanometer in connexion with
the secondary of the induction coil, provided the rise and fall of the current
were equal and in equal periods. The case would, however, be different provided either the rise or the fall were more rapid relatively to the other, and
one might expect under these circumstances that there would be some movement of the needle of the galvanometer, notwithstanding that its period of
oscillation was not synchronous with the pulsations of the current. (de la Rue
and Müller 1879, p. 228)

They found what they had built the device to find: "In every case where the
strata are to the eye or rotating mirror perfectly steady, slight deflections of the
needle are seen. . . . These deflections, though very manifest, do not amount to
more than about three or four divisions of the galvanometer scale." This consti-

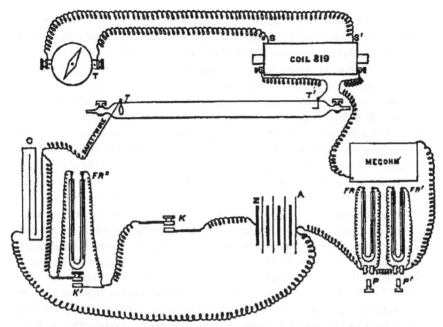

FIGURE 43   De la Rue and Müller's intermittence detector (de la Rue and Müller 1879)

tuted evidence for two things: first, for intermittence itself but, second, for a substantial difference in the rise and fall times of the partial discharges.[9]

Hertz tried to reproduce de la Rue and Müller's result. He accordingly included an induction coil in circuit with the tube and attached either a standard galvanometer or a ballistic galvanometer (dynamometer) to the free coil of the induction device. "In no case," he wrote, "did I obtain a deflection of these instruments," despite de la Rue and Müller's contrary claims. Reproduction had failed, but Hertz offered no explanation. He simply discarded their claims, a move which he implicitly justified by claiming that their device could not possibly have detected what they claimed even if the effect had been present, making their observations an artifact of badly organized apparatus:

> [A]s far as the effect on the galvanometer is concerned, the accepted theory on induction does not indicate that any effect should be expected, even if the current at each separate discharge sinks more rapidly than it rises. I was only induced to perform these experiments by the fact that results to the contrary had been obtained by Warren de la Rue and Müller. Unfortunately [!] I did not succeed in reproducing the phenomenon observed by them. *When the galvanometer had been removed from the direct magnetic action of the coil through which the current flowed,* no permanent deflection could be perceived after the battery current was closed, although the induction impulse on opening and closing the current drove the needle beyond the visible scale. (Hertz 1883c, p. 229; emphasis added)

With these extraordinary remarks, Heinrich Hertz, a young and comparatively inexperienced experimenter, called into question not only the laboratory competence but also the elementary knowledge of two respected British scientists. Their device could not have worked as they thought because they used a galvanometer and not a dynamometer.[10] So Hertz used both. And yet he *did* obtain a deflection, even with a galvanometer, just as de la Rue and Müller had claimed. But Hertz could make it go away just by moving the detector far from the direct magnetic action of the primary coil, which transformed the result from effect into artifact. This is rather nicely disguised by Hertz's formal presentation. He does not actually tell the reader at all directly that he *did* succeed in getting an action. Instead, he wrote that the action de la Rue and Müller were looking for should not be detected with their device, and that it is not. But note his wording: "I did not succeed in reproducing the phenomenon observed by them." Here he has carefully distinguished between an artifact, which is something that can be made to disappear by instrumental adjustments, and a proper phenomenon. Hertz could reproduce something very like de la Rue and Müller's galvanometric deflection; but he could not reproduce their claimed effect. Instead, however, of baldly writing that the British investigators had stupidly failed to make a simple modification, Hertz formally credited their claim and therefore discredited their experimental competence as well. He was learning rapidly not only how to do interesting work but how to deploy the appropriate rhetoric to make his claims convincing in the face of apparently contrary ones. Hertz went on to demolish, he felt, every possible further argument for intermittence. (The details of his experiments and arguments on this point are quite significant for understanding the development of his craft, but the reader may without confusion jump ahead to the next section, which turns directly to Hertz's work on the nature of cathode rays.)

All this did not mean that the very high frequencies that had been assumed by Goldstein and Wiedemann could not be present—only, apparently, that the intermediate frequencies claimed by the British could not be detected. At this point Hertz carefully, methodically divided the problem of demolishing intermittence into distinct parts. He first set up a pair of tube experiments, without capacitance or induction coil. They differed from one another in only one respect: the first contained the Geissler tube, as well as a powerful battery shunted through a large liquid resistance; in the second, the tube, resistance, and battery were all replaced by a Daniell's cell (i.e., a much weaker battery) shunted through a metallic resistance. In addition, both circuits were designed in such a fashion that a galvanometer and a dynamometer could be switched in.

The simultaneous switching of galvanometer and dynamometer into the circuits provided a way to detect indirectly a punctuated current—one whose rise and fall times differ from one another—by comparing the readings on the dynamometers in the two circuits when the resistance in the second experiment is adjusted to produce same galvanometric reading in both of them. Hertz's

apparatus acted, in effect, as a multiplier to bring out the effects of punctuated currents.

> Suppose, for example, that the duration of one of the partial discharges was
> equal to a fourth of the time from the beginning of such a discharge to the
> beginning of the next. While this current lasted it would be four times as
> strong as a continuous [i.e., an unpunctuated] current capable of exerting an
> equal magnetic effect. While it lasted, its dynamometric effect would be six-
> teen times as great, or, on an average over the whole time, four times as great
> as that of the continuous current.[11] (Hertz, 1883c, p. 230)

Resorting to a tactic that was becoming a signature of his experimental craft,
Hertz had found a way to detect *only* the effect he was interested in excising,
the one that de la Rue and Müller claimed to have found: the actual discontinuity of the current over time, which is to say its punctuated existence. His multiplier factor is equal to the (mean) ratio of the duration of a single discharge to
the period between discharges (call this the discharge ratio). As the duration
approaches the period, the current approaches temporal continuity, albeit not
constancy. This experiment is actually independent of the discharge frequency;
it depends only on its discontinuity. But "the dynamometer reading also was
precisely the same as before," evidently telling against punctuation.

These first experiments had, however, a signal disadvantage: they left open
the possibility that the discharge ratio might be too small to detect even if the
period between discharges is not itself extremely large. If the discharge occurred extremely rapidly, producing something like a spike, then even
otherwise-minute intervals between the spikes would not suffice to make the
discharge ratio detectable. In which case, Hertz's apparatus would have failed
to detect a highly punctuated current, one in which very rapid discharges are
separated by comparatively long intervals. His device excluded only high discharge ratios, that is, currents in which the duration of the discharge is of the
same order of magnitude as the interval between discharges. He needed a different way to uncover intermittence.

At some point it occurred to Hertz that he could actually mimic intermittence—produce it artificially—in a way that would put an explicit lower limit
on the discharge's degree of punctuation. Like any competent and creative experimenter faced with the problem of detecting something new, Hertz would
likely have canvassed his previous experience, reading, and craftwork for
something that he could turn to new ends. At first his earlier work in electrodynamics, which had used dynamometers in conjunction with Wheatstone
bridges, probably did not seem relevant to the new situation. Those experimental devices had been designed to produce no dynamometric deflection at all
with changing currents, because that was how to reveal the presence of electric
mass (which would have unbalanced the apparatus). Here the point was precisely to reveal a certain kind of changing current. Hertz realized that one of

his previous devices could be redesigned for his new purposes by putting an induction coil in one branch only of the Wheatstone bridge and adjusting it so that the dynamometer showed no deflection when the current through the bridge was guaranteed to be steady, that is, when the Geissler tube was not included in the circuit.[12]

Hertz's reasoning here depends in the first place upon reducing a punctuated current to the combination of a steady current with an alternating one: "we may regard an intermittent current as composed of a part which flows continuously, and another part which continually changes its direction" (Hertz 1883c, p. 230). Which is to say that a current that frequently and asymmetrically (in time) comes close to vanishing can be simulated by the superposition of an alternating current upon a steady one. This perception links at once to a pragmatic, craft-based truth that scarcely needs theory: namely, that induction-loaded bridges balanced to currents that do not change direction at a high frequency will be unbalanced to currents that do. The single coil in Hertz's device would accordingly unbalance the bridge if the current through its arms did change with time at a rate high enough to bring the coil's self-inductance significantly into play.

Operating the modified apparatus with the tube in the circuit, Hertz again found "no deflection." But even this was not enough to establish the claim that the current is not intermittent, Hertz reasoned, because of what was later termed the skin effect: "if the number of the partial discharges amounts to 100,000 or more per second . . . there is no doubt that the current-variations can only penetrate a small distance into the coil, on account of its large self-induction, and that inside it they must be effaced." Hertz sought a way out of this impasse—a way to test for intermittence without using large self-inductions.

Hertz accordingly designed a new piece of equipment, again based on the Wheatstone bridge (fig. 44), which "consisted of equal liquid resistances of 700,000 S.U. [approximately 400,000 ohms] each"; the bridged portion (*ab*) consisted of a metal-sheathed electroscope instead of a dynamometer. Its gold leaves were connected to *b*, and they were surrounded by a metal case connected to *a*. In this arrangement the leaves will diverge from one another only if *a* and *b* have different potentials. Hertz found that he could detect a potential difference "about one-tenth of that which existed between *a* and *c* when the current was flowing." With the Geissler tube in circuit, the resistances were set such that the leaves do not diverge. Thus far nothing depended on intermittence, because only resistance was involved. Hertz's notion was to replace the function of the coil (with its large self-inductance and accompanying skin effect) in detecting temporal changes with a capacitor, which he connected across arm *ac*—rather as de la Rue and Müller had done, albeit for thoroughly different reasons, since the British scientists switched the resistance out of the circuit to produce the discharge, whereas Hertz deliberately left it in.

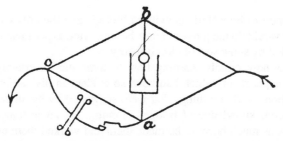

FIGURE 44   Hertz's bridge for testing intermittence (Hertz 1883c)

Hertz used the capacitor to make the resistance across *ac* vanish for alternating currents. He wrote of this only that the condenser "is capable of taking in and giving out the quantities of electricity conditioned by the alternating current without any appreciable change of the potential difference between its coatings," and it is very likely that he did not rely on high-level reasoning at all.[13] Hertz supplied these remarks primarily to justify after the fact what he had done, which was indeed the only thing he could have done if he wished to continue using the bridge. He could not use an inductor, because of the vitiating skin effect. He could not put new resistances or capacitances in series in the arms, because that would simply produce an equivalent configuration. The *only* thing that could pragmatically be done to vary the bridge (without, that is, drastically reconceiving the experiment) was to try adding capacitance in parallel to one of the arms, to rebalance the bridge for steady currents, and then to see whether the bridge was unbalanced for alternating currents. De la Rue and Müller's diagram (fig. 42) already placed capacitance in that way, albeit in their apparatus the detector was not in the bridge. Hertz saw what they had done, and given his desire to continue with the bridge, he may have decided to adapt their circuit to his needs. Having done so, he would then have tried to send a current he knew to be intermittent through this novel configuration to see what happened.

No direct evidence supports this claim for the course of Hertz's work, but he did *in fact* send artificial, intermittent currents through at some point to see whether they did indeed unbalance the new bridge:

> Into the external circuit was introduced a toothed wheel having a large number of teeth, by which the current could be broken artificially up to 2000 times per second. While the current was thus interrupted the gold leaves still remained at rest, provided that the condenser was not in action. When the condenser was introduced they diverged immediately; the divergence increased with the rate of interruption, and was very considerable at the above-mentioned rate. A single opening and closing of the current could be recognised, when the condenser was introduced, by a slight twitching of the gold leaves. (Hertz 1883c, p. 232)

This is a report of how Hertz convinced himself that the only change that he could make would do the job—not how he supported experimentally a conclusion arrived at by some sort of circuit theory.

Hertz was now rapidly learning how to tame the uncertainty and tentativeness of experiment, which had been so evident in his earlier publication on evaporation, by doing things that were suggested by his increasingly well developed craft knowledge of bridged circuitry. He knew how these things behaved just as much because he made them and worked them as because he could provide a theoretical account for them. Indeed, comparing this publication with the one on electric mass, only two years before, shows an immense difference in laboratory acumen. In the past Hertz had tried to calculate as much as he could, to provide analytical models for every aspect of the experiment. Here he scarcely calculated anything at all, and nothing whatsoever from complicated approximations. Experimental craft and know-how, developed over two hard and often disappointing years, had substantially displaced elaborate calculations.

Hertz successfully produced, and reproduced, the result he wanted. Although the circuit was now highly sensitive to alternations, adding a Geissler tube in series with it produced no effect at all. He made several changes to the essential design, and he gave a very simple estimation, which had little to do with circuit theory, that the intermittence would have to reach a frequency of "at least 50 million per second" before his device would fail to detect it. Other phenomena had been raised in the past that argued for intermittence; Hertz disposed of them by designing an electromechanical illustration that could reproduce analogues of the phenomena without punctuation.[14]

10.4.2. *The Divorce of Ray from Current*

Hertz had rid the Geissler-tube discharge of intermittence, a property that Goldstein had used to undergird his belief that the rays in the tube are a species of motion and not a transfer of material particles. Hertz had now blocked that route to propagation, in which he himself believed, and so he had to provide an alternative one:

> Perhaps we shall form a correct conception of the circumstances in question if we admit that the discharge as a whole is continuous [i.e., is not punctuated], but assume that its course along the separate current-lines is a function of time. For example, if the contact of a gas-molecule with the cathode gave rise to an electric disturbance travelling in waves through the medium, the successive production of striae would be easily intelligible without necessitating any splitting up of the discharge into partial discharges. This would still be a continuous discharge in the sense in which we have used the word. (Hertz 1883c, p. 238)

What this says is simple enough, but quite radical in its implications: namely, remove the time dependence from the process of discharge and transfer it altogether to electric waves themselves (which, according to Goldstein, were produced *precisely because* of intermittence). Goldstein had long identified cathode rays (albeit rather vaguely) with waves, but he had nevertheless not conceived of separating the rays from current processes since, for him, the rays (waves) were also open currents that began and ended seriatim down the tube. He saw pulsating open currents traveling as rays (waves) between the tube's glowing striations. Hertz set out, as it were, to liberate cathode rays from Goldstein's punctuated discharges by distinguishing them altogether from electric currents.[15]

Hertz now undertook a subsidiary and a main experiment to break nearly completely the association between ray and current, a task he had begun by removing intermittence. The main experiment involved tracing the current in the tube by means of its magnetic effect to see whether it followed the path of the rays. In order for the resulting map to indicate the path of the *current,* the magnetic needle must not be deflected by anything else, such as, for example, by the cathode rays acting in a nonmagnetic way. In Hertz's words: "it was not improbable that the cathode rays would in any case produce a deflection of the magnet; and this effect might be other than an electromagnetic effect. If such an effect existed, the proposed experiment would be useless" (Hertz 1883c, p. 239).

For Hertz the cathode rays may be entirely novel kinds of things, and they may accordingly have special interactions with other things, such as magnets. The rays might be like currents; or, better put, their interaction with magnets might be precisely the same in nature as the interaction between conduction currents and magnets. Or the rays might interact with magnets in an entirely different way. If so, then Hertz's proposed experiment would not tell anything, because the magnetic deflection would be the result of the electromagnetic action of whatever current the tube carries together with the nonelectromagnetic action of the cathode rays. The latter had accordingly to be excluded, and that was the purpose of Hertz's preliminary experiment.

To do so Hertz assumed that the rays *do* behave just like electric currents, and he fabricated a clever device which because of its symmetrical design would have no external magnetic effect. In Hertz's device (fig. 45) the cathode is a brass disk that just fits the diameter of the Geissler tube. A hole bored through its center carries a thermometer tube, through which the anode, made of nonmagnetic metal, protrudes a bit.

> [T]he current-lines [in the device] must at all events be symmetrical with reference to the axis of the tube; if we suppose the currents replaced by magnetic surfaces, these would be closed ring-magnets which would have no external action. But the cathode rays were fully developed and, according to the

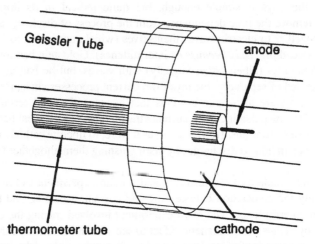

Geissler Tube

anode

thermometer tube

cathode

FIGURE 45   Device to find a nonelectromagnetic action by rays on a magnet

density, filled either the whole tube or a part of it with blue light. If they have
any action peculiar to themselves upon a magnet outside the tube, it would
here exhibit itself apart from any electromagnetic effect. (Hertz 1883c, p.
239)

Since the anode emerges from the center of the disk-shaped cathode, Hertz's
device is completely symmetric about the disk's axis. Currents from cathode
to anode in the tube will, because of the symmetry, form an ensemble whose
magnetic field lines circulate about the disk's axis—and which therefore cannot deflect anything. Any deflection of the magnet would therefore have to be
due to a nonelectromagnetic interaction between it and the cathode rays (provided that the rays are not open currents, because then magnetic deflection
could be construed as constituting evidence for Goldstein's beliefs).[16]

"The tube," Hertz wrote, "was now brought as near as possible to the magnet, first in such a position that the magnet would indicate a force tangential to
the tube, then radial, and lastly, parallel to the tube. But there was never any
deflection—none amounting to even one-tenth of a scale-division in the tele-
scope" (Hertz 1883c, p. 240). Abandoning the first, symmetric arrangement
and placing an anode in such a position that a current flowing from cathode to
anode now parallels the cathode rays produced a very different result:

The strength of the current was from 1/100 to 1/200 Daniell/S.U. [approxi-
mately amperes]. By using a second anode the same current could be made
to traverse the length of the tube; it then produced deflections of thirty to
forty scale-divisions. Similar deflections were obtained when the first anode

was retained and portions of the circuit outside the tube were brought within a few centimetres of the magnet. (Hertz 1883c, p. 240)

In other words, when the cathode-anode paths *paralleled* the cathode rays, Hertz detected a magnetic deflection; when the cathode-anode paths looped symmetrically about the central axis of the cathode and the cathode rays continued straight on down the tube, he found no deflection.

This does not mean that the cathode rays have no magnetic action. On the contrary, they may very well have one because Hertz did detect a magnetic deflection in the experiment in which the rays paralleled the line joining the cathode to the anode. Under these circumstances the rays might very well be at least contributing to the deflection. The symmetric experiment decisively broke the association. Here the path from cathode to anode was utterly different from the path traced by the rays. Any currents following the cathode-anode route could not, because of the symmetry, deflect magnets. Nor, for the same reason, would the rays themselves deflect magnets if they were currents that eventually followed their own return paths to the anode. Hertz had therefore demonstrated just what he wanted; namely, that the rays have *at most* an electromagnetic effect on magnets; there is no special ray-magnet interaction that can be distinguished from current-magnet interactions.[17]

Readers unfamiliar with the full context of Helmholtz's electrodynamics naturally missed the subtlety of Hertz's experiment. George FitzGerald, for example, remarked, in a review of Hertz's papers after the latter's death:

> From experiments on kathode rays projected down a tube, and quite away from both electrodes he [Hertz] deduced that they produce no magnetic action outside the tube, although they are deflected by the magnet. . . . This experiment on the magnetic action of kathode rays is quite inconclusive, and it is very remarkable that Hertz should have attributed much importance to it. Whatever current was carried down his tube by the kathode ray must have come back the tube by the surrounding gas, and these two opposite currents should have produced no magnetic force outside the tube; and this is exactly what Hertz observed. (FitzGerald 1896, p. 440)

FitzGerald's remarks wonderfully miss the entire point of Hertz's experiment. The Irish Maxwellian was perfectly correct in writing that Hertz got just what he should have supposing the rays to be closed currents. But he was entirely wrong in supposing that Hertz was unaware of the possibility. On the contrary, it was precisely because Hertz did think that the rays, if currents, should form closed paths that he was able to use the experiment as he wanted, namely, to exclude the possibility that they can affect magnets in any other fashion than by behaving *like* currents. FitzGerald had missed this critical aspect of Hertz's preliminary experiment because he thought that all of the experiments were designed solely to demonstrate that the rays cannot be "streams of electrified particles." One of them was indeed designed (in part) to do that, but their major import was vastly more general: it was to show that the rays do not carry much of the current in the tube and (as a persuasive though not binding conclusion) that they are not electric currents, whatever currents may themselves be. Rather, in consequence, the rays must be some novel physical entity, or at least an otherwise undetected state of something already known. The purpose of this preliminary experiment was to enable Hertz to interpret his next experiment in just that way by removing the possibility of a disturbing action on magnets that had nothing to do with the currentlike character of the rays.

Having therefore removed the possibility of nonelectromagnetic interactions, Hertz turned to his main experiment, which was designed to demonstrate that the rays carry little, if any, of the current in the tube. The apparatus was designed to separate as much as possible the path between the cathode and the anode from the path taken by the rays. To do so Hertz built an extraordinary "tube," in fact a flat case instead of a cylinder (fig. 46), with the anode and cathode inserted through the legs of a corner—which meant that the path between them had to turn through a right angle. The flat case had to sustain a good vacuum, which caused problems:

> The plates sustained safely the powerful pressure of the air, and could be heated while this pressure was on; but they bent under it so strongly that the curvature could easily be observed on looking at them sideways. Through the brass frame were inserted a tube with stopcock for pumping out, and also several aluminum electrodes; the latter were cemented in glass tubes so as to be insulated from the frame. It was only after several fruitless attempts that the case was made air-tight. One difficulty arose from the bending of the glass; this made any accurate grinding impossible, and every solid cement cracked on pumping out. Another difficulty arose from the fact that no trace of any decomposable organic substance could be allowed inside the case, so that a free use of any fatty substance would have been fatal. (Hertz 1883c, p. 241)

Hertz accordingly designed a special stopcock, which had to be inserted into the case after carefully heating the glass plates and using a special mixture of "four parts rosin and one part of olive oil." This seemed to do the job, providing a flexible, yet reasonably solid seal between the electrodes and the case.

Here's how the device worked:

> After exhausting it to a pressure of a few hundredths of a millimetre, it was placed upon a board covered with co-ordinate paper (squared paper) and provided with levelling screws. Exactly over the zero point of this co-ordinate system hung the magnetic needle which has been already described, at such a height that the exhausted case could be moved about underneath without touching it. . . . When the current was turned on, the magnet was deflected, the deflection depending upon the strength of the current and the direction of its path with reference to the needle. . . . With the help of the square paper

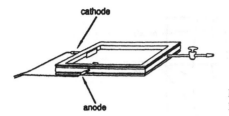
cathode

anode

FIGURE 46   Apparatus for tracing the current
lines (after Hertz 1883c)

the position of the plate with reference to the magnet could be altered and
accurately read off. (Hertz 1883c, pp. 241–42)

This device could therefore be used to map the deflection of the magnet as a
function of its position over the movable case. Referring to Maxwell's *Treatise,*
Hertz noted that the lines of current flow in a plane parallel the magnetic equi-
potential lines, so that a magnetic map can at once be turned into a current
map: the device revealed the distribution of the current.

It was no mean task to move the large plate beneath the delicately suspended
magnet in such a fashion as to measure the latter's deflection, but after practice
Hertz decided to move the case in a way that required plotting the results
graphically, interpolating them, and then measuring "mechanically" the area
under the resulting curve, which gave the relative changes of potential. To minimize the "uncertainty as to the values obtained," Hertz measured along an
orthogonal set of paths, so that "the value of the potential at every point could
be determined in a corresponding number of independent ways." Figure 47
resulted from this procedure. The paths of the current lines in the plate are
markedly, and obviously, different from the paths of the cathode rays. The result was not subtle, and Hertz unhesitatingly concluded that the

> figures show without any doubt that the direction of the cathode rays does not
> coincide with the direction of the current. In some places the current-lines
> are almost perpendicular to the direction of the cathode rays. Some parts of
> the gas-space are lit up brilliantly by the cathode light, although the current
> in them is vanishingly small. Roughly speaking, the distribution of the current
> in its flow from pole to pole is similar to what it would be in a solid or liquid
> conductor. From this it follows that the cathode rays have nothing in common
> with the path of the current. (Hertz 1883c, p. 245)

FitzGerald commented of this that Hertz had "showed that the average flow
at any point was nearly the same as if the whole space were a conductor: that
there was no connexion between the kathode rays and the flow of the current"
(FitzGerald 1896, p. 440). This in fact was all that Hertz insisted he had demonstrated in regard to rays and currents: he did not claim to have shown conclusively either that the cathode rays might not be closed currents or that they
are not streams of electrified particles. But even if the rays are currents, they
nevertheless carry almost none of the current that flows through the tube. Con-

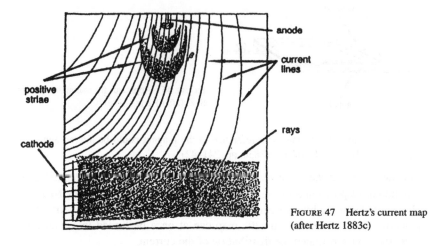

FIGURE 47   Hertz's current map
(after Hertz 1883c)

cerning electrified particles he in fact remarked that these experiments "can quite well be reconciled with the view, which has received support in many directions, that the cathode rays consist of streams of electrified material particles" (Hertz 1883c, p. 253).

Since Hertz knew that his divorce between ray and current was persuasive rather than compelling, he also had to provide an argument to explain how the rays might be affected by a magnet and yet not in their turn affect it. Here he had first of all to break the analogy between a ray's deflection and that of a current-carrying wire. That required positive evidence as well as a countervailing explanation. Hertz accordingly provided three bits of countervailing evidence. First of all, the cathode rays are bent completely no matter how rapidly the discharge occurs, even if it lasts "less than a millionth of a second." But wires take time to bend, and, more important, so does a "gaseous discharge in motion" according to de la Rive, who had actually measured the time taken by such a thing to deflect. If the cathode rays, which after all do occur in regions where a low-density gas sustains a discharge, are similar to such things, then they should also take time to deflect.[18] This disanalogy accordingly refutes the existence of any purported electromagnetic (deflecting) force that might act on the stuff that the rays traverse. Perhaps then the deflection is similar to "Hall's phenomenon," in which the current flow deviates without perceptible delay in a magnetic field, that is, without a deflection of the stuff that sustains the current. "But this analogy again is seen to be defective," Hertz continues, because his experiment had already shown that the path of the current in the tube has nothing to do with the cathode rays.[19] So the only thing that occurs in circumstances similar to the rays and that is also deflected by a magnet (the moving gaseous discharge) behaves in other respects in a very different way, and the only other electromagnetic action that could be used (the Hall effect) presupposes that the rays are indeed currents, for which there is no independent evidence at all.[20]

Finally, Hertz turned to a positive analogy: "it is known," he wrote, that a strong magnet can completely stop the discharge, which begins again after removing the magnet, and this "shows that the action of the magnet upon the discharge cannot be purely electromagnetic" (since a "purely electromagnetic" action of a stationary magnet has nothing to do with starting and stopping discharges). Thus, here we have independent evidence for the existence of a magnetic action that does *not* involve a current. This stopping effect "can only be an action upon the medium through which the current has to pass" because under no other circumstances do magnets ever do anything except deflect currents. But given that this effect does indeed exist, Hertz could argue *positively* that it might also be involved in the rays' deflection:

> Without attempting any explanation for the present, we may say that the magnet acts upon the medium, and that in the magnetised medium the cathode rays are not propagated in the same way as in the unmagnetised medium. This statement is in accordance with the above-mentioned fact, and avoids the difficulties. It makes no comparison with the deflection of a wire carrying a current, but rather suggests an analogy with the rotation of the plane of polarisation of light in a magnetised medium. (Hertz 1883c, p. 246)

Confident that he had argued convincingly for the novel character of rays, Hertz began to insist on embedding the claim in appropriate language. Agreeing with Eilhard Weidemann and Goldstein that the essence of the rays probably involves an "ether-disturbance," Hertz wished them to give up referring to the discharge as though it were the same thing as the rays. "I should," he remarked, "like to see the word 'discharge' replaced by 'cathode rays': the two things are quite distinct, although the physicists referred to do not observe the distinction." In this way Hertz staked out his claim to discovery: *he* alone had broken apart the association between ray and current. He went further and began to think of ways to enforce the dissociation, perhaps even by generating cathode rays without any current processes at all:

> If we could prevent the production of the cathode rays, the gas would everywhere be as dark as it is in the dark intervals between the striae (although the current flows through these intervening spaces). Conversely, if we could produce the cathode rays in some other way than by the discharge, we could get luminescence of the gas without any current. For the present such a separation can only be carried out ideally. (Hertz 1883c, p. 248)

Here was a program awaiting realization, waiting, that is, the production of devices to generate and to manipulate rays without the troublesome presence of the unessential electric current, a program as it were for fabricating electric rays. The first step in the program involved determining whether the rays have any electric properties.

10.4.3. *Electric Effects of the Ray*

"If," Hertz continued, "we admit that cathode rays are only a subsidiary phenomenon accompanying the actual current, and that they do not exert electromagnetic effects, then the next question that arises is as to their electrostatic behaviour." These words have for decades been read as introducing Hertz's final step in demonstrating that the cathode rays are not streams of electric particles. And indeed Hertz did do just that. But that is not all he did with his concluding experiments, nor from his point of view perhaps was it even the most important thing. His work here was intended much more for the eyes of his Berlin contemporaries, of the community within which he was trying to establish an important presence, that for anyone else. Hertz's efforts aimed at producing novelty, not (just) at providing evidence for established and overt Berlin beliefs that he and everyone else already held. He *had* to distinguish himself, to set himself apart within the Berlin community, and the best way to do that was to probe the character of a new state or entity, to bring it, as it were, to instrumental life. A consequence of doing so may very well be to support standard belief—it should *at least* do that—but there should be something more as well, in this case something that embodied Hertz's unique contribution to the common research effort.

"If we admit that cathode rays are only a subsidiary phenomenon accompanying the actual current, and that they do not exert electromagnetic effects, then the next question that arises is as to their electrostatic behaviour." Why is this the next question? Because if we are interested in demonstrating that the rays are not currents of any kind whatsoever, then we must find a way to show this without resorting to the ray-magnet interaction. Hertz already knew that because the rays do not deflect magnets, they are not open currents, but they might still be closed currents (albeit ones that carry little of the tube current). To clinch the issue required using an independent property of such things, and the most salient one was the accumulation of electric charge. Hertz accordingly decided to capture the rays and measure their charge.

But for Hertz a charge-measuring experiment was fraught with difficulty, because in order for it to have any meaning at all required meticulously separating the rays from the tube current that flows from the cathode to the anode and that he had mapped in his previous experiments. This latter (main) current can of course also produce electric accumulations if it is intercepted, and extremely large ones at that. *Measuring* that charge would have no implications for whatever charge the rays might carry. Hertz accordingly designed an instrument in which a perforated anode is placed very close to the cathode; the rays flow past the anode (mostly through its holes, Hertz thought at the time) and emerge into an electrode-free space. These, he remarked, are "pure cathode rays." In this region altogether devoid of impurity (i.e., of main current), Hertz looked for signs of charge.

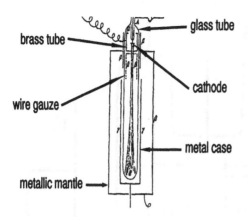

brass tube

glass tube

wire gauze

cathode

metal case

metallic mantle

FIGURE 48   Hertz's ray purifier (after Hertz 1883c)

The apparatus that Hertz built consisted of three major components: a glass tube containing the cathode and anode, a metal case that surrounded the portion of the tube that did not contain the electrodes, and a metallic mantle with a hole in it for inserting the tube (fig. 48–49). The metal case and the grounded mantle were connected to the poles of an electrometer, which therefore measured the case's potential.[21] The device was driven by an induction coil, because by this time (probably late January 1883) the battery he had spent so much effort constructing had succumbed.

The critical part of the design involved the electrodes. The "pure" rays that alone concerned Hertz required keeping the electrodes as close together as possible in order thoroughly to separate the apparatus that detected the charge of the pure rays from the main current. Hertz fashioned an unusual design, in which the anode is a perforated brass cylinder that surrounds the cathode, which lies on the axis and aims at the perforation. Or, to be precise, this is the part of the anode that lies within the evacuated glass tube. It is in metallic connection with a wire-gauze or wire-mesh cylinder that hangs below it and also with the grounded external mantel that encompasses the tube. The mantel in turn connects to a pole of the induction coil.

The space between the enclosed cathode and the perforated anode fills with intermixed current and rays as the coil works. The rays stream perpendicularly toward the perforation; the current presumably follows a broad path to the cylindrical anode. Just past the perforation the space therefore contains primarily rays, but also some current that has leaked through. The mesh captures this last bit of current, so that "the cathode rays are to be regarded as pure after they have passed through the opening in the metal cylinder and the wire-gauze beyond it." The space past the gauze is bathed in the current-free rays. Here, Hertz felt, he had produced a region within which the rays could be manipulated in their purest form. The state of the rays had to be protected from the influence of external electric interactions. The mantel not only carried away to

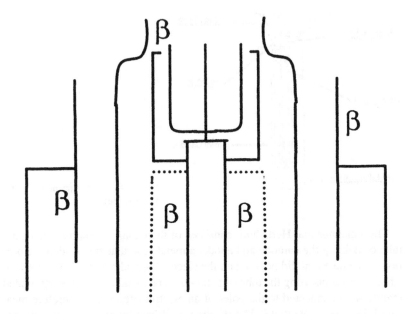

FIGURE 49   Purifier electrodes

ground the polluting main current, but it also shielded the purified rays by acting as a Faraday cage. To detect (but hardly to measure) the charge of these rays was now a simple matter of surrounding the region that contains them with a metal case connected to the opposite quadrants of the electrometer.

In two previous experiments Hertz had faced the critical problem of having no way to calculate the theoretical magnitude of the effect that he was trying to elicit or to erase. His earliest work for Helmholtz, designed to cast doubt on Weber's massy electric particles, could place an upper limit on the mass, but in the absence of a claim on the part of Webereans that the mass should be such-and-such, the experiment's critical potential was severely limited outside the Berlin community itself. In his failed attempt to establish an evaporative limit, Hertz had had no independent means of calculating what the limit might be. Here, however, Hertz did have a way to determine what the charge of the purified rays would be if, in his words, "they consisted of a stream of particles charged to the potential of the cathode." Note the wording: the particles Hertz had in mind are said to be charged to a potential, which means that, whatever else they might be, they are not Weberean electric points, because the latter are not charged to a potential in any meaningful sense at all. Hertz was instead thinking of small bits of metal literally torn away from the cathode. These particles retain all the usual metallic properties, such as conductivity, and are in fact merely tiny pieces of the cathode: either they are not atoms or molecules or else Hertz was prepared to carry conductivity down to the microlevel (which was entirely permissible in Helmholtz's scheme).

Imagine with Hertz a stream of tiny metal pieces flowing out from the cathode, some of which escape through the holes in the oppositely charged mesh. The stream, he evidently felt, would be packed sufficiently densely to constitute in effect a single metallic object the size of the path traced by the cathode rays. The object is in contact at one end with the cathode and therefore has the same potential that it does. If, accordingly, one actually built a metal rod "about the same size and position as the cathode rays" and stuck it onto the cathode itself, then it would be an electric simulacrum of the rays themselves, supposing them to consists of such a dense stream. One could then fire up the ray machine and watch the electrometric deflections produced by this thing, which is what Hertz did. The deflection, Hertz wrote, "was too great to be measured, but could be estimated at two to three thousand scale-divisions." Here is what happened when the cathode rays replaced their metallic simulacrum:

> When the quadrants of the electrometer were connected together and the induction coil started, the needle naturally remained at rest. When the connection between the quadrants was broken, the needle, in consequence of irregularities in the discharge, began to vibrate through ten or twenty scale-divisions from its position of rest. When the induction coil was stopped, the needle remained at rest in its zero-position, and again began to vibrate as above when the current was started. As far as the accuracy of the experiment allows, we can conclude with certainty that no electrostatic effect due to the cathode rays can be perceived; and that if they consist of streams of electrified particles, the potential on their outer surface is at most one-hundredth of that of the cathode. (Hertz 1883c, pp. 250–51)

What Hertz *saw* was entirely unambiguous: the electrometer needle merely "vibrated" through a distance about one-hundredth of the stable deflection produced when the metal simulacrum replaced the rays. Evidently he felt certain that the electrometric accuracy was worse than these tens of divisions, which meant that the experiment provided no evidence for ray charge. As in his experiment for detecting electric mass, Hertz could determine an upper limit of a sort by calculating it from the electrometer's accuracy. Here, however, it was not possible to say without further specification what such a number would delimit, since to know that one had to have a model for the rays as charge carriers. If they did consist of the dense metallic streams that Hertz had replicated with his simulacrum, then, he concluded, the shared potential of the objects that constitute the stream "is at most one-hundredth of that of the cathode."

The critical center of Hertz's experiment lay in its removal of the polluting current by means of the close spacing of the electrodes and the addition of the metallic mesh or gauze. He undoubtedly spent a great deal of time adjusting the spacing and setting the mesh in order to make certain that the rays were sufficiently unpolluted that his electrometer could not tell the difference. But nothing is ever perfect in the practical arts of the laboratory, and it is always necessary to establish a base point for measurement, or a *calibration.* It seems probable that Hertz discovered he had to do this more carefully than he might otherwise have had to, because as a matter of fact the electrometer did detect charge in the purified space amounting to a significant "150 to 200 scale-divisions." But this was a special, *calibrating* kind of effect. It showed up when the machine was started after having long been at rest, and, more to the point, it did not change or go away for a long time even when the machine was shut off, that is, when there were no rays present at all. Hertz could therefore use it to calibrate his electrometer to a new zero point.

But this zero point was not altogether stable, because "while the discharge is on," it can be moved "instantaneously when a magnet is brought near the tube, and the needle remains constant in its new position so long as the magnet is not moved." This had to mean, Hertz admitted, that his purified space was not altogether unpolluted by current, that "as a matter-of-fact . . . electricity does penetrate through the wire-gauze into the protected part of the tube." But this was merely the usual sort of imperfection that all devices manifest—a limitation on one's ability to isolate thoroughly the object of investigation. It had nothing at all to do with the rays proper. Why not? Because

> the passage of [the rays] is in no way influenced when the further penetration of the electricity is prevented; nor, as the first experiment shows, is the amount of electricity in the tube appreciably increased when the cathode rays again begin to enter it. (Hertz 1883c, p. 251)

This requires a bit of unpacking. Hertz first made the point that the magnetic deflection that affects the needle's zero point can deflect the current in the space between the anode and the cathode so that it hits the walls of the anode rather than the hole in it ("further penetration . . . is prevented"). Yet when this takes place, the rays are just as visible as ever. His second point reiterated his earlier statement that the new zero point is not at all affected by the presence of rays since it remains the same (absent the magnet) whether the machine is on or off. So, he concludes, a bit of current must leak past the mesh and, terminating in the experimental space, generate a charge there, but this stops, or rather reaches a steady state, once the potential rises to a certain point ("until its [the current's] entrance is prevented by the rise of potential"). A magnet produces fluctuations in this potential by pressing the current away from its escape route through the anode, thereby upsetting the steady-state maintenance of the potential by the small current that continues to pass through.[22] The cathode rays remain completely indifferent to this process—though, of course, they are themselves shifted about by a magnet.

10.4.4. *Was Hertz's Experiment on the Ray's Electric Properties Ever Replicated?*

Hertz insisted that his sensitive apparatus showed simply no sign of a ray charge within the cleansed experimental space past the polluting electrodes. Although it is no part of my purpose here to provide retrospective explanations for why Hertz obtained the results that he did, this experiment (and one other, considered in sec. 10.4.5 below) did arouse particular interest a decade and a half later when attempts were made to elicit the kinds of properties that Hertz felt he had done away with. The French physicist Jean Perrin, twenty-five years old in 1895, claimed to detect precisely what Hertz asserted could not be detected, the "electrification" of the cathode rays.[23] Hertz had died shortly before, and so one cannot say with certainty what his initial response to Perrin's claims would have been. Much later, however, Perrin's experiment came to be seen either as, in effect, a failed replication of Hertz's or as having elicited something that Hertz, using similar apparatus, had earlier failed to do.[24] By the late 1930s Rayleigh, J. J. Thomson's biographer, was certain that Hertz's apparatus had been badly designed:

> Another important point tested by Hertz was to try whether the cathode rays, fired into a metallic vessel (known in this connection as Faraday's cylinder or Faraday's ice pail) would carry with them an electric charge, detectable by an electrometer connected with the vessel. He failed to observe this effect, but the design of his experiment was open to certain objections which were removed in a later investigation by Perrin in 1895, directed to the same question. Perrin got definite evidence that the rays carried a negative charge. J. J. Thomson, in a modification of Perrin's experiment, showed that if the Faraday cylinder was put out of the line of fire of the cathode rays, it acquired a charge when, and only when, the cathode rays were so deflected by a magnet as to enter the cylinder. (Rayleigh [1942] 1969, pp. 78–79)

More recent commentary reiterates Rayleigh's assertion that Perrin had gotten "definite evidence" for ray charge.[25]

What had Perrin done, then, that Hertz had not, and why was this later seen as having removed objectionable aspects of Hertz's design? Was it the case, as Rayleigh implied, that Perrin had performed the *same* experiment as Hertz, only with better apparatus? Let us put Perrin's apparatus (fig. 51) on the table next to Hertz's (fig. 50) to see how they fit together. They are not the same devices, even though they both use Faraday cages to measure charge. Rayleigh evidently understood this. Whether Perrin's must necessarily be thought an improvement over Hertz's remains to be seen.

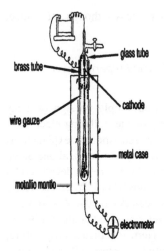

FIGURE 50   Hertz's ray purifier (after Hertz 1883c)

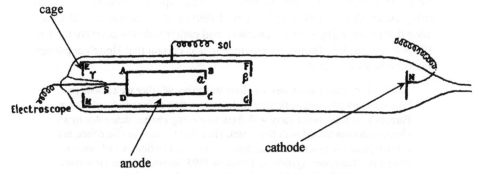

FIGURE 51   Perrin's charge-detection device (Perrin 1895)

The salient characteristic of Hertz's device, which he had labored hard to produce, was its careful purification of the cathode rays by separating them from the current. The device effects the separation by placing the electrodes very close together. Look now at Perrin's apparatus. Here there is no attempt at all to separate the rays from the current, primarily because it never occurred to the young Perrin that this was the crux of the matter. In his apparatus the cathode rays and the current parallel one another all the way from the cathode to the anode; the hole in the anode allows both to penetrate. Perrin's experimental space is, from Hertz's point of view, utterly useless to probe the electric character of the cathode rays because it is irreversibly polluted with current. *Of course,* Hertz would have replied, Perrin measured charge: he measured the charge produced by the current that terminates at the anode, which has nothing to do with the cathode rays. Far from being an improvement over Hertz's design, Perrin's ruins it.

There is no failure of replication here. There is rather the production of a new device altogether. Hertz's experiment was in fact never replicated, because no one subsequently tried to prevent the pollution that Hertz so carefully excised. His Berlin colleagues undoubtedly saw no reason to replicate success; elsewhere, the distinction was simply not made, at least not in the way that Hertz had made it,[26] which was why Perrin could so confidently assert that he had measured the ray charge. Hertz had carefully caught only cathode rays in his metal bucket; Perrin later caught both rays and current.

In 1897 J. J. Thomson perceived this dissimilarity between Hertz's and Perrin's experiments, and he built a new device that, like Hertz's, placed the anode close to the cathode and let the rays pass beyond. In fact, the only difference in the *designs* of Hertz's and Thomson's experiments was that Thomson placed the ray catcher out of the direct line of the rays, using a magnet to deflect them into it. He successfully detected charge when, but only when, the rays were caught. Hertz could have used a deflecting magnet, but if he had, he would have found nothing more than he did. The critical difference between their experiments concerned the degree of evacuation of their tubes.[27] Thomson caught charge when he achieved a low enough pressure.

One could easily make retrospective assertions about why Hertz got the results that he did, given Thomson's low-pressure success, but they would be just that—retrospective comments without the backing of work done directly with Hertz's device. One might ask whether Hertz himself could have done something at the time that would have changed his results, even though he did not, and nobody subsequently challenged him on the point. It seems probable that nothing Hertz could have done would have changed his result, because to do so would have required something very like Perrin's experiment, in which the current path merges with the ray path. Hertz simply did not measure any ray charge with his quite accurate device, undoubtedly because there was not much charge to measure: he had after all (one might say) prevented most of the rays from reaching his experimental space by pulling them into the anode. Whatever luminosity he saw on the walls of that space must have been due to residual rays that were simply too few in number to produce detectable charge.[28]

This might even have been advanced as an argument at the time. If it had been, Hertz would have rejected it, because he was certain that the experimental space was filled with lots of rays: the rays past the anode trap "are none the less vivid; at low densities they cause the glass at B to shine with a brilliant green phosphorescence, upon which the shadow of the wire-gauze is plainly marked." (Hertz 1883c, p. 250) This might puzzle the retrospective analyst, because the "phosphorescence" that Hertz refers to is directly proportional to the intensity of the ray beam. Hertz has sucked away most of the beam (since it goes with the current). And yet the glow on the walls is "none the less vivid." Here one would have it seems simply to assert that since Hertz did not actually produce a comparative *measure* of the glow's intensity, his experiment can be treated as, at best, inconclusive.

But this was not Hertz's only claim concerning cathode rays and electric properties. He also asserted that the rays are not affected by electric forces. This was a distinct question from the rays' charge, but only because the charge test had come up negative. Had it been positive, the second experiment would not have been necessary, because anything that generates an electric action must necessarily be affected by a similar action. The rays do not do so, but it remained possible that electric action could still deflect them—just as they can be deflected by magnetic fields even though they are not currents.[29] If this last experiment did deflect the rays, then it would mean only that electric actions can affect their paths, not necessarily that such actions can directly affect *them*. Of course, the combination of both electric experiments speaks directly to the possibility that the rays are streams of electric particles, as Hertz carefully emphasized.

10.4.5. *Electric Effects on the Ray*

Hertz took care to ensure that his experimental space remained unpolluted by the current that emerges from the cathode along with the rays: he again surrounded the cathode with a punctured anode, followed by wire gauze. Some distance past the gauze he placed a fine wire, which cast a "sharp shadow" on the phosphorescence produced by the rays 12 cm down the tube. He then placed the entire tube "between two strongly and oppositely electrified plates." The wire's image was not moved. There was, however, a possible difficulty: "But here there was a doubt whether the large electrostatic force to which the tube was subjected might not be compensated by an electrical distribution produced *inside it*" (Hertz 1883c, p. 252; emphasis added). Later commentators (e.g., G. P. Thomson 1966, p. 40) have interpreted Hertz as remarking that the walls of the tube itself might have been charged inductively by the external plate, thereby shielding the experimental space from the field. Hertz, however, does not write about the tube's walls; he refers to "an electrical distribution produced *inside*" the tube.[30] Hertz probably meant that the extremely powerful, externally applied electrostatic force might have turned the remaining gas in the tube into a conducting mass, in which case it would act like any conductor and completely shield the experimental space within itself from an externally applied electrostatic force. The effect is on the gas, not on the tube, and it requires a very powerful force to produce, as Hertz emphasizes, because only powerful electric forces can render the low-density gas conducting. By "conducting" Hertz did not have anything at all like an ionic model in mind; he meant that the gas, considered as an object, has acquired the property of conductivity, just as though it were an attenuated metal.

He now placed his metal strips within the experimental space itself, set 2 cm apart. If these strips were sufficient to render the gas conducting when placed outside the tube, they could just as well do so inside the tube. In his first experiment Hertz accordingly used twenty small Daniell's cells (so about 20 V), which produced a much lower field than the "strongly and oppositely electrified plates" that he had first placed outside the tube. "Opening and closing [the connections to the battery]," Hertz wrote, "produced not the slightest effect upon the phosphorescent image"; that is, the rays were not deflected by this *electrostatic* field. Though Hertz did not explicitly make the point, he was undoubtedly quite certain that, under these conditions, the gaseous mass between the plates was not rendered conducting by them—for if it had been, then discharge would have occurred. As far as he was concerned, discharge, and only discharge, signaled conductivity when the stimulating objects (the plates) were placed in the gas itself. This point cannot be too strongly emphasized: for Hertz conductivity remained a property of a substance in bulk; as such it had some value, either zero or finite. And if it were finite, then discharge had to occur. But suppose that discharge *did* occur, that is, that the gaseous mass had been rendered conducting? Under those circumstances the plates would no longer exert an *electrostatic* force on the region between them; instead, the region would be subject to the kind of *electromotive* force that produces currents in wires connected to batteries. What would happen now? Perhaps, Hertz evidently reasoned, the rays might be deflected by a current-generating force even if they are not deflected by an electrostatic action.

Hertz investigated precisely that question by using 240 Planté cells: discharge now occurred, "as soon as the induction coil was set to work and the cathode rays filled the space between the strips." Indeed, once the cathode rays went to work, even twenty or thirty cells could effect discharge 'in accordance with Hittorf's discovery that very small electromotive forces can break through a space already filled with cathode rays." Adding a large liquid resistance to the circuit weakened the discharge enough that it shut off when the induction coil shut off: "during each separate discharge of the induction coil there was now only a weak battery-discharge lasting for an equally short time." Hertz had thereby crafted a machine that produced a discharge across the plates in synchrony with (and as result of) the passage of cathode rays in the space between them. This *guaranteed* the presence in that space of the kind of electromotive force that generates currents. This is what he saw on observing the shadow:

> The phosphorescent image of the Ruhmkorff discharge appeared somewhat distorted through deflection in the neighborhood of the negative strip; but the part of the shadow in the middle between the two strips was not visibly displaced. (Hertz 1883c, p. 252)

Much was made in later years of Hertz's remarks concerning the distorted image, and we shall return to it below. His attention was concentrated on the nondisplacement of the image's center, not on the distortion, and he was certain that this meant that the rays had not been deflected. He held this even though his experimental space was now admittedly filled with current flowing between the plates. As Hertz evidently saw it, the space had become something like a homogeneous conducting mass into which a pair of conductors (the plates) had been immersed. Current passes between the plates, and there was accordingly no reason to think that the electric force between them could possibly vanish. On the contrary, if it did vanish, then discharge would fail, which it did not.[31] And yet the image remained the same, which meant that cathode rays traversing a space across which a current is indubitably flowing, and in which there is accordingly an electromotive force, are even then not deflected.

However, near the negative strip the image "appeared somewhat distorted through deflection." This would not have surprised Hertz, because Goldstein had already noted that neighboring cathode streams seem to deflect one another in the vicinity of the cathode. Goldstein's observation had not been taken as a sign that the rays carried charge, and neither was this one—because both could be attributed to the action on the gas near the strip of the intense electric field at its surface. The image distorts, or the rays seem to deflect, not because a proper repulsion occurs, but because the region through which the rays pass is under extreme stress. Farther from the strip, where the field weakens, the effect naturally disappears.

It was often remarked, in retrospect, that Hertz's plates (presumably in the experiment with Daniell's cells) were probably neutralized by the presence of residual, ionized gas particles.[32] This would not even have occurred to Hertz in 1883, because he understood conductivity as an extensive property of matter in bulk. Electrolytic conduction, he knew (since Helmholtz discussed point), requires a supplementary point of view, into which ions enter, but even there it is not at all manifest that conductivity is to be reduced to ionic mobility, and in any case the concept does not extend to metallic substances or to gases.[33] Consequently, from Hertz's point of view the gas in the tube was either conducting, in which case it behave like an attenuated metallic mass, or else it was not, in which case it behaved like an insulator. He had to examine whether the rays were or were not deflected under both circumstances, which he did.[34]

Hertz's arguments and experimental work were tightly closed, carefully wrought to preclude damaging criticism. Even in retrospect it is not possible to destabilize Hertz's experiments by pointing out that something might have been going on that he did not take into account. That sort of argument would have had no weight at all in Berlin unless it could have been backed up with appropriately crafted experiments that could produce the effects that Hertz had been unable to elicit. Perrin's experiment twelve years later was not appropriately crafted; J. J. Thomson's two years after that was, but then it required a technological device, a pump of novel power, that the electrical industry had not as yet produced in 1883. Once Thomson's high-vacuum devices started to accumulate charge and electrically to deflect the rays, he rapidly became able to manipulate them. Soon he crafted devices that created a world of charged-particle rays, objects that simply had not existed in any tangible form in Hertz's experiments.

10.5. THE IMPRIMATUR

Hertz evidently stopped his experiments sometime in mid-February 1883, having returned to work on the rays on January 20.[35] The huge battery he had built the previous autumn and that he had used to distinguish the current from the ray path had by then become useless from corrosion, and this depressed him. On February 11 he wrote home that he had to begin writing up the experiments of the previous autumn. "Meanwhile," he continued, "I am slowly beginning to rebuild the battery, for it has become quite corroded during the months that I did not use it." Hertz's immersion in ray experiments had only in part been motivated by the twin desires to achieve a proper position in Berlin physics and to bring the ray firmly into being as a novel entity; he also adored constructing his powerful new battery. The tactile manipulation of wires, tubes, chemicals, and plates clearly enthralled Hertz. In the fall he had written home, after finishing the battery and getting it to work routinely, that he would "try to think again like a normal person about the many other things there are in this world, aside from the battery" (J. Hertz 1977, p. 167). But now (mid-February) the battery was dead with corrosion, and the ray experiments on electric interactions still remained to be done. "I am not sure whether I should involve myself in this big job [of rebuilding the battery]," Hertz wrote on February 11, "which would again cost me several weeks, or whether I should look for another subject; but in the meantime I have begun to work, since that is the best way to use my time while I am in doubt, and I hope that the battery will be finished before the doubt is resolved" (J. Hertz 1977, p. 173).

It was not, and he soon put it aside, concentrating instead on producing a written argument for his conclusion about the rays. He was so confident of his work and of its particular suitability within the Berlin environment, that he intended the written piece to serve as his *Habilitation* paper, which was required for a university position. "I find the prospect of qualifying altogether awful," he wrote home on February 17, "like pushing oneself into the turmoil before an overcrowded ticket window, but I do not see any other way of getting a ticket" (J. Hertz 1977, p. 175). Hertz's ticket was simultaneously one of acceptance as a producing member of Berlin physics, a ticket of reputation, and a ticket to the job lists.

We do not know with certainty that Helmholtz saw the details of Hertz's work at this time,but it seems unlikely that he could have remained unaware of what was going on, particularly since Helmholtz had long been closely interested in Goldstein's own work with Geissler tubes. Scarcely two weeks after

Hertz had finished the written report, Kirchhoff came to him with a job offer—not the job Hertz wanted, but a job nevertheless. This was a momentous event in Hertz's career, momentous because it marked not only his proper entry into the profession but also (as we shall see) the divorce between Hertz and laboratory work that rapidly followed. As usual, Hertz wrote a detailed and highly revealing letter home, in which he described how Kirchhoff came to him announcing that a *Privatdozent* was wanted at Kiel, a position with a salary of 500 *Taler* a month from a special fund,[36] and that he and Weierstrass "had thought of" Hertz when asked for a recommendation by the ministry for universities. At first highly enthusiastic at this prospect of employment, or perhaps of *change,* Hertz sobered a bit on speaking with Weierstrass. The mathematician warned him that the professors at Kiel in fact wanted an associate professorship in mathematical physics, but the ministry claimed poverty. Perhaps, warned Weierstrass, the faculty will react by refusing the compromise position since professors "are often strange people." Hertz demurred a bit at this, and Weierstrass promised to obtain an unofficial agreement with the ministry concerning the pay and any possible future candidacy for an associate professorship (J. Hertz 1977, p. 177).

Thus far Hertz had spoken only with Kirchhoff and Weierstrass, even though the ministry had also asked for Helmholtz's advice.[37] Kirchhoff, not Helmholtz, had gone to Hertz, though it was as Helmholtz's assistant that Hertz had spent the last three years in Berlin. Of course, the Kiel position was explicitly in mathematical physics, which fell more closely under Kirchhoff's expertise than Helmholtz's. Hertz went to Helmholtz shortly after his talk with Weierstrass. He described the meeting in a letter home:

> Then, yesterday, I spoke to Helmholtz about it; he said that he knew about it, too, and had also thought of me; however, he could not say whether it would be advantageous for me (nor did he say the opposite); he said the professor who had the chair of physics at Kiel, Prof. Karsten, stood somewhat in opposition to the faculty at Berlin, so that a recommendation from here would not be much use to me; moreover the circumstances at Kiel seemed to be very modest and I should be probably completely cut off from laboratory facilities. Otherwise, he, too, thought that I could be of more use there as *Privatdozent* and promised to recommend me to other professors. As is his way, he spoke very circumspectly within very restricted boundaries and so to speak visibly endeavoured not to exert any influence on my decision, whereas what I wanted was precisely his guidance. Everyone advised me to go to Kiel in person at the beginning of the vacation, and to look around for myself; for whatever the ministry might do would only be done in response to a petition from the faculty there; thus their good will would be the most important thing for me. This much is certain: if I do not receive confidential assurance about the stipend and the candidacy, and if at the same time I do not see that my appointment is in accordance with the wishes of the faculty, and that they regard it to some extent as a substitute for the denied professorship, I shall

not got to Kiel. However, some scruples have arisen as to whether it would not be foolish to agree to the move in any case, even if all this were true. I would gain: time and the prospect of an associate professorship in two years or a little more. I would lose: command of working space, instruments, books, which I have here without restrictions. But there are also other things that concern me when I think about leaving here. Coming from the big world to which I am accustomed, I would have to fit myself into a presumably very small one. (J. Hertz 1977, pp. 177–78)

Here we have a bright spotlight thrown onto the ambitious young physicist's social environment. Hertz had spent three productive but frustrating years as an assistant in Berlin, gaining craft-knowledge, cementing connections within the Berlin community, and trying to find an appropriate way to make his reputation. Just after he had completed the most promising work of his short career, Kirchhoff, who not long before had rewritten Hertz's paper on elastic impact for him, came to him with a job offer, one that he and the mathematician Weierstrass—not Helmholtz—thought would suit him, presumably because of Hertz's mathematically sophisticated doctorate and his paper on impact (not because of his experimental work). Helmholtz kept aloof, and Hertz had to seek him out. When he did, Helmholtz's advice was subtle and complex, which Hertz resented. Watch out for Karsten, Helmholtz was saying; he does not think like us. You will not have a laboratory, and not much money or other facilities. But do not let me dissuade you—probably you would be more "useful" there than here, where you are only an assistant. Helmholtz, as we have already seen, tended not to exert overt pressure, at least on Hertz, but he clearly had his own wishes. He had earlier wanted Hertz to work on the Berlin prize questions; he was now not happy to see him leave for a place far removed from the Berlin research environment, from its attitudes, its imperatives, and also from its facilities. Hertz, he knew, was not likely to flower at Kiel, and he feared that the young man would no longer progress along desirable lines. On the other hand, Helmholtz may also have recognized the rebellious streak in Hertz and what was by then his manifest desire for movement in his career. In the end he gave no advice. Hertz decided to accept the offer, hoping that an associate professorship might come about in a couple of years. He left for Kiel without Helmholtz's explicit benediction. It is perhaps not entirely surprising that within a very short period of time Hertz began critically to examine the foundations of Helmholtz's electrodynamics, to reach out past its overt structure to ponder, not without great difficulty and ambiguity, its wider implications.

At the end of July, when he had already been at Kiel for several months, Hertz received a letter from Helmholtz concerning Hertz's *Habilitation* paper on cathode rays. "Bravo!" he wrote in what must have seemed to Hertz to be unprecedented enthusiasm. "This matter seems to me to be of the widest importance" (J. Hertz 1977, p. 349). Hertz's flattered reaction must have been tempered by the rest of the letter, for Helmholtz at once turned the experiments to his own ends and then raised an objection which shows unequivocally that he had not spent much time studying Hertz's argument. Helmholtz, who had years before squirmed at Goldstein's early insistence that the cathode rays are not streams of particles, now embraced the claim with enthusiasm, but not because he was necessarily averse to the particle scheme (since these need not be Weber's electric particles). Rather, by the early 1880s Helmholtz, as we have seen, had with some initial reluctance married the polar ether to his electrodynamic energy function. This had many effects, among which was the possible presence of electrodynamic entities that do not occur absent the ether, such as waves of compressive polarization. These things exist, he thought, if the disposable constant k in the electrodynamic potential is not zero, given the ether. Consequently, Goldstein's less-explicit understanding of the rays was now turned by Helmholtz into a specific form of electrodynamic entity, one that his scheme could uniquely produce: waves of compression in the ether. Hertz's claim that, taken together, his experiments cannot be reconciled with the stream-of-particles hypothesis fitted Helmholtz's views, providing positive evidence for the only available alternative.

But that had not been Hertz's main purpose, which was to separate rays from the current. Here one must be careful to understand Hertz's motives, implicit though they may have been at the time. He had no desire to dive at once into the midst of the intensely problematic issues that orbited about the ether; his career was not likely to advance by a trek over such marshy ground. Despite this (I think intentional) distancing of his work from the subject of the ether, Hertz nevertheless did write that the view of Eilhard Wiedemann and Goldstein that "the discharge consists of an ether-disturbance" seemed to him "to be based on convincing arguments," provided one replaced the word "discharge" with "cathode ray." But he did not (in print) go beyond this, and he would not have wanted his work to stand or fall on that point.

Hertz had shown that the rays have nothing in common with currents and that they neither exert nor are affected by electric force. He had constituted a novel object. It remained to be seen what that object was. But Helmholtz jumped at once to what seemed important to him, and in so doing he also missed the subtlety of Hertz's argument. As Helmholtz remarked in his letter, there is an objection to the experiments "that could perhaps be removed even more completely than it has been before now and that would have to be discussed in any case": namely, "if cathode rays are (in accordance with the earlier view) electric currents, they would necessarily have to have an invisible return component, possibly in the vicinity of the container wall." He had, he goes on, "often discussed" the point with Goldstein, who, recall, had searched unsuccessfully for evidence of this hypothetical entity, concluding in the end that the rays are "open" currents.

Evidently Helmholtz thought that Hertz's preliminary experiment was designed to show that the rays do not affect magnets, which was how many subsequent readers also understood it. Helmholtz read the experiment in this way because that was what most concerned him: particle streams would be removed, and evidence constituted for the polar ether, by showing that the rays have no magnetic actions. If there were a return current present, then the experiment would of course be inconclusive. But Hertz was well aware of that fact, and his experiment was not designed at all to do what Helmholtz took from it. On the contrary, Hertz's preliminary experiment gained its significance from Hertz's assumption that if the rays *do* have a magnetic effect, then they must be paired with return currents, that Goldstein's open-current view is untenable. Given that conception, Hertz's symmetric experiment showed that the rays cannot have *nonmagnetic* action on the magnet, which Hertz required in order to carry on with his current mapping. The experiment in fact said nothing at all about their possible *magnetic* action.

Helmholtz missed all of this, no doubt because he was preoccupied with a much wider picture than the intensely ambitious Hertz, whose goal was much narrower. Moreover, Helmholtz spoke frequently with Goldstein, and we have seen that Hertz was trying to open a space for himself by disputing one of Goldstein's less-dearly-held convictions. Goldstein may not have pointed this out to Helmholtz, who did not read Hertz's paper with very great care. Hertz's pride in Helmholtz's Bravo! could not have been unalloyed, and this may only have accentuated the physical, and eventually intellectual, distance that Hertz was placing between himself and Berlin.

In his reply to Helmholtz (Koenigsberger 1965, pp. 345–46) Hertz elaborated the point that had gripped his mentor, namely, that there was here support for the rays being longitudinal ether waves. In fact, Hertz wrote back, he himself had already developed rather specific ideas along these lines:

[I] was inclined to think that the cathode rays are produced by the longitudinal waves, which correspond to the transverse vibrations of light. For it seems to me as if the longitudinal waves, in a medium in which the plane of polarization of the transversal waves rotates, must be propagated along curved lines, and thus the direction of rotation for light and for the cathode rays would be

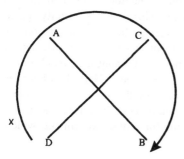

FIGURE 52   Ray displacement (after Hertz 1883c)

identical. Then, if the arrow [x in fig. 52] gives the direction of the positive current, produced by a magnetic field, the plane of polarization of all gases hitherto investigated will be turned in the direction of this arrow,—that is, a force is produced which acts along AB, and produces a displacement at an angle to this, as CD. There must also be longitudinal impulses propagated in a curve that is deflected to the right hand. But an elastic wire in which a positive current was flowing to the cathode would also be deflected to the right hand, and so a confusion between the two phenomena would be possible.

Hertz has here assimilated the deflection of cathode rays by magnetic fields to the Faraday effect, in the following way. In the Faraday effect the plane of polarization of light passing in active media along the line of magnetic force will be rotated in the direction of the currents that produce the magnetic field. Let us suppose, Hertz was suggesting, that this action affects any propagating displacement in the ether, whether it pertains to a transverse wave or to a longitudinal one. In that case a longitudinal disturbance that propagates in the direction of the transverse wave's displacement will itself be rotated in the same direction as that displacement. Thus in Hertz's figure, where x represents the direction of the current that produces the field, a light ray propagating perpendicularly to the plane of the figure (and so along the field lines) will have its polarization, initially along, say, AB, rotated to CD after some distance. A longitudinal wave propagating in the direction AB (and so across the field lines) will accordingly be bent into a curve, ending up after this same distance having turned through the angle between A and C. Of course, precisely the same sort of thing would happen to a flexible current-carrying wire that lay initially along AB, thereby confusing the two effects. Hertz also reported that he had tried to diffract the cathode rays by sending them through a grating "but obtained no result."

At the end of the letter Hertz wrote that he would reflect "upon the objection you pointed out" (presumably on the problem of the return current). "The latter, I think," he obscurely continued, "may be entirely refuted, if we succeed in obtaining more certain proof that cathode rays are possible in the absence of all electrostatic differences." He is apparently saying that one need not worry about whether the rays form parts of closed currents if one could create them without the normal apparatus of current production, since then the rays would necessarily be something entirely novel. This does not, strictly speaking, answer Helmholtz's objection, which Helmholtz thought to concern Hertz's experiment, but since Hertz's experiment was not in fact affected by that objection, Hertz turned it into something else, namely, a demand for explicit proof that the rays cannot be currents (even though Hertz had intended his experiment only to show, and it did in fact show, that the rays cannot carry much of the current in the tube). This made Helmholtz's otherwise rather irrelevant remark into a positive thing, an impetus for a new experiment, which, however, Hertz never performed.

PART FOUR

Studying Books

$$\cdot \; \cdot$$

E L E V E N

Frustration

11.1. THE *PRIVATDOZENT* AT KIEL

Hertz had settled into Kiel by early May 1883. He "paid his calls" on colleagues, but Karsten, whom Helmholtz had warned him about, claimed to be sick. "[S]o that's the way it is," Hertz first thought, "but afterwards at the colloquium [where Hertz was introduced] he was graciousness itself" (J. Hertz 1977, p. 181), although Hertz described the subsequent discussion with Karsten and the astronomer Pochhammer as "a sort of formal talk without content." A few days later Hertz gave his inaugural lecture, which was well attended, though he had characteristically worried whether anyone would show up. On May 8 he began teaching a *Repertitorium,* or review course, and on the ninth he started lecturing on the mechanical theory of heat, with, respectively, eight and seven students in attendance.

For a while things went well, despite the absence of a laboratory. Lecturing became "progressively easier" (at first Hertz had prepared "for 1–2 hours" beforehand). Because of this he found that he had more time for his own work than he had had as a busy assistant in Berlin, where "I was seldom really in the mood for reflection." Now he had a great deal of time to reflect, but as time went on he began to feel oppressed. The signs of depression appeared early. At the end of May, in an otherwise cheery letter, he remarked the "deadly sameness" of his evening walk. He began to walk more and more in the late spring and early summer and to go for swims in the sea. The heat became "oppressive" early in July, but just then he received Helmholtz's approval of the cathode-ray work which he had submitted for his *Habilitation.*

At the end of the summer he was on military exercises, which evidently improved both his physical and his mental state. "I feel completely well at present," he wrote home in mid-September, and by late October, when the semester began, Hertz was busy constructing an electrometer "for my own experiments," which had accordingly turned by then to electricity. At this point he realized, perhaps for the first time, just how badly off he was when it came to laboratory work. "There are all sorts of shortages here in Kiel," he complained, and "God knows how long it takes to run down a suitable piece of platinum wire or a glass tube."

Things went well for Hertz that fall. He began to make friends and develop a social life. He became particularly friendly with a Dr. Petersen, whom he met "often" in the evening. But by mid-December he was unhappy, remarking that his mood matched the dreary whether and short days. "My own work will not progress, and I am dissatisfied with the course as well." Karsten was friendly but unhelpful. Between January and March Hertz thought about "electromagnetic experiments" and tried to perform several with the meager apparatus that he had.[1] His thoughts turned in late January to "electromagnetic rays" and to how such things as dispersion and optical rotation might be dealt with by the "electromagnetic theory of light." In February he wondered about the dynamics of "falling raindrops." Then, between mid-February and early March, Hertz decided to read Kant's *Metaphysische Anfangsgrunde,* as well as the materialist Lotze's *Weltanschauung.* He also took Fechner's *Atomistik* out of the library. He "fetched many books from the library and studied literature."

In early April, terrible news arrived from Hamburg. His younger brother, Otto, with whom he had played every day for half an hour during the years that he had studied at home, was now deathly ill. The next day Hertz's friend Petersen, who must have known the dreadful report that he bore, came to the door with a telegram. Otto was dead. "Home in the afternoon," Hertz tersely wrote in his diary, "in the evening with Otto's body." The funeral took place on April 12, the day before Easter. The diary entry for April 14 reads "Terrible Easter Monday." Two weeks later the unhappy young man was deep into the most abstract and taxing considerations of his career, considerations that began to open wide for him the furthest reaches of contemporary electrodynamics. Hertz was plunged by depression, distance from the laboratory (where he might otherwise have sought solace), and, it may be, his reading of Kant into foundational thinking.

11.2. "FOR THE MOST PART THOUGHT OVER ELECTROMAGNETICS FRUITLESSLY"

Late that spring Hertz wrote what is surely one of the most perplexing, and penetrating, pieces of physics in the nineteenth century. Little known today (though occasionally discussed),[2] it amounted to an attempt to deduce Maxwell's equations without using the ether and without assuming that "force" in the usual sense takes time to travel from one object to the next. This would have been an extraordinary accomplishment if only the result had been coherent. But it was not. Hertz's first excursion into the deeper recesses of theory combined thoroughly incompatible approaches, not merely to electrodynamics but to the very understanding of what it means for one object to affect another one.

Hertz was certainly well aware that Helmholtz had obtained equations that looked rather like Maxwell's as early as 1870 by making the difficult and extremely problematic hypothesis that the ether is a dielectric with very high polarizability, that changing polarization must be included as a current in the continuity equation, and that polarization proper can be generated by the electromotive force due to a changing current. At Karslruhe, where Hertz immersed himself in the experimental investigation of propagation, this aspect of Helmholtz's theory became critically important to him, in major part because he then had the means to pursue aspects of it in the laboratory. But in the Kiel spring of 1884 he was engaged in something even more complex than an attempt to find laboratory traces of a difficult, frail concept, hard though that proved to be. He was concentrating on theoretical primitives, on how the interactions of bodies with one another are to be conceived whether or not polarization in Helmholtz's sense exists. He exerted a formidable (and, he knew, flawed) effort to extract propagation without recurring to the most problematic element in Helmholtz's *theory:* the polarizable ether. Hertz's thoughts on the subject were certainly not entirely coherent at this time, in major part because he was trying to bridge two utterly different ways of conceiving of bodily interactions, Helmholtz's and Maxwell's.

Isolated by force of circumstance from the laboratory, Hertz began to brood over theory, in particular over why there are three forms of electrodynamics that seem to be very different from one another and yet that work equally well in the contemporary laboratory. He thought himself into the laboratory even though he was physically absent from it and tried to find a common thread to unite the disparate theories. Hertz became convinced after intense study that he had uncovered a principle that all theories of electrodynamics had necessarily to accept but that many of them violated. He introduced his idea in a seductively simple manner:

Hertz's Principle I: The Unity of Electric Force

It has perhaps nowhere been explicitly stated that the electric forces, which have their origin in inductive actions, are in every way equivalent to equal and equally directed forces of electrostatic origin; but this principle is a necessary presupposition and conclusion of the chief notions which we have formed of electromagnetic phenomena generally. (Hertz 1884, p. 274)

He elaborated:

According to Faraday's idea the electric field exists in space independently of and without reference to the method of its production; whatever therefore be the cause which has produced an electric field, the actions which the field produces are always the same. On the other hand, by those physicists who favour Weber's and similar views, electrostatic and electromagnetic actions are represented as special cases of one and the same action-at-a-distance emanating from electric particles. The statement that these forces are special cases of a more general force would be without meaning if we admitted that they could differ otherwise than in direction and magnitude. But, apart from all theory, the assumption we are speaking of is implicitly made in most electric calculations; it has never been directly rejected, and may thus be regarded as one of the fundamental ideas of all existing electromagnetism. Nevertheless to my knowledge no one has yet drawn attention to certain consequences to which it leads, and which will be developed in what follows. (Hertz 1884, p. 274)

Both Faraday's (field theory) and Weber's electrodynamics, Hertz asserts, satisfy this principle. Faraday considers the electric field to exist independently of its sources. Weber considers all actions to be "special cases of one and the same action-at-a-distance emanating from electric particles." This must mean that the forces differ solely "in direction and magnitude," that it does not matter how they are produced. In which case Hertz's principle holds for Weber's electrodynamics, it seems, as well as for Faraday's. Hertz termed this idea the "unity" of electric force.[3] Despite the rather innocuous sound of the phrase "unity of electric force," Hertz argued from it that all forms of electrodynamics are inferior to Faraday's because only Faraday's satisfies the principle in execution. Although the other kinds of electrodynamics start with "unity" as their "necessary presupposition," they have all somewhere gone wrong because in their final results they violate it. If Hertz is correct, then every electrodynamics, including his mentor Helmholtz's as well as Weber's, with the sole exception of field theory, must be self-contradictory. This would be a powerful result indeed if it were true.

Hertz's sentences reflect in their structure the convoluted path he followed that fateful spring month of 1884. The "principle" with which he begins—that of the equality between "electric" forces of electromagnetic and electrostatic origins—is, he asserts, a "necessary presupposition and conclusion." But which is it? Does it come first, or does it follow after, "the chief notions which we have formed of electromagnetic phenomena generally"? Or is it perhaps equivalent to these notions? He does not say. But he goes on to give us two particulars in theory, no doubt meant to support the claim that the assumption is a presupposition, and one in calculation (hence experiment), intended to support the claim that it is a conclusion: first Faraday then Weber are mentioned, and finally an assertion that the "assumption . . . is implicitly made in most electric calculations" appears. But is it indeed? And note that Hertz does not feel it necessary to gloss the claim that the assumption holds for Faraday, whereas he adds an additional (and problematic) sentence for Weber. The uneasy mix of these pregnant sentences betrays the incoherence among the conceptions that Hertz was trying to merge.

Until the spring of 1884 Hertz had almost certainly not looked at all deeply at field theory, though he had delved into Maxwell's formidable *Treatise* in order to retrieve useful equations for his doctoral investigation on induction in rotating spheres. We must keep this comparative novelty in mind, because what struck Hertz from the first as a signal characteristic of field theory was its insistence on what he termed the "unity of electric force." Since in field theory a body is acted upon by something that is separate from the body but that exists at the body's locus, the only determinant of the action is the state of the field. To assert, as Hertz does, that electrostatic force and electromagnetic induction are one and the same means in this context that they are shorthand ways of referring to the same state of the field but that conducting bodies can affect the state in two different ways. In which case it is indeed almost axiomatic that changing currents can move charged bodies, because they alter the state of the electric field, which is what moves charged objects. This is Hertz's point of departure, both as the first sentence in his series of examples and, most probably, in his thoughts. As we have seen, it is not at all obvious that electromagnetic and electrostatic forces are equivalent to one another in Helmholtz's system, and it is hardly axiomatic in Fechner–Weber either. This is why, having continued with Fechner–Weber, Hertz must gloss the point for it. He must argue the equivalence on the basis that both forces are merely special cases of "one and the same action-at-a-distance." But this is hardly apropos. Hertz's claim specifies a relationship between the actions of charged bodies and of current-carrying ones, and not one between static and moving electric particles. Since the latter comprise the former in what may be rather complex ways, it is hardly obvious that Hertz's principle holds for them (and indeed it does not).

The only situation, in fact, in which the assumption appears in a stark form involves Ohm's law. Three kinds of electromotive forces appear in the law: those due to chemical, thermal, or mechanical effects, those due to electromagnetic induction, and those due to static actions, the very ones that are necessary but problematic in Helmholtz's electrodynamics. Hertz's opening remark that his assumption of equivalence is a "presupposition and conclusion" consequently parallels the division between Maxwell's and Helmholtz's conceptions. In the former it is a presupposition, but in the latter it is a conclusion drawn from the empirical requirement (first insisted upon by Kirchhoff) that the gradient of the static potential must be included in Ohm's law. Here we have a fundamental asymmetry between the schemes, one that reflects the inherent and irremediable differences between them over the ways in which bodies interact with one another. Hertz tried to relieve this tension, whose deepest meaning he certainly did not as yet grasp, by raising to the status of a principle what, in Helmholtz's electrodynamics, must remain tentative and problematic in order for the scheme to retain its completeness: the actions between charges and changing currents.

The influential argument that Hertz based in part on this principle is pregnant with unstated assumptions and riddled with intricate difficulties, for it challenges the deepest parts of contemporary German electrodynamics, including both Fechner–Weber's and Helmholtz's. But we will also see that these problems remained unclear to Hertz himself for some time. The very statement of principle I violates the nature of Helmholtz's scheme because it asserts as a matter of principle that electromagnetic induction can move electric charge, which is precisely what Helmholtz does not assume and which, if true, presents it with an extremely difficult problem in the absence of a corresponding interaction energy. At this stage in his career, separated by circumstance from easy access to the laboratory, Hertz was looking for a way to understand the connections between field theory, on the one hand, and both kinds of German electrodynamics, on the other. Propagation occurs in field theory; it does not occur naturally in either of the German theories. Evidently Hertz was gripped by the possibility that propagation might somehow be inherent in phenomena that the German theories, lacking all consideration of the ether, at least tacitly accept, but that the phenomenon that implies it must be a marginal one whose effect on the usual laws of electrodynamics is quite small. Hertz knew already that Riemann (in 1858) and Lorenz (in 1867) had made a similar claim but from the other direction: if we assume propagation, then some small change must be made to the usual laws. In effect, what Hertz did was to look for a weak effect that would answer this requirement, and he eventually found it in the largely ignored, but (he proclaimed) tacitly accepted, action of a changing current on a charge.

For a week in mid-May he struggled almost daily with this "problem," until he suddenly saw a solution on the morning of May 19. His diary records the (for him) unprecedented devotion he gave to this purely theoretical issue, one which continued to plague him for weeks after he had reached the solution we will discuss below (J. Hertz 1977, pp. 194–95):

11 May	Hard at Maxwellian electromagnetics in the evening.
12 May	Magnificent sail with Krümmel, Haas, Wendt, to Heikendorf.
13 May	Nothing but electromagnetics.
16 May	Worked on electromagnetics all day.
19 May	Hit upon the solution of the electromagnetic problem this morning. . . .
19 June	Worked on electromagnetics.
26 June	Electromagnetics finished, ready for copying.
27 June	Electromagnetics polished and copied.
28 June	Finished electromagnetics.

The article may have been finished on June 28, but Hertz was still bothered by something:

3 July	Reflected on electromagnetics.
4 July	Electromagnetics, still without success.
8 July	Some work on course, some on electromagnetics.
11 July	For the most part thought over electromagnetics fruitlessly.
14 July	Nothing but electromagnetics.
17 July	Depressed; could not get on with anything.

Hertz, we have often seen, was by nature extremely moody and often depressed. Loneliness, recalcitrant students, or even bad weather could send him into prolonged fits of despondency. More than anything else, he was keenly sensitive to the feeling that his work was going nowhere and that in consequence his career would never rise above the mediocre. Here, on July 17, we find him depressed just after he had failed to make sense of something in electromagnetics. For three months he ignored the subject. In late October he returned to it, only to be thrown again into depression:

20 October	Sent meteorology paper to Köppen. Thought anxiously about electromagnetics.
24 October	Turned back to electromagnetics.
25 October	Thought about electromagnetics.
29 October	Very bad mood.

Twice within a few months Hertz had troubled himself deeply over "electromagnetics." Twice he became depressed shortly afterward. Later he repeated the same pattern of intense work followed by failure and then depression when he investigated electromagnetic waves. But what was it that escaped him and helped send him into despondency? One cannot be certain: not only are there no pertinent records from this critical period in Hertz's career, but there almost certainly could not be any because he probably could not have articulated precisely what disturbed him. Nevertheless, I shall argue that Hertz was greatly, even profoundly, troubled by the ill fit between the structure of electrodynamics as he understood it and its most elementary physical concept, that of the source, and he simply did not know how to resolve the problem.

We turn now to the quandary that Hertz faced that painful and depressing spring, to the problem that he could not even quite formulate, much less solve. The argument that he built upon the fragile foundation of his principle I is difficult, incomplete, and impossible to reconcile at the most fundamental level with German theory of any kind, despite his printed assertion that he had started "from premises which are generally admitted" in electrodynamics and that he had used "propositions which are familiar in it" (Hertz 1884, p. 289). This very incompleteness betrays the struggle that Hertz was just now beginning, a struggle to forge a consistent understanding of electrodynamics out of the contemporary chaos.

11.3. The Dilemma

Although Hertz referred frequently to the unity of electric force, which he had introduced via principle I (according to which inductive actions can produce the same effects as electrostatic ones), his argument was actually based on two principles, only the first of which directly involves the "unity" of force in any meaningful sense. This was fully realized, we shall see, by at least some of his readers (in particular, by Boltzmann), and the fact that Hertz's argument seemed to be incomplete as it stood was apparently quite obvious, though perhaps not at first to him. Indeed, the difficulties Hertz had over the next few months were perhaps prompted by this incompleteness, or by its implications. The second principle, which Hertz did not explicitly state but which he unquestionably used, can be expressed in the following way:

Hertz's Principle II: The Indistinguishability of Sources

If a body A that has a property l exerts a force of type x that moves a body C that lacks property l, and if a body B that has the same property l as A also exerts a force of type x that moves body C, and if C exerts moving forces upon A and B that are the counterparts of the actions that A and B exert upon C, then A and B must, while they possess property l, likewise exert forces of type x that move each other.

This can be stated in a way closer in spirit to Helmholtz's understanding:

If one and the same interaction potential Φ exists between a body C in some state λ and each of two bodies A and B that are in the same state l, then Φ also exists between A and B when they are in state l.[4]

Hertz had in fact begun his article with a form of II rather than I, albeit for magnetic, rather than electric, actions. Ampère's old argument that currents (A and B) must exert forces on one another since they and magnets (C) do involves, Hertz wrote, a principle of the unity of magnetic force. Although Ampère may not have begun with the principle, Hertz continued, "he certainly stated it at the close of his investigations when he reduced the action of magnets directly to the action of supposed closed currents" (Hertz 1884, p. 274). Ampère's reduction is, however, much better expressed as the fact of the indistinguishability of *sources* rather than as a principle of unity of magnetic force, because it depends upon a physical model. The forces are unitary because all *sources* of magnetic action are indistinguishable from one another; in fact, they are physically identical. Ampère certainly sought a unitary foundation for current–current and magnet–current actions, but he himself would not have accepted a priori a magnetic version of Hertz's principle II. Yet for Hertz a general principle is at work here, one that does not rely upon particular models.

The best way to see how Hertz's principles work for electric forces, and why both of them are necessary for his purposes, is to examine his discussion of a special (but highly significant) case, that of two ring magnets, or magnetized toroids, T_1 and T_2. Let us assume that the magnetization of T_1 is changing. Then in all theories this will produce an electromotive force within T_2 unless their planes are mutually parallel. If the magnetization of T_2 is also changing, then it will exert an electromotive force within T_1. Here Hertz's principle I enters. Both T_1 and T_2 must act to move an electric charge according to I since the principle does not allow us to distinguish electromotive forces that move charges from those that create currents. Introduce principle II, in which l represents changing magnetization; T_1, T_2, and the electric charge are, respectively, A, B, and C. The charge lacks the property l, which A and B possess. Now, we assert not only is C moved by the electromotive forces exerted on it by A and B, but so also are A and B moved by one another: since each of them moves C, and C reacts, then, by principle II, they must also move each other.[5] Bodies in which magnetization changes (or, equivalently, in which currents change) must therefore exert mechanical forces on one another that are proportional to the products of the rates of change.

Hertz remarks that such an action is missing from what he calls the "usual" electrodynamics,[6] among which he includes Weber's.[7] And yet, he has argued, the usual electrodynamics at least tacitly assumes the unity of electric force in the sense of principle I. But not only is this claim of Hertz's itself questionable, it is not at all the same as principle II. For the heart of the issue, despite Hertz's own phrase "the unity of force," was not force per se but rather how one must treat the things that generate and are acted on by force in relation to it: how, that is to say, one must conceive the nature of the source in relation to its action. Hertz's principle II, but not his principle I (which alone does not yield Hertz's conclusion) introduces into electrodynamics a physical unity of sources in relation to their actions, and such a unity is completely foreign to both kinds of German electrodynamics. The nub of the principle is its prohibition on distinguishing as objects to be acted on any things that produce electric (or magnetic) forces. All things that exert forces of a given kind must, in Hertz's principle, be acted on by any and all other such objects, whatever their actual physical structure may be. Insofar as these forces are concerned such things do not have direct physical relations with one another. Rather, their mutual connections are determined only in relation to something of a different kind from themselves, namely, that which acts upon them (Hertz's "force"). That is the core, the elementary sense, of Hertz's tacit second principle, a sense that suggests the relationship between sources exemplified by field theory. Here, despite his claim to have used propositions that are "familiar" in the usual electrodynamics, Hertz has actually violated its most elementary concept, the tacit but irrefragable understanding that objects act on objects and that force is a relation between two things of the same kind; force does not stand apart as a thing in itself.

The usual electrodynamics does not—indeed, cannot—detach an action from its source, because forces do not exist independently of sources. Since a force is always an action between a pair of objects, the concept of, say, the value of the force at a point in space usually lacks physical significance: the objects between which the force acts must also be given.[8] Consider again how the usual electrodynamics, based upon electrostatic (U_s) and electrodynamic (U_d) force functions, describes interactions. For the two distinct kinds of physical objects, charges (q) and currents (c), there are two types of interactions: charge–charge and current–current. All theories agree on three points: a charge

q can be moved by another charge; the body containing a current c can be moved physically by another current; a current can be altered in magnitude by a changing current.

Charge on charge: ∇U_s

Current on current: $\nabla \times U_d$ and $-dU_d/dt$

Now it *may* also be the case that a changing current can move a charge and that a charge can generate an electromotive force at a point, but this interaction cannot be directly represented by either of the two potentials U_s, U_d alone, because they are formulated solely for charge–charge and for current–current interactions respectively. Despite Hertz's principle I, the modes of interaction that are involved in these two cases are decidedly different from one another: one is a charge–charge, the other a charge–current, interaction. To represent the latter one needs a third function because the potentials necessarily and directly reflect the physical character of the sources. The third potential may (and Helmholtz assumes that it does) have the same form as U_s for the electromotive force generated by a charge, and as U_d for the force with which a changing current moves a charge, but the interaction is physically distinct from the usual two kinds.

Things are rather different in a theory that, like Weber's, does not begin with a multiplicity of potentials but, instead, postulates that all electrodynamics involves a single physical entity, namely, a particle. Weber's theory can be, and by 1884 certainly was, presented in terms of a potential function, but here the function directly determines the interaction of electric particles only, not of charges with currents or of currents with currents. To obtain these latter, Webereans must introduce Fechner's hypothesis, which specifies the link between Weber's particles and the electrodynamic sources, between, that is, the particles and the charges and currents of laboratory experience. We may then calculate a potential function similar to U_d if we wish, and we may even compute such a thing for interactions between charges and currents. But these functions will be merely secondary representations of the fundamental interaction between particle and particle. Indeed, they may not even be precisely equivalent to the ones that prevail in the usual electrodynamics, since by "usual," Hertz intends Helmholtz's system, wherein U_d is given a priori rather than via a calculation from a more fundamental interaction.

Despite these important differences between Weber's and the "usual" systems, they all share one overriding characteristic: namely, that the actions cannot be divorced from their physical sources. In the usual systems this means that energies must be given for every type of interaction. In Weber's case it means that what are ordinarily thought of as sources, namely, charges and currents, must be computed from the true sources, which are moving electric particles. Interactions do not occur between these calculated representations but between the particles proper. So, for example, it may be true, as Hertz insists it is, that a changing current can move a charge and that the charge cannot tell whether the force acting on it at a given point and at a given moment derives from induction or from static action. But the usual systems must simply postulate this as a fact; Weber can deduce it from the force between his particles. One may say that for these two kinds of electrodynamics, it is essential to know the physical nature of the sources. Interactions occur directly between them, and only between them.

Field theory provided a very different way of thinking from this one. According to it electric and magnetic fields can—indeed, must—be divorced from their sources. The values of the electric and magnetic fields at a given point are in themselves sufficient to determine how an object placed there will behave. What produced the fields, or how the fields were produced, makes no difference at all to their effects: the only important thing is the interaction between the field and its object, not between object and object. Consider Hertz's ring magnets from the viewpoint of field theory. Each of them produces a contribution to the E field through its changing magnetization. The field can move a charge, and the charge should, according to a common view among French and German field analysts after the late 1880s, and certainly among most British Maxwellians during the 1880s, be able to move the sources that contributed to the field that moves it, namely, the ring magnets. The only field that is associated with the charge, the only means by which it can move anything, is its contribution to E. But there is no way to distinguish one contribution to the field from another one. And so it would appear on these grounds that the ring magnets should move each other since each of them also contributes to the E field.[9]

This sort of interaction, it might be thought, could be represented analytically in field theory by energy functions of the very kind that Hertz was considering when he emphasized the conservation principle. Whether two objects (sources) interact is contingent on whether or not there is a term in the field energy that depends on both of them. If there is, then they affect each other. In the example of the ring magnets the energy varies as the square of the E field. Since the total field consists of a linear superposition of all the partial fields, which include the two E_A, E_B from the ring magnets, the energy contains a term of the form $E_A \cdot E_B$, and so the ring magnets should, it seems, interact with one another. An essential feature of this way of thinking is that one cannot distinguish between the E fields produced by the sources however different the sources may be as physical objects. In that sense this may be thought of as a principle of the indistinguishability of the sources as generators of fields.

Hertz's principle II therefore applies at most to field theory, but only to field theory (though even there it is not *necessary:* i.e., the principle makes sense only within the context of field theory, but it is not itself a necessary implication of field theory).[10] And so the otherwise surprising, indeed startling, culmination of Hertz's analysis in Maxwell's equations reflects the very origin of his argument. The next chapter presents Hertz's argument in a way that is designed to bring out its essential features and so to show how, and why, it seems to lead to Maxwell's equations, to the equations of field theory. It will also show that Hertz's argument necessarily ended in confusion because he tried to obtain a result from the usual electrodynamics by importing into it a conception of the source that cannot be reconciled with it.

TWELVE

Hertz's Argument

12.1. GENERATION OF THE NEW VECTOR POTENTIAL

Hertz derived Maxwell's equations by converting electromotive into mechanical forces and then applying an energy conservation principle. Although he gave no references for his method, he indubitably obtained it from Helmholtz, who discussed it in 1870 and, in somewhat greater detail, in 1874 (Helmholtz 1874a, pp. 702 ff.). Neither discussion is entirely lucid, but the method always begins with a generalized Ohm's law and then introduces energy conservation:

Generalized Ohm's Law

(1) $$C = \frac{1}{\kappa}(-\nabla\phi + E_{ind} + E_{ctm})$$

where κ is the resistivity and C is the current *density*. Helmholtz had split the actions that can generate currents into three forces: the electrostatic force$(-\nabla\phi)$,[1] a force due to electromagnetic induction (E_{ind}), and whatever chemical, thermal, and mechanical forces may exist (E_{ctm}). To generate from this an appropriate equation for energy conservation, consider a time interval δt, take the scalar product of C with $C\delta t$, and integrate over all space:

(2) $$\kappa\delta t \int C^2 d^3r = \begin{Bmatrix} -\delta t \int \nabla\phi \cdot C d^3r \\ +\delta t \int E_{ind} \cdot C d^3r \\ +\delta t \int E_{ctm} \cdot C d^3r \end{Bmatrix}$$

Helmholtz interpreted this in terms of energy by first remarking that the left-hand side represents the "work equivalent" of the heat generated by the current system in time δt.[2] Then, he continued, the three terms on the right-hand side must, respectively, be the work equivalents of the electrostatic, electromagnetic, and other heats that are generated in the current system by these actions during δt. Let Q represent the time rate of the second of these three contributions to the energy dissipation. Energy conservation requires that Q be obtained, at the expense of the energy stored in the *current system*, which Helmholtz represented by Φ_0. This energy depends only upon the current states and the distances between the current-bearing objects at any given moment.[3] It can also be changed by whatever internal mechanical forces F_{mech} are engendered by the currents between the bodies in which they occur. If a body on which such a force works has velocity v, the time rate W of this working will be $\int F_{mech} \cdot v d^3 r$, and so, Helmholtz concluded, energy conservation implies

Helmholtz's Principle

$$Q + W = -\frac{d\Phi_0}{dt}$$

(see appendix 3 for details). This principle is the foundation of Helmholtz's (and Hertz's) analysis; from it the existence of electromagnetic induction F_{ind} follows from the Ampère force F_{mech} and vice versa if Φ_0 has the form

(3) $$\Phi_0 = \frac{1}{2} A^2 \int C \cdot A_c d^3 r$$

With Hertz we now generalize our understanding of the conservation equation by interpreting C as any electromagnetic variable that enters into an equation of this form. And then we draw the conclusion, as Hertz did, that the existence of forces which tend to alter the position coordinates of the bodies containing the C, whatever the current may be, implies the existence of forces that tend to · alter the magnitudes of the C themselves, and vice versa.

In particular, variation of Φ_0 with respect to position coordinates entails a force of the form $C \times (\nabla \times A_c)$ that alters these coordinates.[4] Variation with respect to the C proper entails a force $-\partial A_c / \partial t$ that alters the C. Or, one may say, the existence of an energy in the form Φ_0 together with either force implies the other one because of energy conservation (viz., $\delta\Phi_0 = 0$). Because the function A_c that appears in the equation is defined as $\int C/r d^3 x$, whenever we have a situation of this kind (whatever C may actually be), we can deduce these kinds of forces as functions of a vector A_c.

It is very important to understand that the interpretation of these actions is not at all arbitrary. The force that depends upon $\nabla \times A_c$ must always be interpreted as one that affects the positions of the bodies that contain the C. If we begin with the usual electromagnetic system, in which the C are simply currents, we must interpret the forces that depend upon material coordinates as "magnetic" actions because in the usual terminology such forces alter the locations of the bodies containing the currents but not the current magnitudes proper.

Consider such a current system and assume that we find forces $\nabla \times A_c$ that affect the positions of the current-bearing bodies. Then from the energy principle we conclude the existence of a force $-\partial A_c / \partial t$ that alters the magnitudes of the currents. At this point we introduce Hertz's principle II, from which we argue that the force $-\partial A_c / \partial t$, which tends to alter the magnitudes of the C, is also a material force that affects the positions of the bodies. The variables on which this force depends cannot be the usual C because the C do not imply a material action of this kind (since, for C, the mechanical force is entirely accounted for by $\nabla \times A_c$). To make this clear suppose that, as a material force, $-\partial A_c/\partial t$ depends upon some set l of electromagnetic variables. In order to apply energy conservation to this new force we must create a scheme for it that parallels the one that we used for C. Instead of the previous energy expression we must have a new expression—a *new interaction energy*—in terms of the l:

$$\frac{1}{2} A^2 \int C \cdot A_c d^3r \quad \text{becomes} \quad \frac{1}{2} A^2 \int l \cdot A_l d^3r$$

And where before we had a material force $\nabla \times A_c$ we must now have a material force $\nabla \times A_l$ that is equal to $-\partial A_c/\partial t$. This force, that is, tends to alter the position coordinates of the bodies in which the l (as well as the C) occur. Taking this to the next step, we can see that, by energy conservation, there must also be a force that tends to alter the l themselves, and it must have the form $-\partial A_l/\partial t$.

In order to understand what Hertz concluded form this we must be careful to interpret properly these new actions, which do not exist according to the "usual" electrodynamics of his time. Although the position-dependent force $(\nabla \times A_l)$ tends to alter the locations of the bodies containing the l, this does not mean that it is itself a magnetic force like our previous $\nabla \times A_c$. Even though the l occur in the same bodies that the C occur in, and even though the bodies in which the C occur are moved physically by magnetic forces, nevertheless the $\nabla \times A_l$ force derives from electromotive actions and not from magnetic actions. Consequently, its cognate force, $-\partial A_l/\partial t$, cannot itself be electromotive, and so (see below) it must be magnetic. Such a force may have a corresponding vector potential, and this new potential must be added to the old one to obtain a total value for it.

Hertz's procedure may therefore be reduced to this. Treat the electromotive force between currents also as a mechanical force and look for a vector of which this force is the curl. Take the time derivative of this new vector, and interpret the result as a new contribution to the magnetic force between the currents. Then compute an addition to the vector potential that corresponds to this force. With this understanding Hertz's analysis becomes reasonably simple. Begin with the usual electromagnetic induction between changing currents, and interpret it following Hertz's principle II as a mechanical force F^l_{mech} that is derivable from the curl of some function A_l:

$$(4) \qquad E = -\frac{\partial A_C}{\partial t} \Rightarrow F^1_{mech} = -\frac{\partial A_C}{\partial t} = \nabla \times A_l$$

Obviously the first thing to do is to find an expression for A_l. This is quite simple, but it requires assuming that (like its progenitor A_C) the new vector lacks divergence,[5] for then

$$(5) \qquad -\nabla \times \frac{\partial A_C}{\partial t} = \nabla \times (\nabla \times A_l) = \nabla (\nabla \cdot A_l) - \nabla^2 A_l = 0$$

$$\text{whence} \quad A_l = \frac{1}{4\pi} \int \frac{\nabla \times \partial A_C/\partial t}{r} d^3r \quad \text{if} \quad \nabla \cdot A_l = 0$$

This means that our vector l must be $\partial(\nabla \times A_C)/\partial t$, which is just the time derivative $\partial H/\partial t$ of the magnetic force. One may say, as Hertz in fact did, that we have to do here with a magnetic current that behaves in an energy principle like the actual electric current C.[6]

From the energy principle we conclude that there must exist a force F'_{mmf} of the form $-\partial A_l/\partial t$ that tends to alter the l, just as $-\partial A_C/\partial t$ tends to alter the C:[7]

$$F'_{mmf} = -\frac{\partial A_l}{\partial t}$$

$$(6) \qquad \qquad = -\frac{\partial}{\partial t}\frac{1}{4\pi} \int \frac{\nabla \times \partial A_C/\partial t}{r} d^3r$$

$$\qquad \qquad = \nabla \times -\frac{1}{4\pi} \int \frac{\partial^2 A_C/\partial t^2}{r} d^3r$$

The force F'_{mmf} acts upon the variable l, and this, we have seen, must be $\partial H/\partial t$. At this point we can go no further unless we make an additional assumption. We must somehow interpret this force in a way that has empirical significance. All we know thus far is that it alters the l. But we do not know how a force that alters l produces its effect in terms of the actions that are present in electromagnetism, which are only two: electric and magnetic force. Hertz *assumed* that the new force must be "magnetic," and so that it must contribute directly to the total magnetic action.[8] If we accept that all forces must be either "electric" or "magnetic," then his conclusion comes about as follows. The "mechanical" force determined by the A_l is "electric" in nature because it is just $-\partial A_l/\partial t$. Given that there can be only two kinds of forces, the nonmechanical force that this yields through the energy principle must be "magnetic." Consequently, we may now look for a vector potential A' corresponding to this magnetic addition, which we see at once from the last equation is simply

$$A' = -\frac{1}{4\pi} \int \frac{\partial^2 A_C/\partial t^2}{r} d^3r$$

We may express the complete vector potential to this point as

$$A_{comp} = A_C - \frac{1}{4\pi} \int \frac{\partial^2 A_C/\partial t^2}{r} d^3r$$

But this is not the last step. Far from it, for we must repeat this process ad infinitum. We now have a new term in the vector potential to consider. This term was not originally included in computing the E field from $-\partial A_C/\partial t$. And so we must now repeat the process that we just went through, this time for the correction to our original expression. That is, we compute a correction to E and then interpret this as a mechanical force that can be obtained from a curl, and so on.[9] Each step implicates yet another interaction energy for the corresponding states (each of which is the rate of change with time of the previous state). The process leads easily to the following series for the complete potential A_{comp} in terms of the recursively defined functions A_n.[10]

Hertz's Series

$$(7) \qquad A_{n+1} \equiv \int \frac{A_n}{r} d^3r \quad \text{with} \quad A_0 = \int \frac{C}{r} d^3r \quad \text{and} \quad \nabla \cdot A_n = 0$$

$$A_{\text{comp}} = A_0 + \sum_{n=1}^{\infty} \left(-\frac{1}{4\pi} \right) \frac{\partial^2 A_n}{\partial t^2}$$

12.2. Maxwell's Equations without the Ether

To obtain what appears to be a wave equation from the new series is, in execution, quite simple indeed. By definition of the A_n we see that each of them satisfies a Poisson equation with the preceding entry in the series as its source:

$$(8) \qquad A_n = \int \frac{A_{n-1}}{r} d^3r \Rightarrow \nabla^2 A_n = -4\pi A_{n-1}$$

And so for the series as a whole we have

$$(9) \qquad \nabla^2 A_{\text{comp}} = -4\pi C + \frac{\partial^2 A_0}{\partial t^2} - \frac{1}{4\pi} \frac{\partial^2 A_1}{\partial t^2} + \cdots$$

Operating with $\partial^2/\partial t^2$ on A_{comp} itself gives

$$(10) \qquad \frac{\partial^2 A_{\text{comp}}}{\partial t^2} = \frac{\partial^2 A_0}{\partial t^2} - \frac{1}{4\pi} \frac{\partial^2 A_1}{\partial t^2} + \cdots$$

Subtracting this equation from the previous one apparently produces the inhomogeneous wave equation with C as the source:

$$\nabla^2 A_{\text{comp}} - \frac{\partial^2 A_{\text{comp}}}{\partial t^2} = -4\pi C$$

In the absence of C (i.e., in "empty space") this seems to be the same equation for the vector potential given by Maxwell's theory, and it had also been obtained in a manner quite different from that of either Maxwell or Hertz years before by G. F. B. Riemann and L. V. Lorenz.

The vector potentials now show themselves to be quantities which are propagated with finite velocity—the velocity of light—and indeed according to the same laws as the vibrations of light and of radiant heat. Riemann in 1858 and Lorenz in 1867, with a view to associating optical and electrical phenomena with one another, postulated the same or quite similar laws for the propagation of the potentials. *These investigators recognized that these laws involve the addition of new terms to the forces which actually occur in electromagnetics;* and they justify this by pointing out that these new terms are too small to be experimentally detectable. But we see that the addition to these terms is far from needing any apology. Indeed their absence would necessarily involve a contradiction of principles which are quite generally accepted. (Hertz 1884, p. 286; emphasis added)

Hertz's analysis depends upon the recursive property of his series A_n, whose succeeding terms are related to one another as sources to potentials. Combining the propagation equation for the vector potential with the Ampère law ($H = \nabla \times A$) and the Faraday law ($E = -\partial A/\partial t$), Hertz produced for the first time the modern "Maxwell equations" in the form

$$A \frac{\partial H}{\partial t} = -\nabla \times E$$

$$A \frac{\partial E}{\partial t} = \nabla \times H$$

Hertz's purpose was to eliminate any asymmetry between the electric and magnetic forces—to make them, in his words, "interchangeable." They should be symmetrical because they derive in the same manner from infinite series of interaction potentials that have precisely the same form, with a contribution to the electric interaction energy giving rise through Hertz's argument to a new contribution to the magnetic interaction energy and so on.[11]

It is instructive to see in a different way how Maxwell's equations might emerge from such a procedure. Suppose we begin with a zero-order vector potential in the Faraday law and then calculate from its time derivative a corresponding zero-order electric field E_0. We could assume, then, that taking the time derivative of E_0 produces a source for $\nabla^2 A_{\text{corr}}$, where A_{corr} is considered to be a first-order correction to the original A_0. The process would work as follows. Start with the lowest-order term A_0:

Maxwell Faraday Law

$$E_0 = -\frac{\partial A_0}{\partial t}$$

Use the value of $\partial E_0/\partial t$ to correct A_0 by solving the equation given by an "Ampère" law in which we consider that $\partial E_0/\partial t$ behaves as a source for a *cor-*

rection to A_0. If we assume that the correction to the vector potential has, like the original, no divergence, we find

$$\nabla \times H_1 = \nabla \times (\nabla \times A_{\text{corr}}) = -\nabla^2 A_{\text{corr}} = \frac{\partial E_0}{\partial t} = -\frac{\partial^2 A_0}{\partial t^2}$$

What we have done is to posit that the time change of E_0 generates a magnetic field, and we have then used the field's vector potential to calculate an addition E_1 to the original E_0 field.

Maxwell's Equations by Recursion

$$A_0 = \int \frac{C}{r} d^3 r$$

$$H_0 = \nabla \times A_0 \qquad\qquad E_0 = -\frac{\partial A_0}{\partial t}$$

$$\nabla \cdot H_n = 0$$

$$\nabla \times E_n = -\frac{\partial H_n}{\partial t} \qquad\qquad \nabla \times H_n = \frac{\partial E_{n-1}}{\partial t}$$

$$E_{\text{comp}} = \sum_{n=0}^{\infty} E_n \qquad\qquad H_n = \sum_{n=0}^{\infty} H_n$$

Hertz's and Maxwell's electromagnetics therefore yield the same wave equation because of what *appears* to be a structural similarity between them. Both theories might be thought of as building up a total effect by a sort of bootstrap process, with each step generating the next one. In Maxwell's case the procedure seems at first thought particularly apposite because the field relations in radiation were often embodied in the image of orthogonal electric and magnetic fields generating each other by their changes, the net result being propagation. Hertz's recursive production of the equations has the advantage over Maxwell's presentation of them in his 1873 *Treatise* of closely paralleling the physical image of the mutual and sequential generation of electric and magnetic forces.[12]

Hertz had, so it seems, obtained Maxwell's equations. And yet he had neither used the ether nor assumed that "force" propagates. Indeed, force per se is nowhere to be found in Hertz's theory because he continued to employ Helmholtz's taxonomy of interactions. But how then *does* propagation arise? Was Hertz's argument unexceptionable? That is, did he demonstrate that Maxwell's equations (and so propagation) are somehow inherent in principles that were widely accepted among his German contemporaries? Hertz's argument in fact fails critically on two grounds. One ground involves the difficult physical assumptions that are embedded in his principles and that he himself suspected were problematic. The other, which neither Hertz nor, apparently, anyone else recognized until 1966, is purely analytical. That it was entirely missed even during the fifteen or so years when Hertz's article raised living issues reflects contemporary physicists' unfamiliarity with the properties of the inhomogeneous wave equation. The fact is that Hertz's series cannot possibly represent the general *retarded* solution to it because his series contains only even powers of the time derivative, whereas the general solution must contain every order of the time.[13] At the very least Hertz's solution lacks generality, and it is actually inconsistent with traveling waves.[14] Since few physicists understood the properties of such equations well,[15] either then or for many years thereafter, this lacuna has no proper historical significance at all, but it is instructive to pursue it a bit further because it reveals how vastly different the conceptual underpinnings of Hertz's Maxwell equations were from true field relations.

Our recursive version of Maxwell's equations fails in precisely the same way as Hertz's own route to them, which is hardly surprising since they are analytically equivalent: it produces only even-order time derivatives in the several series involved (for *A, E,* and *H*). The procedure fails to recover true field relations as opposed to partial facsimiles of them because it introduces an unphysical artifice: the assumption that the change in a field quantity generates a correction to the field. This is what permitted Hertz (and us) to consider each term in a series as though its effect were limited to its immediate successor. Maxwell's theory does not permit this kind of separation. In it field values must be taken *as indivisible wholes.* The recursive procedure fails in the end to capture the very thing that the field equations so strikingly produce—a propagating wave—precisely because it divides the whole into parts and establishes physical connections between the parts. In Maxwell's physics field quantities can be linked through partial-differential equations to other field quantities, but one cannot physically link a part of given field quantity to another *part* of the same quantity. No Maxwellian would, consequently, ever have generated "Maxwell's equations" in Hertz's fashion. Hertz did so because he was trying to interpret them as the effect of small *corrections* to the usual equations of Helmholtz's electrodynamics—as extra forces that emerge by seeming necessity from the deepest recesses of that scheme.[16]

The second difficulty with Hertz's argument is, unlike the first, historically significant because Hertz himself had an inkling of it; he knew that his argument was far from conclusive and that it rested upon a supposition that might not compel universal assent. He wrote near the end of his article:

> The mode in which we have deduced conclusions from the principal of the conservation of energy clearly marks at each stage the point at which our deductions are only the most fitting, and not the necessary ones. *This mode is the most fitting from the standpoint of the usual system of electromagnetics,* for it corresponds exactly to the accepted proposition in which Helmholtz in 1847 and Sir W. Thomson in 1848 deduced induction from electromagnetic action. But perhaps it may not be the only possible method; for just as in that

proposition, so we have in ours made tacit assumptions besides the principle of the conservation of energy. That proposition also is not valid if we admit the possibility that the motion of metals in the magnetic field may of itself generate heat; that the resistance of conductors may depend on that motion; and other such possibilities. (Hertz 1884, p. 289; emphasis added)

Hertz had indeed pinpointed one locus of difficulty: his use of Helmholtz's energy principle. But he certainly did not see precisely why the principle is problematic, because he apparently thought that the questionable aspects of his argument concerned such ancillary "assumptions" as, for example, that motion per se does not of itself generate heat in a magnetic field (which would require more terms in the energy principle than he had taken into account). Certainly such assumptions are necessary for his argument, but they are not the reason for its ultimate failure. The argument does not fail because of missing auxiliary requirements; it fails because Helmholtz's energy principle is irreconcilable with field theory.

The energy principle assumes and requires a particular conception of the relationship between sources, namely, that they determine an interaction energy which depends only upon their existence and the distance between them. Neither force nor field nor even time intervenes. Hertz's argument fully maintains this conception; indeed, he has added an infinite series of interactions to Helmholtz's taxonomy. but to obtain this series Hertz has had to recur to his two principles, and they are simply not compatible with Helmholtz's energy requirements, for Hertz's principles divorce interactions from sources. Far from being consistent with Helmholtz's energetics, Hertz's principles outrage it. Either sources interact directly or they do not: either Helmholtz's energy principle is correct, or else Hertz's principles are correct, but both cannot be.

The incompatibility arises at the point in Helmholtz's principle where one assumes that the mere existence of the sources at a given moment specifies the systemic interaction energy at the same moment (see appendix 3, requirement 3). This is simply not true in field theory, and not merely because in it effects propagate.[17] Even if interactions were instantaneous, the principle would still not hold in field theory, because the system's energy must be treated as an entity that is only indirectly dependent on the "sources." Consider, for example, Maxwell's dynamical illustration of current–current interactions in the *Treatise*. The mechanism he devised consists entirely of rigid connections so that all interactions are, as it were, instantaneous. The "sources" are loci at which certain motions occur that reflect other, hidden motions of the mechanism. Now as long as the only way to draw upon the energy stored in the mechanism is through these loci, Helmholtz's principle can be applied to it. But if we wish to leave open the possibility that the mechanism's energy can be drawn upon in some other way—if, in other words, we take seriously the assertion that the energy is stored in the mechanism itself—then Helmholtz's system energy will no longer be directly determined by interactions between sources. The interactions occur directly between the sources and the mechanism and only indirectly between the sources proper. One cannot say a priori in field theory that the apparent interaction between two objects depends on them alone, because the intervening mechanism may store energy that is not produced by the actions of these objects on the mechanism but that the objects can nevertheless draw down. The technical result is that the interaction energy between a pair of objects could in principle contain coefficients that are functions of the coordinates of other objects that feed energy into the field.[18] Field actions, in other words, are not necessarily bipartite (though they may be in particular cases), whereas Helmholtz's scheme insists that all actions must be.

It is worthwhile quoting Hertz's concluding remarks:

> In what precedes I have attempted to demonstrate the truth of Maxwell's equations by starting from premises which are generally admitted in the opposing system of electromagnetics, and by using propositions which are familiar in it. Consequently I have not made use of the conceptions of the latter system; but excepting in this connection, the deduction given is in no sense to be regarded as a rigid proof that Maxwell's system is the only possible one. It does not seem possible to deduce such a proof from our premises. The exact may be deduced from the inexact as the most fitting from a given point of view, but never as the necessary. (Hertz 1884, p. 289)

We see from this that Hertz clearly did not assume that the principles he had used cohered with Maxwell's theory; he had, rather, started "from premises which are generally admitted in the opposing system of electromagnetics." There is more than one route to Rome, and there is more than one way to obtain Maxwell's equations. Indeed, Hertz asserted, this way of obtaining field relations can hardly be regarded as a rigid deduction of them, because, supposing field theory to be absolutely exact, the relations with which Hertz had begun cannot possibly be exact. They contain within themselves, as it were, a proof of their own inexactness, but the precise form which this defect takes cannot be *deduced* with certainty from this very imperfection. A comparable situation, Hertz remarked in a note, involves Helmholtz's deduction of electromagnetic induction from magnetic force between circuits. This conclusion contained "tacit assumptions," such as that "the motion of metals in the magnetic field may [not] of itself generate currents." If such a thing did happen, the system energy would clearly not be determined completely by the current magnitudes and the distance between them, in which case the deduction would fail (see appendix 3).[19]

From Hertz's point of view then, his principles stood far apart from field theory. He went on to remark that, in fact, "Maxwell's system [and not Hertz's route to it from his principles] offers by far the simplest exposition of the result [Maxwell's system of forces]" supposing we decide that the results are exact. We must, however, be careful to understand what Hertz would at that time have meant by "Maxwell's system": namely, Maxwell's relations as deduced from

Helmholtz's principles combined with the assumption of a polarizable ether, taken in a certain limit. There is, accordingly, another tension here in addition to the importation of a field conception of the source into the "opposing system of electromagnetics." Maxwell's scheme obtained in Helmholtz's fashion is based directly on system energetics—indeed, on an extreme generalization of it—for the ether is a ubiquitous body whose every part can interact with every material electrodynamic entity as well as with every part of the ether itself. Yet energetics, Hertz felt, actually led in and of itself, sans ether, to Maxwell's system. This must certainly have opened new divisions in Hertz's mind: a division between the system of forces and the two apparent routes to that system, and between those two routes. One could, for example, distinguish Helmholtz's polar-ether analysis from his energetics by taking the electrodynamic and electromotive forces between conduction and polarization currents entirely a priori, setting aside the energetic origins of the forces. That is probably what Hertz had in mind at the time when he asserted that Maxwell's system "offers by far the simplest exposition of the results." This leaves one, however, not with Maxwell's own system, but with a Helmholtzian version of it stripped of fundamental energetics but still based on a non-Maxwellian understanding of object–object interactions—an understanding that Hertz had, however, already violated with his principle II. On the one hand, then, Hertz's perfected vision of Helmholtz's energetics pulled directly towards field theory through its principle II but yielded Maxwell's forces only with difficulty and obscurity. On the other hand, Helmholtz's polar ether, divorced from energetics, forthrightly gave Maxwell's "system of forces" but continued to embrace a nonfield understanding of interactions. When, four years later, Hertz did begin to experiment with effects that he associated with propagation, these conceptual difficulties resurfaced to influence his laboratory work by erecting a rather strong barrier between his goals in context and the global issues raised by propagation. For some time Hertz worked productively with his devices using concepts that could not in fact be reconciled with Maxwell's system. Eventually, we shall see, Hertz in effect returned to the results of his perfected vision by taking Maxwell's system of forces as a given (he introduced at first nothing involving the ether) and by (in his initial reconceptualization) throwing aside altogether anything to do with sources, thereby avoiding the foundational incongruities that arise when a system comes out of Helmholtz's electrodynamics by means of principle II.

12.3. HERTZ'S PRINCIPLES AND APPROPRIATE PHYSICAL UNITS

In the next chapter we will examine a difficult controversy that was stimulated by Hertz's novel claims, but it is extremely important to understand that these claims were, for Hertz, not at all isolated from the immediately practical. On the contrary, in his very next publication Hertz (1884) made use of his principles to argue that, contrary to widespread opinion, the electromagnetic system of units contains precisely the same sort of asymmetry that had often been raised as an objection against the electrostatic system. His argument is particularly fascinating because it depends directly on his principles—and, consequently, could not possibly persuade either Maxwellians or Webereans who carefully probed it.[20]

Units were not a neutral issue. In 1881 Helmholtz had striven at an international meeting in Paris to remove even terminological vestiges of Weberean structures. Units also had a great deal of industrial significance, with, in direct consequence, powerful connections to political economy. Hertz may not have been strongly aware of these latter associations, but he was certainly highly alert to the former. The issue that Hertz tackled concerned an asymmetry between electrostatic and electromagnetic units that seemed to speak in favor of the latter and against the former. The electromagnetic seemingly had a privileged position among systems of units. Hertz wanted to demonstrate that its superiority was in fact only apparent, that it was based on the concomitant, and physically untenable, privileging of electric currents as magnetic sources over magnetic currents as electric sources. Hertz interestingly (and perhaps significantly) framed his argument in the form of a thesis and an antithesis, in the following way.

The thesis, or common argument, used two equally well accepted assertions. Each of them works with the two systems of units, but with an important difference. The first (a) involved the work W_m done on a magnetic pole carried once around a constant current; the second (b) involved replacing a magnetic dipole with a circular electric current.

$a.$ $W_m = k_1 me/t$, where k_1 is (following Maxwell's choice) a dimensionless constant, m is the strength of the magnetic pole, and e is the quantity of charge conveyed by the current in the time t during which the pole circles it. Since the units of work are ML^2T^2 (where M, L, and T denote respectively the units of mass, length, and time), it follows that the dimensions of the produce me must be

$$[m][e] = ML^2T^1$$

$b.$ $m\delta = k_2 ef/t$, where δ is the length of the magnetic dipole of strength m, e/t is the strength of the electric current that is to replace it in terms of charge per unit time, f is the area enclosed by that replacing current, and k_2 is a constant. Clausius, basing his argument on Ampère's requirement that magnets are concealed currents, insisted that the constant must be dimensionless in order immediately to represent the replacement of magnetic moment by a current circulating around an area. Consequently, the dimensions of m and e must satisfy the relation

$$[m] = [e]LT^{-1}$$

The asymmetry between the electrostatic and the electromagnetic system of units arises in the following way. Each system begins with a pole whose dimensions are $M^{1/2}L^{3/2}T^{-1}$; the electromagnetic system assigns this dimension to the magnetic pole $[m]$; the electrostatic system assigns it to the electric pole $[e]$. If we take our two assertions a and b in turn and use each of them to deduce the units for the electromagnetic and the electrostatic systems, we obtain the following:

	Electrostatic: $[e] = M^{1/2}L^{3/2}T^{-1}$	given	Electromagnetic: $[m] = M^{1/2}L^{3/2}T^{-1}$	given
(a)	$[m] = M^{1/2}L^{1/2}$	deduced	$[e] = M^{1/2}L^{1/2}$	deduced
(b)	$[m] = M^{1/2}L^{5/2}T^{-2}$	deduced	$[e] = M^{1/2}L^{1/2}$	deduced

Since assertions a and b were equally well grounded in all contemporary forms of electrodynamics, it seemed to many people that the electromagnetic system had the virtue of giving a consistent dimension to the electric pole, whereas the electrostatic system gave inconsistent dimensions to the magnetic pole.

Hertz set out to prove that the superiority was only apparent. To do so he relied directly on his second principle, according to which no distinctions can be drawn between the sources of a given kind of force. The thesis had been based on magnetic poles moving around, and on replacing magnetic dipoles with electric currents. Hertz now proposed an equally acceptable (he thought) antithesis based on electric poles moving around, and on replacing electric dipoles with magnetic currents—that is, on the entities he had used in his "Fundamental Equations" paper to exemplify his principles. Magnetic currents produce electric forces, and according to principle II, closed magnetic currents can be completely replaced by electric dipoles. Consequently, the assertions a and b can be rewritten in the following way, where in assertion a', W_e represents the work done in carrying an electric pole once round a magnetic current:

a'. $W_e = k_1' em/t$, where k_1' is a dimensionless constant, e is the strength of the magnetic pole, and m is the quantity of magnetic charge conveyed by the (magnetic) current in the time t during which the (electric) pole circulates around it. Consequently

$$[m][e] = ML^2T^{-1}$$

b'. $e\delta = k_2' fm/t$, where δ is the length of the electric dipole of strength e, m/t is the strength of the magnetic current that is to replace it in terms of (magnetic) charge per unit time, f is the area enclosed by that replacing current, and k_2' is a constant. "Theoretically," Hertz continued, "there would be nothing wrong . . . if we started from this equation and made k_2' a pure number" (1885, p. 294), in which case one has

$$[e] = [m]LT^{-1}$$

Hertz could accordingly draw up a new table:

	Electrostatic			Electromagnetic	
	$[e] = M^{1/2}L^{3/2}T^{-1}$	given		$[m] = M^{1/2}L^{3/2}T^{-1}$	given
(a')	$[m] = M^{1/2}L^{1/2}$	deduced		$[e] = M^{1/2}L^{1/2}$	deduced
(b')	$[m] = M^{1/2}L^{1/2}$	deduced		$[e] = M^{1/2}L^{5/2}T^{-2}$	deduced

The electrostatic system is now consistent across the two assumptions, whereas the electromagnetic system is not. "The thesis and antithesis [a nice Hegelian echo] together show that, regarded purely from the standpoint of calculation, neither system has any advantage over the other." To show just how much he refused to privilege electric currents over magnetic currents or vice versa Hertz emphasized that he, following Helmholtz, used the Gaussian system, which requires conversion factors "whenever electrical and magnetic quantities occur together" (1885, p. 295).

Hertz had washed away the asymmetry between the two systems of units by asserting the physical symmetry of electric and magnetic currents. The two propositions (a', b') that do this were, he felt, contained by "every system of electromagnetics." Certainly proposition a' does follow from both Weberean electrodynamics and from field theory, but proposition b' does not follow from either kind of German electrodynamics, because it derives from Hertz's principle II. According to b' electric dipoles "can be completely replaced" by "magnetic circular currents."[21] According to Fechner–Weber a pair of electric dipoles certainly will exert forces on one another, whereas a pair of ring magnets will not. In Hertz's thinking, however, the ring magnets would have to move each other since each can move an electric charge—otherwise, one could not "completely replace" electric dipoles with them. Far from producing something that, as it were, stood regally apart from, and above, contemporary debates, Hertz had inserted directly into them something that was at least as contentious as the asymmetry between the two systems of units. The next chapter shows just how difficult Hertz's principle II proved to be.

• •

T H I R T E E N

Assumption *X*

13.1. CONFUSION OVER HERTZ'S ARGUMENT

Hertz's argument occasioned much troubled reaction among German physicists, the overt sign of which was a series of claims, counterclaims, rejections, and emendations that appeared in the *Annalen* between December of 1884 and the spring of 1887. The last of them was published in the same volume that, later in the year, contained the first of Hertz's experimental papers on electric oscillations. The first to take up Hertz's provocative work was Eduard Aulinger, almost certainly at the direct instigation of Ludwig Boltzmann. Aulinger in fact submitted his analysis of Hertz's argument as his doctoral thesis, completed under Boltzmann's direction, in December 1885.[1] Boltzmann later remarked that he had suggested the fundamental idea on which the thesis was based. He also became a prominent participant in the debate with Hermann Lorberg provoked by Hertz's and Aulinger's claims. Whereas Aulinger and Boltzmann sympathetically received the radical character of Hertz's arguments and attempted to improve them, Lorberg rejected them outright, although it is clear that at first he was rather confused by them. Hertz himself remained significantly silent throughout the debate.

Aulinger remarked (1886) that his goal was to seek the relationship between "Hertz's principle" and Weberean electrodynamics. After quoting Hertz's statement of what I have called his principle I, Aulinger wrote that he found Hertz's interpretation unclear; Boltzmann had brought this lack of clarity to his attention.[2] The vagueness they found concerned Hertz's failure to mention *what* is being acted on, since usually much was thought to depend upon whether the interacting bodies are, say, stationary charges or changing currents. Aulinger proposed to clarify the point by reformulating the principle in the following way:

The Aulinger–Boltzmann (AB) Principle

If, at each point of a finite or infinite space (electromagnetic field), the quantity and direction of the electrostatic force (that is, the force that would act on a unit quantity of electricity at rest at the affected point) and the quantity and direction of the magnetic force (that is, the force that would act on a stationary and unchanging unit north pole) are given, then with that each and every one of the magnetic and electric forces that act in the entire field on the moved and unchanging electricity and magnetism are completely and unambiguously determined whatever the origins of the magnetic and electric forces might be. (Aulinger, 1886, p. 121)

According to this (AB) principle, the "magnetic and electric forces" at any point are determined completely once we know two things: first, the force that acts there on a stationary unit charge and, second, the force that also acts there on a stationary unit north magnetic pole. To draw out the principle's implications, imagine that these two forces are being exerted by some set of sources O_i^1. Suppose that some of the sources in this set are replaced by others, forming thereby a new set O_i^2, but under the proviso that the forces which act at a given point on a stationary unit charge and on a stationary unit north pole are precisely the same for both sets. Then according to the AB principle, whatever object may be placed at the point in question will be affected electromagnetically in precisely the same way by the two sets of sources. But since some at least of the sources in O_i^2 differ in kind from the corresponding sources in O_i^1, electromagnetic actions are completely indifferent to the nature of the sources that give rise to them.

This AB principle is not equivalent to Hertz's principle II, because, for example, one cannot conclude from it alone that variable currents exert ponderomotive actions on one another that depend upon their rates of change. To reach this result, which was essential for Hertz's purposes, Aulinger had to bring in a further "principle of action and reaction," which, in combination with AB, is equivalent to Hertz's principle II, as we can see by applying it with Aulinger to Hertz's ring magnets. Take a space A and in it place an electric dipole layer as well as a ring magnet with changing magnetization. By AB the ring magnet exerts an electric force on the dipole layer, and then by action–reaction the dipole layer exerts a ponderomotive force on the ring magnet.[3] Take a second space B that contains, instead of A's electric dipole layer, an infinitely thin ring magnet that curves round the boundary of the region occupied in A by the dipole layer. Let the magnetization of the ring magnet change at just the correct rate to produce in B the same electric force that the dipole layer produces at the corresponding positions in A. Then the electric force on a stationary charge is the same in B as it is in A, and in both spaces there is no magnetic force. Consequently by the AB principle we must have precisely the same electromagnetic actions in B as in A, in which case it follows that ring magnets with changing magnetizations must exert ponderomotive forces on one another.

Aulinger (and so Boltzmann) have only made explicit what Hertz had already asserted implicitly, as Aulinger himself realized. But, unlike Hertz, they clearly recognized that a significant alteration had been made to the common (in Germany) understanding of fundamental electromagnetic principles, one that could not in fact be satisfied by, for example, Weber's theory or, for much more difficult reasons, even by Helmholtz's itself. Aulinger produced a concrete example that demonstrated the incompatibility between Weberean theory and the new principle. Consider with Aulinger a sphere uniformly charged with positive electricity at its surface. Within the sphere place a wire circle that carries a changing current. Then we can use the Weber law to compute the force between every particle of the current and every particle of the stationary charge on the sphere. Doing so via a binomial expansion, Aulinger determined that the circuit experiences no net translational force from the stationary charge that surrounds it but that it does experience a net torque from it: according to Weberean theory a small wire circuit bearing a changing current that is placed at the center of a hollow, charged conducting sphere will accordingly he twisted about an axis perpendicular to its plane. Now, Aulinger continued, suppose that we had placed a stationary electric charge within the sphere instead of a current-bearing circuit. Then there would be no force at all on this charge, and, of course, there is nothing to exert a force on a stationary magnetic pole placed within it either. Consequently, the AB principle requires that no electromagnetic actions of any kind occur within the sphere, in which case the Weberean torque should certainly not exist. "An experimental proof of the foregoing formula," Aulinger portentously concluded, "would furnish an experimentum crucis for the Weberean theory" (1886, p. 130)[4]

Hermann Lorberg, a Strassbourg *Gymnasium* teacher, had been a specialist in Weberean electrodynamics for fifteen or so years by the time that Hertz's paper appeared, and he would have none of it. Lorberg (1886) reiterated Hertz's argument for a ponderomotive force between ring magnets, concentrating on the derivation of this force from the supposition that changing currents (or, here, the changing magnetization of a ring magnet) can move an electric dipole. Lorberg did not challenge this supposition, because, on Weberean grounds, it is in fact correct: a changing current *does* move a stationary charge, for if it did not, electromagnetic induction could never occur given the Fechner hypothesis. But Lorberg completely rejected the next step.

Certainly, he admitted, it is "in a purely mathematical sense" correct to say that the force exerted by a changing current on an electric dipole has a "potential" with respect to it. But in Fechner–Weber this kind of potential was in itself devoid of physical significance; it could merely represent the resultant actions between groups of electric particles moving in various ways. This "purely mathematical" potential lacks physical significance simply because there is no real electric dipole for it to act upon. A ring magnet with changing magnetization is just that; it is not an electric dipole. So the fact that it can move a dipole certainly does not necessarily imply that it can move another object which is not a dipole, namely, another ring magnet with changing magnetization. In Lorberg's words: "a motion of such a [purely mathematical] electric double layer absorbs no mechanical work since no ponderomotive force acts on it" (1886, p. 669).

And so Hertz's and Aulinger's argument collapses. Nevertheless, Lorberg continued, the conclusions that Hertz had obtained might be correct even though the argument he had used must be rejected. In particular, there might be a force between ring magnets with changing magnetizations, though if there is, it is not because they can also move electric charges. To show that he did not disagree with everything Aulinger and Hertz had written, Lorberg also picked up Aulinger's deduction from the Weber law and generalized it to prove that, even if the conducting circuit within the sphere is not circular, a torque still occurs.

Lorberg's rejection of what had after all been based on his own suggestion brought Boltzmann directly into the fray (1886). But Boltzmann preferred first to emphasize the sole point of agreement between Aulinger and Lorberg, namely, their deductions from Weber's law of the circuit torque. This, Boltzmann remarked in paraphrase of Aulinger (or perhaps Aulinger had first remarked it in paraphrase of Boltzmann), constitutes an "experimentum crucis" for Fechner–Weber, though the experiment would lie at the very limits of technical capabilities since the effect would be extremely small. But then he turned to Lorberg's critique, to which he replied in the following way:

> [The assumption that Lorberg rejects] consists in this, that the electric forces which engender induction currents also act ponderomotively and not merely electromotively. This ponderomotive action is so small that it has to this point nevertheless not been experimentally observed; only it seems to me and obviously also to Hr. Hertz entirely self-evident that a self-closed solenoid will not merely engender an induction current in a closed conducting wire that encircles its central line but will also act ponderomotively on a small, electrostatically charged sphere in the neighborhood. . . . Let this be denoted assumption X. (1886, pp. 598–99)

Boltzmann had thoroughly mistaken Lorberg's meaning. According to Boltzmann the problem was that Lorberg refused to admit that changing currents exert ponderomotive forces on stationary charges. But this is entirely false: Lorberg never questioned *this* action. He had objected only to the next step in the argument, that such an action implies a similar one between the changing currents. Boltzmann, however, felt it necessary to insist that an action between changing current and charge is "self-evident," almost certainly because he was well aware that there were some who did *not* find the action to be "self-evident," even though Lorberg himself had not questioned its existence. Boltzmann was mistaken about this in the case of Lorberg, but he was certainly correct in thinking that someone *would* question the "self-evident" character of the action: Helmholtz's electrodynamics could only regard it as an action sui generis, one that requires the sources involved in it (charge and changing current) to generate an interaction energy that the standard theory does not take explicitly into account. Boltzmann, that is, interpreted Lorberg as though

Lorberg were writing from the standpoint of Helmholtz's scheme, whereas Lorberg's argument was based entirely on Weberean principles.

Since Boltzmann realized that he could not convince anyone who did not already accept the idea that electromotive forces which produce currents must necessarily move charges (because he apparently understood that what he found to be self-evident is not a part of the usual Helmholtz account), he left it to experiment to decide the issue, carefully describing how it might be done. But if we do make this assumption X—and this was the crucial point—"the potential of the electric forces of induction has a precise, real physical meaning just like that of the electromagnetic and the electrodynamic forces." This, however, was precisely what Lorberg, as a partisan of Fechner–Weber, could never accept, though—assuming that the corresponding force does indeed exist—a partisan of Helmholtz's scheme had necessarily to agree with Boltzmann's claim since the potentials are directly given by the very nature of the sources.

Lorberg was quick in reply to what he called Boltzmann's anticritique (1887). He at once pointed out Boltzmann's "misunderstanding" of his position. That a changing current will move an electrically charged sphere (taking Boltzmann's example) is, Lorberg remarked, just as "self-evident" to him as to Boltzmann. In fact it was not, because Lorberg had always to deduce what happens from the interactions between the underlying structures of electric particles. Boltzmann, on the other hand, intended quite literally that the action was, to him, self-evident, by which he meant that it could not be obtained from anything else. In driving home the point Lorberg unwittingly revealed how thoroughly he was himself now misunderstanding Boltzmann's position.

How could Boltzmann possibly have thought otherwise of his position, Lorberg wondered, when he, Lorberg, had even gone so far as to generalize Aulinger's demonstration that the Weber law actually requires an interaction between a charged sphere and a circuit bearing a changing current that is placed within it? This, Lorberg was saying, is obviously a ponderomotive interaction between a changing current and a stationary charge. But, again, Lorberg has to deduce such an action, here through a rather intricate computation indeed, and, more to the point, Boltzmann and Aulinger were in fact arguing that this very action does not occur, because it violates the AB principle. Whereas for Lorberg the existence of a force in the Aulinger example is entirely reasonable and is prima facie evidence that a changing current can act to move a charge, for Boltzmann and Aulinger this particular action should not occur even though they find it "self-evident" that changing currents do indeed move charges. The gulf between what one might call their modified Helmholtz position and Lorberg's Fechner–Weber outlook simply precluded them from understanding one another.

These interchanges between Boltzmann and Aulinger (and the silent Hertz), on the one hand, and Lorberg, on the other, mark the public emergence into German electrodynamics of a fieldlike understanding of the nature of a source, one attended by misunderstandings, lack of clarity, initial hesitancy, and frequent reversion to the previous way of thinking. This new way of thinking about the fundamental nature of physical objects as they relate to one another was vastly more difficult to assimilate than was the idea that forces propagate. One could entertain propagation—as Lorenz and Riemann did—without adopting something like Hertz's principle II, that is, without abandoning the distinct identity of the source. But we shall also see that the new way never did in fact achieve much currency in Germany despite the avidity with which Hertz's theoretical papers were read in the early 1890s. It was soon displaced by a third understanding, which combined the unitary sources of Fechner–Weber, which Hertz himself very much wished to remove from electrodynamics, with the unitary actions of Hertz's and Boltzmann's principle, producing thereby a unique, and deeply influential, structure.

13.2. BOLTZMANN IN THE LABORATORY

Boltzmann and his student Aulinger were unusual among German physicists for their interest in, and knowledge of, certain aspects of field theory, which in Boltzmann's case involved a great deal of experimental work. Between 1865 and 1887 Boltzmann wrote eighteen published pieces on electromagnetism, seven of which were experimental. Many of them concerned aspects of the subject that are particularly significant for field theory, such as the value of the dielectric constant or phenomena like electrostriction. His work clusters in three major groups. The first occurred in 1873 and 1874 at about the time that he became an *Ordinarius* in mathematics at Vienna University. Here he intensively examined dielectric behavior, explicitly considering its relationship to field principles.[5] He did little more in this field until nearly 1879, three years after his appointment to a chair of theoretical physics at Graz University made experimental work easier for him to pursue (Jungnickel and McCormmach 1986, vol. 1:213). For about a year he plunged into magnetization and then into electrostriction. He wrote almost nothing else in electromagnetism for about six years, until 1886, when he published two articles pointing out that the Hall effect undercuts the Fechner hypothesis (Buchwald 1985a, chap. 11).

Boltzmann's earliest work on electromagnetism was published in 1873. He undertook experiments to measure the force between a dielectric and a neighboring, charged conductor. The experiments began in 1872, when the twenty-eight-year-old Boltzmann was in Graz, and continued in Berlin at Helmholtz's institute. We know from Boltzmann himself that Helmholtz had been sufficiently involved (at first only through the mail) to recommend a particular technique. Helmholtz, we also know, was around this time particularly interested in experiments that could be used to determine dielectric capacities, since he had had Nikolaj Schiller use induction currents for that purpose. Boltzmann occupied himself intensively with this and related questions for two years, producing in the end five detailed experimental papers.

Boltzmann's initial aim was to investigate dielectric polarization through its interaction with conductors, and he felt that a careful distinction between polarization and conduction was necessary:

> The phenomena engendered by dielectric polarization differ totally from those called forth by trace conduction. Dielectric polarization cannot call forth an electric current within or on bodies, for as soon as the molecules are completely dielectrically polarized, the electric motion ceases. (Boltzmann 1873, p. 473)

Dielectric polarization differs from the usual current, which (according to Ohm's law) increases along with the "electric forces." But there must in some sense be "electric motion" during the exceedingly small time interval before complete polarization occurs. Boltzmann attempted to explain why this should be so:

> If this hypothesis is correct, that electric force displaces the positive electricity within the molecules of insulators to one side and the negative electricity to the other side (which, by the way, furnishes precisely the same results as if the molecules of insulators were already originally polarized and are only reoriented by electric forces), then an originally unelectrified isolating body, when placed in the vicinity of an electrically charge body, must be pointed out by the latter . . . just as a piece of white steel is pointed out by a magnet. I will call this phenomenon "dielectric action at a distance." (1873, p. 473)

He offers two possible explanations: the first is the old Mossotti hypothesis; the second assumes that the molecules are always polar and that electric force merely aligns them. In either case, Boltzmann takes care to emphasize, nothing like conduction occurs. Indeed, a basic goal of his experiments was to demonstrate that the "dielectric action at a distance" between dielectric and charged conductors cannot be due to some sort of residual conduction in the dielectric.

To pursue dielectrics Boltzmann constructed a simple, clever device. In figure 53 the small (7 mm) dielectric ball L hangs from $ECDGFH$, its weight balanced on the left by a mirror S. L has its center aligned along NM with the much larger metal sphere M; M is itself connected by a thick wire bent through a right angle at N to a smaller metal sphere R. Near R hangs the balanced, grounded rod PQP', whose ends carry two metal balls of the same size as R; the ball P, opposite R, has its center aligned with RN. The center of the rod carries a mirror S. The device is charged by means of the apparatus in figure 54, which contains two Leyden jars. Wires connected to the top of each jar are brought to the metal balls U and V, whose mutual distance can be adjusted by means of a micrometer. Jar Y connects through T to the device in figure 53.

The apparatus works in the following way. Jar Z is charged through V until

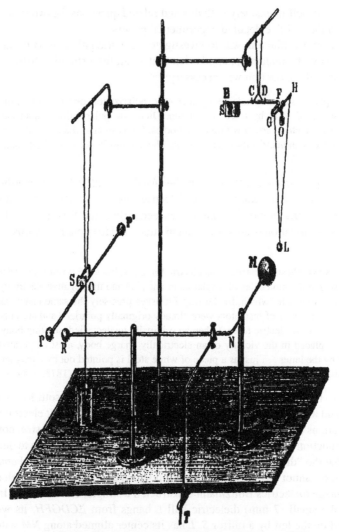

FIGURE 53   Boltzmann's capacity device (Boltzmann 1873)

a spark jumps across the gap between *U* and *V*, which charges jar *Y*—so that the quantity of the charge can be regulated by adjusting the gap. Both *M* and *R* in figure 53 receive charge from *Y*, and they are at the same potential. The spheres *P* and *L* will be charged by induction, and as a result the arms *PP* and *EF* are twisted against the reacting torque of the sustaining wires. The twist can be read by reflecting light from the mirrors *S*. Consequently, the deflection of the metal sphere *P* can be used to establish the potential of the *RNM* system, while the deflection of the sphere *L* is a function of the dielectric capacity.

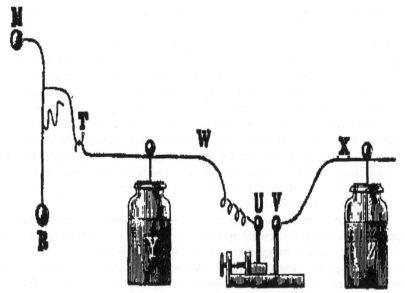

FIGURE 54   Boltzmann's Leyden jars (Boltzmann 1873)

Boltzmann's device can be divided into three quite different kinds of equipment. There is, first, the object that provides the stimulus necessary to generate the effect: the Leyden jars. The second object (the grounded rod PQP') measures the stimulus produced by the first. Both of these—the jars and the rod—remain completely unproblematic for Boltzmann. His concern centered on the third piece of equipment, the object in which the interesting effect is generated: the pendant dielectric sphere. Since he was looking for a difference between this effect and the one that would be generated by a metal sphere of equal size, Boltzmann was greatly concerned to convince his readers that the deflections of the dielectric that he observed could not possibly be due to anything at all like metallic charging.

The device was indeed nicely adapted to distinguishing dielectric from metallic effects, a task that he first approached by coating a second small dielectric (sulfur) sphere with tinfoil. Boltzmann slightly readjusted the device to hang this second sphere on the opposite side of M from the first dielectric (L). The goal was to compare the forces that act on the two small spheres: the one coated with tinfoil experienced only the usual metallic induction, whereas the uncoated one was polarized dielectrically. Boltzmann found that the metallic force was twice as large as the dielectric force, which he found satisfying but not entirely convincing. Suppose that the dielectric had somehow been charged beforehand has a result of a small, residual conductivity which it possessed; this would obviously vitiate any claim to have elicited here an altogether different kind of charging. Boltzmann, still at Graz in 1872, wrote Helmholtz about his problem.

Helmholtz suggested a way out, namely, that Boltzmann should alternate the charge on the electric source.[6] Then the actions of any original metallic charge on the dielectric sphere, should it be present, would on average vanish, whereas the induced effect on it would always be the same. To that end Boltzmann altered the first part of the device—the stimulator—by turning it into something that (in different forms) was just then becoming important in Berlin, namely, into a charge alternator (fig. 55). Here the original pair of Leyden jars are now oppositely charged, and an electromagnetic device oscillates the tuning fork (*a*), which alternately connects the Leyden jars. The results this device generated reproduced the earlier ones but eliminated the possibility of contamination by metallic conduction.

Boltzmann pursued the experiment in elaborate detail, later providing a demonstration that his several approximations were reasonable (1874b), and he also investigated (using an entirely different technique) the polarization of gases (1874a). His published articles are deeply experimental. They are for the most part nearly devoid of higher (problematic) theory, going only so far as to isolate the critical difference between metallic charging and dielectric charging. His laboratory work accordingly was almost completely divorced from the kinds of problems that were by then beginning to coalesce around the conception of polarization. Nevertheless, Boltzmann was not entirely insulated from these currents, for we have already seen that he did offer several brief introductory comments on the possible nature of dielectric polarization (viz., Mossotti's hypothesis and the notion of permanently polar, orientable molecules).

We cannot conclude very much from these short remarks, but they do show that Boltzmann (unlike Helmholtz) was willing to consider models that reduced polarization either to the shift of already charged objects or else to some sort of molecular induction. This almost certainly means that he had not at that time fully sensed the thrust of Helmholtz's new electrodynamics, namely, its abnegation of all reductive models. It would indeed be surprising if Boltzmann's experiments *had* penetrated deeply into Helmholtz's (or, for that matter, Maxwell's) new kind of physics, since he knew little about either. Both were obtainable only through articles—in Helmholtz's case only one article at that (which Boltzmann does refer to, albeit with an incorrect date). We can be quite certain that Boltzmann knew little of Maxwell from two remarks he made many years later. According to the first, Stefan gave him Maxwell's papers to read while he was still a university student and knew no English at all. He must have picked up little indeed from his attempts to assimilate Maxwellian English, as the second remark unequivocally demonstrates:

> I made the experiments on dielectric constants in Helmholtz's laboratory in order to test Maxwell's now-celebrated electromagnetic theory of light. Helmholtz, who did not have the precise formula in his head, told me verbally

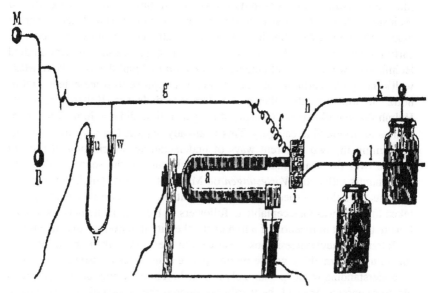

FIGURE 55   Boltzmann's charge alternator (Boltzmann 1873)

that according to Maxwell the index of refraction n must be equal to the dielectric constant D. Since my experiments did not give this, I thought I had completely refuted Maxwell with the firm results from Berlin and already had the notion of publishing it when the idea suddenly came to me to look at Maxwell's formula once more and there found $n = \sqrt{D}$, which worked well. (Boltzmann [1891–93] 1982, pp. 12–13)

So Boltzmann did not even know that Maxwell's theory required the wave speed to vary as the square root of the dielectric capacity, though that is perhaps the most striking result of Maxwell's pre-*Treatise* papers. Helmholtz offhandedly told him that the speed was proportional to the capacity, which was in thorough discord with the measurements on gases that he also undertook in Berlin (Boltzmann 1874a). Evidently unwilling to doubt the great Helmholtz, Boltzmann nearly decided to publish his results until the "idea suddenly came" to check the formula for himself. Helmholtz's own work was not hidden behind a linguistic palisade, as Maxwell's was, but had Boltzmann studied it, he would necessarily have discovered there the very formula, thoroughly discussed, that only impulse gave him. The evidence is unequivocal: before 1873 Boltzmann knew little about either Maxwell's or Helmholtz's novel assertions. Moreover, Boltzmann, unlike Hertz, was already quite well known before he came to Berlin, having published extensively, though not in electrodynamics. He was also nearly a decade older than Hertz was on his (Hertz's) arrival in Berlin, so that Helmholtz's impress upon him was considerably mitigated by maturity.

By the early 1880s Boltzmann had apparently begun a thorough study of

Maxwell's *Treatise,* no doubt in its 1883 German translation by Max Bernhard Weinstein, though his first profound encounter with Maxwell had occurred much earlier, in 1873, when he produced a lengthy commentary on Maxwell's early model for the field. However, there is no evidence that he ever studied Helmholtz's work with quite the same attention to detail that he gave to Maxwell's though he certainly did assimilate, and approve to some extent, Helmholtz's treatment of the ether as a polarizable dielectric. Boltzmann, unlike Helmholtz, but like most Germans of his generation, did not think of polarization as an irreducible category. This is already apparent in his early remarks concerning the two different ways of understanding polarization. It did not change over the years. Indeed, the twelfth lecture of the second volume of Boltzmann's influential *Vorlesungen* on Maxwell's theory, published in 1893, presents the field as a special case of polarization, but interpreted in a way that, taken literally, was unacceptable to followers of Helmholtz, namely, as though it were the shift in location of a hypothetical entity that exerts a distance force.

Boltzmann nevertheless took great care in the *Vorlesungen* not to interpret this thing as something whose motion per se constitutes a current, since such an understanding would presumably tumble him into the embrace of Weberean electrodynamics. Instead, he warily considered the polarization to constitute an "integral current," so that its variation with time is just a "current." This is pure, *uninterpreted* hypothesis, much as in Helmholtz's own theory. The difference is that for Boltzmann the objects themselves are intrinsically affected by electromagnetic actions: his objects are not things that have states, and thus their interactions are not to be understood as implications of a more fundamental energetic relationship.[7] Boltzmann's objects—the stuff of polarization, as it were—accordingly remain ultimately mysterious, whereas in Helmholtz's theory there is no comparable mystery because there are no ultimate Boltzmannian objects: there are only states of objects, which include as a class the ether.[8]

Although Boltzmann's understanding of Helmholtz's scheme might seem to be rather complicated, nevertheless it was (more or less) a common one in Germany, and also in France, in the years immediately following Hertz's experiments with electric waves in the mid to late 1880s. It had the advantage, on the one hand, of avoiding the Weberean trap by resolutely envisioning charge and polarization as strange but closely related things and, on the other hand, of considering the field to be a seat of polarization. The distinction between object and state is weakened, and yet Weberean reduction is neatly bypassed. At the same time the Maxwellian abolition of sources remains a thoroughly foreign idea. Boltzmann, as well as Poincaré and others in the early 1890s, engaged in a delicate balancing act indeed, but one that could scarcely have been avoided without embracing the still-strange conception of the field as an entity sui generis, though Hertz himself suggested just that.

Electric Waves

FOURTEEN

A Novel Device

Nor, indeed, do I believe that it would have been possible to arrive at a knowledge of these phenomena by the aid of theory alone. For their appearance
upon the scene of our experiments depends not only upon their theoretical
possibility, but also upon a special and surprising property of the electric
spark which could not be foreseen by any theory. (Hertz [1893] 1962, p. 3)

14.1. UNHAPPY IN KARLSRUHE

Until he left Kiel early in 1885 Hertz remained fitfully unhappy, depressed by
his lack of access to experimental apparatus and by the obstacles—human,
social, and intellectual—that he saw everywhere around him to the progress of
his career. Two months before he was offered, and at once accepted, a position
at Karlsruhe he wrote his parents:

> 30 November 1884 Concerning the inquiry made of Prof. Ladenburg, I was
> mistaken, for I thought that a professorship had been granted here; actually
> he told me that he had been asked for information by a colleague in Karlsruhe
> and that they had their eye on me. I do not know for what. I am dissatisfied
> once again, for I would rather have one certainty than two possibilities, and I
> view the future through glasses so dark that I am subjectively convinced that
> neither the one nor the other possibility will eventuate, and I torture myself
> with these specters and monsters of my fancy. . . . my feelings are in part
> most irrational, and I could not even defend them myself, if I were not sure
> that what is troubling me is a longing for activity and life and change, rather
> than longing for a silly title. (J. Hertz 1977, pp. 197 and 199)

Rumors about the Karlsruhe position abounded. In the meantime efforts were
being made to keep Hertz at Kiel by securing his entitlement to a professorship
that would not be available until after the following April. During a trip to
Berlin in late December, Hertz met Kayser, Helmholtz, Kirchhoff, and above
all the powerful Prussian Minister of Culture Althoff, who all evidently
pressed him to think seriously about Karlsruhe, which he visited on December
28. He did not at all like what he saw that first day; by the next, after seeing
the laboratory, he had changed his mind, and that evening he accepted. "On
the whole I am doing little besides longing to be away from here [Kiel]," he
wrote home on February 7.

Hertz took up his new position in March. He played around fitfully with various devices and prepared his course over the next months, but the rest of 1885 was a miserable year for him. According to Johanna Hertz, he went through a period of extreme "mental conflict" during these months, which, given his earlier behavior, probably means that he was powerfully depressed. There is evidence that Hertz had an unsuccessful infatuation or love affair during this period. That, combined with the lack of clear progress in his work, would certainly have been enough to drive someone so highly strung into deep despair. The year ended badly. In late December he "struggled hard with ill-humour and lack of hope." On New Year's eve he wrote that he was happy the year was over, and hoped that it would "not be followed by another like it."

Hertz's recovery from depression was probably furthered in no small measure by Elisabeth Doll, the daughter of Max Doll, lecturer in geometry, with whom he became involved sometime during the winter. He was hard at work on building a battery for "future experiments" in February, and he took on a rather uninspiring new laboratory apprentice to replace his first one, Martin, who had left ("First performance: he smashes the glass plate of the large electric friction generator" [J. Hertz 1977, p. 213]). On April 12, while on an excursion to Bühl and Windeck Castle, Hertz proposed marriage and was accepted. They were married the following July 31, and by the next fall Hertz's energies and enthusiasm were at high pitch. Elisabeth worked with him in the laboratory at least occasionally, though no record apparently remains of what she did there. Later, we know, she helped him produce diagrams, and it may very well be that she was involved in his early discoveries, perhaps as a witness to subtle effects.

14.2. The Invention of the *Nebenkreis*

Among the devices that Hertz found for his lectures in Karlsruhe were a pair of Riess, or Knochenhauer, spirals. According to his (much later) recollections Hertz first produced what he thought to be a novel electrodynamic effect while playing with these objects. In the introduction to his *Electric Waves* Hertz reminisced:

> In the collection of physical instruments at the Technical High School at Karlsruhe (where these researches were carried out), I had found and used for lecture purposes a pair of so-called Riess or Knochenhauer spirals [fig. 56]. I had been surprised to find that it was not necessary to discharge large batteries through one of these spirals in order to obtain sparks in the other; that small Leyden jars amply sufficed for this purpose, and that even the discharge of a small induction-coil would do, provided it had to spring across a spark-gap. *In altering the conditions I came upon the phenomenon of side-sparks which formed the starting-point of the following research.* (Hertz [1893] 1962, p. 3: emphasis added)

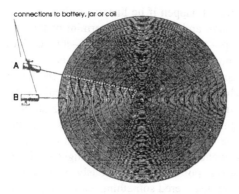

connections to battery, jar or coil

A

B

FIGURE 56   The Riess, or Knochenhauer,
spiral (G. H. Wiedemann 1885)

"In altering the conditions . . ." This phrase holds the clue to the origins of the rapidly evolving laboratory research that Hertz pursued for three years beginning in the spring of 1886. Our goal in this chapter will be to uncover the beginnings of that sequence, with its original signification for Hertz.[1]

Begin with the Riess spirals. Hertz used these devices to demonstrate electromagnetic induction to his students. They were a pair of spirally wound conductors, each of which terminated in metal balls. Placing one spiral directly above the other produces, in effect, an induction coil: the upper spiral, say, would act as the primary element, in which a current would be created and interrupted, while the second, lower spiral would then act as the secondary element, in which a current would be induced by the changing current in the primary (upper) element. This had the advantage for purposes of demonstration that the induced current was made visible by sparking across the metal balls. Hertz began to play with the devices; he stimulated them with other, less-powerful sources than interrupted batteries, at first "small Leyden jars." These, much to his initial surprise (given the comparative weakness of the current they can directly produce), also worked very well: sparks appeared in the lower spiral when the Leyden jars were discharged across the upper one. He went further and tried an even more unlikely candidate as a current source: he hooked the secondary element of a small induction coil across the termini of the upper spiral. In playing, Hertz found that the upper spiral could make the lower one spark even when driven by a small coil, provided that a spark always jumped between the coil's termini (which the design of the Reiss spiral made easy: since it terminated in a pair of knobs, the stimulating coil could produce the necessary spark across them). This effect—the spark's appearance when a small coil's termini discharged to one another, but its absence when this discharge did not take place—puzzled Hertz. He pursued it.

The upper spiral now underwent a topological transformation as Hertz molded it into a different form in order to probe the effect. In a critical move he evidently decided to see what would happen if he linearized the spiral by unwinding it, breaking it at the center, and attaching the termini of the coil there, bringing the critical spark gap—the focus of his interest—directly into the device's middle (fig. 57). This unusual device—an open linear circuit with a central spark gap—remained capable of producing induction in the lower spiral. Hertz began to probe his laboratory novelty, looking in particular for a way to explain the oddly stimulating ability of the spark gap. In its absence induction did not occur, which might mean that under those circumstances the rate of change of the current in the device was simply not great enough to effect sparking through electromagnetic induction. The gap might (and this would certainly have been the simplest possibility) enormously increase that rate, in which case Hertz would have discovered something about the properties of spark gaps under these circumstances, which he would certainly have considered to be worth pursuing to publication.[2] But how was he to elicit a telling response from the device, one that would bear upon this issue?

Hertz's published article begins its discussion of the experiments at this point, with a convoluted explanation of how he eliminated the several alternative sources for the gap's power—convoluted because Hertz did not completely efface the traces of his earliest thoughts even after the device had evolved for him into a stable probe of what he thought, and certainly hoped, to be unexplored electrodynamic terrain. The immediate question concerned what takes place along the discharging wire as a result of the presence in it of the spark gap. Whatever the precise cause of the gap's inducing power, it had of course to work via electromotive force. To examine the electromotive force required a device that could reveal differences in potential between close points along the wire (e.g., between C and D in fig. 58)[3] Such an object did exist: the Riess spark-micrometer. This was, in its simplest form, a pair of metal termini (1 and 2 in the figure) separated by an adjustable air gap. One could probe the potential between two points by connecting them respectively to these knobs and closing down the gap until a spark just occurs across it. The sizes of the gap for different positions of the connecting wires along the circuit being examined could then serve as measures of the relative potential differences.

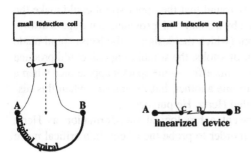

FIGURE 57   Transformation of the spiral into a linear device

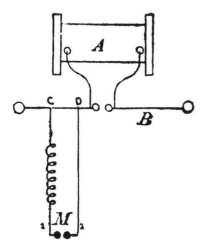

FIGURE 58   Probing the potential along the discharge circuit with a Riess micrometer (after Hertz 1887a)

Hertz evidently discovered that sparking occurs so intensely no matter where along the discharge wire the probe's connections are placed that he could not, as he had hoped, use the gap size to measure the potential differences. The micrometer was, as things stood, useless. One way to modulate the sparking would be to bridge the gap with a low-resistance shunt, forcing most of the potential drop across the Riess circuit between 1 and 2 to occur across the shunt (fig. 59). "It might be expected," Hertz wrote of this configuration, "that, if the metallic shunt were only made short and of low resistance, the sparks in the micrometer would then disappear." But they did not: "Even when the two knobs of the micrometer are connected by a few centimetres of thick copper wire sparks can still be observed, although they are exceedingly short" (1887a, p. 30).

Hertz wanted to make these micrometer sparks go away; only then could he begin to examine the properties of the gap in the primary circuit. But he could not make his instrument work. The Riess micrometer refused to stop sparking, no matter how low Hertz made the resistance between its knobs. This must at first have been intensely frustrating; unwanted and unavoidable sparking blocked his attempt to probe the primary gap's novelties. At some point Hertz decided that he was not going to be able to make the sparking go away, and he started looking around for an explanation. He learned that a similar effect was not only "already known" but had been turned to technical use as a lightning protector for telegraph wires: if they are bridged by spark gaps, lightning surges harmlessly across the gap rather than damagingly along the wires themselves.[4] But why could he not make the effect disappear?

Hertz does not tell us, but one likely possibility was that the current was so intense—the potential gradient along the discharge circuit being so large—that the resistances which he had available were simply not low enough to decrease the drop along the micrometric circuit sufficiently to obliterate spark-

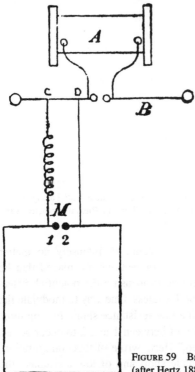

FIGURE 59   Bridging the Riess micrometer with a metallic shunt
(after Hertz 1887a)

ing. There was a simple way to test this possibility: remove one of the two
connecting wires between the micrometer and the discharge circuit. The first
figure in Hertz's printed article concerns precisely this situation: there is no
wire between knob 2 on the micrometer and the discharge circuit (fig. 60).
Despite the absent connection (and therefore the absent *current* through it)
sparks *still* jump vigorously between the micrometer knobs.[5]

Here was a second novelty. Not only did the gap in the discharge circuit have
the ability to make that circuit act inductively even with weak coils stimulating
it, it apparently did not do so in the most likely way, namely, by producing
huge currents (and so huge drops in potential). Hertz pursued what must have
been an unanticipated, and initially very puzzling, discovery, concentrating
now on the singly attached micrometer, which he dubbed the side-circuit, or
Nebenkreis. He described how the side-sparks increased in length with more
powerful sources, and how they depended as well on what terminated the gap
in the discharge circuit. His remarks on this point are particularly compelling
because they seem to convey his energetic, active probing of the best conditions to make the *Nebenkreis* perform its magic:

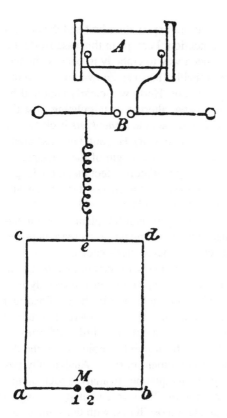

FIGURE 60   Hertz's diagram for demonstrating that resistance does not account for the device's behavior (Hertz 1887a)

> If [the sparking in the discharge circuit] takes place between two points, or between a point and a plate, it only gives rise to very weak side-sparks; discharges in rarefied gases or through Geissler tubes were found to be equally ineffective. The only kind of spark that proved satisfactory was that between two knobs (spheres), and this must neither be too long nor too short. If it is shorter than half a centimetre the side-sparks are weak, and if it is longer than 1 1/2 cm. they disappear entirely. (Hertz 1887a, p. 32)

Hertz became quite specific: the best things to use were "sparks three-quarters of a centimetre long between two brass knobs of 3 cm. diameter." But the effect was extremely capricious: "the most insignificant details, often without any apparent connection, resulted in useless sparks instead of active ones" (i.e., instead of ones that would stimulate the *Nebenkreis*). In the end only experimental craft could make things happen: "After some practice one can judge from the appearance and noise of the sparks whether they are such as are able to excite side-sparks."

Crackling, jagged white sparks stimulate the *Nebenkreis,* but why does it react at all? The published article presents an explanation first and then offers one experiment as evidence for the contention. Perhaps in this case Hertz did first entertain a reason and then construct a way to embody it, because he had certainly already excluded the most probable (because known and common) factor, namely, high current. But perhaps not; Hertz was clearly captured by the *Nebenkreis* at this point; he was probing, altering, manipulating, first the discharge circuit (to stabilize the effect) and then the *Nebenkreis* itself (to see what properties of that device determine its behavior). He found that its "thickness and material" had no important effect "on the length of the side-sparks" that it sustained. But something else did very markedly affect them: the length of the *Nebenkreis*. The longer it is, the longer are the sparks that can be sustained in the micrometric gap. That property, which Hertz would certainly have found (if he did not already suspect it) during his careful, planned manipulation of the device, meant that the cause of the sparking had to increase in power with the length of the wire that connects the micrometer's knobs.

Although we do not, and probably will not, know for certain, Hertz may at this point have begun to think about the entire apparatus in a markedly new way. Circuitry and devices for electrodynamic induction in the confined and controlled space of the laboratory were for the most part considered to produce essentially homogeneous currents, in the sense that it made little difference to their operation that the current might vary from point to point *along* the device's circuitry as well as from moment to moment in *time*.[6] Analytically this is reflected in the use of ordinary differential equations to analyze such things. Even circuits that were deliberately built to oscillate were treated as doing so in the same way that, for example, a pendulum oscillates, with the value of the charge at any point within the circuit being analogous to the position of the pendulum: the change in charge value from point to point in the circuit at a given instant was not important for these kinds of devices, which were small pieces of laboratory equipment.[7]

Hertz's device would in this respect not have seemed to him to be any different from these common ones, which was why he would initially have sought the power of the discharge gap in some effect on the current in the circuit, naturally assuming that the current, and so the electromotive force that drives it, has pretty much the same value throughout the circuit. But the *Nebenkreis* was not behaving in an ordinary way, and its extraordinary behavior was exacerbated, not weakened, by its length. At some point Hertz began to look at the *Nebenkreis*, and the discharge circuit to which it was connected, in a new way. He transmuted the effect's dependence on *length* into a dependence on *time*. The apparatus ceased for him to be substantially like other pieces of electrodynamic equipment (with an interesting, but limited, property associated with the spark gap) and became instead an indicator of something that had always been treated as utterly unimportant: the variation from point to point within a laboratory circuit of its electrodynamic state at a given moment, and in particular the propagation of that inhomogeneous state through the circuit. The *Neb-enkreis* had become a detector of an *electric propagation* that had been engendered by the extraordinary spark gap in the discharge circuit.

Hertz now understood the behavior of the device in the following way. For some unknown and thoroughly unanticipated reason the spark gap in the discharge circuit resists breakdown long enough, but finally gives in swiftly enough, that the potential in that circuit propagates through it in a way that has dramatic effect.[8] When the disturbance reaches the point at which the *Neben-kreis* is attached to the discharge circuit, it begins to propagate through it as well, first reaching knob 1. At that moment there will be a large potential difference between the two knobs (since the effect that alters the potential has not as yet reached knob 2), engendering a spark. The sparking of the *Nebenkreis*, in other words, was due entirely, Hertz now felt, to the finite rate at which the disturbances must propagate through it, and this in turn depended on the surprising and unanticipated behavior of the spark gap in delaying breakdown long enough to build up a great deal of charge, and then breaking down quickly enough to send its effect out as a pulse. The *Nebenkreis* sparking depends on the length of the side-circuit because that length alters the time taken by the disturbance to move through it; it does not depend on the thickness or material of the *Nebenkreis,* because that changes only the resistance of the wire, which has in these circumstances of comparatively small capacitance a negligible effect on the speed, as Hertz remarked.[9]

This point of view suggested an immediate and obvious research goal: to see whether one could control this novel effect, this propagation through the *Nebenkreis.* Hertz could indeed do so by moving around its point of connection to the discharge circuit:

> [I]f we shift the point of contact on the side-circuit, which we have hitherto supposed near one of the micrometer-knobs, farther and farther away from this, the spark-length diminishes, and in a certain position the sparks disappear completely or very nearly so; they become stronger again in proportion as the contact approaches the second micrometer-knob, and in this position attain the same length as in the first. (Hertz 1887a, p. 33)

In manipulating the position of the contact point on the *Nebenkreis,* Hertz believed that he was controlling the locus at which the electric pulse enters it and divides to follow the two paths to the micrometer knobs. If the paths have the same length, both pulses arrive at the same time, equalizing the potential at the knobs and obliterating the sparking. The greater the difference in paths, the stronger the sparking, providing, as it were, a visual indicator of the difference in arrival times.

During these manipulations, however, Hertz discovered that he could make the sparks reappear, after obliterating them by creating equal path lengths, by playing with the micrometer knobs—by touching to one of them another con-

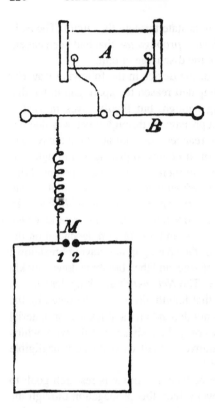

FIGURE 61 Reflections in the *Nebenkreis*
(Herb 1887a)

ductor.[10] This could obviously not affect the initial time of arrival since the conductor lies beyond the knob to which it is attached. Hertz decided instead that it must be affecting subsequent events, due to effects that persist after the initial arrival—in fact, to reflections from the knobs: "The waves do not come to an end after rushing once towards *a* and *b* [fig. 61]; they are reflected and traverse the side-circuit several, perhaps many, times, and so give rise to stationary oscillations in it." No sparking occurs because the reflection from the one knob arrives at the second at the same time as the reflection from the second arrives at the first. Hooking another conductor to a knob obliterates the reflection from it. As a result the reflection from the other knob, passing back around the circuit, will produce sparking. Given his overall interpretation of the device's working, Hertz could now consider that he had *discovered* the ability of a stimulated *Nebenkreis* to sustain electric oscillations for what might be comparatively long periods of time. He had not anticipated this. On the contrary, when Hertz first created the *Nebenkreis* by removing the second connection to the discharge circuit, he was looking only for something that spoke to conditions in the latter; this broadened "when I had discovered the existence of a neutral point in the middle of a side-conductor, and therefore of a clear and orderly phenomenon" (Hertz [1893] 1962, p. 2).[11]

From Hertz's point of view he had managed to fabricate two pieces of equipment: first, a *generator* (the gap-divided discharge circuit) whose electrodynamic power depends directly upon an unforeseen electric pulse; and, second, a *detector* (the *Nebenkreis*) that reveals the pulse's motion. His researches had by this point moved far from their original stimulus, which had been the strange ability of a Riess spiral stimulated by a weak induction coil with a spark gap to induce current in another spiral. Hertz now reached back for a connection to this original impetus by querying the power of the generator.

14.3. How the Discharge-Circuit Became a Spark-Switched Oscillator[12]

Hertz considered the sparking of the *Nebenkreis* to involve a species of "self-induction," in the sense that it resulted from the electromotive effects of changing current in the device. There was, however, an important difference between inductive effects in traditional laboratory apparatus and the ones that Hertz believed he was detecting. In the former the effects are the same throughout the device because the current is treated as effectively homogeneous (i.e., the process was, for instrumental purposes, quasi-static). In Hertz's apparatus, as he now understood it, induction acts by effecting a pulse rather than by producing a uniform, in situ electromotive back-force. Although Hertz was convinced that the behavior of the *Nebenkreis* amply supported this contention (as opposed to a belief that, say, the sparks result from resistance-produced voltage drops), he nevertheless recognized that it was not going to be easy to convince his contemporaries that these comparatively powerful and quite visible effects—sparks—resulted from the weak currents that small induction coils would presumably produce in his device. He admitted that "if we consider that the induction-effects in question are derived from exceedingly weak currents in short, straight conductors, these appears to be good reason to doubt whether these do really account satisfactorily for the sparks" (1887a, p. 36).[13]

To settle *that* question, and not anything wider, Hertz decided to manipulate his apparatus into yet another form, one that would altogether divorce the *Nebenkreis* from the currents that pass through the discharge circuit and that might otherwise be thought somehow to account for the sparking. He removed the last wire connecting the two together. Or, rather, he left a vestigial *Nebenkreis* in place and built additional spark-gap circuits which he did not connect to the discharge circuit. Despite the lack of physical connection, these sparked "when brought sufficiently near" with a side "parallel to the conductor [discharge circuit]," and even (though less vigorously) when brought near the vestigial *Nebenkreis* itself, where, Hertz was certain, he had been able to witness the effects of propagation directly. "Thus it is scarcely possible," Hertz obscurely argued, "that our conception of the phenomenon is erroneous."

Despite Hertz's remark, the new apparatus differs critically from his previous one, and if both work by induction, nevertheless they must work in radically different ways, as Hertz undoubtedly knew. The original, attached *Nebenkreis*, according to Hertz, sparks because the wire that connects it to the discharge circuit sends a disturbance into both of the branches that emerge from the connection. Sparking reflects the different times it takes the disturbance to reach the knobs. But Hertz had now removed the wire. There were no longer any points of divergence to establish path differences. Yet sparking still occurs, and in the detached circuits the only known effect that can produce it (excepting electrostatic influence, which Hertz separately disposed of) was electromagnetic induction. His otherwise obscure remark reflects his implicit contention that the sparking in both cases must be due to a single cause, which in the case of the detached circuit had of course to be induction. But here the induction could not work in the same manner as it did on the attached *Nebenkreis* (i.e., by effecting pulses that diverge from a stimulating point); instead, the entire detached circuit was now acted on as a unit by the discharge apparatus.

What began as an effort to bolster the view that the attached *Nebenkreis* is worked by rapid propagation (induction in effect) accordingly turned into an exploration of the relations between the detached circuit and the discharge circuit "to study the phenomena more closely," to study, in particular, this extraordinarily powerful induction effect "between two simple straight lengths of wire, traversed by only small quantities of electricity." To do so Hertz in effect removed the spark gap from the original, attached *Nebenkreis*, producing the wire *igh* in figure 62, leaving a free end (*h*)—that is, *igh* was a vestigial *Nebenkreis*. Hertz's intention was evidently to take advantage of his recent observation that sparking in a detached *Nebenkreis* occurs when it is placed near its attached brother; he removed the latter's spark gap because it only complicated the phenomenon and was not necessary to witness the induction, which was now to be observed in the detached circuit.

Bringing one side (*cd*) of the latter parallel to *gh* at first produced little sparking. He pulled off "an insulated conductor" from "an electrical" machine and stuck it (*C*) onto *h*, which he knew would pull more electricity through the wire by increasing the capacitance. As he hoped, the effect intensified: "sparks up to two millimetres long appeared in the micrometer." But now a possible disturbing effect occurred to Hertz, one that was to be a constant issue for him in his subsequent experiments with this device and its descendants: the possibility that the sparking he observed in the detached *Nebenkreis* was not due to electrodynamic, but rather to electrostatic, effects; in particular, that it was due here to the charge that the comparatively high capacity terminal *C* accumulates. Since *C* is not at the same distance from the knobs 1 and 2, it

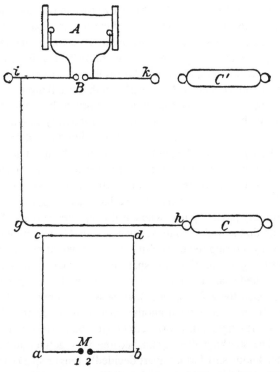

FIGURE 62   Exploring induction effects with a detached *Nebenkreis* (Hertz 1887a)

could induce different charges on them, perhaps enough to produce sparking. Calculation could certainly say nothing about this; but experimental manipulation could. Hertz took *C* off of *h* and reattached it at *g*. The sparking ceased, even though *C* would certainly have been charged just as well at *g* as at *h*. But at *g* it would reflect the disturbance, forbidding the wire *gh* from acting electrodynamically at all. In this way, which set a pattern for his subsequent work, Hertz could make disturbing electrostatic effects disappear by altering the device rather than by trying to calculate something impossibly difficult.

Hertz felt that he could thereby convince readers that the effects were indeed due to electrodynamic induction between "simple straight lengths of wire," but the question remained *why* they were so unusually powerful. The answer, Hertz suspected, "lay in the fact that [the electric disturbance] did not consist of a simple charging-current, but was rather of an oscillatory nature"—of, that is, a regular oscillatory nature, which is to say one that undergoes a reasonably large number of equally spaced alternations before disappearing. It is important to understand that nothing in the design of the discharge circuit, as it now stood, compelled such an assumption. Regular, oscillatory currents were reasonably well known, having been produced and detected by Feddersen in the years between 1857 and 1866, but they arose in devices with large capacitances. Hertz's discharge circuit was at this time terminated by comparatively small knobs, and it was divided by the oddly behaving spark gap, whose resistance during discharge might easily vary wildly.[14] Therefore, why should such a thing produce anything like a *regular* oscillatory discharge?

Hertz probably conceived fairly early on that oscillations of some sort exist in the discharge circuit since the circuit is in principle similar to a discharging Leyden jar, if the spark gap breaks down sufficiently rapidly and lowers the resistance enough. But because Hertz at first thought that any such processes would be highly irregular, he paid no attention to them per se but concentrated instead on the speed of the moving pulse rather than on the accumulated action of possibly several passages back and forth. As his experiments progressed, however, he became convinced that the effect was too powerful to be explained by an irregularly passing pulse, whereas the repeated actions of a regular oscillation would do the trick.[15] Manipulative as ever, Hertz sought to witness these oscillations by altering the device in a way that would reinforce them and that would accordingly exacerbate the spark in the detached *Nebenkreis*.

He first pulled apart the latter's knobs until "sparks only passed singly." That is, he weakened the effect as much as possible without making it disappear altogether. He then connected another insulated conductor (C'), this time to the end k of the original discharge circuit (fig. 62 misleadingly shows C' unattached): "The sparking then again became very active, and on drawing the micrometer-knobs still farther apart decidedly longer sparks than at first [i.e., in the absence of C'] could be obtained." This successful manipulation constitutes the invention of the linear, spark-switched oscillator, a device that Hertz was to use again and again as the instrumental backbone of his subsequent work in electrodynamics. He explained its working:

> This [renewed, powerful sparking] cannot be due to any direct action of the portion of the circuit ik, for this would diminish the effect of the portion gh; it must, therefore, be due to the action of the conductor C' upon the discharge-current of C. Such an action would be incomprehensible if we assumed that the discharge of the conductor C was aperiodic. It becomes, however, intelligible if we assume that the inducing current in gh consists of an electric oscillation which, in the one case, takes place in the circuit C—wire gh—discharger, and in the other in the system C—wire gh, wire ik—C'. It is clear in the first place that the natural oscillations of the latter system would be the more powerful, and in the second place that the position of the spark in it is more suitable for exciting the vibration. (1887a, p. 38)

As Hertz understood it, introducing the second large knob C at the terminus of the discharge circuit itself would exacerbate the inductive sparking by increasing the frequency of the oscillations, supposing them to occur in a significantly regular way at all. The sparking did become much more intense, which Hertz decided would be "incomprehensible" if the discharge were irregular because he was certain that C could have no other action that would produce the same sort of effect.

When Hertz decided that his experiments provided exceedingly strong evidence for regular oscillations, he also began to think about the discharge circuit in a new way, one that provided him with the tools for explaining the odd behavior of the spark gap within it. In the first place, the discharge circuit, modified by the additions of C and C', truly became for him a spark-switched oscillator. It operated in the following way. The induction coil first charged these high-capacitance termini until the electromotive force between them was sufficient to break down the air gap. At that point the discharge circuit became, Hertz believed, independent of the induction coil's circuit, despite the fact that they remained connected.[16] The function of the spark, as Hertz now saw it, was to behave as a fast-acting switch (as noted by Bryant 1988, p. 17), transforming the discharge circuit into a self-subsisting electrodynamic entity.

This was not all. In order for oscillations to occur, very special conditions had to be satisfied during the breakdown process at the air gap, and these conditions could not have been predicted ahead of time. They involved the questions that had first stimulated Hertz's investigations, and he felt that he could now provide an answer based on the special conditions required by the presence of oscillations:

> [I]t is not sufficient that the spark in this circuit should be established in an exceedingly short time, but it must also reduce the resistance of the circuit below a certain value [else the response would be deadbeat], and in order that this may be the case the current-density from the very start must not fall below a certain limit. (1887a, p. 39)

When he went back over his papers for their publication in *Electric Waves*, Hertz added clarifying notes, and among these he included a particular remark concerning the special requirements for the spark gap to work:

> I expect that the action of the induction-coil partly depends upon the fact that directly before the discharge it allows the potential to rise very rapidly. Several accessory phenomena [again unspecified] lead me to believe that when this rapid rise takes place, the difference of potential is forced beyond the point at which sparking occurs when the difference of potential increases slowly; and that for this reason the discharge takes place more suddenly and energetically than when a statical charge is discharged. ([1893] 1962, p. 270)

Hertz emphasized in his introduction to *Electric Waves* just how unexpected the behavior of the spark gap was: it depends *"upon a special and surprising property of the electric spark which could not be foreseen by any theory"* (p. 3; emphasis added).

Hertz now concentrated on his new and powerful device for producing electrodynamic induction, the spark-switched oscillator, and tried "to arrange a modification of the experiment which appeared interesting." This modification

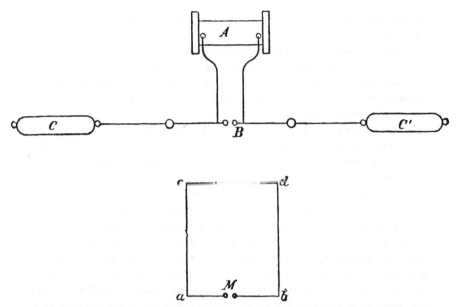

FIGURE 63   The spark-switched oscillator with detached *Nebenkreis* (Hertz 1887a)

was intended to amplify the effect by bringing both arms of the spark-switched oscillator, instead of just one of them, into inducing action (fig. 63). He removed the modified *Nebenkreis, igh,* with its high-capacitance terminus, *C,* and stuck the latter directly onto end *i* of the discharge circuit.

Such a configuration should considerably increase the inducing power and indeed it did. Hertz placed the termini *C* and *C'* 3 m apart and used 2-mm-thick copper wire; the detached *Nebenkreis* was 120 cm by 80 cm. In his first experiments with detached circuits (originally 10–20 cm on a side) he had placed them only 3 cm from the inducer. When using *C'* Hertz had detected power by showing that sparking could occur, at this same distance, over a wider gap. Now he varied for the first time the distance between the inducer and the detector, increasing it to 50 cm, more than sixteen times the distance in his previous experiments, where he had produced sparks 2 mm in length. With this new device the sparks were just as long and, what was more, sparking could be produced at distances up to a meter and a half. Walking between the two pieces of equipment "in no way interfered" with the sparking, which, however, could be stopped by taking away "one or both halves" of the oscillator. The oscillator was powerful, and it was stable, for it could even be used to spark another gapped linear oscillator placed parallel to it.[17] "I believe," Hertz wrote in a sentence undoubtedly designed to catch Helmholtz's attention, "that the mutual action of rectilinear open circuits which plays such an important part in theory is, as a matter of fact, illustrated here for the first time."[18]

14.4. How the Detached *Nebenkreis* Became a Resonator

We have seen that, for Hertz, regular oscillations in the discharge circuit served to explain its powerful inducing action and were evidenced by the addition of capacitance to both termini of the circuit. He had not initially suspected regular oscillations, nor had he immediately turned his attention to them as a centrally interesting effect. "At first," he wrote in the introduction to *Electric Waves*, "I thought the electrical disturbances would be too turbulent and irregular to be of any further use." However, "when I had discovered the existence of a neutral point in the middle of a side-conductor, and therefore of a clear and orderly phenomenon, I felt convinced that the problem of the Berlin Academy was now capable of solution." It is tempting to read this as a statement that Hertz's discovery of null points or nodes in the *Nebenkreis*, which suggested a route to the Berlin problem, was closely bound in his mind to the presence of oscillations in the discharge circuit, which would in turn suggest that a regular oscillatory discharge existed very early for Hertz as a critical, central effect. That, in other words, from the beginning, as later, oscillations constituted the core of Hertz's work. Temptation must, however, be resisted because context suggests a considerably different, and more complicated, situation.

Although the printed article (1887a) does not mention "the problem of the Berlin Academy," we can be quite certain that Hertz would have had it in mind once he discovered the neutral point in the *Nebenkreis*, for such a thing could be disrupted by the presence of other inducing objects, like dielectrics with changing polarization. But it is very important to separate this later discovery, which Hertz emphasized in the introduction to *Electric Waves*, from the entirely distinct question of oscillations in the discharge circuit. His considerations in the early 1880s of the Berlin problem for dielectrics had not involved oscillations at all, and it is likely that his first thoughts here did not either.

The apparatus that Hertz had conceived in the early 1880s sought to detect the electromotive action of changing polarization on the current in the inducing coil itself. The circuit containing the latter was designed so that, in the absence of a coupled dielectric, it would not oscillate at all. The presence of the dielectric, Hertz had found by calculation, superadds a very high frequency oscillation to the otherwise purely damped current in the stimulator, with the result that the oscillations' effect over the lifetime of the inducting current averages to zero and cannot therefore be detected. These were the only oscillations that Hertz had earlier considered, which undoubtedly means that he would not have had the stimulation of oscillations in mind as a way to attack the Berlin problem. Nothing in his early précis for Helmholtz of methods for solving it suggests such things, nor would they likely have been considered because there was nothing about high-frequency oscillations to suggest that they might be useful in producing strong inductive effects.

Hertz certainly did not begin his experiments on the Riess coils expecting that he was on the way to finding something useful for solving the Berlin questions, nor did he necessarily link the latter to discharge-circuit oscillations once he realized that a path to the problem had been opened by the presence of nodes in the *Nebenkreis*. His route was not predetermined either by a set of specific initial goals or by the inherent properties of his apparatus. Every step that we have thus far examined depended entirely on its immediate context, a context of highly localized problems linked to extremely specific bits of equipment. Hertz's goals and his apparatus kept changing together. Discharge-circuit oscillations became important to him as a way to understand the circuit's inducing strength on a detached *Nebenkreis*, which itself had been built in order to establish that the attached-*Nebenkreis* sparking could not be the result of a noninductive process. And that determination had originally been important because it provided a route for probing the odd behavior of the spark gap in the discharge circuit, to wit, its ability to make the latter effect induction when driven even by very weak coils.

Hertz always sought to probe a new effect as far as he could, and the course his *Nebenkreis* experiments took once he became convinced that the discharge circuit oscillates is therefore hardly surprising. He brought the oscillations to vivid instrumental life by constructing a new piece of equipment that would work appropriately only if the discharge circuit did in fact oscillate, a piece of equipment that, more importantly, could be used to witness the oscillations indirectly. The device originated when he conceived the notion that the detached *Nebenkreis* itself, which is driven inductively by the discharge circuit, might actually be set like a pendulum into sympathetic vibrations with the discharge circuit, thereby reinforcing through resonance the mutual action of the two circuits. This, more than any other single step, profoundly directed the course of Hertz's subsequent investigations. He described his idea in the following words:

> [I]t seemed to me that the existence of such oscillations [in the discharge circuit] might be proved by showing, if possible, resonant[19] relations between the mutually reacting circuits. According to the principle of resonance, a regularly alternating current must (other things being similar) act with much stronger inductive effect upon a circuit having the same period of oscillation than upon one of only slightly different period. If, therefore, we allow two circuits, which may be assumed to have approximately the same period of vibration, to react on one another, and if we vary continuously the capacity or coefficient of self-induction of one of them, the resonance should show that for certain values of these quantities the induction is perceptibly stronger than for neighbouring values on either side. (1887a, p. 42)

Context is particularly critical here. In his introduction to *Electric Waves* Hertz had remarked that he first thought to use the apparatus for the Berlin questions when he had discovered "the existence of a neutral point in the middle of a side-conductor," in fact in the middle of an attached *Nebenkreis*. We will see below that the experiment he eventually performed did use the apparatus to establish inductive equilibrium in a detached *Nebenkreis,* an equilibrium that was upset by the approach of other conductors or—and this was its main point—of dielectrics. However, he need not at first have been altogether certain that the primary discharge circuit *oscillated* in order to conceive the null-balance device, because the latter depended only upon the observed existence of a null point in the *Nebenkreis.* Hertz may have been perpetually on the lookout for ways to attack the Berlin questions, but there is little evidence either in the introduction or in his earlier, unpublished manuscripts for Helmholtz (SM 245) that he had ever thought that oscillations in an inducing circuit would be useful for doing so. He did design equipment that made use of these sorts of oscillations, but his initial conception of the experiment had not perhaps considered them at all.

Once, however, Hertz had decided that the discharge circuit must be a spark-switched oscillator, which took place after he had observed a node in the *Nebenkreis,* and therefore after his first thought that he had a way to attack the Berlin questions, *then, and only then,* his understanding of the detached *Nebenkreis* also changed radically. On first observing a node, he had considered the attached *Nebenkreis* to be stimulated rather like a bell struck by a hammer. "We must . . . imagine the abrupt variation which arrives at [the center point of the attached *Nebenkreis* from the discharge circuit] as creating in the side-circuit the oscillations which are natural to it, *much as the blow of a hammer produces in an elastic rod its natural vibrations*" (1887a, p. 34; emphasis added). These "natural vibrations" only modified the spark-producing effect, which arose primarily from the initial division of the incoming pulse and its first arrival at the knobs. Detaching the *Nebenkreis* did not at once imply that this earlier image, according to which standing waves are stimulated by shock, had become entirely inappropriate. However, the effect must at least now arise from the inductive action of the pulse that traverses the discharge circuit instead of from its divided physical presence in the *Nebenkreis,* which meant that what had been a secondary effect in the attached *Nebenkreis* (the stimulation of its natural vibrations) now becomes the central feature.

According to this way of thinking, these newly central oscillations of the *Nebenkreis* might in fact be the only regular part of what might very well be an otherwise highly irregular process (the production of pulses in the discharge circuit). That is, the discharge circuit proper did not have to involve persistent oscillations. But if the discharge process were itself fairly regular, as Hertz was now convinced, then the detached *Nebenkreis* must be thought of as akin to a harmonic oscillator rather than a struck bell. It has, in other words, become a *resonator.*

We may recapitulate Hertz's route to the resonator in four major steps. First, and before he thought much about oscillations, he found a node in the attached

Nebenkreis, which surprised him because he thought the discharge process would have been too irregular to produce such things. This alone suggested a route to the Berlin problem, independently of other factors. Second, he conceived in consequence that the attached *Nebenkreis* rings like a bell struck by a hammer under the impact of the impulses sent along the wire to it from the discharge circuit. Third, he decided that the discharge circuit oscillates in a regular fashion. Fourth, he conceived that the *Nebenkreis,* which he had first detached in order to substantiate the inductive origin of the sparks in it, must in consequence be thought of like a driven pendulum rather than a struck bell. As such it had become an altogether novel form of electrodynamic apparatus, one that could in particular be used to witness the newly conceived oscillations in the discharge circuit by manipulating the circuits until they came into synchrony, whereupon the sparking power of the resonator would be at its strongest. This complex route, which was not predetermined in any simple way either by theory or by a well-laid plan, produced the resonator, a tool for investigating oscillations.

Hertz produced synchrony in the simplest possible way. He scarcely calculated anything (except loosely to ensure that the secondary circuit would oscillate a bit more slowly than the discharge circuit). He manipulated the inducing oscillator, or the resonator, to change its period: he added conductors to the discharge poles (changing the capacity) and altered the distances between the poles of the discharge circuit (changing its coefficient of self-induction):

> The easiest way of adjusting the capacity of the secondary circuit was by hanging over its two ends two parallel bits of wire and altering the length of these and their distance apart. By careful adjustment the sparking distance was increased to 3 mm., after which it diminished, not only when the wires were lengthened, but also when they were shortened. That an increase of the capacity should diminish the spark-length appeared only natural;[20] but that it should have the effect of increasing it can scarcely be explained except by the principle of resonance. (1887a, p. 43)

Hertz varied these manipulations, now exacerbating, now weakening, the sparking in ways that corresponded to increases or decreases in the resonance. In another series of experiments Hertz altered the self-induction of the resonator by changing the lengths of the side-wires; here he estimated the spark lengths and produced a pair of graphs to show the marked effect of resonance.

Given Hertz's habitual practice of pursuing a new effect as fully as possible, one might suspect that he would not have stopped with experiments that witnessed resonance and therefore that provided an instrumental indicator of oscillations in the discharge circuit. We have come to expect him to go on to employ such a device in a novel way, which he did: he shifted his attention away from the discharge circuit and used resonance to manipulate harmonics in the resonator by detecting *and changing* the nodal points in it. The rectangle,

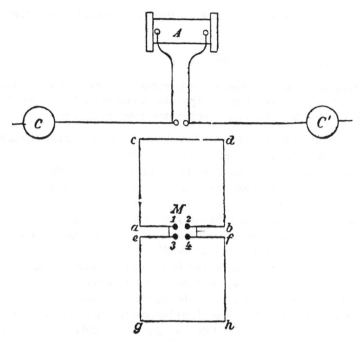

FIGURE 64   Manipulating harmonics (Hertz 1887a)

for example, must have a node at the midway point between the knobs. Reasoning that the resonance would not be affected by touching the wire near a node, Hertz found the sparking to be unaffected near the central point but to be lessened by touching "at the side-branches." Nodes could be detected. Could they also be manipulated?

Hertz decided, in his customary manner, to alter the device in a way that would give him the ability to manipulate these new effects: he added a second resonator connected to the first (fig. 64) at their corresponding knobs (*1* and *3*, *2* and *4*). This system should have two nodes, one between *c* and *d* in the upper circuit, the other between *g* and *h* in the lower circuit. Moreover, the additional circuit should not in any substantial way change the sparking between *1* and *2* that takes place in its absence. The nodes were certainly there, as always in comparative measure, but the *1–2* sparking dropped dramatically from 3 mm to 1 mm. This seems not to have concerned Hertz at all, and he did not pursue the point any further. Instead, he decided to cut the connection between knobs *2* and *4*. This, he remarked, should still leave the two nodes where they were before, but in addition it makes possible a deeper vibration "with a single node between *a* and *e*," which implies that sparking should occur to some degree between knobs *2* and *4*. Indeed, "we find that there is feeble sparking between them."[21] Hertz had thereby created a *resonator* in which he could detect two harmonics.

14.5. THEORY ADAPTED TO APPARATUS

The results already described were only obtained by careful attention to insignificant details; and so it appeared probable that the answers to further questions would turn out to be more or less ambiguous. (Hertz 1887a, p. 49)

Throughout all of the many complex changes that we have examined, Hertz made very few calculations and used only the most elementary (in contemporary context) theory. Indeed, it would be extremely misleading to say that Hertz's creation of the spark-switched oscillator and the resonator involved a theory in any important sense at all. Hertz did not discover the efficacy of the discharge circuit from theory; he was surprised by it. He did not initially suspect that the discharge circuit oscillated; experimental manipulation convinced him that it did. He did not produce the *Nebenkreis* for deep theoretical purposes; he built it to examine the puzzling inductive efficacy of the discharge circuit. And he did not detach the *Nebenkreis* to examine oscillations; he built it to provide powerful evidence for the inductive origin of the discharge circuit's action. The detached *Nebenkreis* became for him a resonator when he wanted to convince himself that the discharge circuit did oscillate. Afterward it became an interesting piece of apparatus in itself. At nearly every moment Hertz thought and worked in an experimental, not a theoretical, space, and the devices that he built and mutated emerged from practice informed by broad, commonly held conceptions (such as, that changing the capacitance or the inductance of an oscillating circuit alters its resonant frequency). Precise (or, rather, quantitative) considerations were at best irrelevant and at worst pitfalls. They did not provide much if any insight into these novel devices, and the measures they did give were not reliable.

Hertz did not want to recur to theory, in the sense of prior computation, at all. "It is highly desirable," he wrote, "that quantitative data respecting the oscillations should be obtained by experiment." But "as there is at present no obvious way of doing this, we are obliged to have recourse to theory, in order to obtain at any rate some indication of the data." He accordingly tacked on a theoretical end-section to his published report, one that caused him considerable embarrassment a few years later when Poincaré pointed out that it contained an elementary mistake (Poincaré 1890, chap. 8). Hertz wanted to provide some measure for the frequency of the spark-switched oscillator, and to do so he used the apparently unproblematic formula $\pi\sqrt{(PC/A)}$, where P, C, and A are respectively the oscillator's coefficient of self-induction, its capacitance, and the electromagnetic constant. To calculate P Hertz ignored the terminating spheres and considered the oscillator to be a straight wire. Including for generality Helmholtz's disposable constant k, he obtained (without providing details)[22]

$$P = 2L \, [\log \text{nat} \, (4L/d) - 0.75 + (1 - k)/2]$$

where L and d are respectively the length and diameter of the (cylindrical wire). The value of k (assuming it to correspond to one of the three alternatives Helmholtz had considered) did not affect P very much, which Hertz calculated to be about 1902 for his wire length of 150 cm.

To calculate the capacitance Hertz simply took the capacity of either sphere considered in isolation, that is, its radius. He erred at this point, and he would at once have corrected the mistake had anyone pointed it out to him at the time. The circuit's capacitance is equal to the charge on either sphere divided by the voltage difference between them. If the voltage difference is 1, then with respect to space at zero potential the one sphere has a potential $+1/2$, whereas the other has a potential $-1/2$. The charges on the spheres are then $+r/2$, $-r/2$, where r is the radius, and so the circuit's capacitance must itself be $r/2$, and not r as Hertz had it.[23] This had repercussions for calculations in his subsequent experiments. The error lowered the frequency from 177 MHz to 126 MHz, but all that Hertz was concerned with at this time was "the order of magnitude," which was a hundred times higher than any oscillations that had previously been fabricated. Hertz calculated as well an upper limit for the resistance that would permit oscillations, concluding from it that the damping would probably limit their number to tens, whereas in the resonator there would be "thousands of oscillations."

By the end of November 1886, Hertz had developed a novel device consisting of two unorthodox components, and he had learned how to make it work. On December 5 he wrote Helmholtz a letter describing his production. "I have succeeded in demonstrating quite visibly the induction effect of one open rectilinear current on another rectilinear current, and I may hope that the way I have now found will in time enable me to solve one or another of the questions connected with this phenomenon" (J. Hertz 1977, p. 215). Hertz emphasized the completely unexpected production of extremely high frequency oscillations in the discharge circuit and his creation of resonance between it and the *Nebenkreis*. Hertz was telling Helmholtz that he had at last succeeded in the very area that Helmholtz had long wanted him to pursue: he had produced a new, unorthodox piece of equipment with which he could create and manipulate processes that not only had never (he thought) previously been fabricated but that could be used to answer questions that were of burning importance for Helmholtz himself. That letter marks the day that Hertz felt he had gained appropriate entry to the Berlin-centered physics community.

FIFTEEN

How the Resonator Became an Electric Probe

15.1. THE STIMULATING EFFECT OF A DEVICE THAT DOES NOT WORK

According to entries in his diary, Hertz had begun probing the unusual effect of the spark-mediated discharge on or about October 4, 1886. Three weeks later (October 25) he began to use the Riess micrometer, introducing the first "short metal loop" and thereby creating the *Nebenkreis,* the next day (October 26). The apparatus mutated rapidly thereafter, with "induction between two open circuits" stabilizing on November 13. During the next two weeks the *Nebenkreis* changed into a resonator, until on December 2 Hertz "succeeded in producing resonance phenomena between two electric oscillations" (J. Hertz 1977, p. 215). The effect became more stable—stronger—during the next day, and on December 3 he found "wave nodes" as well as a "peculiar effect on sparks." On December 4 he felt secure enough to work the device for a colleague, Dr. Schleiermacher, with sufficiently good results that the next day he wrote Helmholtz.

Early in December, Hertz, ever attuned to novel effects, had noted certain odd sparking behavior that, he felt, depended on whether or not the resonator's spark gap could see the gap in the the oscillator. Evidently, though, he did not at once follow this (disruptive action) up but tried first to use the apparatus as a tool for the purpose of answering the Berlin questions that he had long ago decided could not be answered on the basis of then-existing devices. But he now had a new device, and he apparently tried to use it to elicit dielectric effects. There is, however, an apparent contradiction here between entries in his diary and remarks in his introduction to *Electric Waves.*[1]

The diary records that on December 8 Hertz "poured a paraffin ring for dielectric experiments" (otherwise unspecified), which he performed "without success" the next day. On December 10 he "conducted experiments on the induction effect of dielectric motion, apparently without success." He tried a "sulfur rod" on December 11 with "no success." He did more work on the twentieth and gave up altogether on the twenty-first. Hertz did not explain in his diary just what he was trying to do at that time or how. However, in the *Electric Waves* introduction (p. 7) he remarked that he tried to attack the second Berlin question, which was to polarize dielectrics using electromagnetic induction, by means of "closed rings of paraffin" *after* he had finally succeeded in detecting a dielectric's electromagnetic actions (the first Berlin question), which took place in November of 1887. "But while I was at work," he continued, "it struck me that the centre of interest in the new theory [Maxwell's theory] did not lie in the consequences of the first two hypotheses" but rather entirely in the third, which realization (he asserted) accordingly led him to the propagation experiments (see chap. 16 for a discussion of these experiments).

If Hertz had not thought to pursue the second Berlin question until after succeeding with the first, then why did he record in the diary that he was making a paraffin ring in early December of 1886? It seems likely, in fact, that he may actually have attacked both Berlin questions in December of 1886.[2] In the absence of further documentation it is difficult to be certain concerning Hertz's apparently inconsistent dating for these experiments, but it seems probable that (in the public introduction) he was trying to make the course of his work seem to be more straightforward than it had in fact been. In retrospect, given his eventual success, it looked better—more logical—to have tried to use the device, albeit unsuccessfully on first design, to seek out the action of dielectric currents and then, having eventually succeeded there, to have continued directly with trials on polarizing dielectrics, which he stopped when he decided that a finite rate of propagation would answer both questions.

Having tried electromagnetically to polarize dielectrics without success on December 9, Hertz may have turned as early as December 10 to electromagnetic induction *by*, instead of *on*, dielectrics, if we can judge by his reference to the "induction effect of dielectric motion." In which case the experiment that Hertz described in the introduction to *Electric Waves* would in fact have been done in December of 1886 rather than in November of 1887 (as he there asserted). One of Hertz's earliest experiments motivated by Riess spirals could be modified very easily for this purpose: his use of the detached *Nebenkreis* to pick up induction effects from a wire connected to the body of the oscillator (or, to be precise, to the vestiges of the original, attached *Nebenkreis*). In that device (fig. 65) the detached *Nebenkreis* was brought near the wire *igh*, whose end *h* was not (of course) connected to the end *k* of the oscillator. The point was to see if this open end could effect sparks. Now we saw earlier (chap. 6) that, in the experiments which he designed for Helmholtz but which he never carried out, Hertz had proposed to manipulate dielectrics precisely as though they were current elements. He had here a device that was superbly crafted to pick out current-element (or "open-circuit") actions. To use it with a dielectric was (it certainly seemed to him at the time) as simple as could be: just attach a pair of plates to *k* and to *h* and put a dielectric slab between them (fig. 66). The primary device will continue to oscillate very rapidly (since its conducting path is still blocked, albeit now by a dielectric) and in so doing will drive polarization currents in the interposed slab (since the large plates at the primary's ends bear alternating charges). The resonator, brought near the dielectric, should react to the dielectric just as though it were an element of a current-

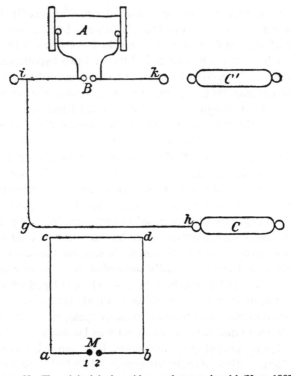

FIGURE 65   The original device with a gap between *k* and *h* (Hertz 1887a)

bearing wire, supposing that the dielectric does work electromagnetically. Hertz described the experiment and its results this way:

> [I] expected that when the block [of paraffin or sulfur] was in place very strong sparks would appear in the secondary, and that when the block was removed there would only be feeble sparks. This latter expectation was based upon the supposition that the electrostatic forces could in no case induce a spark in the almost closed circuit *C* [see Hertz (1893) 1962, p. 41] for since these forces have a potential, it follows that their integral over a nearly closed circuit is vanishingly small. Thus in the absence of the insulator we should only have to consider the inductive effect of the more distant wire *ab*. The experiment was frustrated by the invariable occurrence of strong sparking in the secondary conductor, so that the moderate strengthening or weakening effect which the insulator must exert did not make itself felt. ([1893] 1962, p. 5)

The device had failed to work in the way that Hertz had expected, whether or not the dielectric does act electromagnetically. Something was wrong with its design. But what? In crafting the dielectric experiment Hertz had conceived that, in the absence of anything (conductor or dielectric) between the plates *A*

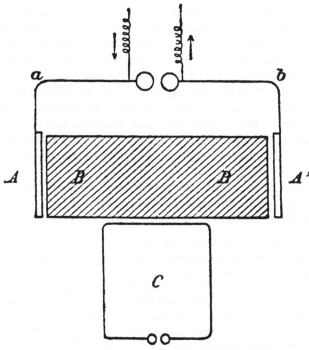

FIGURE 66   The original device modified by a dielectric bridging the gap (Hertz [1893] 1962)

and A', the secondary would spark feebly at best, because there would be no current to act by induction on it, excepting the "distant wire ab." The large electrostatic charges on the plates would not do anything, because, he had tacitly assumed, the secondary is in effect a closed circuit, albeit closed by sparking. And closed conducting paths cannot have currents generated in them by electrostatic forces, he knew, since the latter derive from a potential (and he had another, perhaps more influential reason for thinking this to be so).[3] But sparking took place whether or not a current device was present, and this, Hertz eventually decided (perhaps not until the following autumn), could only mean that electrostatic actions were at work.

> It only gradually became clear to me that the law which I had assumed as the basis of my experiment did not apply here; that on account of the rapidity of motion even forces which possess a potential are able to induce sparks in the nearly closed conductor; and in general, that the greatest care has to be observed in applying here the general ideas and laws which form the basis of the usual electrical theories. These laws all relate to statical or stationary states; whereas here I had truly before me a variable state. ([1893] 1962, p. 5)

It was not that the *experiment* had failed. Not at all. The experiment had never been done. It was not at first obvious to Hertz what the problem could be, precisely because neither he nor anyone else had ever worried about electrostatic effects on closed circuits. They were simply negligible. But they were also the only things that could be working here, because there were only two kinds of electromotive force. Nearly a year later, during the autumn of 1887, as Hertz pondered and probed, he revised his understanding of the nature of the apparatus. Instead of being only the locus of unusually rapid, but otherwise conventionally active, electric processes, it now became the locus of something altogether novel: processes in which static forces, as well as electromagnetic ones, could induce current in closed circuits. The enormous speed of the processes vitiated conclusions drawn from apparatus that deployed incomparably slower ones.

During the course of his work with the device in early December of 1886 Hertz had observed a peculiar effect on the spark length of the resonator, which he eventually described in the following words:

> The [resonator] spark was not very luminous; in the experiments its maximum length had to be accurately measured. I occasionally enclosed the [resonator] spark in the dark case so as more easily to make the observations; and in so doing I observed that the maximum spark-length became decidedly smaller inside the case than it was before. On removing in succession the various parts of the case, it was seen that the only portion of it which exercised this prejudicial effect was that which screened the [resonator] spark from the [oscillator] spark. (Hertz 1887b, p. 63)

Hertz decided after typically careful probing that the effect was due to the emission of ultraviolet light by the oscillator spark. He did not go beyond this conclusion, in part, certainly, because the discovery was for him a by-product of other researches, and also because he felt that the conditions determining the effect "are very complicated" and should be studied without an induction coil (presumably using batteries to avoid the induction coil's complicating oscillations).

Hertz's production of what was later called the photoelectric effect nevertheless gave him for the first time a definite claim on novelty. He wrote his parents on May 1 that he was "keenly at work," that "the lean time is over and that some fruit at least is ripening, even if it is only plums and not pineapples" (J. Hertz 1977, p. 221). He continued experimenting on the effect until early July, but, as he wrote in explanation to his father on July 7, though proud of his discovery he would have been happier had it been "less puzzling" (and therefore more easily integratable into the questions that really gripped him). Hertz did feel strongly that the discovery would be quite important for his reputation. He rushed it as quickly as possible into print in order to maintain his "considerable lead" and not to have it hanging over his head as a perpetual interruption to his work with the oscillator and resonator.[4]

15.2. The Evolution of the Resonator into a Stable Instrument

15.2.1. *How the Resonator Became a Differential Analyzer and a Force Mapper*

During August and early September Heinrich and Elisabeth vacationed in the Herrenalp to escape the summer's heat. On September 7 he was back in the laboratory, and on that day he turned back to electric experiments. On the eighth he "[e]xperimented with a movable induction conductor and quickly came to the heart of the matter" (J. Hertz 1977, p. 229). Two days later (September 10) he could record "success." But five days further on he was greatly troubled by "peculiar disturbances." On the sixteenth he decided to change the experimental venue and moved the "apparatus to the auditorium, where the experiments gave better results." "Better" became "beautiful and complementary" on the seventeenth—so good, in fact, that Hertz began "very eagerly" to sketch out what were probably new apparatus configurations and designs on the eighteenth; the next day he began to play with "the relative positions of the circuits," which are what he probably also sketched that day. On the twenty-first he had progressed far enough to consider bringing in dielectrics in a systematic fashion.

According to the diary, then, it took Hertz almost exactly two weeks to move from a tentative exploration of the electric force in the neighborhood of the oscillator, designed to find appropriate loci for testing dielectric effects, to re-introducing the dielectric proper.[5] In the *Annalen* paper (1888c) written in February and concerning in part results he reached between September 7 and October 10, Hertz described the apparatus and presented a very brief "analysis of the forces acting on the secondary circuit." This was a qualitative, but very powerful, way of determining what forces the resonator was responding to in particular orientations. Following this he described the results of his experiments, or, better put, he described how the resonator responded in various positions and then provided for each case a brief explanation, in terms of his qualitative force analysis, of why this was so. Significantly, Hertz almost apologized for providing the analysis, but not because it was qualitative rather than quantitative. He apologized for providing it at all, offering it solely "for the sake of convenience," insisting "that the facts here communicated are true independently of the theory, and the theory here developed depends for its support more upon the facts than upon the explanations which accompany it" (Hertz 1888c, p. 81).

Hertz's statement probably reflects quite directly *how* he had actually gone about his work. He had literally *discovered* through manipulation how to understand the resonator's behavior; he had not first generated a quantitative analysis and then confirmed it. On the contrary, the analysis was the result, not the forerunner, of the experiments that he undertook.[6]

The apparatus consisted as before of two devices: the (primary) oscillator and the (secondary) resonator. Hertz described it quite carefully:

> The primary conductor consisted of a straight copper wire 5 mm. in diameter, to the ends of which were attached spheres 30 cm. in diameter made of sheet-zinc. The centres of these latter were 1 metre apart. The wire was interrupted in the middle by a spark-gap 3/4 cm. long; in this oscillations were excited by means of the most powerful discharges which could be obtained from a large induction-coil. The direction of the wire was horizontal, and the experiments were carried out only in the neighbourhood of the horiontal plane passing through the wire. This, however, in no way restricts the general nature of the experiments, for the results must be the same [on account of symmetry] in any meridional plane through the wire. (Hertz 1888c, p. 81)

The oscillator, then, was not substantially different from what it had been nine months before. But the secondary was different: he now used a circular resonator:

> The secondary circuit, made of wire 2 mm. thick, had the form of a circle of 35 cm. radius which was closed with the exception of a short spark-gap (adjustable by means of a micrometer-screw). (Hertz 1888c, p. 81)

He changed the secondary's form because he very rapidly discovered (on September 7, in fact) that the sparking depends critically upon the position of the spark gap in the secondary's own plane: rotating the secondary about an axis normal to it and through its center generally produces tremendous changes in sparking. To analyze the effect required eliminating all asymmetric actions due to changes in distance, which required a circular form:

> Now the choice of the circular form made it easily possible to bring the spark-gap to any desired position. This was most conveniently done by mounting the circle so that it could be rotated about an axis passing through its centre, and perpendicular to its plane. This axis was mounted upon various wooden stands in whatever way proved from time to time most convenient for the experiments. (Hertz 1888c, p. 82)

Hertz's new design made it very simple to explore the sparking around the oscillator. The axis mounted through the center of the resonator could be set vertically or horizontally. For simplicity call P the plane defined by the oscillator and the center of the resonator, and suppose that this plane lies in the horizontal. Then with its axis in the vertical the resonator lies in P, and with its axis in the horizontal the plane of the resonator (call this R) is perpendicular to P. The resonator could be rotated about the axis, thereby altering the position of its own spark gap in R. In addition, the axis could itself be pointed in any direction when in P (which is always the horizontal plane), so that Hertz could examine how the sparking changes when, with R normal to P (i.e., with the plane of the resonator normal to the plane through its center and the oscillator), the position of R changes from, for example, cutting the oscillator normally to paralleling it. Note that the radius of the resonator is over twice the radius of the oscillator's spheres, which meant that portions of it extended quite far above and below P.

Hertz soon discovered how sensitive the device could be to the position of the resonator's spark gap (call this S) in R. He looked for how the sparking changed as he altered two significant parameters: p_1, the position of S in R, and p_2, the direction in which the axis points when it lies in P. There can be little doubt that the experimental results which Hertz listed in his published paper represent fairly well what he discovered in the laboratory, and not what he had expected ahead of time. It is quite probable that Hertz's understanding of the forces at play developed along with his reduction of the sparking to rule.

Hertz performed two series of experiments. Both are mentioned in his laboratory notes. In the first (series A), which is the most extensively discussed in entries dated September 7 and October 10, R is perpendicular to P; in the second (B), which is just mentioned on October 10, R parallels P. He achieved the following results, which are described in detail in Hertz 1888c:

Series A: R Normal to P[7]

1. If S (the resonator's spark gap) lies in P, no sparking occurs. Call these loci of no sparking the (two) null positions.
2. If S does not lie in P, sparking does occur, starting from a small effect near the null positions and growing to apparently equal maxima (sparks 2–3 mm long) when S lies vertically above or below P. Call such loci the maximal positions.
3. When S is in one of the maximal positions and the resonator's axis (which lies in P) is turned around, the sparking goes twice to a maximum and twice to very weak minima during a complete turn.

Series B: R Coincident with P[8]

1. When the center of the resonator lies along an extension of the oscillator's own axis O (recall that the resonator's plane contains that axis), sparks disappear when S itself is turned onto the oscillator's axis. Apparently equal maxima (2.5 mm long) occur when S is turned into the normal to O.
2. When the center of the resonator lies along a line through the oscillator's center and inclined at about 30° to O, two unequally long maxima occur; there are null positions, but they no longer lie along a line that is normal to the one which contains the maximal positions.
3. When the center of the resonator is turned through another 30°, to lie at a 60° angle to O, the two null positions appear to merge into a single one, and one of the maxima also vanishes. The remaining maximum produces 4-mm-long sparks "opposite to a very extended tract of extinction."

4. When the center moves even closer to the normal to O, the null positions vanish altogether, leaving a maximum and a nonzero minimum. The spark lengths now range between 1.5 mm and 5.5 mm.

5. When the center lies directly on the normal to O, a 2.5-mm minimum occurs when S lies on the part of O that points away from the oscillator's spark gap, and a 6-mm maximum occurs when S is nearest the gap (i.e., directly below the minimum).

Let's try to envision what Hertz would have done as he observed the sparking. First of all he would have rapidly decided to look for two important positions: those in which the spark lengths are longest (where the effect is the strongest) and those in which they are weakest or vanish altogether. A varying spark length was what the apparatus provided, and it was obtained by turning the micrometer in the resonator until sparking just began (or just stopped). Move the device; point its axis; turn the spark gap around; adjust its width until sparking just starts (or stops); write the width down when it looks like the spark is longest or shortest. These operations (though not always in this order) provided the observations that Hertz distilled into his two compact, carefully chosen series.

By early November Hertz had decided that series A separated actions due to the resonator's openness from electrodynamic induction, which affects the resonator as a closed circuit.[9] This distinction is revealed in the following way. Experiments A.1 yield *diametrically opposite* null positions. In general, Hertz felt, both electrostatic and electrodynamic forces cause the sparking; the former does so only because the resonator is open; the latter can in principle produce effects (albeit different ones) whether the circuit is open or closed. If we have null positions, either the forces cancel one another there or else they both vanish altogether at those points. If the electrodynamic force is the usual one that governs a closed circuit, it cannot depend upon the position of S; but the force that acts because of the resonator's openness certainly can depend upon the position of the spark gap in it. This latter force must (obviously) be periodic in the length of the resonator. Consequently, if the open–dependent force cancels a (presumptive) closed-circuit electrodynamic force at one of the null positions, then it cannot do so at the other one, because it has by then changed sign, whereas the closed-circuit force (by assumption) remains constant. The conclusion was obvious: when the plane of the resonator is perpendicular to P, the electrodynamic force that usually drives a closed circuit vanishes altogether. Here, then, Hertz had found a configuration that could be used to probe *only* the forces that arise because the resonator is open. There was more.

Experiment A.3 provided a method for mapping the direction of this active, open-circuit force, for here there are directions in which the sparking almost completely ceases but in which it can be made to recur by rotating the plane of the resonator. First, place the resonator anywhere, but with S at the high point (i.e., from A.2, at the maximal sparking position), and then turn its plane around until the sparking vanishes, or at least until it nearly does so. If such a position exists, Hertz concluded, then there can be *no* force at all in the resonator's plane, in which case he had here found a direction, specified by the normal to the resonator, that must point *in* the force's direction (since he knew the force was in general not zero because sparking could be effected in other orientations). Hertz could accordingly *map* the total acting force in this way.

The mapping eventually produced, or at least worked with series B to produce, more detailed information about the contributors to the open-circuit force.[10] Hertz felt quite certain that electrostatic action predominates here. But it was certainly not the only action present. The arrows in figure 67 represent force direction.[11] These, Hertz remarked in the published article, are "similar" to the lines that map the electrostatic force between oppositely charged spheres, thereby indicating that, indeed, static action predominates in governing the open-circuit force. However, it was also apparent to Hertz that static force was not the only force present, because the lines are, in his words, "pushed away" from the axis of the oscillation. Were static force acting alone, the lines would crowd much more closely about the axis. This necessarily meant that something causes them to bend away from the axis. Decomposing the force at a point into components parallel and perpendicular to the axis revealed that the bending arises from a nonelectrostatic contribution that lies along the axis but that opposes the (considerably larger) component of the static force in that direction. The only other kind of force is electrodynamic in origin, and so Hertz had found that the open-circuit force has in general two contributions: electrostatic, which follows the usual pattern for such a force between oppositely charged spheres, and electrodynamic, which points from the negative to the positive sphere along the axis. That is, the open-circuit electrodynamic action parallels the oscillator's axis and always opposes the component of the static action in that direction.

Hertz could establish a great deal about the open-circuit force by setting the device in such a way that the closed-circuit electrodynamic force cannot affect it. But what about configurations in which both open and closed actions are effective? How do they combine to affect the sparking? Series B addressed that question. Here the plane of the resonator contains the oscillator. Sparking was examined by first placing the center of the resonator in some position and then turning the spark gap around to find maximal and null (or minimal) positions. Hertz certainly began the series thinking that both open and closed forces must be active here—the latter because the resonator is perpendicular to the position in which the closed force has no effect at all (and because it was obvious in all theories that in this configuration the oscillator, considered as a circuit element, would generate electromotive force around the resonator). Following the convention that Hertz established in his printed article, call the

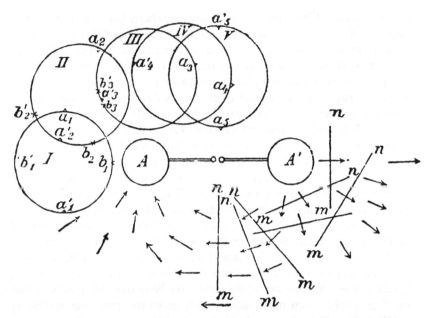

FIGURE 67   Hertz's depiction of the direction of the force near the oscillator (Hertz 1888c)

closed-circuit force α. Let β denote the maximum value of the open-circuit force (due to both static and dynamic causes) along the resonator for a given position of the resonator's center. Then a simple conjecture might be that the sparking due to the open-circuit force varies as βsinθ, where θ denotes the angular position of the spark gap in the resonator with respect to the direction of the electromotive force in the resonator's plane. Then the combined action for a given position of the spark gap and orientation of the resonator's plane will be α + βsinθ. Here, it seems likely, Hertz may have conjectured such a dependence (given the results of series A) and used series B to convince himself that it worked reasonably well. At the same time, series B provided information concerning the comparative magnitudes of α and β in various positions.

Consider, for example, experiment B.2. Here Hertz had found two *unequal* maxima at *diametrically opposite* positions of S, which therefore correspond to the sum of α + β and the difference α − β. Hertz also found two minima much closer to the smaller of the two maxima. This immediately told him that in this position β must be larger than α, which would yield a spark-length curve looking something like figure 68.[12] Here the proximity of the two null positions to the smallest maximum is clearly apparent. Consider next experiment, B.3, in which the null positions merge into one, and one maximum vanishes altogether. Here we must accordingly have near-equality between α and β (since then there obviously will be a single maximum at θ = 90°, and the two null points both occur at θ = 270°). In B.4 the null points vanish altogether, in

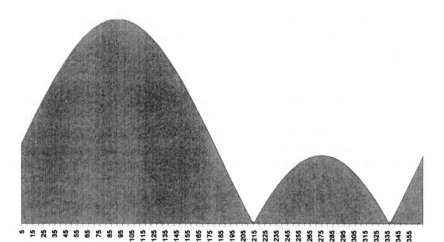

FIGURE 68   Hertz's conjectured force curve

which case the closed-circuit force α must now have grown larger than the open-circuit force β.

When used as in series A the resonator revealed the direction of the total open-circuit force. When used as in series B the resonator distinguished the comparative contributions of the forces to the sparking by responding to their combined action in a way that depended upon its precise setting. Depending on how it was used, it was either a *mapper* of one kind of force or a *differential analyzer* that distinguished between the two kinds of force. Only the most qualitative considerations were necessary for these purposes, and it seems highly likely that Hertz developed the appropriate ones as he experimented. But when he made his results public, he rather reluctantly prefaced them with an attempt to ground them in a brief analysis of the resonator's behavior. This is worth considering because it illustrates just how deeply qualitative Hertz's considerations were *and had to be.*

15.2.2. Why Hertz Did Not Produce a Theory for the Resonator

Hertz did not in any profound sense ever develop a theory for the resonator's behavior. Rather, he early provided an argument-sketch in order to justify his working rule that the spark length can be represented by an expression of the form $\alpha + \beta\sin\theta$. He provided, in other words, a rationale for something that was itself grounded directly in experiment. Hertz did not argue for his basic assumption that the activating cause can always be split into a part that depends upon the circuit's orientation (the closed-circuit term) and parts that depend upon this as well as upon the position of the spark gap in the circuit (the open-circuit terms). Instead, he concentrated solely on explaining why he felt it reasonable to reduce the two possible low-order open-circuit terms to a single one.

Hertz assumed straightaway (1888c, p. 82)—and this was a considerable assumption—that the total activating force Σ at some point of the resonator can be written in the following way:

$$\Sigma = A + B \cos(2\pi s/S) + \ldots + B' \sin(2\pi s/S) + \ldots$$

in which A is the closed-circuit term, s is distance along the resonator from the spark gap, and S is the resonator's circumference. There are, he admitted, higher-order periodic terms, but he felt justified by his laboratory experience in neglecting them, which allowed him to ignore "weak sparks" at inconvenient locations. His goal was to explain how this point-by-point force can build into a spark-producing action at the gap such that the spark lengths α and β can replace the force terms A and B, and why that action can reasonably be reduced to the first two terms of Σ.

Hertz came right to the point. First of all the A term is unproblematic because it is constant: if it is effective then it integrates to an electromotive force across the gap and so translates directly into a proportionate spark length. The sine and cosine terms, however, require some thought, because their action varies from point to point along the resonator. Hertz decided to conceive of the resonator as being broken into four quadrants, with the gap lying at the top, because the *sign* of the force remains the same for each of the two terms respectively throughout a quadrant. Figure 69 depicts Hertz's argument: the directions of the arrows represent, for each of the two terms (sine and cosine), the corresponding directions of the force in a quadrant, and the thickness of the arrows represents the force intensity at the corresponding point. From the diagram we see, as Hertz argued, that the sine term cannot effect an oscillation: the force at every point in the two right-hand quadrants is countered by an equal and opposite force at the corresponding point in the two left-hand quadrants. Since the forces are completely symmetrical about the gap, sparking cannot occur at it. That leaves the cosine term, which does not produce symmetrically opposed forces on either side of the gap. It suffers from the apparent problem that the forces in the quadrants located on either side of the gap are countered point by point by the forces that act in the two lower quadrants. However, the forces in the lower quadrants certainly could effect an oscillation if they were not countered by their opposing numbers in the upper quadrants, if, that is, proximity to the spark gap makes a difference.

Hertz asserted that the forces far from the gap are more effective in producing an oscillation than are the opposing forces near it, because, he argued, the current terminates at the gap's boundaries. This is, as he knew and remarked, a "somewhat brief statement," but he came to it by analogy with the vibration of a string tied at both ends. If, he argued, the forces near the center of the string are opposed to those near its tied ends, then the former will govern its motions if the central forces oscillate in synchrony with the "fundamental note of the string." Here the current corresponds to the string's displacement, and

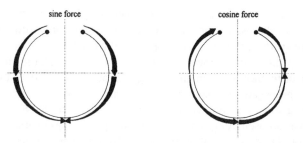

sine force   cosine force

FIGURE 69   Representation of Hertz's account of the forces that act on the resonator

the gap ends (where the current ceases) correspond to the string's termini. The cosine term can therefore excite the "fundamental note" of the resonator as a result of the forces that act on parts *opposite to* the gap. The cosine term furthermore easily leads to Hertz's working rule that the spark length varies as $\alpha + \beta\sin\theta$.[13]

Is this a "theory of the resonator"? Hardly: it is a justification for relations that emerged from experiment and, moreover, has very little behind it, since Hertz in no way attempted to derive the periodic terms from, say, inductive relations at high frequencies. Could he have done so? Both Henri Poincaré, using field relations, and Paul Drude, who avoided fields, provided theories for the resonator in 1894. However, neither analysis is entirely straightforward, and more to the point neither one at all yields the simple relations upon which Hertz relied.

Both Poincaré and Drude (who had read Poincaré's analysis) took as their task the derivation of a differential equation for the current at an arbitrary point in the resonator; both of them inserted a driving term to represent the external force. Hertz, recall, had concentrated his attention on the form of that term, arguing in the first instance that it can be expanded in a sine and a cosine Fourier series and then throwing away the sine terms on the grounds that they cannot produce an oscillation. Poincaré and Drude, by contrast, were not interested in such a simple argument. Indeed, both of them just ignored Hertz's sine–cosine expansion and expanded the applied force in a straightforward Fourier sine series. Each of them, in his own way, then bracketed corresponding terms in the resonator's differential equation. As a result they could not provide anything like Hertz's simple expression for the spark lengths (viz., $\alpha + \beta\sin\theta$). They could corroborate Hertz's claim that the force opposite the gap has the most effect, but they did not neatly separate the actions in the way that he had. This is hardly surprising. By 1894 both Drude and Poincaré were rapidly becoming well versed in field concepts, and it would not have seemed appropriate to either of them to split up the force in Hertz's fashion into parts that are open- and closed-circuit dependent.

There is an even deeper difference that thoroughly distinguishes Hertz's pre-

1888 work from what others later produced. Drude did provide separate expressions for the oscillations of a rectangular resonator (in different orientations) that are due primarily to the electric field and those that are due primarily to the magnetic field. Although they are superficially similar to Hertz's distinctions, the difference between them is significant. Drude assumed that the external force propagates in plane waves and simply took it for granted that, for example, a wave traveling in the z direction in which the E field oscillates along x has an oscillating (and electric force producing) magnetic field in the y direction. Drude's separation of electric from magnetic effects accordingly corresponds to the resonator sitting in such a position that it responds to an E whose integral around it nevertheless vanishes, or else to an E whose circuit integral does not vanish. In either case—and this is the critical difference between Drude and Poincaré, on the one hand, and Hertz in 1887–88, on the other—there is only one cause that produces sparking, a unitary E field. Hertz in 1887 and even in early 1888 was not thinking in these terms. For him the unitary character of electric force was rather a perplexing fact, as we have seen, than a sign of something fundamental about the nature of interactions. The absence of a resonator theory was for Hertz not only an important laboratory convenience but also a profound conceptual necessity.

15.2.3. The Interference of Forces from Different Objects

Hertz's use of the resonator to explore the pattern of electric force around the dipole depended critically on his separation of the actions upon it into separate parts that worked together to effect sparking. These parts might be said to *interfere* with one another, in the loose sense that the final effect depended upon how they combined. There was nothing difficult or problematic about this in Hertz's initial understanding: forces from different sources simply combine to produce a net resultant. The purpose of probing the forces had been to find loci that could be used to reveal electromagnetic induction produced by dielectrics. Hertz's first effort in that direction, we saw above, was simply to stick a dielectric between a pair of conducting plates, in effect using it to replace a piece of wire. The dielectric, if it worked at all, would exert inductive force just like a metal wire placed in its position, and then the resonator could pick up the effect. This design had failed because the device sparked even in the absence of anything at all between the plates. At some point, perhaps quite early in his September work, it struck Hertz that he might succeed if, instead of looking for the action of a single object—the dielectric—he sought instead for the resultant action of two objects (the oscillator and the dielectric). He invented, that is, an investigative technique based on the *interference of forces from pairs of objects*.

The overtones of such a procedure are striking, in that they bring to mind Hertz's 1884 discussion of the unity of force. There, recall, Hertz had considered the interactions between three objects. Two of the objects were in the same state, the third was in a different state, and by assumption the first two interacted with the third. Hertz's aim there was to use his principles to conclude that the first two objects must also interact with one another. If we consider only the topology, as it were, of Hertz's 1884 arrangement, then it is strikingly similar to the one he conceived in 1887, for in both cases a pair of objects a and c interact with a third object b. Of course, in 1884 Hertz was not concerned with the *simultaneous* effects of a and c on b. His 1887 experience with simultaneously acting forces from a single entity, it may be, combined with the earlier image of the triad of interacting objects to suggest a novel arrangement, one that would detect the (hypothetical) action of b (the dielectric) on object c (the resonator) by playing it off against the (unhypothetical) action of a (the oscillator) on c.[14] The apparatus to detect dielectric actions, one may say, incarnated precepts of Helmholtz's electrodynamics.

Hertz would not have struck out in this new direction with a dielectric.[15] That would have made little sense, because the dielectric was to be subject of his investigation, and he had first to learn how to manipulate the device in a way that would with security and stability reveal interference between the forces generated by different objects. To do so he would most likely first have used a conductor instead of a dielectric as the second object because there was no doubt that the conductor, when stimulated by the oscillator, would produce an inductive effect. Hertz had to learn how to detect the interference between the effect and the one produced by the oscillator itself. And, in fact, Hertz's communication (1887c) to Helmholtz does proceed by first establishing the effects of a second conductor and then drawing a relation to effects obtained when using a dielectric instead.

Figure 70 shows the oscillator configuration that Hertz designed.[16] Three characteristics stand out. First, the oscillator now has flat plates instead of spheres at its ends (the plates are each 40 cm to a side and are connected by a 70-cm-long copper wire, 0.5 cm thick and broken by a 0.75-cm spark gap).[17] Second, above it sits another conductor (C) with flat plates but with a large metal fin joining the plates instead of the wire with spark gap that joins the oscillator's plates. Finally, the same resonator that Hertz had used to explore the force distribution now sits with its plane normal to the oscillator plates and with its center on the line that is normal to and bisects the oscillator's spark gap. Why did Hertz change from spherical to flat capacitance loading?

Although Hertz nowhere explained the alteration, it is not hard to understand why he made it in view of the goal he was aiming at. The ultimate purpose was to investigate dielectric electrodynamics; the proximate one was to use a second conductor (here C) to see if the former goal was practicable. Hertz could certainly have replicated the spherically loaded dipole for the second conductor C, but if he had done so, he would not have been able to control the effect very well. Spheres placed near each other have considerable variation in the distances between their parts. Flat plates can, however, be placed parallel to one another. If Hertz had used his previous oscillator and had produced a separate dipole using it as a model, his observations would have depended in a rather complicated way upon the distance between the oscillator and the stimulated dipole. With flat plates distance only weakens the effect; there are no complicated changes due to configuration. His change to flat plates was accordingly for purely practical reasons: to make it easy to examine what happens when a stimulated conductor is moved about near an oscillator.

The resonator, oriented as in the figure, could be rotated around its central axis to bring the spark gap into any position about or below the plane of the oscillator's flat plates. If the oscillator alone works, then the sparking is strongest with the gap at b or b'. Null points occur at a, a'. This much Hertz had already seen in his series A experiments. His immediate aim was to separate static from inductive action. With the resonator in the symmetrical position Hertz already knew that it was responding primarily (though, as we saw, not exclusively) to the oscillator's static action, since the closed-circuit force of induction vanishes. But if the resonator is shifted downward, so that more of it lies below than above the plane of the oscillator, the symmetry breaks; the positions of the null points and the strengths of the maxima change. The "sparking distance increases at the highest point and diminishes at the lowest point . . . [the null points] appear to be rotated downwards through a certain angle on either side" (Hertz 1888d, p. 98). This effect must, Hertz argued, be due primarily to the production of a nonvanishing (closed-circuit) electrodynamic induction, and not to a change in the electrostatic action, because the latter has not altered very much, whereas the former now has a nonzero value because of the broken symmetry. Consequently, shifts in null points and changes in the strengths of the maxima, when produced in this fashion, bring out closed-circuit inductive action. Hertz's initial aim was to see whether he could replicate this shift by keeping the resonator fixed and bringing near a stimulated conductor C.

He succeeded:

> When [C] is lowered towards the primary conductor AA', we observe the following effects:—The spark-length has decreased at the highest point B, and has increased at the lowest point b'; the null-points have moved upwards, i.e. towards the conductor C, whereas there now is noticeable sparking where the null-points originally were. (Hertz 1888d, p. 99)

Hertz knew that he could get the *opposite* effect—a downward shift in the null points—in the absence of C by moving the oscillator upward relative to the resonator. That same effect can be produced, Hertz argued, by introducing "above AA' a second current having the same direction as that in AA'." In C's presence the opposite happens, and so, Hertz concluded, "this effect is naturally explained as being to an inductive action proceeding from C," supposing that the current induced in C is opposite to that in the oscillator. That is, the null points are shifted upward as a result of *interference* between the closed-circuit induction proceeding from the superior conductor C and the open-circuit, predominantly electrostatic action emanating from the oscillator (since the closed-circuit induction from the latter remains zero).

Hertz accordingly thought of C as being akin to a rectilinear resonator: like the resonator, it was driven by the forces of the oscillator. C therefore should show resonance effects that could be altered by changing its electromagnetic characteristics. In the case of the circular resonator Hertz effected changes simply by altering its radius. He could not do that here, but he could instead change the bridge between C's flat plates. He could use wires of different lengths and diameters to bridge the plates; the longer the bridge, and the thinner the wire, the greater C's self-induction and so the lower its frequency. In this way Hertz could bring C's natural frequency into any relation relative to the natural frequency of the oscillator. Since, he supposed, C is governed by the equation for a damped, driven harmonic oscillator, in which the driving force has the frequency of AA', it followed that[18]

1. if the frequency of C is lower than that of AA', C leads AA' in phase;
2. if the frequency of C is greater than that of AA', C lags AA' in phase;
3. if C and AA' have the same frequency, they differ maximally in phase, the difference being 90°; and
4. the amplitude of the oscillation in C decreases with increasing difference in frequency between C and AA'.

The resulting effects require subtle interpretation. Hertz found the following. As the frequency of C was gradually lowered, at first the null points moved "farther and farther upwards, but at the same time became more and more indistinct"; the sparking at the high point, initially much smaller than at the low point, began to grow again "after the disappearance of the zero-points." Then, "at a certain stage the sparks in the highest and lowest positions again became equal, but no null-points could be found anywhere in the circle; in all positions there was vigorous sparking." After that the low-point maximum began to decrease, and null points formed around it, which gradually moved toward the line aa', replicating in reverse order the initial changes.

These effects, Hertz thought, corresponded to moving from a higher frequency in C than in AA', through resonance, and then to a lower frequency. The argument depended upon the sparking in the resonator being seen as the result of interference between the action of C (itself produced by a current induced in C by AA') and the action of AA'. As the frequency of C approaches that of AA' its amplitude grows, as does the phase difference between its oscillation and that of AA'. The effect on the resonator results from the combination of amplitude with phase growth: as C's amplitude increases, its action becomes stronger, which is reflected in an increase in the high-point sparking and in an

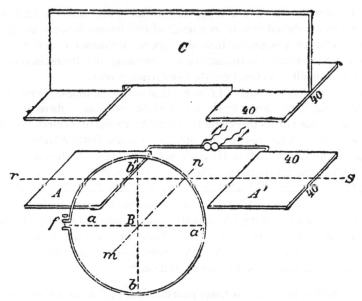

FIGURE 70   Apparatus for detecting the effect of neighboring conductors on sparking (after Hertz 1888d)

upward motion of the null points. However, because the phase difference is approaching 90°, the interference between the actions decreases, which makes the upward-moving null points begin to spark weakly as the forces begin to fail to cancel each other. At resonance, where the phase difference is 90°, the failure is complete, and the null points vanish altogether. At the same time the amplitude of C is at its strongest and so therefore is the sparking everywhere around the circle (since, at resonance, no interference from AA' decreases the effect of the closed-circuit induction from C). Past resonance, the effects reverse.[19]

The diary records that on September 10, two days after he had begun playing with "a movable induction conductor," Hertz had "experimented, and with success." This may refer to a first indication of downward-moving null points with a movable conductor. But five days later he was stymied by something: "Peculiar disturbances in the experiments, great trouble," he recorded, without specifying the nature of the "trouble." Given the experimental design the only likely trouble was that Hertz could not reliably distinguish the displacement of the null points, that the displacement was hard to produce reliably. Frustrated by the problem, and apparently unable to do anything about it by monkeying with the apparatus. Hertz decided to use the device in a different place. On September 16 he took it out of the small room he had been using and moved it "to the auditorium, where the experiments gave better results." Apparently the small room and the apparatus did not like one another; the auditorium and the apparatus got along very well, which no doubt gave Hertz a great deal of confidence in the stability of that experimental locus, which, we shall see, had important consequences several months later.

The apparatus, then, could stably and reliably elicit the effects of the interference between the closed-circuit induction of the induced current and the (predominantly static) force of the inducing oscillation. Suppose now that we replace the metal device C with a block of dielectric material (or, rather, that we place the block beneath, instead of above, the oscillator; see fig. 71). Since the oscillator lies entirely in a single plane (having flat plates), the charges upon it will, by static induction, elicit mirror polarization charges near the surface of the dielectric—that much was entirely unproblematic. This image will track its progenitor. If, consequently, changing polarization can also produce electrodynamic induction, then the oscillator's image in the dielectric mirror will interfere with its progenitor's action on the resonator in precisely the same way that the metal circuit C did. Here, however, there was no way to control the image's self-induction, or indeed even to speak of its natural frequency, for that completely lacked meaning. However, the image, which must always be in phase with the current in AA', will nevertheless be weaker in strength (since the polarization charges are much smaller than the inducing, conduction charges of the oscillator). Thus, the effect should be to pull the null points toward the dielectric. This is precisely what Hertz eventually found:

> I made the first experiments with a material which lay ready to hand, namely, paper. Underneath the conductor AA' I piled up books in the form of a parallelepiped 1.5 metre long, 0.5 m. broad, and 1 m. high, until they reached the plates A and A'. It was shown without doubt that sparks now appeared in those positions of the circle which before were free from sparks, and that in order to make the sparks disappear the spark-gap f had to be turned about 10° towards the pile of books. (Hertz 1888d, p. 101)

Hertz preceeded to refine the experiments and to increase markedly their demonstrative power. On October 5 he employed the "pile of books" just referred to as well as 800 kilograms of "unmixed asphalt." The effects were quite pronounced: the high-point sparking was greatly increased, and the null points moved downward 23°. "We have here all the effects of a conductor of small period of oscillation" (Hertz 1888d, p. 102). Now he sought to remove any likely disturbing factors. Because the asphalt contained "a large amount of mineral matter," which might be conducting, the effects could be attributed to actual conduction, rather than to polarization, currents. So Hertz had another block made, this time of "artificial pitch," which had no similar impurities and which reproduced the previous results, albeit somewhat more weakly. But the pitch did contain "free carbon in a very fine state of division," and this would certainly be conducting.

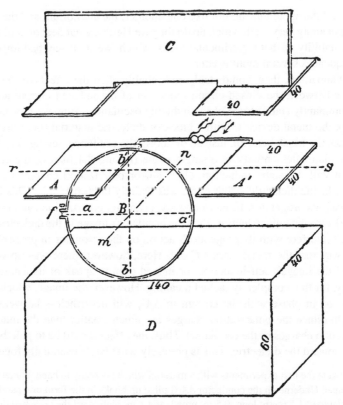

FIGURE 71   Hertz's apparatus for detecting dielectric induction (Hertz 1888d)

Hertz could not afford any more thousand-kilogram dielectric blocks, so he scaled the apparatus down by a factor of 2. In the end he "investigated altogether eight substances" and succeeded in generating null-point displacements ranging between 8° and 30°, with most being closer to the smaller value. With such a wide variety of materials all producing displacements, Hertz felt there could be scarcely "any doubt that the action is a real one, and that it must be attributed to the substances themselves, and not to [conducting] impurities in them."[20]

These were heady days for Hertz, though not precisely because he had answered the first Berlin question by demonstrating the action of dielectric currents. Röntgen had already done so (albeit more crudely) in 1885, and Hertz was well aware of the fact. Rather, Hertz was excited by the apparent power of his device as a sort of universal electric-force explorer, able to probe aspects of electrodynamic processes that were particularly important in Helmholtz's electrodynamics but that had hitherto eluded its manipulative grasp. Certainly success in answering a Berlin question was of the highest importance to him, even if it had already been done—because it was important to Helmholtz. On October 30 he rather solemnly wrote his parents:

> The work that I will complete, with God's help, in the next few days, is actu-
> ally the solution of the problem that was set by the Berlin Academy in 1879
> but that has remained untouched.[21] Helmholtz urged me to work on it at the
> time after I had solved the university problem [on electric inertia], but I held
> back because I saw absolutely no practicable way of doing it. Now I have
> succeeded in solving it, almost like child's play, in a way that naturally could
> have been scarcely thought of at the time. Therefore this result is a sort of
> personal triumph for me. (J. Hertz 1977, p. 233)

He sent his paper to Helmholtz for the Berlin Academy on November 5, emphasizing that "it deals with a topic that you yourself once urged me to tackle some years ago, *and that I have therefore always kept in mind,* but that I have in the meantime found no way of approaching, until now, with any sort of prospect of unequivocal success" (J. Hertz 1977, p. 235; emphasis added).

In many ways the fall and the early winter of 1887/88 were the happiest, the most exhilarating of Hertz's short life. He had an apparatus in hand that he could make produce all sorts of interesting new effects and that could answer old questions and raise new ones. The laboratory was not the only birthplace of importance to Hertz at the time; his daughter Johanna Sophie Elisabeth was born at 2:45 a.m. on October 2, in the midst of Hertz's playing with asphalt blocks. On November 5 Hertz notified Helmholtz that "the electric oscillations that I have used could become very useful for the electrodynamics of open circuits; I have already taken some steps towards further applications" (J. Hertz 1977, p. 235). Open circuits, he knew, were ever Helmholtz's central concern, and Hertz felt that he was now on the verge of bringing them strikingly to heel. Little wonder that his wife wrote his parents on November 9 that "[Heins] simply pulls these beautiful things out of his sleeve now!"

SIXTEEN

Electric Propagation Produced

16.1. WHY HERTZ LINKED THE BERLIN QUESTIONS TO PROPAGATION

Hertz's diary records that on November 7, just two days after sending the dielectric paper to Helmholtz, he "found standing electric oscillations 3m in wavelength in wires stretched straight."[1] Three days later he was "producing interference between the direct effects and those transmitted along wires." Two days after that (November 12) he "set up experiments on the velocity of propagation of the electrodynamic effect." The obvious questions are why Hertz made wire waves in the first place, and why he used them to look for "the velocity of propagation of the electrodynamic effect."

His laboratory notes unfortunately provide little help here, except to confirm the diary entries. According to the historical introduction to *Electric Waves,* Hertz turned to wires and propagation experiments after the dielectric success when he realized that though he could not produce a practical way to polarize dielectrics electrodynamically (to answer the second Berlin question, having succeeded with the first), a successful demonstration that "a finite rate of propagation" occurs in air would positively and simultaneously answer both questions. The problem with the introduction's account, as we saw above, is that the "closed rings of paraffin" that Hertz there refers to, and which he had cast to investigate the second Berlin question, had in fact been made for that purpose a year before. It is highly unlikely that Hertz redid these unsuccessful experiments sometime between November 5 and 7, in which case the introduction's account is in this respect strongly misleading. It does not, however, follow that the introduction also misleads in regard to the course of Hertz's *thoughts.* If we take his remarks there as referring to what he was thinking about, rather than to what he did, then we can come to some understanding about why he was experimenting with wires in the first place on November 7, which would otherwise remain something of a mystery.

He had by then successfully produced the electrodynamic action of dielectrics; polarization of dielectrics by electrodynamic action seemed out of experimental reach. Thus far Hertz had done nothing that Röntgen had not already accomplished, albeit (as he wrote his parents on October 30) he, Hertz, had done it "almost like child's play," whereas Röntgen had done it much more crudely. It would have been entirely in keeping with Hertz's intensely competitive character for him to have thought long and hard about how to move decisively beyond Röntgen. The introduction to *Electric Waves,* it seems quite probable, tells us *how* he thought to do so, though it does not accurately relate the sequence that convinced him he *had* to do so.

Hertz's reasoning here, though apparently obvious in retrospect, would not have seemed so to him at the time. The full account in the introduction reads:

> The first assumption [the electrodynamic action of dielectrics] was now shown to be correct. *I thought for some time of attacking the second. To test it appeared by no means impossible; and for this purpose I cast closed rings of paraffin.* But while I was at work it struck me that the centre of interest in the new theory [Maxwell's theory] did not lie in consequences of the first two hypotheses. If it were shown that these were correct for any given insulator, it would follow that waves of the kind expected by Maxwell could be propagated in this insulator, with a finite velocity which might perhaps differ widely from that of light. I felt that the third hypothesis contained the gist and special significance of Faraday's, and therefore of Maxwell's, view, and that it would thus be a more worthy goal for me to aim at. I saw no way of testing separately the first and the second hypotheses for air; but both hypotheses would be proved simultaneously if one could succeed in demonstrating in air a finite rate of propagation *and waves.* (Hertz [1893] 1962, p. 7; emphasis added)

Put aside for a moment the last two words in this account. The other emphasized passage is the one that cannot properly represent the sequence of Hertz's laboratory work. But the unemphasized passages can, and probably do, correctly represent the thoughts that he had on or just after November 5, precisely because the argument they advance would not have been entirely obvious at the time and therefore owed much to inspiration ("it struck me"). To understand Hertz's difficulties here we must turn back for a moment to Helmholtz's electrodynamics.

We have repeatedly seen that Helmholtz's electrodynamics enforced a complete separation between effects of different kinds, requiring appropriate states and interaction potentials to represent them. Hertz was more than aware of this separation; he had been living and breathing it for nearly a decade. The Berlin questions that had long ago defeated him nicely embodied this separation. They asked for experimental evidence

> for or against the existence of electrodynamic effect of forming or disappearing dielectric polarization in the intensity as assumed by Maxwell

and

> for or against the excitation of dielectric polarization in insulating media by magnetically or electrodynamically induced electromotive forces.

These two questions refer to different interactions. The first concerns interactions between a dielectric in a state of changing polarization and a conductor in a current state. The second concerns interactions between a dielectric as a polarization-capable body and conductors with changing currents or magnets in motion. These involve (in principle) different interaction potentials, just as, for example, the ability of a charged conductor to generate a current in another conductor involves a distinct potential from the one that enables a conductor in a state of changing current to do so. In Helmholtz's electrodynamics there is little reason to connect these two effects directly.

In the parts of his 1870 foundational paper that examined propagation, Helmholtz had just ignored the difficulties by making two assumptions: first, that changing polarization can appear in any equation where conduction current appears (except for Ohm's law); and second, that anything that can effect a current can also effect polarization. We have already seen how, in 1884, Hertz had attempted to come to grips with the difficulties inherent in this approach by inventing two principles, the second of which was in fact thoroughly incompatible with the spirit of Helmholtz's scheme (and was never explicitly stated). These, when coupled to Helmholtz's version of the energy principle (which was itself based on assumptions concerning the existence of interaction potentials), yielded (Hertz thought) propagation precisely like Maxwell's, and so in the same form that Helmholtz had apparently obtained as a limiting case from his own structure. This has an important implication for Hertz's thoughts in 1887 and indeed suggests just why he might have put together the Berlin questions in the way that he then did.

The Berlin questions implicitly incarnate the division of interactions that Helmholtz's electrodynamics could not avoid, but they leave the difficult issues hidden. Troubled by these difficulties, Hertz had decided to avoid them by taking an extraordinarily radical step: instead of ignoring the interaction potentials by hiding them beneath a request for experiments, Hertz had brought them directly into the open by inventing a novel axiom that infinitely multiplied them. Where the distinct Berlin questions reflected Helmholtz's divided, but limited, potentials, Hertz's 1884 argument united an infinitude of such things in a single result: the propagation equation for the vector potential. For Hertz propagation was accordingly the inevitable result of an underlying, infinite set of interactions.

This raises an interesting question. If the Berlin questions contain only two potentials, and if an infinite set is necessary to obtain propagation, then it would seem that a positive answer to the questions cannot have much to do with propagation. How then did Helmholtz obtain waves in 1870? Appendix 13 gives full details, but the essential answer is quite simple: to obtain propagation Helmholtz had to assume that the polarization current must be included in the continuity equation along with the conduction current. This is entirely separate from the question of whether the polarization current can generate electrodynamic effects; it might very well do so and yet not appear in the continuity equation. Helmholtz had obtained propagation, as it were, by using continuity to avoid answering the very difficult questions raised by the introduction of his two novel interactions. Hertz, on the other hand, did not use continuity at all. He instead obtained propagation directly from the interactions themselves by multiplying them indefinitely.[2] Helmholtz the originator had certainly been attempting to follow Maxwell's route in spirit (by treating the polarization current as equivalent to a conduction current in every respect except for Ohm's law), thereby producing an inevitable set of unanswered problems. Hertz the radical student perceived the difficulties and sought to avoid them by making the foundation explicit, and in the end by avoiding problematic polarization currents altogether.[3]

Given this, we can see why Hertz would not have easily connected propagation with a positive answer to the Berlin questions, why, that is, it cost him some thought to do so. It was not at all clear from Hertz's 1884 analysis alone that the effect hypothesized by the second Berlin question—the polarization of dielectrics by electrodynamic action—should in fact take place, for it appears no where at all in the argument, not even implicitly. Even the first Berlin effect—the electrodynamic action of polarization currents—appears at most implicitly (and even then only for empty space) in Hertz's deduction of the "Maxwell equations." It is certainly not explicitly present in the infinite set of interaction potentials, because these hold only between n^{th}-order currents *in conductors*. Polarization appears nowhere in any of this. If, therefore, propagation emerges in Hertz's 1884 fashion as a result of infinite interactions between conduction currents, then there is apparently no necessary reason to link it to the Berlin questions, unless one returns to Helmholtz's original, and widely known, way of obtaining propagation.

Until Hertz was faced with his inability to answer the second Berlin question (the one that no one had in fact answered as yet) experimentally, he did not put back together what he had broken apart in 1884, namely, propagation with dielectric effects. His aim was to succeed in the laboratory in answering the two questions posed by the master, questions that had long ago defeated him. It cannot be sufficiently emphasized that for Hertz these questions stood by themselves as queries for the laboratory. Answers to them possibly had wider implications—hence their originally compelling interest—but they could be treated as objects of concern in their own right. Hertz had now succeeded with one query, but so had someone else; the second continued to resist his attempts. At some point between November 5 and 7 Hertz must have thought back to the master's original paper, in which the two effects were used, together with continuity, to generate propagation, and he then decided that, *faute de mieux,* and despite his own 1884 considerations, he would use the latter to claim success with both of the former, just as he tells us in the introduction to *Electric Waves.* And he felt that he might be able to detect propagation because his method for observing the interferences between the forces exerted by different objects, which he had just used extensively for the electrodynamic action of dielectrics, could naturally reveal other aspects of interference as well (should they occur), in particular, phase differences due to the finite rate at which the electrodynamic interaction might propagate. Propagation emerged initially as a tool to be used for other purposes, but it emerged in a way that began to separate Hertz in the laboratory from the deepest reaches of Helmholtz's physics as Hertz had come to understand it. The question now was how to detect propagation.

16.2. How to Manifest Wire Waves with the Resonator

Near the day (November 5) that Hertz sent his paper on dielectrics to Helmholtz he had connected a wire to a plate which he had put parallel to one of the oscillator's own plates. The wire was run from the plate over the (horizontal) oscillator's spark gap and then drawn out along the normal to the oscillator. On November 7 he succeeded in using the resonator to detect the standing electric waves that this arrangement should, he thought, produce in the wire. Both his laboratory notes and his diary record this success, which he refined over the next two days. There can be little doubt that Hertz undertook these experiments, and the ones that immediately followed, specifically in order to detect a propagation of the "direct action" that, emanating from the oscillator, drives the resonator because both the notes (dated November 11–12) and the diary (dated November 12) record his intention to do so. Moreover, the laboratory notes leave no doubt that Hertz had a reasonably clear conception of what to do from the beginning of these experiments, although we shall see that his attempts were hardly untroubled.

The essential idea probably occurred to Hertz as he reflected on his dielectric experiments, which had utilized interfering actions to displace loci of null sparking. Here the actions were assumed to propagate instantaneously; they interfered only in the sense that the force emanating from the driven dielectric combined with the force emanating from the oscillator to produce a compound effect. But suppose that action propagates. In that case the resultant would presumably depend on the respective distances of the resonator from the two drivers. Hertz did, we shall see, soon construct an experiment of this kind, but it was not the first one that he tried. In the absence of specific remarks one cannot be certain, but experimenting with two drivers might not have appealed to him, because he knew it would not yield anything like a clear indication given the vagaries of the spark indicator. At best he might have detected a near-vanishing of the spark as one of the drivers moves away and then an increase in sparking of some modest amount. This would hardly constitute persuasive evidence.

To be convincing he needed to play off the oscillator's direct action against some effect whose propagation could be independently and persuasively demonstrated. Then he could transform the interference of two presumptively un-propagated actions (as in his dielectric experiments) into the interference of one action (*w*) that he *knew* to be propagated by another action (*a*) that *might be* propagated. Hertz's reasoning on this point probably differed little from his published remarks (1888e, p. 117): namely, if *a* is not propagated at a finite rate, the interference between it and *w* will vary in lockstep with *w*'s own oscillations.[4] In which case spark behavior will reflect *w*'s periodicity. If, on the contrary, *a* is propagated, one of two things may happen. If *a* travels at the same rate as *w*, the interference will be the same no matter where it is measured because the actions have everywhere the same phase difference. Spark behavior under interference would accordingly not vary markedly with distance. If, instead, *a* and *w* travel at different rates, the phase difference between them will depend upon where the spark measurement is done. In which case the spark behavior under interference will change noticeably with distance and will show a variable periodicity.[5]

The immediate problem was how to generate *w*, the propagated action. Very likely Hertz was again aided here by his recent driven-oscillator experiments, in which the one device produced oscillations in the other without direct connection between the two. In an inspired moment Hertz thought to generate standing electric waves by connecting an unterminated wire to the driven-oscillator plate. This was in itself an entirely novel thing to do. No one before had written about generating wire waves in this fashion. More to the point, Hertz now knew quite well how his resonator responded in different orientations to the oscillator's direct action, and this had immediate implications for how it could be used to provide a graphic map for the wire waves.

Hertz mapped the wire carefully on November 8, according to the laboratory notes, and then tried different metals and thicknesses on the ninth. He used the resonator in the following way (fig. 72). He knew that the resonator responds particularly strongly to the solitary oscillator when both the resonator and its spark gap are parallel to the oscillator. Call *P* the plane that contains the center of the resonator and the oscillator. The oscillator in these experiments always lies along the horizontal. Move the resonator until *P* is also horizontal, whereupon its center lies at the same height above the floor as the oscillator, and then rotate the resonator until its plane is vertical and parallel to the oscillator. Following Hertz's terminology, we will call this orientation of the resonator position 2. Finally, turn the spark gap around until it lies horizontally at the top. (Under these circumstances it reacts to both the static and the dynamic open-circuit forces.) The resonator responds strongly here, as it also does when its plane contains, and its gap points directly toward, the oscillator.

Easy application to the wire follows, because the wire is itself a long oscillator. Instead of orienting the resonator in position 2 with respect to the oscillator, orient it in a vertical position 1 parallel to the wire. Set the spark gap vertically above or below. Hertz further required the gap to face the wire in these stationary-wave experiments, and in practice he probably set the resona-

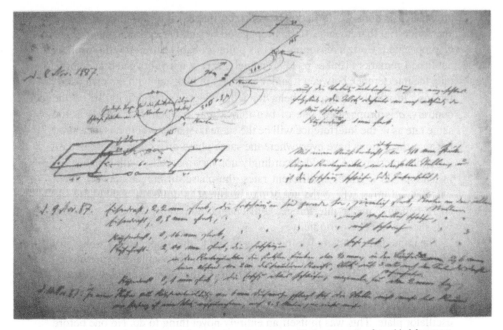

FIGURE 72   Hertz's diagram for mapping wire waves with the resonator, from his laboratory notes. By permission of H. G. Hertz.

tor vertically below it. We will call this orientation position 1_{so}. (For later purposes Hertz also defined a position 3, in which the resonator's plane is set horizontally to contain both the wire and the oscillator, and in which the resonator's spark gap points directly toward the wire.) To map the stationary wave set the resonator in position 1_{so}, move it along the wire, and measure the spark lengths by adjusting the resonator's micrometer screw until sparking just stops or just begins.

Hertz's published account provides a compelling image of how he worked, manipulating the resonator to find, and the wire to produce, appropriate waves:

> [W]e allow the wire to end freely at some distance from its origin,[6] and bring near to it our secondary conductor in such a position that its plane includes the wire, and that the spark-gap is turned towards the wire. We then observe that at the free end of the wire the sparks in the secondary conductor are very small; they increase in length as we move towards the origin of the wire; at a certain distance, however, they again decrease and sink nearly to zero, after which they again become longer. We have thus found a nodal point. If we now measure the wavelength so found, make the whole length of the wire (reckoned from the point n) equal to a complete multiple of this length, and repeat the experiment, we find that the whole length is now divided up by nodal points into separate waves. If we fix each nodal point separately with

all possible care, and indicate its position by means of a paper rider, we see that the distance of these are approximately equal, and that the experiments admit of a fair degree of accuracy. (Hertz 1888e, p. 112)

Hertz's bits of paper marked the wire wave, but note that he provided no explicit limits on the reliability of his measures—just that they admitted "of a fair degree of accuracy." His experiments with different kinds of metals and different wire thicknesses yielded "nodes in the former places." The laboratory notes for November 8 and 9 do not explicitly record the wavelength. In the evening of the tenth he used a "small apparatus at half the scale," and here he explicitly recorded a "wavelength" of 1.55 m, which he noted was "half" the previously measured one (which therefore would have been about 3.1 m).[7]

Hertz's characteristic unconcern with precision measurement is strikingly evident in these experiments and in the ones that he undertook afterward. In his published account he did provide six numbers for the locations of the "paper riders," but he nowhere tried to estimate just how accurately the resonator could be used to mark these positions. He was confident that the numbers were good enough and that they had meaning, because he could strengthen or weaken the nodes at will, simply by manipulating the wire's length. Hertz, one might say, substituted skill in handling and changing his apparatus for the rigid and dead numbers of exact measurement. His laboratory notes catch him in action:

> If the wire at the small circle [resonator] is gradually shortened by rolling it up, then the nodes always become indistinct, until the wire length is again shortened by half a wavelength, whereupon they are again sharp. So it changes from 3 to 2, from 2 to 1 nodes, and very good conspicuous appearances are obtained. (p. 2b)

Hertz was evidently quite taken with this new-found ability to manifest wire waves. Elisabeth wrote his parents on the ninth that he "had again succeeded in the most beautiful experiments" and that "it makes him very happy, and me as well, when he tells me about it with a radiant face" (J. Hertz 1977, p. 235). The next step was to use the waves to examine the propagation of the "direct action" from the oscillator (as he referred to it by December 15 in the laboratory notes and in his published accounts). On November 11 he succeeded in producing a detectable "interference" between the wire waves and the direct action.[8] Hertz's success did not breed further elation.

16.3. THE RESONATOR BECOMES AN INTERFERENCE DETECTOR BUT FAILS TO PRODUCE NOVELTY

The entry in his laboratory notes for November 11–12 reads, "Fruitless experiments to determine the velocity of propagation through air."[9] The diary entry for November 12 is somewhat more precise: "Set up experiments on the velocity of propagation of the electrodynamic effect. Contrary to expectations, the result is infinite propagation" (J. Hertz 1977, p. 235). Neither the diary nor the notes contain any further information on these experiments, but we can discover what Hertz tried to do at the time by extrapolating backward from the technique that he deployed five weeks later.[10]

The concept of the experiment was quite simple. Since Hertz's metal wire was drawn out along a normal to the oscillator, the resonator's position 2, which responded strongly to the oscillator, was always perpendicular to its position 1, which responded strongly to the wire. In each position, Hertz already knew, the resonator responded scarcely at all to the other stimulator: that is, in position 1 and in the absence of the wire, the resonator hardly sparked whereas in position 2 it sparked (strongly) in essentially the same way whether the wire was present or not. Choose position 1 so that the center of the resonator lies at the same height as the oscillator. If we now rotate the plane of the resonator between these two mutually orthogonal positions—between being parallel to the wire and being parallel to the oscillator—then we change the resonator from a wire-wave detector into an oscillator-action detector and back. In between these two positions the resonator responds in some measure to both stimulators, and, therefore, the two can be played off against one another— each "interferes" with the other's effect on the resonator. On November 11 Hertz did something that, he was convinced, showed this interference.

In Hertz's published diagram (fig. 73; 1888a, p. 198, and 1888e, p. 108) the direction of the resonator B is specified by the unfeathered arrow drawn normal to its plane.[11] In position 1 this arrow points directly toward (or directly away from) the wire; in position 2 it points directly toward or away from the oscillator. The first thing Hertz did was to calibrate the device by setting it so that in a chosen position the sparks are equally strong in both positions. He did this simply by adjusting the distance between plates A and P. Suppose that we begin with the resonator in position 1, with its spark gap at the highest point. We rotate the resonator's plane so that the arrow's tip turns away from the pair of plates AP (P being the one to which the wire is connected) until it points along the extension of the wire (marked in the figure by the feathered arrow). Call this position L_2 (fig. 74). Then, Hertz found (for his chosen position), the sparking increases in power. Suppose, conversely, that again beginning from position 1, we turn the tip of the arrow toward the plates AP. Call this position L_1. Then the sparks at first disappear and only reappear when the resonator's spark gap is "considerably shortened"; that is, the sparking intensity has markedly decreased. If we do this same experiment, but with the resonator's spark gap at the lowest point rather than at the highest, the spark intensity changes in reverse, being most powerful when the arrow points toward AP and weakest when it points away.

What did this mean? Hertz, we shall see in a moment, characteristically provided (in print) a qualitative theory, but he did not need much at all to be

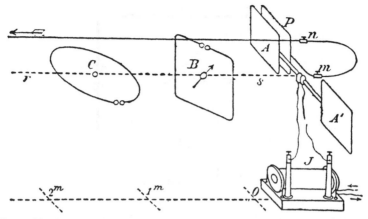

FIGURE 73   Hertz's published diagram for his interference experiments (Hertz 1888e)

convinced that his device was picking up some sort of interference between the wire and oscillator actions. The near-vanishing of the sparking when the arrow points toward *AP* (with the resonator's gap lying vertically above) vividly indicates that (always at this particular distance of the resonator from the oscillator) the wire and the oscillator are working against one another since the sparking *picks up again* as the resonator continues to turn toward the oscillator. In the wire's absence there is, Hertz knew very well, no *minimum* point, just a gradual increase of sparking. Start from the minimum locus of the arrow, say toward *AP*, with the spark gap accordingly at the top. Turn the arrow toward the wire and then away from *AP*, until extremely powerful sparking occurs. Past this orientation the sparking decreases, until it reaches the usual intensity for position 2. Here, then, we have detected a sparking *maximum*, again indicating some sort of interference. This was all that Hertz needed: his device was, he felt certain, picking up a compounded wire–oscillator effect.

But why was there interference of this kind at all? Its presence unquestionably meant that the direction of the net resonator-driving force due to one, but not both, of the two stimulators—the wire or the oscillator—must reverse direction in going from position L_1 to L_2. Obviously, the reversing action had to be the one exerted by the oscillator, since that is the one that goes to zero when the resonator is in position 1, halfway between L_1 and L_2. The wire's action, in contrast, reaches a maximum at position 1.

In his published account Hertz explained, purely qualitatively and using what he certainly already knew by October, why he thought this should happen. Recall first that Hertz's September–October experiments (series A) had likely shown him that, at least not terribly far from the oscillator, the total force exerted by it is dominated by electrostatic action, which is opposite in direction to the oscillator's (open-circuit) electrodynamic force,[12] and so the oscillator's

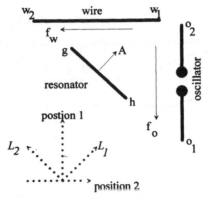

FIGURE 74   Hertz's understanding of the resonator's response

force (f_o in fig. 74) is essentially parallel to it and directed at any given moment from the positively charged end (say o_2) to the negatively-charged end (o_1). The force exerted by the wire (f_w in the figure) parallels the wire and is entirely electrodynamic. These forces affect the top and the bottom of the resonator, but Hertz already knew that the force which activates the portion of the resonator opposite its gap governs sparking. Consequently, he could determine the resonator's response simply by projecting the wire force and the oscillator force onto a directed line that represents the resonator's active portion. Figure 74 illustrates the various forces in relation to the resonator's responsive part, to which the arrow A is normal.

From the figure we can see, in general, how the resonator behaves. Suppose first that A points toward position L_1, in which case end h of the resonator lies closest to end o_1 of the oscillator. Project the forces onto gh in this position. Now turn gh so that A points toward L_2. In this orientation end h of the oscillator now lies closest to end o_2 of the oscillator. Again project the forces onto gh. In the first orientation the projection of the oscillator force pointed from g toward h. In the second orientation the projection reverses: it now points from h toward g. On the other hand, the projection of the wire force points in both cases from h to g. Hence in going from L_1 to L_2 the oscillator's effect on the resonator reverses direction, whereas the wire's action does not.

Suppose that we again began with the arrow A pointing toward L_1, but that instead of rotating it to L_2 we rotated it 90° in the other direction, ending up therefore with A pointing directly opposite to L_2. If we do this, then the wire's action on the resonator, but not the oscillator's, reverses direction (changing from h to g in position L_1 to g to h when A points directly away from L_2). An experiment that detects differences between deviations toward L_1 and deviations toward L_2 is an *oscillator-reversal;* an experiment in which the second deviation is toward a direction opposite to L_2 is a *wire-reversal.*[13]

On November 11 and 12 Hertz tried to determine "the velocity of propagation through air." He did not specify in the notes how he went about doing so, but his method was no doubt similar to the one he carefully deployed five weeks later and was in fact specified by the structure of his apparatus. The essential idea was quite simple: move the resonator along the wire and observe whether (and if so, how) the interference between the wire wave and the direct action changes. Evidently at this time Hertz found that the interference does seem to track the wire wave, indicating that the direct action is not propagated: the diary entry for November 12 remarks, "Contrary to expectations, the result is infinite propagation" (J. Hertz 1977, p. 235). One cannot of course be certain, but it seems highly likely (in view of his work a month later, when he initially replicated this result) that Hertz found something like the following.

In his later work Hertz divided kinds of interference into three sorts, which he designated $+$, 0, $-$. A $+$ interference indicates that the resonator sparking was smaller with the resonator oriented toward L_1 (i.e., with its normal tending toward AP) than with it oriented toward L_2 (i.e., with its normal tending toward the wire's end). A $-$ interference designates the reverse. A 0 indicates that there is no difference in sparking between deviations toward L_2 and deviations toward L_1.[14] If the direct action propagates at infinite speed, then whatever type of interference ($+$, 0, $-$) is found at a given point along the wire should reoccur at regularly spaced intervals that measure the length of the wire wave. If the direct action propagates at the same rate as the wire wave, then only one kind of interference should occur wherever the measurement is made along the wire. The case of infinite speed would, therefore, ideally produce something like table 2, where the distances are given in terms of the wire-wave's length (of course, the first mark might be any of the three). As we have repeatedly seen, Hertz's device was not designed to make precise measurements of spark length, and his decision to distinguish three general kinds of interference reflects that character. The device's qualitative nature meant that transitions between marks could be hard to discern, and Hertz eventually settled on a 50-cm distance between observation points. His November observations were "futile" because they evidently produced a table of marks in which the symbols seemed (as in table 2) to repeat at nearly regular intervals (which, we shall see, occurred again five weeks later).

Hertz was not inordinately discouraged by whatever initial results he obtained, because the next day (November 13) he wrote an enthusiastic letter to his parents concerning his work:

> This week I have again had good luck with my experiments, and though there were some mishaps where I had counted on almost certain success, I must confess that I was almost impertinent in my wishes and that I must be quite spoiled if I do not take sincere pleasure in what has been accomplished. However, I have never before been on such fertile soil, prospects are opening right and left for new, interesting experiments, and I take my pick from among the plenitude. (J. Hertz 1977, p. 237)

TABLE 2   An ideal scheme for wire-wave sparking in the case of infinite air speed

$\lambda/4$	$\lambda/2$	$3\lambda/4$	λ	$5\lambda/4$	$3\lambda/2$	$7\lambda/4$	2λ	$9\lambda/4$	$5\lambda/2$
+	0	–	0	+	0	–	0	+	0

One of the "mishaps" Hertz referred to was surely his failure the previous two days to detect a finite velocity. Within five days (November 18) Hertz had given up, or put aside, the wire apparatus for a very different kind of interference experiment.[15]

This new trial used two mutually parallel oscillators. Given Hertz's previous experiments, one of the oscillators was probably driven synchronously by the other. According to the entry in his laboratory notes for November 18 Hertz put the resonator in the horizontal position 3 between the two drivers, all three devices therefore lying in the same plane. In position 3 sparking always occurs, whatever the orientation of the resonator's spark gap, in the presence of either of the two drivers by itself. If, in this configuration, the currents in the two oscillators are in phase with one another, then they cancel one another's actions since (Hertz probably already knew—he certainly knew it four weeks later) the electrodynamic action here predominates. Hertz's idea was to "regulate" the driven oscillator (he did not specify how) in such a way that interference does not take place. By manipulating the driven oscillator he was able to effect a 90° phase difference between the current in it and that in the driver, thereby obliterating the mutual cancellation of their actions on the resonator. This calibrated the device by setting appropriate phase relations between the currents themselves.

Having balanced the device in this way, Hertz now sought to unbalance it. If direct action propagates, then moving the driven oscillator away from this position will introduce new phase differences between its action and that of the oscillator at the resonator's locus, thereby regenerating interference at some point. If the direct action does not propagate, then the calibrated setting of a 90° phase difference between the oscillator currents will ensure that the oscillators can never cancel one another's effects however they may be arranged. To judge from the diagram in his laboratory notes, Hertz took the driven oscillator and placed it on the same side of the resonator as the other oscillator, but 1.5 m farther away. However, he could not manifest anything interesting: at a 1.5-m remove there was still no interference.[16]

Hertz, then, could not make either set of apparatus produce a finite speed for the direct action. He had on his hands an unfortunate reminder of the kind of experiment that had dominated his early career and that lacked the cachet of novelty: a negative experiment, one in which something does not happen that would occur if the effect in question existed. Both of Hertz's devices were demonstrably capable of detecting interfering actions: the wire device could detect interference between the wire waves and the oscillator's direct action; the twin-oscillator apparatus could detect interference between the two direct actions. Hertz could make them show such things stably and reliably. But the first device, it seemed, locked the interference to the wire wave; the second locked it to the phase relations between the currents in the twin oscillators. Nothing novel here, other than the considerable, but already deployed, novelty inherent in Hertz's devices themselves.

Little wonder that Hertz's mood turned somewhat sour, as always in the fate of recalcitrant devices. According to the diary he had (unspecified) "thoughts about Maxwell's theory" on the eighteenth. Eight days later he "experimented, without real enthusiasm." At the end of the month he immersed himself in his lectures. On December 2 he turned to "theoretical work"; the next day he pondered "oscillations according to Maxwell's theory." On the sixth Hertz "developed theory of propagation of waves in wires."

Hertz would at this time still have thought of wires as objects with very high conductivity and little or no polarizability, because the classes of conductors and dielectrics were kept quite separate outside Maxwellian circles. The wire wave would for him still be a wave of current, and it would accordingly propagate much as in the non-Maxwell theories, namely, at something like the speed of light.[17] On December 8 Hertz wrote to inform Helmholtz about the successful production and detection of wire waves. In the letter he remarked that he had calculated the wave speed to be 200,000 km/sec from the measured wavelength and a frequency computed from the oscillator's "potential and capacity." "However," he continued, "perhaps it would be better to take the velocity of propagation in the wires as equal to the velocity of light [viz., 300,000 km/sec], since nearly all theories point to that result" (J. Hertz 1977, p. 239). *Nearly* all theories are not *all* theories; the only exception to the rule was Helmholtz's own 1870 calculation of the wave speed in a cylindrical wire. He had there shown that the wire wave will travel at the speed of light only if the wire's radius is vanishingly small or if the general constant k is zero.[18] If k is not zero, or if the wire's radius is not very small, the wave speed will differ from light speed in some complicated way that Helmholtz did not explicitly calculate.

Hertz, who was thoroughly familiar with Helmholtz's electrodynamics, may have tried to derive a general expression for the speed of the wire wave in the case that the radius cannot be ignored, in order to find out what effect k might have. We will see that he certainly did later ponder possible effects of this kind, though only after he had carried out a further series of experiments. His initial failure to detect a finite speed for the direct action evidently led him to think more closely about what "Maxwell's theory" (i.e., Helmholtz's equations in the Maxwell limit) had to say about "oscillations" (i.e., about the propagation of the direct action) and also about what the general value for the speed in the wire might be, regardless of whether the Maxwell limit (which, recall, included $k = 0$) obtained or not. That way he could make "comparisons and calculations" for more general situations. In particular (though he wrote nothing on this point), he might have wondered whether the wire speed could be very much slower than the speed of light, in which case his experiments would certainly have failed even if the direct action did propagate at light speed.[19] Hertz's device, in other words, could detect a finite speed for the direct action only if it did not differ by some extraordinary amount from the speed of the wave in the wire.

Nevertheless, as Hertz wrote Helmholtz on the eighth, there was a fairly general consensus that the speed of the wire wave was not an order of magnitude different from that of light, and indeed measurements done in 1850 by Fizeau and Gounelle for iron and copper wires and by Siemens in 1875 for iron gave values ranging between 100,000 km/sec and 260,000 km/sec (i.e., at the worst, about 30% of the speed of light). Hertz might at first have thought that even such a comparatively small difference might have vitiated his device, though he gives no such indication in either his diary or his laboratory notes. Yet the possibility of a detectable, influential difference between the wave speed in wires and that of the direct action was probably very much in Hertz's mind quite early on.

16.4. An Appliance for Showing Interference

On December 15, nearly a month after he had failed to make his apparatus perform in the way that he had hoped, Hertz returned to his experiments in order to make certain that they were secure—to, as he put it in the diary, "fill in the gaps" (J. Hertz 1977, p. 239).[20] In his introduction to *Electric Waves* Hertz reminisces about the intervening month:

> Disheartened, I gave up experimenting. Some weeks passed before I began again. I reflected that it would be quite as important to find out that electric force was propagated with an infinite velocity, and that Maxwell's theory was false, as it would be, on the other hand, to prove that this theory was correct, provided only that the result arrived at should be definite and certain. I therefore confirmed with the greatest care, and without heeding what the outcome might be, the phenomena observed. (Hertz [1893] 1962, p. 8)

During these weeks, it seems, Hertz talked himself into accepting what seemed to be an unavoidable *negative* result, namely, that he could not detect a finite speed with his device. We have repeatedly seen that he was not happy with negative consequences. His training placed a tremendous premium on producing positive (i.e., new) effects, but there seemed to be no way out. He accordingly decided to write a paper detailing his results, one that would count as a negative answer to the second Berlin question and so also against Maxwell's theory. But first he had to check his results and, in particular, to make his apparatus utterly stable as an interference detector, since everything depended upon its not being subject to doubt.

Hertz knew that he had to convince an audience just introduced to very high frequency oscillations that a detector built to respond to such things was a stable, reliable indicator of the composite force exerted by these oscillations and by equally high frequency waves in wires. He had to be able to argue persuasively that this entirely new device, which he had just recently invented and with which he had already produced novel and complicated effects, could now be deployed as an instrument of such eminent trustworthiness that it could detect something that lay at the extreme boundaries of contemporary theoretical speculation, and certainly beyond the boundaries of accepted instrumentation. Hertz accordingly worked to turn his device into an appliance for showing interference. He did so in the following way.

On December 15 Hertz set his resonator for an interference experiment using a mutually orthogonal wire and oscillator, which was located at a point along the wire where it showed no detectable interference, that is, where a 90° rotation of the resonator's plane produced no change in sparking (call this position of the resonator Q). This occurs, he remarked, because the "effect of the wire and of the direct action there have a 1/4 wave phase difference." This locution presumes that there is only one wavelength involved, the wire's, which is what Hertz at the time thought to be the case, since his first attempts had yielded no finite length at all for the direct action.[21] To change the interference required altering the phase of the spark curve, which Hertz easily did simply by adding an additional 150 cm of wire between the wire's origin at plate P and the point at which the wire passes the oscillator (between points n and m in fig. 73). Since (presumably) the wire wave had a length of about 600 cm, this altered the phase by an additional quarter-wave, which should therefore have produced interference. It did.[22] Hertz's confidence that he was reliably producing phase differences here was, it seems, greatly furthered by his ability to reverse the nature of the interference (using the extra 150 cm of wire) by rotating the resonator's spark gap from top to bottom, but only with the resonator fixed in position Q. In the first setup, without the additional wire, where interference did not occur with the gap at the top, it still did not arise with the gap at the bottom. In other words, Hertz could manipulate the kind of interference where he should be able to do so and could not manipulate it where he should not be able to do so. This impressed him greatly since he here placed exclamation marks around his comments.

Hertz's confidence in his device as an interference detector was now extremely high, and two days later (December 17) he began a numbered series of experiments that were designed to turn the apparatus into a thoroughly unproblematic tool that he could publicly deploy to substantiate the (negative) result that the direct action does not propagate. These experiments, numbered from 1 through 35 and performed from December 17 through 21, divide into two sets; in all thirty-five the resonator sits at the same distance from the oscillator. In the first twenty-one experiments the resonator is itself manipulated in various ways, or else the plate to which the wire is attached is shifted from one side of the oscillator to the other (altering which oscillator plate drives it). In the remaining fourteen experiments the wire is lengthened backward by increasing amounts, thereby altering the phase of the wire wave at the resonator.

None of these experiments has anything at all to say about the propagation velocity of the direct action, because the resonator never changes its distance from the oscillator. Whether or not the direct action propagates, its phase at the resonator's locus remains the same. These experiments were designed to specify the resonator's interference-detecting behavior by indicating what was significant, and what was not, in its use. They were to provide, as it were, an instruction manual that showed how to ensure the device was properly constructed and worked. They may be characterized in the following ways.

Experiments 1–21 (begun on December 17) pursue the resonator's sparking behavior with respect to

1. the position of the resonator's gap (vertically above or below);
2. a 180° rotation of the resonator's plane;
3. the resonator's placement near the oscillator plate that drives the wire or near the other plate of the oscillator;
4. the effect of replacing a circular resonator with a rectangular one;
5. the effect of raising or lowering the center of the resonator, which usually sits in the horizontal plane that also contains the wire;
6. effects when the spark gap is turned into the horizontal plane and the resonator is raised or lowered with respect to the wire and the oscillator; and
7. what occurs when the resonator's plane is horizontal (and therefore contains both the wire and the oscillator).

Experiments 22–35 (performed on December 21) specify how the device responds to changes in the phase of the wire wave by delaying it through the addition of progressively longer bits of wire between the position of the resonator and the point at which the wire attaches to its driving plate.

The first set of experiments was carefully designed to circumscribe the device's behavior. Let's follow Hertz in stabilizing the apparatus (refer to Hertz's published diagram, fig. 73). We begin, in experiment 1, with a calibration. We disconnect the wire from its driving plate and position a circular resonator, its plane vertical and its spark gap at the highest point, so that no sparking occurs. This places its plane visually parallel to the oscillator and so visually perpendicular to the wire. We do not, in other words, simply begin by setting the resonator physically in this locus by eye or ruler as best we can. On the contrary, we calibrate the device by establishing the locus through its sparking behavior. From this position we now go on to see what changes cause spark-ing—we go on to see what makes the device work.

We first hook up the wire to its driving plate, whereupon sparking begins. We now seek to change this sparking. To do so we slightly rotate the plane of the resonator about a vertical axis so that its normal (which points initially along the normal to the wire) points a bit toward the nearest plate of the oscilla-tor (which is the direction shown for the rectangular resonator in the diagram). The sparking weakens.

The next step is to establish what happens when the resonator's plane devi-ates in the other direction (whereupon the oscillator's action reverses direc-tion). This, one might think, would be easy: just turn the resonator back, so that its normal points to the wire, and then continue turning a bit. Evidently the device did not work for Hertz when operated in this fashion. It needed instead to be recalibrated between each trial. His notes specify that the wire must again be disconnected from its driving plate, and the resonator then set (as before) in the null position. Then, and only then, can the effect of a devia-tion in the opposite direction be reliably examined. When this is done the sparking strengthens.

There is no mention of this in Hertz's published paper, where he in fact leaves the impression that the experiment can be done by turning the resonator from one position, through the calibration point, and then onward to the other, watching the spark continuously strengthen along the way. This would cer-tainly be an effective visual way to compare the resultant actions in the two positions. Hertz did not, however, work the device in this fashion. He did not determine spark strength by visual intensity; he determined it by the maximum gap at which sparking could still be induced for a given orientation of the reso-nator.

This illustrates just how sensitive Hertz's device was, and why therefore he had a great deal of confidence in it. In the situation Hertz has described the interference between the two actions drops the force that drives the resonator by no more than several percent for a slight deviation. This is, however, evi-dently sufficient to lower it below the point necessary to effect sparking if the device is adjusted in such a way that the resonator's spark gap at the calibration point is set to the widest size possible at which sparking still occurs. After the spark vanished Hertz might have narrowed the gap to make it reappear, thereby affording a certain indication that the sparking had been weakened (and not that the device had, say, been rendered useless by deviation). This would then have required a subsequent resetting in order to reacquire the appropriate gap width at the calibration point. From there a deviation in the opposite direction allows the spark-permitting gap to be increased.[23]

Having determined how the device could be reliably worked, Hertz com-bined and recombined the seven types of experiments in the first group to see just what the apparatus was sensitive to and whether he could substitute a rectangular for a circular resonator (he could, and thereafter usually did).[24] In the second series Hertz determined with great care just how the sparking behaves when more and more wire is added between the resonator's position and the driving plate (we will hereafter refer to these phase-changing additions as the back-wire). In these experiments (unlike the ones shown in fig. 73) the wire was placed to one side of an oscillator plate, and it was driven by the other oscillator plate.[25]

Hertz started with a 35-cm back-wire, introduced a 100-cm one, and then kept adding wire, in 50-cm or 100-cm increments, until he reached 9 m, for a total of twelve trials. In each case he wanted to see whether a deviation markedly weakens the sparking, markedly strengthens it, or has no noticeable effect. Hertz then drew up a table of these back-wire lengths, placing a 0 next to lengths in which the interference was not marked (i.e., where deviation had little effect on sparking), a + in the two positions where deviation strengthened sparking, and a − in the two positions where sparking was weakened. Since the phase of the direct action remained constant here, the three distances between the contiguous + and − loci directly measured half the length of the wire wave, for which Hertz obtained 2.66 met. (meters); that is, the wavelength was 5.32 met.[26]

These wire-phase experiments constituted a second sort of calibration, in which the operation of the device as an interference detector was checked against its already-stable function as a wire-wave detector. Previously Hertz had measured the length of the wire wave by putting the resonator parallel to the wire and moving it along, checking the sparking as he went. This way he was able, as it were, to examine the wave directly. And now he had obtained the wavelength in a very different way. First of all, he did not move the resonator; he moved the wire by increasing its length before it reached the resonator. This alone, however, is not the important factor in his new measurement, because he could still have observed in the old way had he wished, namely, by leaving the resonator undeviated and adjusting its gap to see how the sparking varied. But he did not do this; instead, he twisted the resonator a bit to see whether the sparking strongly weakened or strengthened. In the absence of the oscillator the deviated resonator would *always* spark less strongly, so that in those circumstances this kind of measurement would have told Hertz precisely nothing. The oscillator's interfering action makes the difference, and if Hertz's device were working correctly, a wavelength obtained in this new fashion had to match fairly close the wavelength obtained in the previous way. In Hertz's opinion this evidently worked, since the standard deviation among his three measurements here puts the wavelength close to what he had previously obtained. Hertz believed that wavelength measurements of either kind were accurate to about 10%–15%.

With these calibrations behind him, Hertz could now turn to the job, as he saw it, of using the device to demonstrate that the direct action propagates vastly more rapidly, if it propagates at all, than the wire wave. He began experimenting to that end on December 22. His laboratory notes speak here of experiments "more accurate than before on propagation in air."[27] The diary refers both to "phase effect in the wire" and to "radical experiments on the velocity of propagation of the electrical effects"—experiments that, he was unhappily convinced, would not produce a finite propagation.

16.5. HERTZ CONFIRMS HIS UNHAPPY RESULTS

On December 22 Hertz carried out experiments 36 through 43. Using a 100-cm-long back-wire, Hertz for the first time measured the sparking sequence at six distinct points. The first, or zero point, constituted the reference standard for the sequence; the sparking there had to be unambiguous. Hertz chose the point so that powerful interference occurred when the resonator (its gap pointing upward) deviated toward the wire's driving plate. As a calibration test he made certain that the interference was genuine by turning the resonator's gap downward, whereupon the sparking intensified when the resonator deviated away from the driving plate. He quickly found that 3 m down the wire again showed strong interference, albeit with a reversal of the effects at the zero point.

The observations were not easily made. Hertz had to distinguish three kinds of sparking behavior: a comparatively powerful sparking increase for resonator deviation away from the wire and toward the driving plate (designated +), a correspondingly strong decrease for deviation away from the driving plate (designated −), and a neutral case, in which deviation made no difference (designated 0). This inevitably meant that qualitative judgments of sparking strength had to be made given the meaning that Hertz sought to extract from his +, −, 0 markings. To see what he faced we shall begin with table 3, which gives the marks that represent the five trials that he carried out on December 22. This table in fact forms part of one that appears in his laboratory notes just past the list of experiments for that day and that includes, we shall see, marks obtained on the twenty-third.

To obtain these marks Hertz attached the corresponding length of back-wire and then found an appropriate zero point, where the sparking increased powerfully with a + deviation. He then put the resonator at each of the sixteen or seventeen points along the wire, recalibrated it, and tried a deviation. Except for remarking that the 8- and 8.5-m marks for the 100- and 400-cm back-wires were difficult to observe, Hertz's laboratory notes contain nothing about the sparking intensity at each position. Nevertheless, Hertz concluded from his observations that the interference remained in step with the wire wave and thus that the direct action had an immeasurably higher speed. We know this from a letter he wrote the next day, certainly in the morning, to his parents, to whom he had not previously explained his new work on propagation.[28]

TABLE 3 Hertz's marks for December 22

	0 m	0.5 m	1 m	1.5 m	2 m	2.5 m	3 m	3.5 m	4 m	4.5 m	5 m	5.5 m	6 m	6.5 m	7 m	7.5 m	8 m	8.5 m
100 cm	+	+	0	−	−	−	−	−	0	0	0	0	+	+	+	+	+	0
200 cm	+	−	−	−	−	−	0	+	+	+	+	+	0	0	0	0	0	
250 cm	0	−	−	−	−	0	0	+	+	+	+	0	0	0	0	−	−	
400 cm	−	−	+	+	+	+	+		0	0	0	−	−	−	−	−	−	−
500 cm	−	0	+	+	+	+	0		−	−	−	−	−	−	−	−	+	+

[W]hat is the unexpected and to me displeasing result of my endeavors? The velocity [of the direct action] is not that of light, but certainly much greater, perhaps infinitely great, at all events not measurable. Even if it were three times as great, it could still be measured, but everything has its limits. Now, there is no arguing with nature; it must be as it is, but I should have certainly liked it better to obtain a clear, positive result than this more negative one. And the "most promising electrical theories" about which I wrote in my last paper [on the dielectric effect] all of a sudden no longer seem so promising. Certainly, caution is indicated here, but once again the experiments seem all too clear to me. (J. Hertz 1977, p. 241)

Hertz extracted these negative results, which he was by this time expecting since he was now looking to confirm his earlier failure to elicit a "positive" response, from the marks in the table. In the laboratory notes a calculation done beneath the table that appears after the listing of the experiments for the twenty-second (the table that also contains marks obtained on the next day) averages the five numbers 4, 2.8, 2.8, 3.4, and 2.5, obtaining 3.1. The day before he had checked his device with interference experiments that yielded 2.66 met. for the half-length of the wire wave; experiments with the wire alone had previously given 3.1 met. for that length. This calculation accordingly seems likely to represent the mean of the distances, in all but one case (at 400 cm) from 0 to 0,[29] that Hertz divined from his table, in which event these successive-interference experiments quite precisely returned the half-length of the wire wave,[30] allowing for about a 15% inaccuracy.

Turn back to Hertz's table and try to extract these numbers from it. It is hardly simple to do so, no matter how the numbers are mapped to the five experiments or how the distances are set off. To obtain the numbers that he wrote down, Hertz had to have had more information than he recorded in his notes; he had to know just how good a particular 0 mark was, and whether he ought to shift its locus some fraction of a meter forward or backward. Assume for a moment that he wrote the averaged numbers down in the order of the experiments and note the corresponding separations from the table (excluding for a moment the 400-cm experiment):

	First 0 to 0	Second 0 to 0	Hertz has
100 cm	3	3	4
200 cm	3	—	2.8
250 cm	2.5	2.5	2.8
500 cm	2.5	3	2.5

It is hard to extract Hertz's numbers from these separations even by re-arrangement, or, say, by counting the locus as the center of a group of similar marks. We could map his 2.5 to the 250-cm experiment, and one of the two 2.8 numbers can be mapped to 500 cm. The other 2.8, the 4, as well as the (here unlisted) 3.4 do not have easy homes, nor is there an obvious separation to be extracted from the 400-cm experiment (unless it be the first + to −, which produces 2.5). If we average the numbers obtained directly from the table, then we find 2.71 met., with a standard deviation of close to 0.3. The upper limit of this mean puts it precisely at the 3.1 that Hertz had obtained in his earliest measurements of the wire wave.

Hertz very much wanted a positive result, though he was by this time convinced that he would not get one, so there was certainly no reason at all for him to have massaged the data in a way that yielded nearly exactly the result he did *not* want, namely, that the interference paralleled the wire wave. He may perhaps have had some way to adjust the separations appearing in the table. The measurements could be difficult. For example, when he tried to make the 100-cm measurements over again with a circular resonator, he found that above 3 met. he could not continue, because he "lacked a magnifying glass and the screw [controlling the resonator's gap] wobbled"—because, that is, the sparking had by that distance become quite weak. The weaker the spark, the more difficult it was to tell just what the effect of the resonator deviation was. Hertz may have used the first set of measurements to gain proximity to the significant points (where the sparking remained truly indifferent to deviation); once in the vicinity he could then have refined the observation.

Whatever the origins of the numbers that Hertz averaged may have been, he certainly did think that his marks strongly implied that the interference keeps step with the wire wave, and indeed they do: even if we use the separations as given directly by the table, the result of 2.71 met. has a standard deviation of about 11% (whereas the numbers explicitly given by Hertz deviate among one another by close to 19%). In his published article, which reproduces the table (including the next day's results), Hertz remarked that "it might almost appear as if the interferences changed sign at every half wave-length of the waves in the wire" (Hertz 1888e, p. 118). Whatever modifications Hertz may have essayed, these marks at first spoke strongly to him of infinite propagation. But he had to be "cautious"—to be certain that he had enough trials to support his claims before going public with such a powerful "negative" assertion. Of course, those who did not think that electric action propagated outside wires would hardly have thought the result to be negative, but Helmholtz, ever Hertz's first audience, certainly would have. Hertz had already warned him about this on December 8; he had now to provide thoroughly persuasive evidence.

To do so Hertz continued the next day, December 23, with six further trials of the same kind, this time with back-wires 150, 300, 600, 550, 450, and 350 cm long. Table 4 gives the marks for these trials (in the order that Hertz performed the experiments). Beneath his table in the laboratory notes Hertz pro-

TABLE 4 Hertz's marks for December 23

	0 m	0.5 m	1 m	1.5 m	2 m	2.5 m	3 m	3.5 m	4 m	4.5 m	5 m	5.5 m	6 m	6.5 m	7 m	7.5 m	8 m	8.5 m
150 cm	+	0	−	−	−	−	0	0	0	0	0	+	+	+	+	+	0	0
300 cm	−	−	−	−	0	+	+	+	+	+	0	0	0	0	−	−	−	−
600 cm	+	+	+	+	0	0	−	−	−	−	−	0	0	0	+	+	+	+
550 cm	0	+	+	+	+	0	0	0	0	0	−	−	0	0	0	0	+	+
450 cm	−	0	+	+	+	+	+	+	0	0	−	−	−	−	−	−	0	0
350 cm	−	−	0	+	+	+	+	+	+	0	0	0	−	−	−	−	−	−

vides another computation for an average. This one finds the mean of fourteen numbers, obtaining 3.4 (with a standard deviation, which he never bothers with, of about 0.7 among them). Four of the numbers also appear in his average (located physically just to the right of this one beneath the table) of the five that had produced a mean of 3.1 (with a slightly smaller standard deviation of 0.6). If we again draw the separations directly from the 0 to 0 distances in these experiments of December 23, we find 3 for the mean, with a standard deviation of about 0.3. If we include the distances from the previous day's results, this changes to 2.9, with a standard deviation of about 0.3 once again.

Neither the 3.4 we find in Hertz's undiscussed average nor the 2.9 obtained directly from his table easily sustains a *positive* argument for finite propagation. On the contrary, the 2.9 result falls right between the 2.66 and 3.1 values for the wire wave's half-length that he had previously found, and the 3.4 result barely misses the first, and includes the second, value if we allow the inaccuracies to be no larger than the standard deviation among the measurements. If, as Hertz would certainly have had to admit, the inaccuracies are in fact somewhat larger than this, then his numbers cannot be instrumentally distinguished from one another. These results, as they stand, accordingly nicely sustain Hertz's anticipated *negative* outcome. But because Hertz later decided his experiments had positive results, in his published articles he had to argue the apparent implication of the table away, which he did in two ways: first, by saying that the alternation is not at all precise (which might, however, speak just as well to the experiment's accuracy as to its implications) and, second, by capitalizing on what might otherwise appear to be inaccuracy in observation, namely, the frequent appearance of four or more zero points together at distances past 3 m.

Hertz could have used these results to argue convincingly that the interference keeps step with the wire wave; he had argued negative points on the basis of evidence little stronger than this in the past. But a great deal was at stake here, given the importance of the issue for Helmholtz. Moreover, the marks at greater distances did seem to produce broad regions of neutral interference, which at the least raised questions about their reliability. Some time later on this same day, December 23, 1888, Hertz put the resonator in a different orientation, one in which it lies in the plane containing the wire and the oscillator. He had checked the behavior of the device in this orientation in his experiments 18 through 21.

16.6. A REPRODUCTION OF HERTZ'S EXPERIMENT

Just how hard was it for Hertz to make his new apparatus perform? What sorts of difficulties troubled it, and what kinds of essential adjustments had to be made that did not make their way into print? One cannot know for certain precisely what Hertz did, but his experimental device can in fact be reproduced quite easily and his operations carried out to the extent that he described them. Naum Kipnis and I did so on June 3 and 4, 1993, at the Bakken Museum in Minneapolis. Kipnis, who has much experience in producing instructional experiments with historical significance, set up a Ruhmkorff coil with a mechanical interrupter that operated at about 20 Hz. He provided as well a rectified voltage source driven by mains power. Building and using the kind of micrometer-adjusted spark gap that Hertz deployed would have taken some time. Kipnis had the idea to use instead a small, fast-acting neon bulb that would respond to very weak voltages.

On the first day, June 3, we used as oscillator plates aluminum-covered cardboard squares (instead of Hertz's brass ones) with much smaller dimensions than the ones Hertz fabricated. Kipnis initially fashioned a resonator out of a circular wire with the neon bulb inserted into it. With this device we could very easily detect the differential action of the oscillator according as the resonator was parallel or perpendicular to it. We made no attempt to calculate in any way the frequency of the oscillator or to put the resonator in synchrony with it. The sound of the coil was loud enough to preclude normal conversation. The strength of its action on the resonator was quite irregular, varying from moment to moment, and the air soon filled uncomfortably with ozone's acrid smell. The effect was, however, unmistakable and easily repeated.

We attached a 6-m wire to another aluminum-covered square in Hertz's fashion and let the wire hang freely. We also fashioned a squared-off resonator by wrapping wire around a piece of cardboard. The first 4 m of wire ran straight from the oscillator; beyond that point we turned the remaining wire through 90° to make the run as long as possible. We were able almost at once to see the wire's independent effect on the resonator. The neon bulb brightened when the resonator was set parallel to the wire and normal to the oscillator—a position in which, absent the wire, the bulb had stayed quite dark. We were, however, unable either to detect nodes along the wire or even to find the node that should exist at its freely hanging (i.e., ungrounded) end.

The next morning we decided to rebuild our device to Hertz's considerably larger specifications and to carry the wire straight on without any bends. This was soon accomplished, and we again set to work looking for stationary wire waves. At first we still had trouble. The wire's free end, in particular, did not seem to be a node at all, because the neon bulb could be made to light quite easily there. Further, the bulb did not seem unambiguously to decrease and then to increase in brightness as we moved it along. We did, however, notice that it was quite sensitive to its distance from the wire. This suggested a way out of our predicament.

Instead of looking for a node at the wire's free end, we decided to *assume* that a node did occur there. That way we could calibrate our resonator for wave measurements as follows. We stuck the resonator on a plastic stick that could push out a distance past the bulb. Keeping the resonator's plane vertically below the wire (where, we found, it was always most sensitive), we brought the neon bulb up to the free end until it barely lit. Pushing the adjustable stick out past the bulb to just touch the wire set a calibrated distance which we maintained as we moved the resonator down the wire toward the oscillator.

Although the bulb's weak light meant that (like Hertz) we had to work essentially in the dark, we were now easily able to detect a pronounced region, about 10–15 cm wide, in which it was at best dimly lit but outside of which it brightened quite noticeably. With the 6-m wire we were then using we failed to detect another such point near the oscillator. The wire was extended straight on several more meters, and this time we could detect two clear nodes. Measuring the distance between them without taking much care gave somewhat more than 2.7 m for the half-wavelength, very close indeed to what Hertz himself had found. Could we now detect interference between the oscillator's action and the action of the wire?

Time did not permit a thorough investigation, but the results were clear to both of us. First, the resonator did in general respond differently when it was in Hertz's position 1 and his position 2. We did not adjust the gap between the driven and driving plates in Hertz's fashion, but we did witness changes in brightness that, however, varied over time as the coil's rather irregular sparking continued. Anything reliable required letting the coil stabilize for a while after turning it on and then not letting it stay on for too long. Within these quite essential limitations we found, unmistakably, that the intensity change between Hertz's positions L_1 and L_2 *did not depend very much, if at all, on the distance of the resonator from the oscillator.* We, unlike Hertz, apparently found air waves that propagated at the same speed as the wire waves using his original technique.

The point is not to determine what in retrospect looks like a pathology (but which, we shall soon see, Hertz understood in a very different way). It is rather to gain access to Hertz's tactile, manipulative understanding, to the kind of thing that rarely appears in print. We learned that the Ruhmkorff coil is disturbingly loud, that it is irregular, and that it eventually makes the air both uncomfortable and, perhaps, too conductive for reliable results. We had no window to open since our room was in a basement. It was hard to work in the dark without our eyes tiring and becoming unreliable; Hertz had considerably worse problems looking at actual sparks. All trials required calibrating the resonator every time the oscillator was turned on, though it could not be left on for too long without our observations becoming irregular. Nevertheless, we developed expertise even in the short time we had. We could make the same kinds of things happen time after time; we could make effects go away by doing things that should make them disappear, and we could make them reappear in various ways.

16.7. Propagation Brought Unexpectedly to Life

In experiments 18 through 21, performed between December 17 and 21, Hertz had monitored the device's behavior for a configuration (position 3) in which the plane of the resonator contains both the metal wire and the oscillator. On the afternoon of the twenty-third[31] he used this configuration with a 100-cm-long back-wire. Three trials were made, numbered 50 through 52. Trial 50 constituted a new calibration, for in this configuration the device has to be manipulated in a different way from before. In the previous experiments the resonator was itself perturbed by rotation about a vertical axis. In this new configuration the plane of the resonator must remain undisturbed (else the configuration itself is altered), and the position of its gap also remains fixed. Instead of manipulating the resonator, Hertz now disturbed the metal wire, moving it from one side of the resonator to the other, with the resonator's spark gap remaining fixed (fig. 75). At the zero point (Hertz's experiment 50) sparking takes place with the wire placed as in the figure; flipping it to the other side of the resonator obliterates the effect.

It is important to understand what Hertz had in mind here. In this configuration, with the center of the spark gap lying along the normal to the oscillator through its gap, and with the resonator's gap perpendicular to the oscillator, Hertz had ascertained that (in the wire's absence) the resonator was governed entirely by the oscillator's closed-circuit, electrodynamic action.[32] Since the relative configuration of the oscillator and the resonator remains fixed in these experiments, the oscillator's inductive action is also fixed in direction and magnitude. Suppose, for example, that it runs clockwise around the resonator. The effect of the wire *relative to the effect of the oscillator* depends upon which side of the resonator it passes by. In the initial position sketched in the figure, where powerful sparking takes place, the wire's inductive action must drive a current clockwise around the resonator since we have specified that the oscillator's action does so. Flip the wire to the other side of this fixed resonator, and the current produced in it by the wire must therefore run counterclockwise around it. Since the current produced by the oscillator remains in the same clockwise direction as before, the two actions now work against one another, and the sparking is obliterated or weakened.

Suppose that the direct action, whether electrodynamic or electrostatic, propagates vastly more rapidly than the wire's action. Hertz's wire wave had a half-length, we have repeatedly seen, somewhere between 2.7 and 3.1 m. At Hertz's zero point the oscillator and the wire interfere constructively with one another. About 1.5 m past this point the interference should begin to change sign, until sparking occurs very strongly in an opposite fashion at about 3-m distance. In terms of Hertz's device, under this assumption he should have seen something like the following. At the zero point a wire flip kills sparking. Move the resonator down the wire several hundred centimeters. Sparking should

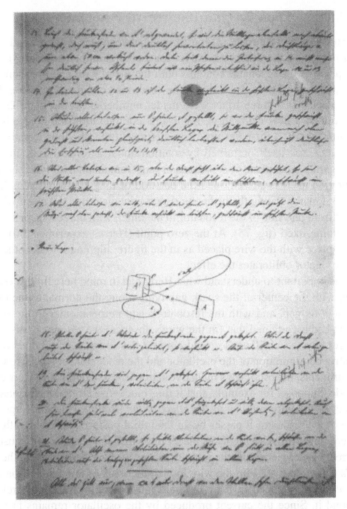

FIGURE 75   The critical arrangement. By permission of H. G. Hertz.

weaken, but a wire flip should still obliterate the effect. Near 1.5 m the sparking should be very weak, and a wire flip should have nearly no effect. Several
hundred centimeters further on weak sparking should occur with the wire in
its previously flipped position, with obliteration now taking place when the
wire is on the previously reinforcing side of the resonator. At about 3 m the
effect must be *very* marked and indeed just as unmistakable as the strong
sparking and obliteration or weakening that take place at the zero position, but
now reversed.

In experiment 51 Hertz placed the resonator 3 m away from the null point.

Leaving the wire in the same spark-strengthening position as in experiment 50, Hertz found *no difference from the null point*. In his words, strong sparking took place with the wire "on the same side as when close [to the oscillator at the null point]." The original reads, with emphatic punctuation, *"Also von derselben Seite wie in der Nähe!"* He immediately pursued the discovery in experiment 52: at 1, 2, 3, and even 4 m from the null point the effect remained the same in direction, though it was sufficiently weak at 4 (and 5) m that things became "doubtful."

This completely unambiguous, and before December 23 certainly unexpected, result could in Hertz's conception of interfering forces have only one interpretation: the direct action must itself propagate at a speed detectably different from that of the wire wave. Hertz was tremendously thrilled and pursued the effect further after Christmas, when he wrote his parents with the exciting news: "I received a great gift in my work on the night before Christmas, at least that's how it seems; nature does seem to be somewhat more kindly disposed towards me than she previously appeared to be" (J. Hertz 1977, p. 241). On December 26 Hertz performed three numbered experiments, 53–55, the first of which repeated 52. In experiment 54 Hertz used a 400-cm back-wire, and in 55 he tried a 250-cm one. The 400-cm wire reproduced the effect; the 250-cm effect was "indistinct." The next day Hertz began a "systematic investigation."

As before Hertz used +, −, and 0 signs to designate the type of interference, this time corresponding to effects that occur when the wire is flipped across the resonator. His trials ran to 5 m from the zero point using back-wires differing by 50 cm in length and ranging from 100 to 650 cm. Table 5 reproduces the table in his laboratory notes and in his published articles.[33]

None of the rows shows anything like the changes between 0 and 3 m that would occur, in Hertz's understanding, if the direct electrodynamic action

TABLE 5   Hertz's table establishing finite propagation for the direct electrodynamic action

	0 m	1 m	2 m	3 m	4 m	5 m
100 cm	+	+	+	+	0	?
150 cm	+	+	0	0	0	?
200 cm	0	0	0	−	−	−
250 cm	0	−	−	−	−	−
300 cm	−	−	−	−	−	−
350 cm	−	−	−	−	0	?
400 cm	−	−	−	−	0	?
450 cm	−	−	−	0	0	?
500 cm	−	−	0	0	0	?
550 cm	−	0	0	0	+	?
600 cm	0	+	+	+	+	?
650 cm	+	+	+	+	0	?

propagated at a speed very different from that of the wire wave. Indeed, the table at once permitted Hertz roughly to calculate the ratio between the two speeds. He calculated that the direct action propagates 1.5 times as fast as the wire wave. Although he did not specify how he came to this result, it is easy, in the light of remarks made in the printed articles, to see what he did. From table 5 (in particular from the rows for 100, 350, 400, and 650) it looks as though the interference will change sign at about 8 m. If we take the wire wave's half-length to be the 2.66-m mean that Hertz obtained in experiment 35, then a simple formula yields about 1.5 for the ratio of speeds. Further, the table also shows that increasing the length of the back-wire pulls a given phase of interference toward the zero point, which indicates that the direct action is the faster of the two (see appendix 14 for the reconstruction of both formulae).

Hertz pressed further on December 28. First, it was now extraordinarily important to obtain as precise a value as possible of the length of the wire wave in order to specify the ratio between its speed and that of the direct action. Experiment 57 resulted in a value of 2.75 m for the half-length; Hertz performed three other experiments (58–60) of this kind, and thereafter he used 2.8 m. Having pinned this important value down, Hertz now sought, for pressing reasons we will come to in a moment, to establish a relation between the speeds of the two direct actions: the electrostatic and the electrodynamic. The discovery experiment is nicely adapted to that end, in principle, because in this configuration the resonator is in fact responding to both static and dynamic forces, though the latter predominates.

Hertz's experiments 61 and 62 probe this situation by removing the wire; 61 uses a circular resonator, 62 a rectangular one. With the resonator in a given position, the signed value of the driving force will be $\alpha + \beta$ with the resonator's gap facing the oscillator, and $\alpha - \beta$ with its gap facing in the opposite direction, where α is the (closed-circuit) dynamic force, and β is the maximum value of the open-circuit action, which is here entirely electrostatic. Set the resonator in a given position and see what happens to the sparking when the gap turns from facing to pointing away from the oscillator. Move it a distance away along the baseline (a line normal to the oscillator's gap and lying in its plane) and repeat the trial. Let + indicate increased sparking, − decreased sparking, and 0 indifference. Hertz obtained essentially the same marks in the two experiments, but he felt that the ones done with the rectangular resonator (62) were much more reliable (table 6). If the half-length of the induction wave is about 4 or 4.5 m, as Hertz had already determined, then these results (which, however, lose observational certainty past about 6 m, where the sparking has become very weak) indicate that the static action must propagate vastly more rapidly because otherwise the marks should not change up to 4 or 4.5 m. These two experiments are also Hertz's first measures of the direct action's propagation that do not rely upon setting it off against a wire wave of known length.

TABLE 6 Experiments to examine the interference between the static and dynamic actions

−0.5 m	0 m	0.5 m	1 m	1.5 m	2 m	2.5 m	3 m	3.5 m	4 m	4.5 m	5 m	5.5 m	6 m	6.5 m	7 m	7.5 m	8 m
+	+	+	0	0	0	−	−	−	−	−	−	−	0	0	0	0	0

They are, however, much less reliable than the wire experiments because the sparking is very much weaker past a few meters.

These two experiments resulted in a question that Hertz now had to answer: why had his earlier experiments, with the resonator plane set vertically, not produced a finite speed for the direct action? The reply Hertz provided in his printed articles would certainly have been clear to him at this point; it has two parts. Consider first the experiments done up to about 3 m from the null point. These quite clearly indicated an infinite speed for the direct action because the interference changed character. Hertz had, however, now experimentally uncoupled the static from the dynamic direct action, a possibility that was in fact strongly suggested by Helmholtz's (and ipso facto Hertz's) interpretation of the general (Helmholtz) equation for propagation in a dielectric through the constant k. The static action falls off with the square of the distance, whereas the dynamic action falls off directly with distance. Thus, in regions closer to the oscillator the static action predominates, and one would expect to find that, in these regions, the interference tracks the wire wave because the static action has such a vastly higher, perhaps even infinitely higher, speed. At points farther away the dynamic action, which is only 50% or so faster than the wire wave, begins to predominate, and in these regions one would expect to find that the tracking grows less distinct, that, to use Hertz's printed words, "the retardation of phase proceeds more rapidly in the neighbourhood of the origin than at a distance from it" (Hertz 1888e, p. 118).[34]

Hertz now had two kinds of experiments. The first kind, in which the resonator responded to both static and dynamic actions, was useless at less than 5 or 6 m from the null point. This kind of experiment had, however, the advantage that sparks could be observed (albeit weakly) at quite large distances. The second kind of experiment, in which the resonator was dominated by the closed-circuit dynamic action, provided strong results up to 3 or 4 m but was useless past that point. Hertz accordingly decided to put the two kinds of experiments together: to use the second kind for regions below about 4 m and to extend the first kind up to about 12 m. He put the results together in the final table of his printed articles (table 7). These combined results gave Hertz what he wanted, including an estimate of about 7.5 m for the distance at which the interference changes sign. From that, and using 2.8 m for the half-length of the wire wave,

TABLE 7   Hertz's final table

	0 m	1 m	2 m	3 m	4 m	5 m	6 m	7 m	8 m	9 m	10 m	11 m	12 m
100 cm	−	−	−	−	0	0	0	+	+	+	+	+	0
250 cm	0	+	+	+	+	+	0	0	0	0	−	−	−
400 cm	+	+	+	+	0	0	−	−	−	−	0	0	0

Hertz asserted in print that the ratio of the speed of the oscillator's electrodynamic action to that of the wire wave is 75/47 (1888e, p. 121).

Hertz "followed the effect throughout the auditorium" (J. Hertz 1977, p. 247) on the thirtieth, not only observing the character of the interference but seeking also to determine the direction of the action, which has by this time matured into a "wave," a term he used for the first time in his laboratory notes for the twenty-seventh (Doncel 1991, p. 22). He described other results he obtained at this time in the article he later wrote to explain the essential working of the device:[35]

> One remarkable fact that results from the experiment is that there exist regions in which the direction of the force cannot be determined; in our diagram each of these is indicated by a star [fig. 76]. These regions form in space two rings around the rectilinear oscillation. The force here is of approximately the same strength in all directions. (Hertz 1888c, p. 92; see also 1888e, p. 122)

At these points the sparking did not change no matter how the resonator was turned around. The total force cannot, however, "act simultaneously in these different directions," because if it did, it would not do anything at all. The only other possibility was that its direction changed rapidly over time. According to his printed remarks, Hertz tried to understand how this might happen by resorting to neglected terms in his expression Σ, but he was unable to find a solution in this way. At some point it occurred to him, probably through analogy to polarization in optics, that rotation over time will occur if two causes act in different directions and if the ratio of their magnitudes changes with time. Hertz had two causes: the static and dynamic forces. But if the ratio does change with time, then the two forces do not remain in step with one another; they must, that is, have a phase difference. This was accordingly further evidence for propagation: one at least of the two forces must propagate at a finite rate since in the starred regions the static and dynamic forces are nearly orthogonal. In these regions whatever phase difference had built up between them would show itself most strongly (in optical analogy, the elliptical polarization of the resultant force would be the most pronounced).

It is, however, important to understand that for Hertz this elliptical variation was a proof of the very distinction between the static and dynamic forces exerted by the oscillator that field theory denies. He was at this point operating entirely within Helmholtz's framework. Indeed, it seemed to Hertz that one major virtue of his experiments was that they could be used to provide information about Helmholtz's general constants. He did not go into details (in the manuscript he sent to Helmholtz on January 21, 1888) but he did provide them in the paper describing how his apparatus works that he sent to Wiedemann for publication in the *Annalen* on February 17, along with new versions of his dielectric and discovery papers (Doncel 1991, pp. 5–6). Hertz emphasized two results in particular.[36]

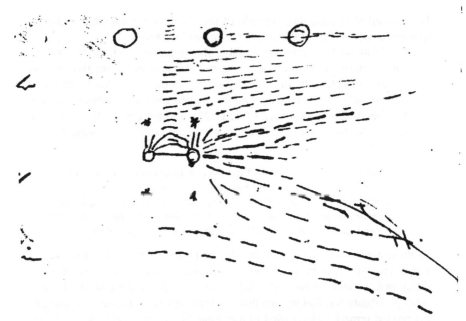

FIGURE 76   Hertz's diagram for regions (*) where the force directions cannot be determined

[1.] It is worthy of notice that, in the direction of the oscillation, the action becomes weaker much more rapidly than in the perpendicular direction, so that in the former direction the effect can scarcely be perceived at a distance of 4 metres, whereas in the latter direction it extends at any rate farther than 12 metres. Many of the elementary laws of induction which are accepted as possible will have to be abandoned if tested by their accordance with the results of these experiments.

[2.] Along one pair of straight lines the direction of the force can be determined at every point. The first of these straight lines is the direction of the primary oscillation itself; the second is perpendicular to the primary oscillation through its centre. Along the latter the magnitude of the force is at no point zero; the size of the sparks induced by it diminishes steadily from greater to smaller values. In this respect also the phenomena contradict certain of the possible elementary laws which require that the force should vanish at a certain distance. (Hertz 1888c, p. 92)

What did Hertz have in mind here? What "elementary laws" were contradicted by results 1 and 2, and why? We have repeatedly witnessed him distinguishing between the electric force that arises from static causes and the one that arises from dynamic action. The static force is produced by an electric dipole (the oscillator) and has a component parallel to the oscillator; indeed, along the two particular directions singled out here by Hertz the static action is entirely parallel to the oscillator. If we take the usual Neumann form of the electrodynamic potential (in which the constant k in Helmholtz's general expression is unity), then the corresponding electric force due to changing current in the oscillator will also parallel the oscillator. Since the static force decreases much more rapidly then the dynamic force, the total force that drives the resonator will, at least at some distance from it, weaken at about the same rate along a line normal to the oscillator as along a line parallel to it, whereas Hertz's observation 1 requires a much greater rate of change in the former than in the latter direction. Further, because the electrodynamic force opposes the static force at points such that lines drawn through them and normal to the oscillator lie between the latter's poles, the action along the perpendicular direction should at some point vanish (inasmuch as the dynamic force decreases less rapidly than the static, whereas the static force predominates near the oscillator). This however violates Hertz's observation 2.

Consequently, the Neumann form of the potential could not be correct, nor could any form that led to a similar expression for the force. Other forms remained possible, given Helmholtz's general expression, because the latter con-

TABLE 8   The sequence of Hertz's dipole experiments

Date	Experiment
October–November 1886	Creation of the dipole oscillator and the resonator
December 3, 1886	"Peculiar effect on sparks"—the photoelectric effect
December 1886	Experiments on electromagnetic polarization of dielectrics
December 1886	Experiments on electromagnetic induction by dielectrics (?)
September–October 1887	Experiments to probe the dependence of resonator sparking on the position of the resonator's spark gap and use of a "moveable induction conductor" to manipulate sparking by interference
October 1887	Successful use of a dielectric instead of a moveable conductor to affect resonator sparking by interference
November 7, 1887	Detection of waves in wires
November 11–12, 1887	Fruitless experiments using wire-wave interference "to determine the velocity of propagation through air"
December 17, 1887	Numbered experiments begun to show that electromagnetic force does *not* propagate
December 23, 1888	Detection of a finite velocity using a new orientation of the resonator
December 27, 1887	First reference to waves
December 30, 1887	New observations up to 12 meters
late February–March 1888	Reflection experiments

tained the otherwise undetermined constant k.[37] Here, then, one had for the first time a way to probe the troublesome constant k. Hertz at this time did not conceive that the existence of propagation per se invalidated conclusions drawn in the usual way for circuit interactions. For some time he continued to distinguish the existence of propagation from the question of what terms appear in the underlying potential law. Experiments could bear on both points, but they were to be distinguished from one another. At this time, and indeed well into 1888, Hertz was still thinking in essentially Helmholtz's terms. This changed as he begun to play with his new effect; as he altered his apparatus into stable, *wave-manipulating* equipment his conceptions altered as well. This took place during the following spring, as we shall see in the next chapter.

SEVENTEEN

Electric Waves Manipulated

17.1. THE NECESSITY OF PERSUASION

"Weary from experimenting," Hertz wrote in his diary on the last day of the year, which he understandably looked back on "with pleasure." The next day he wrote his parents, complaining that he had not had enough Christmas greetings and telling them that his "results are clearer and more important than I should have ever dared to hope six months ago," though six months before he had had no notion at all of demonstrating propagation. But as was usual with him, he was now somewhat depressed, anticipating the distasteful risk of carrying on with additional experiments, "because one can only lose by them"— as, of course, he had years before in his work on evaporation. A week later he began writing up his work, which he found "terribly difficult." He was troubled by occasional doubt: the results were "too far reaching" and might be "figments of the imagination." But by January 21 he had put aside his doubts and had written the article, which he sent to Helmholtz, who immediately wrote back "My heartiest congratulations on this success too."

The paper (Hertz 1888a) was received by the *Sitzungsberichte* on February 2. On February 5 von Oettingen from Dorpat visited Hertz, and he was apparently the first colleague to witness Hertz's trials. Von Oettingen, who reacted enthusiastically, had, Hertz remarked to his parents, already heard about the experiments from Helmholtz in Berlin but "had not understood them there." Little wonder: to understand Hertz's experiments required very careful attention to novel and therefore unfamiliar instrumental details. Indeed, it is extremely unlikely that at this point, and even after reading Hertz's two *Sitzungsberichte* papers (this one on propagation and the immediately preceding one on the dielectric effect), anyone could possibly have grasped what Hertz had actually done without paying him a visit. To understand his claims required thoroughly understanding his apparatus, and as yet he had not published an account of how it worked. He had described its progenitor in his paper on rapid electric oscillations, and he had written about its use in detecting the electrodynamics of dielectrics, but he had as yet nowhere explained how or why it worked as a detector. Details were missing, and they were essential in order for Hertz's discovery claim to be validated. The same was not true for Hertz's dielectric experiments, even though they used the new device, because in them all he had had to show was that the dielectric produces the same sorts of effects on sparking as a neighboring conductor.

This was one of the reasons for Hertz's taking advantage of an opportunity afforded by a letter from Wiedemann, the editor of the *Annalen,* which he had received shortly before January 24. Wiedemann wanted to print his "last paper" (J. Hertz 1977, p. 251), which was the one concerning dielectric effects (Doncel 1991, p. 3, n. 5). Hertz wrote back asking instead that Wiedemann print a trilogy, including reworked versions of his *Sitzungsberichte* reports on the dielectric effect and propagation, as well as a new one. This last, which was therefore written well after the discovery events of December, appeared as "On the Action of a Rectilinear Electric Oscillation on a Neighbouring Circuit" (Hertz 1888c).[1] Hertz wrote this paper with two goals in mind. First, he wanted to make it possible for readers to understand how his new device could be relied upon to stably produce novel forms of behavior and how such behavior was to be interpreted. But, second, the paper was also designed to make Hertz's own route to propagation seem to be much more straightforward and indeed inevitable than it had been. The paper is presented as a complete account of the trials he had performed before he succeeded in producing propagation. In particular, the paper's penultimate section discusses "the forces at greater distances." The last of the four points in this section concerns the regions at which the direction of the force could not be determined and concludes with the remark that "we probably have before us here the first indication of a finite rate of propagation of electrical actions." This observation was, however, almost certainly made after Hertz had already obtained propagation.[2] The other two papers were rewritten to refer to this new one, giving the entire trilogy the character of a more logical progression than had in fact been the case. Hertz's "On the Action" paper served as a sort of manual of operations for his device. From it one could begin to learn how to set the apparatus up, how to move the resonator around, what to look for, what to ignore, what to adjust, what not to adjust, and so on.

February passed uneventfully as news of Hertz's work spread among his German colleagues by word of mouth and through the *Sitzungsberichte* account. Hertz recorded no other visits or correspondence until the beginning of March, when he received the first concrete indication that he had finally succeeded in reaching the front ranks among his peers. On March 4 Röntgen of Giessen, Hertz's most creative competitor in electrodynamics, the very man who had bested him in first demonstrating the electrodynamic force exerted by moving, charged dielectrics, wrote, in Hertz's triumphant words, "with congratulations on my splendid papers, saying that he considered them the best achievement in the field of physics in recent years. That is a great compliment, considering that he has also produced papers himself and is in many respects a competitor" (J. Hertz 1977, p. 253). Apparently no one else visited in March, nor did Hertz record any other noteworthy congratulations. He was, we shall see, busy trying to make his newly produced propagation behave like proper waves. He spoke about "the propagation of electromagnetic waves" on March 10 to the Mathematics Club in Karlsruhe shortly before the spring vacation. His new experiments had thoroughly stabilized by March 17, and on March 19 he wrote Helmholtz about them. By the end of the month he had begun to think about and to calculate with "Maxwell's theory."

Hertz's rapid and tremendous success evidently took several of his colleagues quite by surprise, for it seems that among some of them at least his abilities were not overly esteemed. In early April, for example, Hertz, on a walking tour, stopped at Tübingen to visit Ferdinand Braun, professor of physics, "at the castle." There he also met Leo Graetz, then *Privatdocent* for theoretical physics at Munich and himself a student of both Helmholtz and Kirchhoff. Braun had preceded Hertz at Karlsruhe and had evidently included him on the list of three that he had recommended as his successor (Jungnickel and McCormmach 1986, vol. 2:85). They had lunch. The diary entry for that day mentions the lunch and then just notes that Hertz "returned home in the afternoon." Six months later, by which time Hertz was achieving widespread notice, he recalled the visit, remarking that when he "was at Tübingen at the beginning of April, I had all the material that has now been published ready, and would have loved to talk about it, but no opportunity arose, since no one else brought up the matter, although by then enough had been published to make a physicist eager to ask questions" (J. Hertz 1977, p. 265).

Neither Braun nor Graetz said a word, though both must certainly by then have known about Hertz's claims for dielectrics and for propagation from the *Sitzungsberichte* at least. Braun's biographer printed a letter written in February from Braun to Graetz which is worth quoting:

> I am no longer impressed with Hertz's effect of ultraviolet light on electrical discharges, after having determined recently that the effect as good as disappears in the electrostatic case, when magnesium illumination alters the discharge potential by no more than 2–3%, too low to be measured. H. should have tried that and described it! But it is a dreadfully exaggerated representation. That's just like his elastic floating plate—the whole thing can be stated in a couple of words, only then everyone would know that a piece of paper with a bulge in it can support a greater weight the more it bulges. (Kurylo and Susskind 1981, p. 68)

Braun was only seven years older than Hertz. He had studied at Marburg and then in Berlin starting in 1869, where Magnus's reign was soon to end. He assisted Quincke at the latter's laboratory in the local *Gewerbeakademie* and wrote his dissertation on acoustics; Helmholtz was one of the examiners in 1872 (indeed, Braun was the first student at Berlin to be examined by Helmholtz [Kurylo and Susskind 1981, p. 19]). Braun published in electricity, but he just missed becoming part of the new intellectual order that Helmholtz established at Berlin.

Little wonder that neither Braun nor Graetz said anything to Hertz, much less offered congratulations. Braun for one thought that Hertz's previous work was "dreadfully exaggerated." The ultraviolet light effect, Braun claimed, could not be fundamental because it can easily be made to go away when it ought not. Braun would almost certainly have had much the same to say about Hertz's work on cathode rays and on evaporation: too exaggerated, too overstated. Or, to put his reaction into contemporary context, what Braun objected to was Hertz's constantly claiming to have sought and to have elicited novel effects, when all he had done was to find a few limited, minor, and really unimportant physical baubles. Braun himself had not absorbed the Berlin ethos that impelled a constant search for thoroughly novel effects; he had come too soon for that. Hertz, he knew, had powerful support from Berlin, but he, Braun, could not see that Hertz had done anything to deserve it. If Graetz disagreed, he apparently kept his own counsel.

Hertz was particularly put out by Braun's and Graetz's silence because he had just succeeded in making electric waves come to life. Until March Hertz had not been experimenting with waves; he had been trying to show that electrodynamic forces propagate. The difference is subtle but crucial, both for Hertz's own emerging views and for the electric contagion he sought to breed among his contemporaries.[3] To uncover it let us for a moment retrace Hertz's work.

In the fall of 1887 Hertz conceived the idea of using his new device to show that electric force propagates in order to provide a compact answer to the Berlin questions. Propagation was for him just as much a tool as it was an end in itself. He did not, in other words, sit down sometime in the summer of 1887 and decide that he was going to turn the German physics community upside down by making them believe that electric action takes time to travel. He decided, *faute de mieux,* to bring propagation to a trial because he could not devise a workable way to show that electrodynamic induction can polarize dielectrics.

The December propagation trials were, not surprisingly, molded by Hertz's route to them. Recall his recently completed experiments with dielectrics. These experiments had a bipartite structure: stabilize the apparatus as a detector of unproblematic induction, and then use it to try problematic induction. Hertz's propagation trials in December had the same structure. First, he had to persuade his audience that the device could pick up an unproblematic effect, namely, wire waves. Then he had to use the device to show that the direct propagation will add its effect on the device to that of the wire wave in a way that indicates propagation. This was hardly an easy task and at first he thought he had shown that the effect did not indicate propagation. Hertz's entire attention was at that time concentrated on one thing only: the instrumental outcome of a contest between an action everyone believed to propagate and an action that might, or might not, do so.

Given this, we can see why Hertz would not have been thinking much, if at all, about waves. Indeed, it is entirely possible that Hertz did not at this time think that the oscillator actually did send out waves, properly speaking; certainly he did not use the word at all in his laboratory notes until the end of December.[4] The presence or absence of waves was not at this time directly important to him, for the following reason.

As the oscillator vibrates it varies the charge on the metal plate to which the wire is connected, and this wire is in consequence the site of a wave; that is, the value of the current at a given point of the wire depends both upon the time and upon its distance from the oscillator. Let us for a moment ignore the succession of phases in the wire as they pass a given point and concentrate instead on a given phase as it moves along. Suppose next that the direct action also propagates. At a given moment the oscillator will therefore do two things: it will send out a given phase along the wire, and it will also send out a given phase of direct action (which differs from the wire phase by a quarter-period). If these two phases of induction-capable actions travel at the same rate, their combined effect, whatever it may be, will be the same throughout their journey down the wire. If, however, they travel at different rates, they will separate from one another, and they will cease to work together, giving rise to different effects at different points. Here we concentrate entirely on the relationship between twinned loci, each with a given phase, as one of the pair moves through the wire and as the other one travels through the air. We need not think further than this, namely, to the phase relationship between succeeding loci in air, because we are at this point interested only in what our detector, the interference-driven resonator, will reveal as it seeks to determine whether the twinned loci do or do not remain in step, or indeed whether there is, properly speaking, such a locus in air at all. All we need to know thus far is this: if the loci travel together, the resonator always shows the same response; if they travel at different rates, the resonator responds differently at different points, but not in step with the wire wave, which we have otherwise examined; if the direct action does not propagate at all, the resonator tracks the wire wave. No need here to think beyond this, to the succession of direct actions proper, because the behavior of such a succession was not itself an issue: the wire wave's presence obviates having to think about it.[5]

This way of thinking began to change by the time that Hertz calculated the speed of the direct action on December 27, when he first refers to waves, because the calculation otherwise makes no sense. Until then, however, Hertz was not concerned with the direct action *as a* wave; he was concerned with it as a propagating effect. Things that have yet to come alive in the laboratory, things whose very existence is at issue, are often instrumentally and intellectually amorphous; what they can do to other things, what other things can do to them, what properties they have, all remain for the investigator to construct. Hertz had convinced himself by the end of December that propagation took place, and by then he was thinking in terms of waves; but would his work, in which waves appear as hidden modifiers rather than as robust presences, be thoroughly convincing to all others?

Röntgen was apparently convinced; so was Helmholtz and probably others within the orbit of Helmholtzian influence. But, if we are to judge by Braun's and Graetz's silence, neither of whom was in that orbit, then Hertz's argument was not entirely persuasive. It could be resisted (though it was not, in print at least) by querying his novel device and its working in this *indirect* trial of propagation. To understand what Hertz had done here, as von Oettingen's remark to him on February 3 makes clear, was no mean feat, and it was best done by visiting the trial site itself, where Hertz could show "experiments all afternoon." Hertz needed, and he was quite aware that he needed, a way to bring propagation to bawling life; propagation had to cry out its presence, not just whisper it. Hertz made essentially this point in the first paragraph of the article in which he claimed to have done just that:

> I have recently endeavoured to prove by experiments that electromagnetic actions are propagated through air with finite velocity. The inferences upon which that proof rested appear to me to be perfectly valid; but they are deduced in a complicated manner from complicated facts, and perhaps for this reason will not quite carry conviction to any one who is not already prepossessed in favour of the views therein adopted. (Hertz 1888f, p. 124)

Hertz needed to produce conviction, to make people like Braun and Graetz believe in propagation. He had already performed experiments to do so before his unsatisfying lunch with them in early April. That meeting no doubt reinforced in him the certainty that his new experimental way was much more than a supplement to his earlier trials; it, and it alone, carried the possibility of enforcing consensus.

17.2. ALONE WITH NATURE?

A month after he had demonstrated propagation Hertz began trying to detect waves. In the introduction to *Electric Waves* Hertz reconstructed his thoughts and actions at the time in the following way:

> While investigating the action of my primary oscillation at great distances, I came across something like a formation of shadows behind conducting masses, and this did not strike me as being very surprising. Somewhat later on I thought that I noticed a peculiar reinforcement of the action in front of such shadow-forming masses, and of the walls of the room. At first it occurred to me that this reinforcement might arise from a kind of reflection of the electric force from the conducting masses; but although I was familiar with the conception of Maxwell's theory, this idea appeared to me to be al-

most inadmissible—so utterly was it at variance with the conceptions then current as to the nature of an electric force. (Hertz [1893] 1962, p. 11)

Further, in discussing here his early belief that the air wave travels faster than the wire wave Hertz remarked that "while performing the experiments, I never in the least suspected that they might be affected by the neighbouring walls" (p. 9).

We can date precisely when Hertz first saw something that he has here described as shadow formation. The entry in his laboratory notes for December 29, when Hertz first followed the action to great distances, mentions that he then "observed the protective action of a zinc sheet." Although Hertz was certainly by this date thinking that the direct action propagates as a wave, he did not yet think that it might behave like a wave in other respects as well, such as reflection. The "protective action" he found at the end of December accordingly seemed, at the time, to be the same sort of thing as the shielding action of a conducting mass from electrostatic force and as such was hardly to be paid much attention. According to the introduction Hertz soon after observed a "peculiar reinforcement," which he at first thought might be due to a kind of electric reflection, but he dropped the idea as ridiculous for some time. There is no mention of any such observation in the laboratory notes, but the last entry in them is for December 30. Hertz might have seen some such thing during January or February.

The diary entry for February 27 reads: "Prepared for new experiments. Made metal shields." On March 2 he tried to reflect the "electromagnetic waves" (this is the first such phrase in the diary) from a zinc plate, "with obscure results." By then, clearly, Hertz had decided that he might be able to manipulate his electric waves as though they had properties like waves of sound or light, but he had trouble making them behave. Three days later he could make "shadows by electromagnetic beams," and on that day he had his laboratory assistant, Amann, make "a large concave mirror." On March 10 Hertz notes that the mirror did not work, though there was "a hint of positive results." He apparently decided to put the mirror aside at the time—and little wonder that he did, because he was trying to focus 9-m waves with a mirror that could have been no more than a quarter their length, which may again indicate that he had not as yet completely assimilated these new kinds of waves to the old kinds in his thinking. But then, on the twelfth, he thought that he had "detected standing electromagnetic waves by reflection in the hall." He built a flat mirror of zinc and pursued the experiments on the fifteenth, when he "grew more certain about standing waves." Over the next two days he must have become extremely adept at playing with these novel things. On the seventeenth he wrote his parents a supremely confident letter, one that reflects his skill in fabricating and manipulating entirely new entities in ways that did not require expert understanding of arcana to appreciate:

I am enjoying my vacation very much. I began it very well, for I had the large chandeliers removed from the auditorium to give me the largest possible empty air space, and yesterday I conducted some new experiments in the space thus obtained, and these if I am not mistaken (for I have done them only once so far), were brilliantly successful. They yield only the same results as my last experiments, but much more directly and clearly, and seem to be a beautiful confirmation. . . . You yourself would be able to appreciate the importance of these experiments if I took the trouble to make them plain to you. In scientific papers there is no reason to do so, since the reader may be assumed to be able to draw the obvious conclusions himself. In my work I now have the comfortable feeling that I am so to speak on my own ground and territory and almost certainly not competing in an anxious race and that I shall not suddenly read in the literature that someone else had done it all long ago. It is really at this point that the pleasure of research begins, when one is, so to speak, alone with nature and no longer worries about human opinions, views and demands. To put it in a way that is more learned than clear: the philological impulse ceases and only the philosophical remains. (J. Hertz 1977, p. 255)

While Hertz was trying to manifest propagation, skeptical colleagues stood looking over his shoulder. Once he could make his effect behave in a wavelike way, his fear that he would not be able to argue convincingly, or that he might be anticipated, lessened. He now felt "alone with nature" precisely because his new "philosophical" (curiosity-driven) experiments were so well adapted to "philological" (persuasive) purposes.

During the first two weeks of March Hertz reconstituted his apparatus as a generator and a detector of electric standing waves in air. He had already produced and manipulated standing waves in conducting wires, and these air experiments may have been undertaken with the wire trials in mind. There was, however, a major difference between the two cases, as a result of which Hertz developed a new way to manipulate the detector. In the earlier experiments with wires Hertz had set the resonator with its plane containing, but with its spark gap facing directly away from, the wire. To map the wave he had then moved the resonator down the wire, observing the sparking along the way. This technique could not be directly applied to air waves because the configurations are essentially different.

Hertz did not discuss the point, and he might have discovered it by direct trial. On the other hand, he may never have tried any such experiment if he was certain that the wire-wave experiments had to do only with the old, standard inducing force of one wire on another one close and parallel to it, whereas the air-wave experiments had to do with a propagating force. This has a direct, and obvious, instrumental effect. In the case of the air waves, to achieve strong sparking still requires the resonator's gap to lie parallel to the inducing wire (which is here the distant oscillator), but the resonator must now move along a line that is perpendicular both to the acting wire and to its own gap. Wire-wave experiments could not accordingly function as models for air-wave ones.

A new approach had to be found, and Hertz produced one by relying directly on the finite size of the resonator in respect to the wavelength that he was trying to detect. In the diagram that he drew for his published article (fig. 77) the oscillator is not even present: it lies far to the right, and the distances are given with respect, not to it, but to the reflecting wall A. The trial consisted in placing the resonator at some point, with its plane parallel to that of the oscillator and with its spark gap either directly facing or directly opposing the reflecting wall. Observe the sparking in one orientation, and then turn the gap to face the other way. Observe the sparking in the new position. The sparking changes markedly, and Hertz mapped these changes to the state of the electric standing wave that he envisioned in the experimental space before the wall.

Let us map the wave with Hertz. Put the resonator in position I, with its spark gap facing the wall. If we assume for a moment that the standing wave's length is comparable to that of the wire waves that Hertz had been investigating, then the 35-cm diameter of the resonator will place its gap near a part of the standing wave that has a much lower amplitude than the part near the side of the resonator opposite the gap. Hertz drew arrows at these points, whose lengths represent the relative strengths of the wave there. What will be the net effect on sparking? The sparking behavior of the resonator, Hertz knew, was governed by the force that acts on the part opposite its gap. In his previous experiments the force at the gap itself was, he felt, entirely inconsequential. But now he exploited it. In position I we have, in Hertz's words, "a stronger force acting under favourable conditions [i.e., opposite the gap] against a weaker force, which acts under unfavourable conditions [i.e., at the gap]." Sparking of a certain magnitude occurs. Now turn the gap to face away from the wall. Here "the stronger force now acts under unfavourable conditions against a weaker force, which in this case is acting under favourable conditions." Whatever the precise outcome of the contest, the sparking had to be weaker than in the previous orientation.[6]

We come now to the core of Hertz's new method. Move the resonator away until a position is found, if it can be, in which the sparking behavior seen at I is reversed; to a position, that is, in which the sparking is stronger with the gap facing away from the wall than toward it. This corresponds to Hertz's position II, a quarter-wavelength farther away, for here the loci of the strong and weak forces have switched from one side of the resonator to the other. "Our explanation," Hertz continued, "carries with it a means of further testing its correctness." In positions V, VI, and VII the sparking should not differ between these two orientations, but the current in the resonator must reverse direction as the gap rotates from one side to the other, in which case the sparking must vanish somewhere between the two.

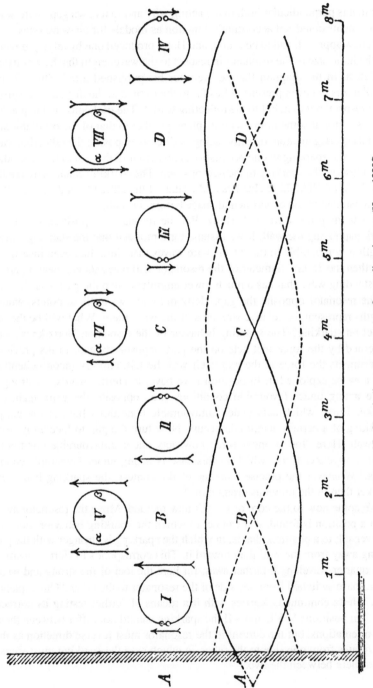

FIGURE 77 Hertz's direct measurement of a standing air wave (Hertz 1888f)

Performing the trial, Hertz mapped positions I, II, and III, which are, respectively, 0.8 m, 3 m, and 5.5 m from the wall. "At distances lying between those mentioned," that is, at B and C, "both sets of sparks under consideration were of equal strength." Further, near C the sparking behavior changes more rapidly with distance than near B. Point B, the locus of an antinode, could be directly found, or it could be estimated by locating the place to its left where no sparking occurs when the gap faces away from the wall, and the place to its right where no sparking occurs when the gap directly faces the wall; B lies halfway between the two, putting it at 1.72 m. This, Hertz wrote, "agrees within a few centimetres with the direct observation." Since the node C lies 2.4 m from B, the origin of the reflected wave would be 0.68 m behind the wall, with the half-wavelength then being 4.8 m. Hertz had "by an indirect method" previously found 4.5 m for the half-length, and he thought it best to take the mean (4.65 m), in part because it seemed to him unlikely that the origin should lie so far behind the metallic wall.

Hertz tried the experiment with a resonator half the diameter, using an oscillator of appropriate dimensions. The effect was much weaker but quite determinate. He then tried to detect a traveling wave by placing the oscillator between the wall (but initially near it) and the resonator, so that the resonator would pick up a wave formed from the interference of a direct wave from the oscillator and a reflected wave from the wall. To do so he placed the resonator 14 m away and moved the oscillator between trials. The sparking varied in the resonator with the oscillator's locus in such a way "that the wave-lengths already obtained are in accordance with the phenomena."

The standing-wave experiments, unlike Hertz's discovery trials, did not depend upon a subtle understanding of the forces at work; they provided powerful, visible indications of the presence of a wave. Hertz knew that these trials would be widely influential and widely replicated, since "no great preparations are essential if one is content with more or less complete indications of the phenomena. After some practice one can find indications of reflection at any wall." On March 19 he wrote Helmholtz of his success. The letter concluded with an extremely illuminating remark:

> It seems to me that the most important point is now the certainty that propagation through conducting wires is perceptibly slower than in air, whereas all previous theories, including Maxwell's, led to the conclusion that propagation in wires took place with the velocity of light. (J. Hertz 1977, p. 255)

The "most important point" was not that Hertz could now manipulate waves nearly at will, that speculative physical entities had been incarnated in the laboratory, though this was how his work was received in Great Britain. It was that he had found something that apparently conflicted with "all previous theories, *including Maxwell's.*" This was a powerful claim to fame indeed, to have used an effect no one before had produced, though many had assumed, to undercut the very structures that implied it. What several years later appeared to Hertz himself to have been some sort of strange artifact of the particular circumstances under which the trials were first performed, namely, the difference between wave speeds in wires and in air, seemed at the time to be something very different indeed.

Precisely because Hertz was now convinced that he was on the track of an entirely novel effect that might upset every contemporary theory of electrodynamics, he decided that he had to understand as thoroughly as possible just what Maxwell's theory had to say about the waves produced by his oscillators and about the speed of waves in wires. By the middle of October he had succeeded in calculating maps of dipole radiation. The paper that he wrote, entitled "The Forces of Electric Oscillations, Treated According to Maxwell's theory," was immensely influential, constituting the basis of field theory for physicists outside Great Britain. Yet its penultimate paragraph describes something that Maxwell's theory "is unable to account for."

17.3. THE RADIATION FIELD REALIZED ON PAPER

The article that Hertz and Elisabeth, who drew the field maps, prepared in October (Hertz 1889a) was, for those outside Britain, the most influential discussion of the electromagnetic field ever written. Widely read and frequently referred to, Hertz's account served at least four distinct functions. First, it provided a compact version of field equations together with an appropriate, abbreviated physics that avoided some of the deeper problems that distinguished it from other structures, thereby easing its absorption. Second, it fashioned the elements of an entirely new instrumental domain: what we might in retrospect call antenna physics. Third, it reconstituted Hertz's earliest air–wire interference experiments in the new context, showing the sense in which some at least of former methods (such as those deployed by Hertz himself several months before) had now to be thought "approximate." Finally, Hertz used his newly developed analytical techniques to deepen his claim to have detected something novel and incompatible with the "limiting" form of Helmholtz's electrodynamics that was embodied in Maxwell's equations: the wave behavior of wires. In this single article Hertz brought together an introductory lesson for the untutored, the construction of a new physical regime, and a claim to discovery.

The foundational equations in Hertz's seminal article are the same in form as the ones he had obtained five years before for Maxwell's theory. It is worthwhile writing them out just as he did, in full component form, because this was what many German physicists saw in first coming seriously to grips with field physics:

$$A \frac{dL}{dt} = \frac{dZ}{dy} - \frac{dY}{dz} \qquad A \frac{dX}{dt} = \frac{dM}{dz} - \frac{dN}{dy}$$

$$A \frac{dM}{dt} = \frac{dX}{dz} - \frac{dZ}{dx} \qquad A \frac{dY}{dt} = \frac{dN}{dx} - \frac{dL}{dz}$$

$$A \frac{dN}{dt} = \frac{dY}{dx} - \frac{dX}{dy} \qquad A \frac{dZ}{dt} = \frac{dL}{dy} - \frac{dM}{dx}$$

$$\frac{dL}{dx} + \frac{dM}{dy} + \frac{dN}{dz} = 0$$

$$\frac{dX}{dx} + \frac{dY}{dy} + \frac{dZ}{dz} = 0$$

In more familiar form these are

$$A \frac{\partial H}{\partial t} = -\nabla \times E$$

$$A \frac{\partial E}{\partial t} = \nabla \times H$$

$$\nabla \cdot H = 0$$

$$\nabla \cdot E = 0$$

These determine the "forces in free ether," and Hertz used them to make his novel device a creature of field physics. Or, rather, he carefully built a special structure that made these equations useful for his purposes while avoiding the difficulties that Maxwellian conceptions of sources posed for the uninitiated (such as Hertz himself). This is worth pursuing to understand just how novel Hertz's construction was.

Maxwellians thought of objects like Hertz's dipole as specifying regions where field structures stopped, not as objects that controlled or produced fields. Among Maxwellians devices of this kind posed problems because they were not amenable to simple boundary analysis; it was much too hard, except in the most elementary of cases (e.g., a sphere) to specify what went on at their peripheries, where the field structures were rapidly altered. Hertz had, how-ever, not as yet thoroughly grasped this quintessential characteristic of field physics. For him the dipole was the object that produced the electromagnetic field, not (or not just) the place where the electromagnetic field underwent severe changes. He could not analyze precisely what went on at the dipole's surface in a useful way, anymore than the Maxwellians could, but he was able to do something that they found very difficult: he was able to avoid the problem altogether by surrounding the dipole with a closely fitting spherical surface into which his analysis did not enter. For him it did not matter just what hap-pened at the dipole's surface; for Maxwellians, at least until they read Hertz's work (and even afterward), this disregard was difficult to accept.

Having transformed the immediate neighborhood of the dipole into terra

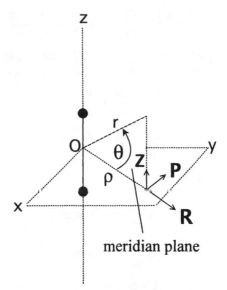

FIGURE 78   Hertz's mathematical dipole

incognita, Hertz turned to his time-tested methods of handling equations of this sort: he sought an appropriate potential to provide a useful physical solution. The potential, which he designated Π, was chosen to satisfy a wave equation; the fields were then defined in terms of the potential, given the axial symmetry of the dipole, in such a fashion that they satisfied the Maxwell equations. This required introducing an auxiliary quantity, Q, which, we shall see in a moment, had tremendous physical significance. Figure 78 shows the coordinate system Hertz used. The dipole lies along the z axis, and because of symmetry the distribution of force will be independent of orientation about the dipole axis. He accordingly introduced a cylindrical system (ρ, z) for analyzing the forces given any such "meridian plane": ρ represents the distance of the field point from the origin (which lies at the center of the dipole) projected into the xy (or equatorial) plane, and z is its distance from the latter.

This symmetrical system is based on the assumption that the electric field at a given point lies entirely within the meridian plane that passes through that point. The electric force may accordingly be decomposed into two components: one, R, lies along the intersection of the meridian and equatorial planes (i.e., along the direction of ρ), and the second, Z, lies along the z axis. The magnetic force is similarly assumed to have no component normal to the dipole (i.e., along ρ), so that it can in principle be specified by two components: P is perpendicular to the meridian plane, and N parallels the dipole.

No trace remains of the hard trials that Hertz must have gone through before he succeeded in fitting this scheme to a potential that satisfies the wave equation. He was, however, a past master by this time in building potential functions, and he eventually discovered that the following set of relations would satisfy all of his Maxwell equations:

$$A^2 \frac{\partial^2 \Pi}{\partial t^2} = \nabla^2 \Pi$$

$$Q = \rho \frac{\partial \Pi}{\partial \rho}$$

$$\rho Z = \frac{\partial Q}{\partial \rho} \qquad \rho R = \frac{\partial Q}{\partial z} \qquad \rho P = A \frac{\partial Q}{\partial t} \qquad N = 0$$

As usual Hertz did not bother with questions of uniqueness, just as he had ignored such things years before in writing on elastic impact. He at once pointed out the central physical significance of the function Q, which, like Π, depends on ρ, z, t only.[7]

Choose any meridian plane, and, in it, draw a line such that, along this line, Q is a constant. Form a series of such lines, that is, lines along which Q is constant. Then draw the orthogonal trajectories to this series of lines. These trajectories will themselves be orthogonal to the lines of electric force, which therefore means that the lines of constant Q are lines of electric force: map Q, and you have also mapped the electric field.[8]

To produce mappable values for Q Hertz began directly with the spherically symmetric solution to the wave equation for Π:

$$\Pi = El \frac{\sin(mr - nt)}{r}$$

He immediately turned to the limiting case in which the radial distance r to the field point is much smaller than the wavelength ($2\pi/m$). Here, he easily found, Π reduces to $-El\sin(nt)/r$, and the E field can be expressed as $\nabla(\partial\Pi/\partial z)$. Consequently, the electric field appears close to the dipole as the gradient of the function $-El\sin(nt)\partial(1/r)/\partial z$. This is precisely the same in form as the field due to a static dipole of strength $-El\sin(nt)$, that is, one with an oscillating moment. In this region, Hertz had found, a static regime holds sway insofar as the electric field is concerned, and the magnetic field obeys the Biot–Savart law, so that the dipole behaves magnetically like a current element of magnitude $AEln\cos(nt)$.

Elsewhere the situation was considerably different, and the function Q had to be brought directly into consideration. From his solution for Π Hertz determined Q to be[9]

$$Q = \rho \frac{\partial \Pi}{\partial \rho} = Elm \left[\cos(mr - nt) \frac{\sin(mr - nt)}{mr} \right] \sin^2\theta$$

Before using Q to map the field, Hertz employed it to distinguish four regions. The first, which corresponds to the modern near-field, specifies an essentially nonradiative regime, as we have already seen. In the second, along the dipole axis proper, the E field lies along the axis and drops off initially as $1/r^3$, further along as $1/r^2$. In the third, which lies in the equatorial plane, the E field behaves as in the second case but drops off farther away as $1/r$. In the fourth, and final, case, which corresponds to the modern far-field and which Hertz defined essentially as a regime in which the distance is sufficiently large that the sine term in Q can be neglected, the E field is everywhere orthogonal to the radius vector from the dipole's center and varies as $Elm^2\sin(mr - nt)\sin\theta/r$.

Hertz's goal now was to provide a physical feel for the behavior of the field over time by using Q to map it at each quarter-period throughout an entire cycle, though he remarked that the "best way of picturing the play of the forces would be by making drawings for still shorter intervals of time and attaching these to a stroboscopic disk," reproducing, as it were, the motion of the field lines. This was not an easy task; even today it is not entirely simple even with the aid of a powerful computer. To map the field he separated a given value for Q into two parts ($\sin^2\theta$ and its factor) and found the values of r or θ that corresponded to a given choice for the appropriate factor. "On setting about the construction of these curves," Hertz remarked in considerable understatement, "one perceives many small artifices which it would be tedious to exhibit here." The maps he generated in this way (fig. 79) were reproduced again and again in subsequent years, often literally, until they eventually came to define for many people what a Hertzian dipole was: it was whatever generated this sort of radiation pattern.

17.4. Reconstructing Previous Results

Although Hertz's analysis powerfully taught how to use and to interpret field equations, nevertheless the bulk of his article was concerned with two other aims that were particularly significant to him. The first, which we will examine here, concerned the interference experiments of the previous winter in which he had first convinced himself that propagation takes place. The second concerned wave speed in wires, an issue of prime importance to Hertz, and which we will also examine below. The interference experiments, recall, had taken their meaning under a particular interpretation: namely, that the "electrostatic" and electromagnetic forces which emanate from the dipole must be treated independently of one another, with only the electromagnetic force traveling at a detectably finite speed; the electrostatic force propagates vastly more rapidly. This is explicitly permitted, indeed suggested, by Helmholtz's equations except in the extreme cases that either the susceptibility of the ether is infinite or that the constant k is zero. These were of course precisely the two conditions that were widely taken, in Germany, to constitute the Maxwell limit of Helmholtz's

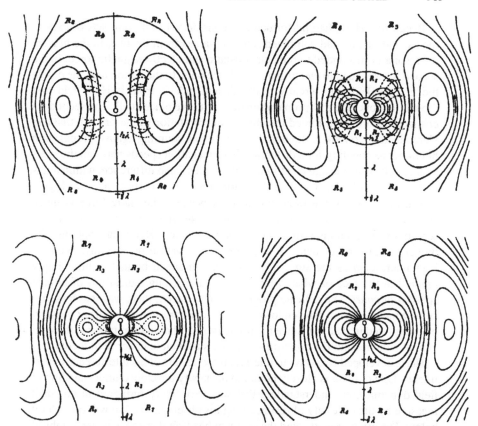

FIGURE 79   Hertz's field maps. The sequence runs clockwise from the upper left. Initially the
current is a maximum and the dipole is uncharged (Hertz 1889a).

general expressions, and so Hertz's original claim to have detected a difference
in speed between the electrostatic and the electromagnetic forces constituted
a claim that, whereas spatial polarizations do take place, nevertheless the Maxwell limit proper does not govern them.

Hertz did not begin his interference experiments with the expectation that
the general bipartite *Gesamtkraft,* or total force, would reveal its split nature to
his manipulations. There is no evidence that he expected this to happen. However, the notion of a *Gesamtkraft* continued at that time to form a basic element
in his physical understanding, an element that he had early absorbed from
Helmholtz. He was, as we have seen, already well aware that the concept loses
its clarity, perhaps even its meaning, in the Maxwell limit, and he here made
the point, and therefore the problem raised by his earlier interpretations, explicit.[10] To see just how important this problem now was for Hertz, turn back
for a moment to his initial understanding.

Recall that Hertz had performed two considerably different sets of experiments, and that he had combined selected values from the two. In the first set the resonator responded to both electrostatic and electromagnetic force; in the second it responded to electromagnetic force only. In the region near the dipole the first set had been greatly troubled, and indeed from it Hertz had first thought to conclude that propagation does not take place at all. The second set changed his mind but forced thereby a reinterpretation of the first. To that end Hertz decided to look at the initial results not as counting against propagation but rather as counting for a difference in propagation speeds between the independent electrostatic and electromagnetic forces. That difference, within a few meters of the dipole, would, if the static force propagates at a vastly higher speed, nearly mask the finite propagation because the static force is much stronger than the magnetic in that region. These were, however, marginal observations that, judging from his laboratory notes, Hertz had not initially thought to provide evidence for the finite propagation of anything. What he later interpreted as the sign of a difference in speed he had at first simply ignored in the face of what seemed to be quite strong indications of infinite speed *tout court*. The delicate problem he now faced, then, was that these results, which he had in print endowed with evidentiary status, could not possibly be interpreted in that way since the foundation of that interpretation—the concept of a compound *Gesamtkraft*—was not sustainable under field theory.

Hertz developed two ways to interpret these results of his interference experiments. One of them is closely bound to his newly developed understanding of field dynamics, an understanding that was very likely furthered in no small measure by the necessity of reconceiving his observations. We will consider it below. The other way attacked the problem head-on through a clever graphical method for representing the interference between the wire and the air actions. Suppose for a moment that both actions propagate as waves, though not with the same speeds. Draw a diagram in which the abscissa represents distance along the wire and in which the ordinate represents the time for constructing the value of the phase at a given point. Since we assume that both actions propagate as waves with constant, but different, speeds, we may represent their respective phases as straight but differently inclined lines: the less inclined to the horizontal a line is, the faster the action propagates.[11]

In figure 80, for example, δ_a represents the phase of an air wave. Following Hertz's procedure, construct a line $\delta_m + \varepsilon$, which is to say a line that differs by the amount ε from the phase of the wire wave. Suppose now that at some point the air and the wire waves differ in phase by this quantity. Use that point as a reference and draw a baseline along which measurements are taken. Next construct a series of lines parallel to $\delta_m + \varepsilon$ but differing from it by increments of $-\pi/2$. Where these lines intersect δ_a determines the loci on the baseline of a series of points such that the phase difference between the wire and air waves

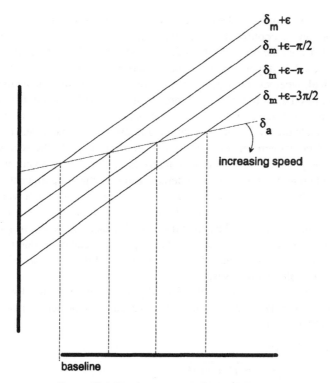

FIGURE 80 A Hertzian nomogram for constant speeds

increases at each step by a constant $\pi/2$. These loci can then be correlated with Hertz's three types of observed interference (designated $+$, 0, $-$).

Hertz's nomograph provides a graphic way to visualize interference under different assumptions concerning wave speed. First of all it is quite obvious from the figure that if both wave speeds are constant, the series of points will be equally spaced along the baseline. Hold the speed of the wire wave constant and increase that of the air wave by tilting its line toward the baseline. Then the distance between successive points decreases until the air wave ceases to be a wave altogether (having no phase), and the interference simply replicates the wire wave's phase at intervals of $\pi/2$. Conversely, moving the air line up increases the distances between the equispaced points, until the air line parallels the wire line, and the phase difference has the same value everywhere. This could be used to reinterpret Hertz's first results by constructing a phase curve for the dipole's electric field.

Two situations had to be considered, corresponding to Hertz's two different sorts of interference experiments. In the one, Hertz had originally thought, the resonator responds to both electrostatic and electromagnetic forces; in the other it responds only to electromagnetic force. That distinction could not be maintained in the context of the unitary field. It was, however, possible to distinguish between the ways in which the resonator could respond to the electric field. In Hertz's first type of interference experiments the plane of the resonator was vertical, whereas the dipole was horizontal; further, the resonator's spark gap was placed at its top. Hertz had previously assumed, in using the *Gesamtkraft*, that under these circumstances the resonator was driven by whatever force parallels the oscillator; he continued to assume that here, leaving therefore the Z component of the field to activate sparking.[12] In the second case, wherein the plane of the resonator contains the dipole, the resonator would be stimulated entirely by the E field generated through changing magnetic induction, that is, by $\delta P/\delta t$ (since P is orthogonal to a meridian plane, it would here lie in the plane of the resonator). Hertz could therefore compute these two fields from his solutions and try to find a phase for each that would be meaningful for his nomograph. It was here, within a specific, detailed problem, that the differences between the concept of *Gesamtkraft* and the demands of the field strikingly revealed themselves.

Take the first case, in which the Z field governs the resonator. Here, Hertz found after some standard manipulation that Z can be represented in an apparently wavelike form:

$$Z = B \sin(nt - \delta_1)$$

$$B = EL \frac{\sqrt{1 - m^2r^2 + m^4r^4}}{r^3}$$

$$\tan\delta_1 = \frac{\sin(mr)}{mr} + \frac{\cos(mr)}{m^2r^2} - \frac{\sin(mr)}{m^3r^3} \bigg/ \frac{\cos(mr)}{mr} - \frac{\sin(mr)}{m^2r^2} - \frac{\cos(mr)}{m^3r^3}$$

A major peculiarity of this expression is that the distance-dependent phase does not behave at all like that of a plane or spherical wave until r becomes sufficiently large that the phase reduces to mr. In the vicinity of the dipole the phase behaves quite strangely, and this is what Hertz capitalized on.

In Hertz's nomograph (fig. 81) the baseline is drawn thickly on the bottom, and the curved line that dips below the upper horizontal represents the phase δ_1. The straight lines 1, 2, 3, 4 represent the succession of metal-wire phases mentioned above. To find the loci of interferences that differ in phase by $\pi/2$ then (as above) requires finding the intersection of these lines with the curved line for the air phase. Close to the dipole the points are by no means equispaced, which, Hertz argued, fits the results he had obtained: "the phase does not increase from the source; its course is rather as if the waves originated at a distance of about $\lambda/2$ in space and spread out thence, partly towards the conductor, and partly into space" (Hertz 1889a, p. 152). The indications of near-infinite wave speed that occurred below about a wavelength are due to the dip

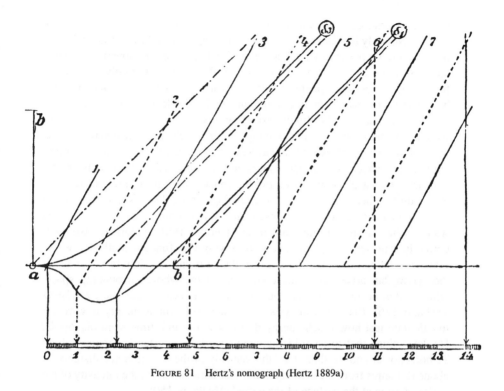

FIGURE 81   Hertz's nomograph (Hertz 1889a)

in the phase curve: at the curve's minimum point its tangent is horizontal, which corresponds to an infinitely rapid propagation. But the width of the minimum is sufficiently narrow that there should be contraindications for infinite wave speed, and according to Hertz's interpretation there were in the closer spacing of his marks.[13]

This marked a stunning, and it must have been to Hertz deeply revealing, difference from *Gesamtkraft* requirements. The latter's explanation of the odd behavior near the dipole, Hertz remarked just before showing how to build the nomograph, can only serve as "an approximation to the truth." But in what sense was it an approximation? Only in the very limited one that if you think distinct, independently propagating forces are at work here, then the observations can be fitted to them. But you simply could not move from the *Gesamtkraft* conception to the field explanation, because in the older way the oddly phased wave that Hertz had now produced on paper lacked meaning altogether.[14] Nothing in the *Gesamtkraft* conception suggested such an idea; nothing even allowed for it as long as it was possible to distinguish between the forces.

Moving to the Maxwell limit was much more than a mathematical procedure, for this kind of "wave" was an entirely new creation on Hertz's part, and he took special care in explaining what he certainly knew to be a puzzling novelty. Not only had Hertz fabricated something new in the laboratory; he had now done so on paper as well. Unlike any kind of waves previously fabricated, these *Hertzian waves,* as we shall now call them, did not simply move away from their source. The sphere $Q = 0$ marks at any given moment the effective boundary within which the lines terminate on the dipole (or, given Hertz's decision not to deal with the dipole proper, on a small spherical surface encompassing it), and so within which the dipole mimics electrostatic behavior; here, in a region that never exceeds a wavelength in radius, the wave has an apparently infinite speed. This sphere moves outward at speeds considerably greater than $1/A$, the wave speed proper, but the sphere is not itself a line of force: it is a boundary between lines of force, distinguishing those which continue to develop out of the dipole from those which have become detached from it and which radiate away. "In the sense of our theory," Hertz remarked, "we more correctly represent the phenomenon by saying that fundamentally the waves which are being developed do not owe their formation solely to processes at the origin, but arise out of the conditions of the whole surrounding space, which latter, according to our theory, is the true seat of the energy" (Hertz 1889a, p. 146). To cap his analysis Hertz used the Poynting energy-flow vector to calculate just how much energy flows away per unit time from his dipole. He determined that to keep his apparatus working constantly required about 22 horsepower and, further, that "the intensity of the radiation at a distance of about 12 metres from the primary conductor corresponds to the intensity of the sun's radiation at the surface of the earth" (1889a, p. 150).

The first kind of interference experiments could, then, be reconstituted. What of the second kind, in which the resonator responded as a closed circuit? Hertz computed a phase line for it, represented by the curve labeled δ_3 in his nomograph (fig. 81). This line, unlike the curve for δ_1, is not inflected, and so it will not entail apparently infinite wave speeds at any point. But all was not entirely coherent with Hertz's experimental results, as he quickly pointed out. The δ_3 curve does not dip to a minimum but is quite markedly curved within 3 m or so of the dipole. Projecting its intersections with the straight lines for the sequence of wire-wave phases indicates that the interference should have changed character at about 2 m, and then have stayed the same in type until about 4.5 m. Instead, Hertz had found no change until about 3 m.[15] He did not now regard this anomaly as a serious one, because he felt that it might be attributable to approximations made in supposing the conductors to be "vanishingly small," as well as perhaps to an incorrect value for the wavelength. But had Hertz in his discovery experiments found what his calculations now indicated that he should have found, then on his earlier interpretation, based on *Gesamtkraft,* he would have concluded once again that propagation does not occur! *The very observations that convinced him force propagates should not have done so.* He of course did not make this point explicit, since it would have undercut the force of his discovery document.

More to the point, Hertz now had another significant problem to point out, one that he had recognized from the very beginning: namely, that the air wave must, in his experiments, travel considerably more rapidly than the wire wave, since otherwise the tangents to both δ_1 and δ_3 beyond about 6 m should parallel the phase lines for the wire wave, yielding no further changes in the character of the interference, which was indubitably not what Hertz had found. Nor was this the only such anomaly for Maxwell's theory that Hertz uncovered. He had two others.

17.5. ANOMALIES EXPOSED

The previous spring, on March 19, Hertz had written Helmholtz telling him about his success in reflecting air waves. He concluded the letter with the following remark, which is worth quoting again:

> It seems to me that *the most important point* is now the *certainty* that propagation through conducting wires is perceptibly slower than in air, whereas all previous theories, including Maxwell's, led to the conclusion that propagation in wires took place with the velocity of light. (J. Hertz 1977, p. 255; emphasis added)

There is not a trace of doubt about the point here: wire waves travel more slowly than air waves, and neither Maxwell's theory nor any other can explain the fact. Hertz's certainty, one might even say his enthusiastic belief, had not in the least changed by the fall. Still, he was not until then utterly certain just what Maxwell's theory did entail concerning wire waves, nor had he as yet thoroughly probed their intricacies on the basis of Helmholtz's electrodynamics.

In the interestingly titled section "Waves in [*sic*] Wire-Shaped Conductors" (1889a) Hertz describes a way to investigate wire-wave processes without initially requiring them to have any particular velocity. He deployed his solution techniques to generate force expressions that became singular along the z axis, which Hertz interpreted as being the locus of a "very thin wire." [16] This procedure left a free parameter available to adjust the speed of the wire waves to any desired value less than or equal to $1/A$. In the particular case that the wire waves travel at the speed $1/A$, however, the electric field lines, which terminate orthogonally on the wire and which in general form curved loops, straighten out. In this case, therefore, the field has no component tangent to the wire at its surface. According to Maxwell's definition, Hertz continued, a "perfect conductor" is one within which only "vanishingly small forces" can exist, and this naturally entails that the E field tangent to the surface of such a thing must vanish. Closing the argument, it follows that in Maxwell's theory, as in most others, waves "in" wires "of good conductivity" travel at the speed of light ($1/A$). These waves, moreover, are purely transverse since they move parallel to the wire and the field lines are both straight and orthogonal to the wire.

The question that Hertz pondered, and that he had been pondering almost from the moment that he had succeeded in producing waves, was what could account for this difference in speeds. "For some time," he wrote, "I thought that it might be affected by the constant k, through the introduction of which Mr. H. v. Helmholtz has extended Maxwell's theory; but further consideration led to the rejection of this idea." Most people familiar with Helmholtz's equations for the polar ether considered that two distinct conditions had to be fulfilled to reach Maxwell's theory: first, that the ether's susceptibility had to be effectively infinite, and, second, that the constant k in the general (Helmholtzian) expression for the electrodynamic potential had to vanish. However, it was also thought that the purpose of the second condition was to forbid longitudinal waves. The first condition captured the essential physical aspect of Maxwell's theory: its effective replacement of conduction charge as the preeminently active entity by the ether's polarization charge, as well as Maxwell's relationship between wave speed and dielectric permeability. One might conceive of separating the two conditions, in particular of retaining the first (and thereby holding onto the physical core of what Germans thought to be Maxwell's theory) and allowing the second to lapse, thereby permitting longitudinal waves. Hertz did so.

The resulting longitudinal waves, Hertz thought, might then alter the composition of force near wires in such a fashion as to produce an *apparent* change in the transverse wave speed itself. This change would, however, not be due to the effect of the wire wave on the transverse wave but rather to the appearance in the wire of a new wave form, and thus, Hertz would in fact have detected the combined result of these two distinct waves.[17] One can see this from Hertz's conception of the wave forms that might exist if Maxwell's criterion for a "perfect conductor" were relaxed so that the electric force curves back into the wire seriatim along it, instead of remaining everywhere normal to it: a longitudinal force along the conductor now combines with the force that is orthogonal to the wire to produce curvature (fig. 82). For this to be possible the two wave speeds must have the same order of magnitude.

On "further consideration," which must have cost him some effort, Hertz realized that the ether's infinite polarizability, which he had not hitherto questioned, and which by itself yielded Maxwell's equations, made all appropriate forms of longitudinal wave vastly faster than transverse ones. In which case the transverse wave could not couple to a suitable longitudinal wave, whatever the value of k might be. Hertz drew what he thought to be the most likely conclusion: Maxwell's equations are not correct, at least for devices like his dipole and near conducting regions. In Hertz's words, it "seems rather doubtful whether the limiting condition [i.e., the ether's near-infinite polarizability] is

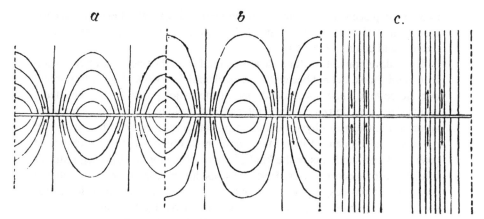

FIGURE 82   Possible forms of wire waves (Hertz 1889a)

correct for rapidly alternating forces." This primary condition for Maxwell's
theory was now itself questionable as a general requirement, and this result
(unlike a lapsed condition on k) called into question core conceptions of the
theory. Nor was this all. Hertz had decided (from nodal observations) that he
could greatly slow wire waves down by spiraling the wires, and this too Maxwell's theory "is unable to account for."

Hertz nevertheless concluded his paper by praising Maxwell's theory as better than "any of the older theories" except von Helmholtz's supplemented by
the "ether as dielectric." He was of course damning with faint praise, because
Hertz was now convinced that the limiting case of Helmholtz's equations that
yielded Maxwell's was not correct, that, in fact, the ether, while certainly present, was not infinitely polarizable. The ether had proved its electric character,
but in Helmholtzian, and not Maxwellian, form. If Hertz had been able to sustain this claim, to have it demonstrated in other laboratories, then he could have
laid claim to the laboratory realization of an entirely novel electromagnetic
regime, one taken into account neither by proponents of the Maxwellian field
nor (of course) by Wehereans.

Hertz's excited conviction lasted for about two years. The apparatus, or
rather attempted reproductions of it and replications of his wave-reflection experiments, rapidly escaped his exclusive control. Problems arose that cast
doubt on the integrity of his conclusions and on the stability of his device,
problems that deeply engaged him as he fought through correspondence to
preserve and to extend his claims. By this time Hertz had moved to Bonn. His
success in the laboratory had enabled the Helmholtz-centered network to place
him in a position that might eventually have led to Berlin and Hertz's crowning
as Helmholtz's heir in power, prestige, and authority had he not died on January
1, 1895. The remaining four and a half years of Hertz's career, and his life,
differ profoundly from his earlier struggles to establish himself, for he was

now firmly placed at the center of an international network of electromagnetic practitioners. His device soon mutated in ways and in circumstances that he did not, and could not, control.

These developments differ radically in kind from the ones that I have examined in this book, but not because Hertz had until then worked in solitary splendor. He certainly had not. Indeed, one of the major points I have repeatedly emphasized was Hertz's constant, intense, and prolonged concern with his position in the German physics community, which reached deep to the core of his work and influenced it in many ways, both obvious (as in his choice of problems) and subtle (as in the kind of approach he took). However, until his fabrication of electric waves, Hertz was not himself, as it were, a center of contagion in physics, a place from which a way of thinking and doing was propagated. He had longed to become just that from the beginning of his career, but once he had succeeded and moved from the (comparative) periphery to the center, he faced a different kind of pressure from what he had previously been accustomed to. He had to propagate his views, to defend and enhance his claims, to canonize his device, indeed to invent a new physics built around it, and above all to establish powerful and binding alliances both within and outside the German physics community. The story of how he did all of this, and of what happened to his device as it was seeded in different and not always sympathetic loci, will be the subject of a succeeding volume.

EIGHTEEN

Conclusion: Restraint and Reconstruction

This book has taken the reader on a long and perhaps disorienting ride through unfamiliar territory. It could not have been otherwise. Its aim has been to help the reader think as much as possible as Helmholtz and Hertz did during the 1870s and 1880s—and so to see symbols, devices, and effects in what are now profoundly unfamiliar ways. To that end I have here refrained from doing two things. First, I have not attempted systematically to contrast Helmholtz's electrodynamics, or for that matter Hertz's, with modern theory.[1] Second, and perhaps even more disconcerting, the many experiments that fill the chapters are not generally rendered familiar by means of modern accounts. Both of these efforts at restraint, which will surely have disturbed many readers who understandably want a clear path through the wilderness, merit discussion and even justification. To do so requires first of all engaging issues raised by one of my major themes, which appears primarily through the pervasive impact of Helmholtz on Heinrich Hertz.

That deep intellectual bond between the master and the apprentice runs through many pages of this book. And yet Helmholtz's way of doing physics, with which Hertz grappled long and hard, is historically elusive because it was not explicitly articulated either by Helmholtz himself or by others. My arguments for its very existence, beyond electrodynamics at any rate, depend on the presence of a common, persistent, and unifying pattern in many different aspects of Hertz's work, on rips that Hertz perceived within that pattern, on the structure of Helmholtz's electrodynamics and the experiments concerning it that were carried out in his Berlin laboratory during the 1870s, and finally on aspects of Eugen Goldstein's understanding of cathode rays. This raises two obvious and pertinent questions which I have not directly addressed until now. First, could I simply dispense with Helmholtzianism—these unarticulated substructures—altogether and nevertheless tell essentially the same story about Hertz? Second, why don't more people show direct evidence of its influence?

If I neglected altogether the deeper characteristics of Helmholtz's physics, then I would tell a story in which Hertz's experiments on (among other things) hardness and elasticity, evaporation, cathode rays, coupled circuits of all sorts, spheres rotating in the earth's magnetic field, not to mention his otherwise odd

(1884) attempt to generate propagation without an ether or the sequence through which he eventually fabricated electric waves, would all appear to be distinct, essentially unrelated events. Further, I would no longer be able to explain the character of Hertz's practice, and indeed the motivation behind many of the specific things that he did. It is precisely the unexpressed aspects of Helmholtz's physics that strikingly unify the young Hertz's work. Any one instance can of course be questioned, taken in isolation, as an exemplar of Helmholtzian practice, but it is precisely the unity across instances that constitutes the power of the conception. That unity might not have existed—it might simply have been impossible to tease it out of the sources.

It is not only Hertz's early work that gains unity and coherence in this way. Helmholtz's own, at least in electrodynamics, does so as well. Where, in the absence of this unarticulated scheme, Helmholtz's claims concerning his ability to deduce electrodynamic force from electromagnetic induction, or vice versa, seem odd even in contemporary context, they now gain coherence and force. Where, otherwise, it is extraordinarily difficult to understand Helmholtz's overarching claims for his electrodynamic potential, we may now better understand why he devoted such effort in the 1870s to teasing out its possible implications in the laboratory. Finally, and perhaps most important of all, Helmholtz's potential was, despite his claims for it, not a universal representative of contemporary electrodynamics; it had its own, unique implications. In which case, either it can be taken as a pure hypothesis, as a sort of deus ex machina, or else (like Weber's or Maxwell's schemes) it had a foundation in a deeper and broader conception.

This brings me to what I believe to be the primary reason, not for the absence of an overt statement of Helmholtz's deeper conceptions (on which more below), but for their elusiveness: Helmholtz did not succeed in bringing them to laboratory life. By the time Hertz arrived in Berlin Helmholtz's laboratory had consistently failed to produce appropriate effects. Worse (for uncovering something that would even in the best of circumstances be difficult to unearth), Helmholtz produced articles that left a great deal to the imagination as he provided explanations for the problems his laboratory encountered in eliciting what he wanted—until he reached in the end for the difficult and dangerous polar ether as a way to save the scheme itself.

Helmholtz's intense concentration during the 1870s on bringing his system alive in the laboratory marks a significant difference between his scheme and the underpinnings of Maxwellianism, or at least of that form of Maxwellianism practiced by Cambridge-trained physicists in the 1880s. Here tacit Maxwellian knowledge (which specified how to construct and manipulate fields) bound to a particular, troubling problem—the nature of conductivity—to produce a great deal of paper activity but very little laboratory work (Buchwald 1985a). Maxwellians had a reasonably clear, coherent sense of being engaged in a progressive problem-solving activity—in, that is, an activity which generated solutions that the community found to be congenial. Helmholtz's electrodynamics had nothing comparable because it was actually designed from the beginning to avoid precisely the sorts of questions that powerfully gripped Maxwellians, and that gave them the opportunity to produce satisfying paper work. This is worth briefly pursuing since it helps explain why Helmholtz's system would not likely have embedded itself in widespread practice on anything like the scale, in Britain, of Maxwellianism.

The essence of Helmholtz's system resided in its assumption that the states of a pair of objects, and the distance between them, completely determine an interaction energy from which all physical effects can be deduced. As a result the scheme poses only two problems: to identify appropriate and completely unreducible object states and to construct corresponding interaction energies. Maxwellianism, in considerable contrast, was very deeply concerned with reducing the states of objects to states of something else (the bearer of fields, the ether), which intervenes between the objects. This provided Maxwellians with a purchase for problem generation that Helmholtz lacked (since, at best, Helmholtz's ether was just one object among others). Helmholtz could not have produced a community of paper problem solvers analogous to the one in Britain because his scheme did not lend itself to that kind of activity. Consequently, the sort of skilled practice that scientific schemes typically produce could be realized for Helmholtzian electrodynamics *only* in the laboratory.

Helmholtz certainly did his best to generate that practice but he (or, rather, his assistants and visitors) were unable to make his new devices work. Skilled practice did not develop very far in the Berlin laboratory, because (unlike Maxwellian paper practice) satisfying success proved to be elusive. Had this not been so—had Helmholtz succeeded in producing potential-revealing apparatus—then Helmholtz's practice could have flowered into widespread activity. More to the point, a major part of this book has argued that until 1887 Hertz attempted to accomplish what Helmholtz had failed to do, although not entirely (or even primarily) within electrodynamics—namely, to make Helmholtz's interaction physics come to life.

The justification for my first form of restraint—my refusal to draw a detailed contrast between Helmholtz's electrodynamics and modern conceptions—lies precisely in the older system's lack of development on paper. To lose restraint would require discussing something considerably beyond Helmholtz's electrodynamic potential; namely, his ether. That would, however, be an exercise in futility because the conception never did give rise to much problem-solving activity, primarily because Helmholtz's ether was, unlike Maxwell's, not thoroughly integrated to an entire physical outlook. Ether problems were peripheral ones that Helmholtz eventually felt forced to engage, whereas his true concern lay with the interaction physics symbolized by his electrodynamic potential, which itself never generated much paper work beyond the initial deduction from it of the novel effects produced by inhomogeneous currents.[2] Hertz's attempt in 1884 to elicit propagation without the ether shows just how alien it was felt to be.

Refraining from modern explanations of experiments requires a different justification. The experiments done in Helmholtz's laboratory during the 1870s need no (modern) explanation at all, because they resulted in consensually acknowledged failure at the time, which is just what modern theory demands. Hertz also performed experiments of this kind, in which he attempted to elicit a specific evaporative power of a substance. Here again modern theory does not raise its eyebrows, because Hertz himself decided that the experiments failed, just as modern theory requires. On the other hand, Hertz did perform at least one set of experiments, with cathode rays, that upsets modern understanding, and yet here too I have refrained from elaborate explanation. More precisely, I have mentioned what many people after the fact took to be the reasons for Hertz's having obtained the results that he did, but I have also attempted to show why Hertz himself would not have countenanced an early version of the explanation (due to FitzGerald), and why one cannot in fact say much at all about what was presumably going on in Hertz's tubes without doing what no one has apparently ever done: replicating his device and manipulating it in ways designed to substantiate modern explanations for the nonelectric behavior of cathode rays in it. Even if someone did just that, it would hardly conduce to an understanding of Hertz's laboratory skills and conceptions.

Undoubtedly the most difficult experiments for us to grasp today are the very ones through which Hertz first convinced himself that electric waves in air could be made and detected. Here, however, it is more than desirable, it is utterly essential, to keep modern notions as far away as possible, for otherwise the very tight connection between Hertz's increasingly skilled manipulations and his developing understanding of just what he was manipulating would be lost. Further, a major part of my story involves the displacement of these discovery experiments from their originally central position in Hertz's own work to the distant periphery when they became problematic and difficult in the light of Hertz's subsequent analysis of the radiation field. Here we do in fact have modern theory—in this case, Hertz's own (later) theory—with which to look at the experiments, but it tells us little about the circumstances of their original performance.

Suppose that Hertz had not gone into print until after he had reflected and refracted his new waves and had provided for them a new (and essentially the classical) radiation theory. Suppose further that we had only a few extracts from his laboratory notebooks, say brief descriptions of early experiments and tables of results, and that we had no way of knowing whether Hertz had generated his radiation theory before or after he had begun experimenting. He would surely never have printed these early results because they only contradict his radiation theory. Historians might then have interpreted the bits and pieces from the laboratory notebooks as evidence for *failed* experiments, whereas these were originally *successful* experiments. To understand that, I have tried to seize them as much as possible with Hertz's discovering hands.

Appendixes

Appendixes

APPENDIX ONE

Waveguides and Radiators in Maxwellian Electrodynamics

J. J. Thomson's *Recent Researches* of 1893 contains extensive sections on electromagnetic radiation, but these deal for the most part with two situations: the coaxial cable (a cylindrical conductor surrounded by a concentric metal shield with a dielectric sandwiched between them) and an isolated cylinder or sphere that is not initially in electric equilibrium. These are the only two situations for which Thomson provides a careful, quantitative analysis. Neither of them corresponds directly to the Hertzian oscillator, for which Thomson offers only a qualitative discussion. This contrasts markedly with Hertz's extensive mathematics for his dipole vibrator (1889a). The absence of a mathematics for the dipole in the later *Recent Researches* is particularly striking given Thomson's extensive qualitative discussion of it and (in part at least) reflects the difference between Maxwellian and Hertzian attitudes toward sources. To illustrate the point requires a brief explanation of Thomson's analyses for the two situations that he does treat quantitatively, and in particular of the important coaxial cable.

In his quantitative analyses of waves whose lengths are comparable to macroscopic dimensions Thomson treats conductors as objects that lack capacity in comparison to their immense conductivity; the surrounding or sandwiched dielectric lacks conductivity. As a result no charge can ever appear within the conductor proper, in which case the internal E field has no divergence. Consequently, the system (conductor + dielectric) requires solving the following set of equations:

Within the conductor: $\nabla^2 E = (4\pi\mu/\rho)\partial E/\partial t$ and $\nabla \cdot E = 0$
Within the dielectric: $\nabla^2 E = \mu\varepsilon\partial^2 E/\partial t^2$
At the boundaries: H_{TAN} and E_{TAN} continuous

The inner conductor of a coaxial cable cannot in general be treated as an object within which nothing at all occurs—that is, as though it were an infinitely thin cylinder or wire—because this would lead to inconsistent boundary conditions. Nevertheless, waves do not propagate through the inner conductor, but only around it, traveling through the embracing dielectric; within it they merely diffuse. The coaxial cable accordingly acts as a prototype for Maxwellian wave motion at long wavelengths in that it exemplifies the general problem as Maxwellians saw it by the mid-1880s, one in which propagation in a dielectric must be linked via boundary conditions to diffusion within an embedded conductor.[1]

Since Thomson was primarily interested here in developing a precise theory of signal transmission by cables and suspended wires, he had little interest in the wire as a radiating object. He considered it solely as a guide for signals whose origin remained unspecified. Rather than sending out radiation into the dielectric (which would hardly be desirable in telegraphy or telephony), the wire, as it were, simply pilots it along. However, in section 301 Thomson turned to electric oscillations on a "metal cylinder surrounded by a dielectric," and here he considered an initial state of electric disequilibrium and the production of radiation as equilibrium is reached. "In this case," he remarked, "the waves starting from one portion of the cylinder travel away through the dielectric and carry energy with them, so that the vibrations will die away independently of the resistance of the conductor." Here, then, the object (cylinder) does not pilot the radiation; it lets go of it, sending it out into the surrounding dielectric.

To analyze this problem, which alone introduces true radiation, Thomson again applied his differential equations and boundary conditions, continuing to assume that the surrounding dielectric lacks conductivity and that the conducting cylinder lacks dielectric capacity. He further takes the special case that the electric distribution on the cylinder does not vary along its length but only from point to point in any section of it normal to its axis. For wavelengths on the order of the cylinder's diameter Thomson finds that the radiation of energy is so rapid that it practically vanishes after only one or two complete oscillations. The same holds for oscillations on a sphere and, by extension, for any closed conductor that, in modern parlance, is topologically equivalent to a sphere. In other words, a long train of Hertzian waves cannot be generated merely by starting a conductor in such a state that the potential varies over its surface: oscillations will indeed occur on it, and radiation will certainly take place, but the radiation is emitted as a pulse rather than as a sequence of oscillations.

What now of Hertz's dipole? In chapter 5 Thomson turns to Hertz's experiments but he does not provide a quantitative theory for the dipole, because the geometry of the object is too complicated. He is willing to attack only problems that can be solved using boundary conditions and suitable approximations. Isolated cylinders and spheres can be treated, albeit with some difficulty, in this way. But two spheres connected by a cylinder—the Hertz radiator—cannot be. Nevertheless, since the objects that compose the dipole would not by themselves emit more than a few complete oscillations, Thomson was quite certain that the dipole must also radiate its energy away very rapidly indeed:

> The Hertzian vibrator is one in which there will be considerable radiation of energy. In consequence of this radiation the decay of the oscillations in the

vibrator will be very rapid, indeed we should expect the rate of decay to be comparable with its value in the case of the vibrations of electricity over the surfaces of spheres or cylinders. We have seen, however, that for spheres and cylinders the decay of vibrations is so rapid that they may almost be regarded as dead-beat. We should expect a somewhat similar result for the oscillations of the Hertzian vibrator.

Taking Thomson at his word, we see he thought that the Hertz dipole does not allow more than half a dozen or so complete oscillations before it must be reset by the induction machine. The phrase "Hertz's oscillator" is, for Thomson, altogether inaccurate.

Yet Hertz's experiments unquestionably show something that looks very much like interference. Since the oscillator cannot generate persistent waves, Thomson concluded that the interference must be due to the resonator itself. The vexing role of the resonator will be examined in a succeeding volume. Here we need only remark with Thomson that the Hertz resonator, unlike the dipole, must continue to oscillate for a very long period indeed in order for it to be responsible for interference: "The rate of decay of the vibrations in the resonator compares favourably with that of pendulums or tuning forks, and is in striking contrast to the very rapid fading away of the oscillations of the vibrator." Bjerknes performed experiments in 1891 which did indicate that the Hertz oscillator only vibrates about a dozen times before its amplitude becomes utterly insignificant, whereas the resonator vibrates about 500 times. And the reason for this marked difference is that Hertz's oscillator loses its energy almost entirely by radiation, whereas his resonator radiates very little and decays primarily because of its resistance to conduction.

This alone is nearly sufficient to explain why Maxwellians did not discover freely propagating electric waves through experiment. From FitzGerald's earliest theoretical investigation of waves generated by oscillating currents to Lodge's experiments with wires and Leyden jars, the Maxwellians had always used circular vibrators or configurations that are radiatively equivalent to them, as Thomson in effect points out for Lodge in section 332. Such things simply cannot generate any significant radiation, because they act as pilots for it, as waveguides. They do not let it go. Yet where radiation does occur—in spheres, cylinders, and topologically equivalent objects—wavetrains, Maxwellians knew, do not also occur. Hertz's signal accomplishment in their eyes was therefore to have discovered how to use an otherwise undetectable pulse from something like a cylinder to stimulate persistent oscillations over a nearly closed loop, or waveguide (the resonator), of such a magnitude that they could actually be observed. This was why Maxwellians tended to concentrate their remarks on Hertz's resonator rather than on his oscillator, even though only Hertz had invented a method, with his spark-switched dipole, to generate the raw power that was necessary to produce truly detectable radiation at all.

One scarcely exaggerates in remarking that field theory deterred Maxwellians from discovering electric waves. Unlike Hertz, who used his dipole directly to control the radiative source, Maxwellians ignored sources almost completely and thought about radiation in terms of initial values and constraints imposed by boundary conditions. Many of them apparently knew that objects which produce significant amounts of radiation emit only (or very nearly only) pulses and that such things cannot, it seems, be detected. But waveguides, which prototypically require Maxwellian tools like boundary conditions, very nicely bounce disturbances around through the field, disturbances that remain permanently linked to the conducting guide. Thus, it is hardly surprising that Maxwellians like Thomson, FitzGerald, and Lodge were quite taken with guidance but had comparatively little interest in radiation, where boundary conditions cannot readily be deployed and where in any case waves properly speaking can hardly be generated. Because Hertz was not a Maxwellian he was initially unaware of the distinction between guidance and radiation, and he had no notion at all when he began his experiments that near-pulses rather than lengthy wavetrains were all that could be generated at the time. His control, or rather assumed control, over the source gave him implicit faith that he could generate a detectable train.

This difference between Hertz and the Maxwellians is nicely illustrated by Oliver Lodge's discussions in a lengthy series on field theory that he wrote for the journal *Nature* shortly before Hertz's experiments. Indeed, the series ends with a breathless announcement of the discovery of electric waves. In August 1888 Lodge remarked that he knew of no practical means to generate a significant amount of radiation, but he certainly did not learn this from mathematics, even though Maxwellian analysis had come right to the edge of the question as early as 1883.[2] If we may judge from Lodge's remarks in 1888, he, at least, intuited it:

> We have seen that to generate radiation an electrical oscillation is necessary and sufficient, and we have tended mainly to one kind of electric oscillation, viz. that which occurs in a condenser circuit when the distribution of its electricity is suddenly altered—as, for instance, by a discharge. But the condenser circuit need not be thrown into an obviously Leyden-jar form: one may have a charged cylinder with a static charge accumulated mainly at one end, and then suddenly released.
>
> In a spherical or other conductor, the like electric oscillations may go on, and the theory of these oscillations has been treated with great mathematical power by Mr. Niven and by Prof. Lamb.[3]
>
> Essentially, however, the phenomenon is not distinct from a Leyden jar or condenser circuit, for the ends of the cylinder have a certain capacity, and the cylinder has a certain self-induction; the difficulty of the problem may be said to consist in finding the values of these things for the given case. The period of an oscillation may still be written $2\pi\sqrt{LS}$; only, since L and S are both very small, the "frequency" of vibration is likely to be excessive. And when we

> come to the oscillation of an atomic charge the frequency may easily surpass
> the rate of vibration which can affect the eye. The damping out of such vibra-
> tions, if left to themselves, will be also a very rapid process, because the
> initial energy is but small. (Lodge 1888, p. 591)

Lodge was in essence asserting that conducting spheres and cylinders damp
extremely quickly—and so cannot produce much radiation—because their capacitance is too small to store much energy; what energy they do have "flashes"
away, to use his word.[4]

These remarks appeared in the issue of *Nature* dated October 10, 1888. The
last section, dated January 31, 1889, concludes with a rather breathless discussion of Hertz's discovery, which Lodge described as follows:

> The step in advance which has enabled Dr. Hertz to do easily that which
> others have long wished to do, has been the invention of a suitable receiver.
> Light when it falls on a conductor excites first electric currents and then heat.
> The secondary minute effect was what we had thought of looking for, but Dr.
> Hertz has boldly taken the bull by the horns, looked for the direct electric
> effect, and found it manifesting itself in the beautifully simple form of micro-
> scopic sparks.
>
> He takes a brass cylinder, some inch or two in diameter, and a foot or so
> long, divided into two halves with a small sparking interval between, and by
> connecting the halves to the terminals of a small coil, every spark of the coil
> causes the charge in the cylinder to surge to and fro for about five hundred
> million times a second, and disturb the ether in a manner precisely equivalent
> to a diverging beam of plane-polarized light with waves about twice the
> length of the cylinder. (Lodge 1889, p. 321)

Notice that Lodge limits Hertz's "step in advance" to "the invention of a suitable receiver" and describes the dipole oscillator as a divided "brass cylinder."
Lodge was steeped in the lore of wave guidance and, in particular, in experiments that illustrate guidance by producing a resonant effect between circuits
(Aitken 1985, pp. 90–93). These circuits are waveguides because their quality
as oscillators reflects (to Maxwellians) their ability to contain and guide fields.
Only a small amount of radiation can escape through the open termini of the
circuits; otherwise, they would simply not be good oscillators. The rapid,
flashlike radiation from cylinders was obviously quite different from the persistent oscillations that alone seemed to offer the possibility of detection, and
so Lodge simply did not think to link the two processes.[5] Instead, to judge
from the remarks just quoted, he had thought to detect radiation from a waveguide by the cumulative amounts of Joule heat that will be generated when it
strikes a conductor.[6] Although Lodge was well aware that radiating electric
oscillations occur on cylinders, and despite the fact that he was already using
spark discharges as fast-acting switches in his guidance experiments, he never
thought to use the latter to stimulate oscillations on the former that could be
detected by an independent waveguide (resonator).

There is also another factor at work here, though perhaps not one that directly influenced Lodge. For Maxwellians who were familiar with Helmholtz's work, the question was not whether radiation occurs at all but whether it is Maxwell's or Helmholtz's kind that does. In the fall of 1888, just before he learned of Hertz's experiments on reflection, but already aware of Hertz's previous work, J. J. Thomson found Hertz's experiments, and in particular his detector, "interesting" but utterly inadequate for a proper test of Maxwell's theory. Instead, and perhaps influenced by Hertz's own remarks, he sought a way to determine whether or not the electrostatic potential propagates at infinite speed, since (probably following Helmholtz here) this translates directly into the question of whether or not all circuits are closed, that is, into the question of displacement currents. He wrote Richard Threlfall on October 3.

> After a full trial of the methods for testing Maxwell's theory I have come to the conclusion that the most hopeful way is to test whether or not the electrostatic potential is propagated with a finite velocity or not. If it is then if you have concentric spheres the potentials of which change their signs millions of times in a second, the electric force will be finite outside the outer and inside the inner spheres instead of vanishing as it does if Maxwell's theory is true. By using electrical oscillations you can get the required rapidity of reversal (see Hertz, Wied. Ann. 1887) and the only thing is to get some way of detecting an electrical force which is continually changing its direction. Perhaps this might be done by the glow produced in a highly exhausted space such as an incandescent lamp, or since the force is not uniform by the mechanical force it exerts on a small unelectrified conductor. If you care to undertake it, it is a thing eminently worth doing though very difficult. Most of the difficulty would vanish if you could get a delicate way of detecting an alternating electromotive force. I have suggested the one I think most hopeful but perhaps you may hit upon something better.
>
> You should read Hertz's papers; they are very interesting. *He has measured the velocity of propagation of electrodynamic action, but this is not the real point which is whether all circuits are closed or not;* if they are then the electrostatic potential is propagated with an infinite velocity, but if they are not then the velocity is finite and could be tested in the way I suggest. (Rayleigh 1969, p. 30; emphasis added)

In other words all Hertz had done so far was to find finite speed, but as it stood this did not unequivocally support Maxwellian, rather than Helmholtzian, propagation. That required a different kind of experiment from any that Hertz had performed—though Hertz in fact at first thought that his initial *failure* to detect electrodynamic propagation in the region near the oscillator did bear directly on a Helmholtzian version of the question posed by Thomson.

Lodge's initial description of Hertz's discovery—his emphasis on the resonator, which was something he understood—neglects the problematic operation of the oscillator. Before Hertz's experiments, Lodge (if he thought much about it at all) probably conceived of the radiative flash produced by a cylinder reaching electric equilibrium as several complete oscillations with a fixed period and very rapidly decreasing amplitude. This, certainly, is the picture implied in the remarks quoted above. However, following Sarasin's and de la Rive's experiments, it was clear that the oscillator could not be thought of in quite this way, and Lodge's subsequent work in producing wireless apparatus was directed in major part at overcoming the broadband effect of rapid damping, though precisely why a broadband effect occurs puzzled Lodge and many others at the time, including Hertz.[7] All of them knew it had something to do with damping, but just what remained a point of confusion and controversy.

APPENDIX TWO

Helmholtz's Derivation of the Forces from a Potential

2.1. THE PRINCIPLE OF ACTION–REACTION AND LINEAR CURRENT ELEMENTS

One of the objections that both J. J. Thomson (1885) and Helmholtz raised against existing laws for the ponderomotive force between current elements was that many of them violated the principle of action–reaction. That is, according to several laws the force on an element $d\sigma$ was not equal and opposite to the force on an element ds. This was not acceptable, from Helmholtz's position, because it would violate the conservation of linear momentum. J. J. Thomson could not reject a law a priori on these grounds, because it was possible that the missing momentum might be assigned to the ether. Any acceptable Helmholtzian interaction had accordingly to satisfy the action–reaction principle, including, of course, the one for electrodynamics.

Recall, however, that the electrodynamic potential function yields a set of forces that apparently *do not* satisfy the principle; namely,

$$F_{ds} = A^2 ij \frac{ds}{ds} \times \int_\sigma d\sigma \times \nabla \frac{1}{r}$$

$$F_{ends} = A^2 ji_{end} \int_\sigma \frac{1}{r} d\sigma + \frac{1-k}{2} A^2 i_{end} \sum_\sigma j_{end} \nabla r$$

These are the forces that, according to the Helmholtz potential, are exerted by an element $d\sigma$ that carries a current j on an element ds that carries a current i. We can easily see that, as it stands, the net force that acts on ds is not equal and opposite to the net force on $d\sigma$. Element $d\sigma$ exerts a total of seven distinct forces on ds. Label the respective ends of the elements $\sigma_{a,b}$ and $s_{a,b}$, and the centers σ_c and s_c. Then the seven forces are as follows:

Force	Source	Object	Direction of force
F^a_1	σ_a	s_a	end to end
F^a_2	σ_b	s_a	end to end
F^a_3	σ_c	s_a	center to end
F^b_1	σ_a	s_b	end to end
F^b_2	σ_b	s_b	end to end
F^b_3	σ_c	s_b	center to end
F_{cc}	σ_c	s_c	center to center

If we now calculate the forces that ds exerts on $d\sigma$, we can see that the forces on ds do not all have equal and opposite counterparts to the forces on $d\sigma$. The end–end forces (which are functions of k and which vanish if either of the elements forms part of a closed circuit) do balance because they are proportional to the product of the changing charge densities at the interacting ends. But the remaining forces do not balance. Although the centers of both elements do exert forces on the ends of the other, the ends of the one element do not exert forces on the center of the other; moreover, the center–center force obviously violates action–reaction since in general the forces are neither equal nor opposite. In terms of the table, F^a_3 and F^b_3 have no counterparts: centers act on ends, but ends do not, as it stands, act on centers.

Something was clearly wrong with this analysis because, Helmholtz knew, the net forces given by the electrodynamic potential had to satisfy action–reaction, for the simple reason that the potential is a function of the distance between the centers of the elements.[1] In 1874 he provided the missing forces. The analysis that yielded the forces listed in the table, recall, had originally divided the potential into two parts. One part has the form of the Neumann potential; the other part contains a factor of $(1 - k)/2$. In computing the forces that arise from the k-dependent part, Helmholtz had had to partially integrate over both elements, since the corresponding force vanishes if either one of them is part of a closed circuit. As a result this part of the potential yields forces between ends. However, in calculating the forces from the Neumann potential, Helmholtz had partially integrated over only ds. He had not done so over $d\sigma$. Consequently, as it stands, the forces are incomplete; in particular, the center–center force requires an additional partial integration over $d\sigma$.

When Helmholtz carried this out, he immediately discovered that the ends of $d\sigma$ exert forces on the center of ds, and that these are precisely equal and opposite to the forces that the center of $d\sigma$ exerts on the ends of ds. That was not all. There is, of course, still a center–center force since the integral does not vanish even when both circuits are closed. This, it turns out, has precisely the same form as Ampère's original element–element force law.[2] Action–reaction is, as it must be, indeed satisfied by Helmholtz's electrodynamic potential. This marks a signal difference between Helmholtz's scheme and the Maxwellian, no matter what limits are taken in the former—and whether or not an electrodynamically polarizable ether exists. This difficult point demands explanation.

Helmholtz's potential determines the interaction *at a given instant* between any two current-bearing elements in the universe, no matter how far apart they may be. If we introduce the ether in addition to material circuitry and dielectrics, then all we have done is to increase the objects that interact with one another. A given ether element may sustain a polarization current at some instant; if it does, then *at the same instant* it must be interacting with every other element in the universe that also sustains a current. Helmholtz's potential, in other words, is utterly nonlocal; currents light-years apart are, according to it, just as much in interaction with one another as currents millimeters apart (albeit with different strengths due to the different distances). Maxwellian field energy is utterly local; it specifies the field energy contained in a differential volume element. That energy may be there as a result of current processes that took place eons ago in far away places; or it may result from processes that took place a moment ago between neighboring elements.

This has the following effect. In Helmholtz's scheme, whether supplemented by the ether or not, current-bearing material circuit elements necessarily interact with one another in a way that conserves momentum. In the Maxwellian scheme, they do not: the ether must be brought into consideration. However, and this is a critical point, in the laboratory Helmholtz's elements will *appear* to violate momentum conservation (action–reaction) because they are subject to forces exerted by the hidden currents in the ether with which they are also interacting. Unless the ether can be removed from a space, the material interactions cannot be separated from the matter–ether interactions, and material elements will apparently violate momentum conservation.

2.2. The Mechanical Force on Currents for Extended Bodies

2.2.1. *General Results*

Helmholtz (1874a, pp. 729–36) examined only the "Neumann" part of the potential, namely, $A^2(C' \cdot C/r_d)d^3rd^3r'$ (where r_d is $r - r'$), which meant that he was here considering effects that require one of the interacting objects to form part of a closed circuit. Then the mechanical force F_{mech} on C can be found by displacing the body in which C occurs, and whose position is determined by r, through δr:

$$\int F_{mech} \cdot \delta r d^3 r = -\delta \phi_0$$

The variation, which must be done with great care, is constrained by requiring the linear current to remain constant during the virtual displacement of r, which is itself completely arbitrary. This means that energy must in general be fed into the system from an external source. Accordingly Helmholtz had to perform the following variation:[3]

$$\delta\phi_0 = -A^2 \int_{r'}\!\!\int_r - \delta r \cdot \nabla \frac{1}{r} C' \cdot C d^3 r d^3 r' - A^2 \int_{r'}\!\!\int_{r'} \frac{1}{r} C' \cdot \delta(C d^3 r) d^3 r'$$

The first step is to find $\delta(C d^3 r)$. If i is the absolute current intensity, then we may write with Helmholtz[4]

$$\frac{C_x d^3 r}{dx} = \frac{C_y d^3 r}{dy} = \frac{C_z d^3 r}{dz} = i$$

Since $\delta(i)$ must vanish under Helmholtz's requirement that the filamentary current intensity remain constant during virtual displacement, we obtain for $x_j = x, y, z$

$$\frac{d^3 r}{dx_j}\delta C_j + \frac{C_j \delta(d^3 r)}{dx_j} - \frac{C_j (d^3 r)\delta(dx_j)}{(dx_j)^2} = 0$$

The variations δx_j and $\delta d^3 r$ are simply found:

$$\frac{\delta(d^3 r)}{d^3 r} = \nabla \cdot (\delta r) \quad \text{and} \quad \delta(dx_j) = \nabla(\delta x_j) \cdot dr$$

Consequently, for δC we have

$$\delta C = (C \cdot \nabla)\delta r - C[\nabla \cdot (\delta r)]$$

And so for $\delta(C d^3 r)$,

$$\delta(C d^3 r) = d^3 r[(C \cdot \nabla)\delta r]$$

Then for the variation of the potential we have

$$\delta\phi_0 = -A^2 \int_{r'}\!\!\int_r [\delta r \cdot \nabla \frac{1}{r_d} C \cdot C' + \frac{1}{r_d} C' \cdot (C \cdot \nabla)\delta r] d^3 r d^3 r'$$

To proceed further requires a partial integration, that is, an application of Green's theorem with a single boundary at infinity. Helmholtz does not give the details, but they are easily reconstructed. Take C' outside the square brackets and then apply the following vector identity:

$$[(C/r_d) \cdot \nabla]\delta x_j = \nabla \cdot [(C/r_d)\delta x_j] - \delta x_j \nabla \cdot (C/r_d)$$

The Green theorem eliminates any term involving only a divergence so that the equation may now be written

$$\delta\phi_0 = -A^2 \int_{r'} C' \cdot \int_r \left[\delta r \cdot (\nabla\frac{1}{r_d})C - \delta r(\nabla \cdot \frac{C}{r_d}) \right] d^3r d^3r'$$

Helmholtz next defines a vector U' and introduces the continuity equation to link current C with charge density ρ:

$$U' = \int \frac{C'}{r_d} d^3r'$$

$$\frac{\partial\rho}{\partial t} + \nabla \cdot C = 0$$

Combining these relations with the previous equation and comparing with the basic variational equation for F_{mech} yield the mechanical, or "ponderomotive," force on the current C:

$$F_{mech} = A^2\left[C \times (\nabla \times U') + U'\frac{\partial\rho}{\partial t} \right]$$

The presence of $\partial\rho/\partial t$ in the mechanical force arises from the term $\nabla \cdot (C/r_d)$ via the continuity equation whatever the value of Helmholtz's constant k may be: it is a special implication that Helmholtz drew from the "Neumann" form of the electrodynamic potential. This term should appear in the corresponding expression for the force that is implied by Maxwell's equations. It does not, despite Helmholtz's early claim that he could reach Maxwell by setting k to zero, because field theory's continuity relation is entirely different from Helmholtz's. The quantity that, in field theory, should correspond to Helmholtz's C (given the expression for the energy of a current system) is the sum of conduction and displacement currents. Consequently, the continuity equation should simply require $\nabla \cdot C$ to vanish, and this eliminates the second term in the last equation.[5] This makes it clear why, when his attempt to detect the effect failed, Helmholtz recurred to polarization currents in the surrounding air and required also that the susceptibility must be quasi-infinite: only in this way could he produce an analytical simulacrum of Maxwell's displacement current and thereby explain the absence of the novel effect in the laboratory. Helmholtz also, in a much simpler analysis, obtained the force between the elements of current-bearing linear circuits.

2.2.2. Gleitstellen

The analysis just detailed applies to every case of conduction, whether or not the conductor is moving, because in Helmholtzian electrodynamics the interaction energy between objects never depends directly upon their motions, as it does, for example, in Fechner–Weber. Indeed, this is a cardinal principle of the theory upon which Helmholtz always insisted. Consequently, he had to develop a way of accounting for motional effects as secondary results of the fundamental interaction, which depends upon the distance between the interacting objects and not upon their relative (or absolute) speeds. Consider, for example, what must happen, in Helmholtz's system, when a conducting wire is free to slide along two fixed, mutually parallel wires that are connected to one another by a third wire, also fixed, when the device exists in a homogeneous magnetic field that is normal to its plane.

Helmholtz develops a long argument about this prototypical situation, whose conclusion amounts to this: the regions between the ends of the movable wire and the parallel, fixed wires which it glides along must be thought of as *Gleitstellen*, or loci of sliding, regions that are themselves conducting but whose velocities change continuously from the velocity at the ends of the movable wire to zero at the fixed wires. To find out whether a force tends to displace the sliding wire requires assuming it to move and then examining whether a change occurs in the electrodynamic potential due to the current flowing through the changing circuit that it forms with the fixed wires (Helmholtz 1874a, pp. 736–42). The constraint that governs the variational computation is the same one that we used above, namely, that the linear current intensity must always be the same—that, in other words, the same quantity of charge flows per unit time through each elementary length of the conductors, whether or not that element is itself in motion. If, again, C is the current density, then this requirement, as we saw above, leads to the constraint

$$\delta C = (C \cdot \nabla)\delta r - C[\nabla \cdot (\delta r)]$$

To see the effect we shall use the example that Helmholtz gave. Suppose that the (plane) ends of the movable wire are parallel to the yz plane and that between them and the fixed surfaces of the stationary wires the *Gleitstellen* have speeds v_y equal to $x\beta/\lambda$ (where the fixed surface is at the origin, and the moving surface is at λ). We may now apply the equation of constraint. First, we assume that the *Gleitstellen* do not involve condensations or rarefactions, so that the divergence term in the constraint may be ignored. Second, δr reduces to $\delta y e_y$, where e_y is a unit vector along y, because that is by assumption the direction in which the slide occurs. Then the constraint implies that the current density acquires an additional component δC_y equal to $C_x(\partial/\partial x)\delta y)$ (since all derivatives with respect to y and z vanish). But the δy generated in time dt is just $v_y dt$, and so we find that the effect of the motion in the region immediately adjacent to the moving wire is to produce in time dt an addition to the current density in the y direction equal to $C_x \beta dt/\lambda$ (which is Helmholtz's result, though he barely gives the analysis).[6]

To see what effect this has Helmholtz considers the layer to be extremely thin so that he can use this expression to represent the complete effect of the sliding. Suppose the surface area of the layer is Q and that the electrodynamic potential has the values U, V in the x, y directions respectively at the layer.

Then the interaction energy of the current in the layer with the external sources of the magnetic field is $-(UC_x + VC_x\beta dt/\lambda)Q\lambda$.

Unless the rotation is extremely rapid, the current I through the layer is very nearly QC_x, and so during a time dt the interaction energy will change by $-VI\beta$. Consider now the end of the sliding wire, which has a speed υ_y equal to β. An electrodynamic force Y acting on it will, in time dt, perform work equal to $Y\beta dt$. This must be equal and opposite to the change in the potential, so we find at length with Helmholtz that the movable end of the wire will be acted on by a force equal to IV. In his words:

> This is simply the value of the force that acts on the end-surface of the displaced layer as if it were an end-surface of the current. The entire electromagnetic action reduces to this end-force, which the remainder of the currents present exert on the displaceable layer if it is infinitely thin. (Helmholtz 1874a, pp. 740–41)

2.3. The Electromotive Force on Currents

2.3.1. Electromotive Force due to Time Changes in the Potential

The electromotive force $-\partial U/\partial t$ contains an additional term in comparison to the corresponding action given by Neumann's potential function:

$$\text{additional term} = \frac{1-k}{2}\nabla\frac{\partial\Psi}{\partial t} \quad \text{where} \quad \Psi(r)$$

$$= \int(C' \cdot \nabla_{r'}r_d)d^3r'$$

$$= \int r_d\frac{\partial\rho'}{\partial t}d^3r' \quad \text{(continuity)}$$

This term can only affect open circuits since it involves the gradient of a potential function. It corresponds to a current-generating electromotive force from circuit termini where the second derivative of the charge density with respect to time does not vanish.

2.3.2. Electromotive Force in Cases of Motion

It is important to understand how electromotive actions arise in moving conductors in potential theory. As usual all actions must derive from changes in potential. As an object moves from one point at which it has a certain potential to another point where it has a different potential, electromotive actions arise directly as a result of the corresponding change in potential. On this basis Helmholtz was easily able to demonstrate that an object moving with a velocity

v in a potential U will have produced within it an electromotive force E_{ind} such that (Helmholtz 1874a, p. 745)

$$E_{ind} = -A^2\left[\frac{\partial U}{\partial t} + v \times (\nabla \times U) + \nabla(U \cdot v)\right]$$

(The two terms in v both arise from a single term $(U \cdot \nabla)\delta r$ in the temporal variation of the energy.) Despite this form, in which E_{ind} appears, as it were, on its own, it is important to understand that *in application* the equation can be very difficult, because in all laboratory situations E_{ind} acts on something that does not exist by itself but forms a part of a circuit.

Indeed, understanding just where E_{ind} has an effect can be quite difficult. Consider a conducting arm that pivots about one end and, at the other, swings over a metallic arc. Connect another radial arm to the pivot of the first arm, and fasten its other end to the arc. As the first arm swings freely over the arc, it forms a varying, wedge-shaped circuit composed of the swinging arm, the fixed arm, and the part of the arc between the two.

Now apply a homogeneous magnetic field that is normal to the plane of the circuit. The currents that form this field must themselves flow in circular paths that are parallel to, and concentric with, the metallic arc of our device (albeit located in another plane)—for otherwise the magnetic field could be neither homogeneous nor perpendicular to the circuit. The parts of any current that exists in the radial arms will (stationary or moving) have *the same* interaction energy with the external currents whatever the position of the arm in its swing may be, and so no electromotive force can possibly appear in it. In terms of Helmholtz's equation for E_{ind}, the term in the curl of U must cancel the term in its gradient. However, as the arm moves the metallic arc changes in length, so that its interaction energy with the magnetic field will change continuously. But this will occur only at the point where the arm slides on the arc, since that is where the circuit length changes. (And at that locus the term in E_{ind} that produces the effect involves the gradient, not the curl, of U.) Here, and only here, the interaction energy changes with time, and this entails an electromotive force. Consequently, in Helmholtz's theory electromotive forces generated by motion appear only at loci where the effective length of the conducting path itself changes. The generation of electromotive force by motion is an end-effect.

APPENDIX THREE

Helmholtz's Energy Argument

It is particularly important to grasp the sense of Helmholtz's deduction of the Ampère force from induction and vice versa as it was assimilated by Hertz and others who worked with Helmholtz.[1] To do so, begin with Ohm's law, which connects the current to the electromotive force (E_{ctm}) provided by a battery or equivalent external source and to the electromotive force ($\nabla\phi$) provided by electrostatic sources, and with Joule's law, which determines the rate Q at which heat is generated in a circuit.[2]

Ohm's Law

$$\kappa C = E_{ctm} - \nabla\phi$$

Joule's Law

$$Q - \kappa \int C^2 d^3r = 0$$

Next, using the Joule law write the equation for the energy balance in the system when the current-bearing circuits are in motion (velocity v). We shall allow for a possible mechanical force F_{mech}, and we shall also assume that the current system may contain an internal energy Φ_0 that can vary as a result of processes that occur at a fixed point in space or as a result of changes in the system's configuration (or both):[3]

Electromagnetic Energy Balance

$$0 = -\int\kappa C^2 d^3r + \int E_{ctm} \cdot C d^3r - \int \nabla\phi \cdot C d^3r + \int F_{mech} \cdot v d^3r + d\Phi_0/dt$$

If we take Ohm's law as it stands above, then the first three integrals in the balance equation cancel one another, but the other terms remain. At this point we cannot conclude anything about either inductive forces or mechanical ones, because the term in the mechanical force contains a velocity, whereas the term in Φ_0 does not seem to.

Consider, however, that the term $d\Phi_0/dt$ is by assumption a total derivative, which therefore has contributions both from the rate of change of the system energy Φ_0 at a point and from the alterations produced in the energy as the configuration of the system changes:

$$\partial\Phi_0/\partial t + (v \cdot \nabla)\Phi_0 = d\Phi_0/dt$$

Consequently, the energy balance may be rewritten in the following way:

$$0 = -\int\kappa C^2 d^3r + \int(E_{ctm} - \nabla\phi) \cdot Cd^3r + \partial\Phi_0/\partial t + \int(\nabla\Phi_0 + F_{mech}) \cdot vd^3r$$

Consider a situation in which the *system energy*, supposing it to exist, does not change with time ($\partial\Phi_0/\partial t = 0$). Then the terms involving C in the balance equation must together account completely for the Joule heat since they do not in any way involve the velocity,[4] and so we are left with only the last term, which involves v. If we know from some other evidence that Φ_0 must exist, then we can also conclude that a mechanical force equal to $-\nabla\Phi_0$ must occur. But what can tell us that the system energy does indeed exist?

Suppose we know from experiment that electromagnetic induction (E_{ind}) occurs and that it has the following form:

$$E_{ind} = -\int\frac{\partial C/\partial t}{r}d^3r$$

This electromotive force must also appear in Ohm's law if it is to generate currents, in which case the Joule law implies that there is an addition δQ of the following form to the heat rate:

$$\delta Q = \int(E_{ind} \cdot C)d^3r$$

$$\text{total heat rate} = Q + \delta Q = \int(E_{ctm} - \nabla\phi + E_{ind}) \cdot Cd^3r$$

The only term in the rewritten equation for the energy balance that can correspond to δQ under all circumstances (i.e., even when the velocity is zero) is $\partial\Phi_0/\partial t$, so we must accordingly have

$$\delta Q = \int E_{ind} \cdot Cd^3r = -\frac{1}{2}\iint C(r') \cdot \frac{\partial C(r)/\partial t}{|r - r'|}d^3r'd^3r = \frac{\partial\Phi_0}{\partial t}$$

If we assume that the value of Φ_0 *at a given time* t *is a function solely of the currents and their mutual distances at that time,* we can at once integrate this to find (excepting a term that is independent of the time)

$$\Phi_0 = -\iint\frac{C(r') \cdot C(r)}{|r - r'|}d^3rd^3r'$$

And so we now know that such a function Φ_0 exists. From its form we can calculate $\nabla\Phi_0$ to be $C(r) \times [\nabla \times \int C'(r')d^3r']$, which is the Ampère force. Here, then, we have moved from electromagnetic induction through Ohm's law to a modified form of the Joule heat; we then saw that the only term in the energy balance equation to which the new heat addition could correspond in all circumstances involved the (partial) time derivative of a presumptive system energy, which we accordingly conclude must exist, and whose form we can now compute. From that we in turn deduced the corresponding mechanical (Ampère) force. Suppose that we instead begin with the mechanical force expressed as the gradient of an unknown potential Φ_0 and integrate to find the latter. We may then compute its time derivative, and from it we can deduce the existence and form of E_{ind} if we assume that the potential function from which the mechanical force derives through a gradient must represent an internal energy of the system. What we would find is that a new force must be included in the expression for the Joule heat.

Although these results amount to considerably less than the bald claim that, given energy conservation, induction can be deduced from the Ampère force and vice versa, nevertheless they are hardly vacuous and may be expressed in the form of four requirements:

Helmholtz's Energy Requirements

1. If an electromagnetic system possesses an internal energy Φ_0, the energy can be altered as a result of changes that may occur while the system is fixed in position, as a result of mechanical forces that act upon the system during its motion, or both.
2. If the internal energy exists, at any time t it is a function solely of the electromagnetic variables and their mutual distances *at that time*.
3. If internal mechanical forces exist, they must be obtained as the gradients of a potential function that represents the system's energy.
4. If electromotive forces due to internal system processes exist, they must be obtained from the partial derivatives with respect to time of a potential function that represents the system's energy.

Given these four conditional requirements, the Ampère force is indeed an implication of electromagnetic induction and vice versa.

APPENDIX FOUR

Polarization Currents and Experiment

In order to obtain an effect in his experiment in which a metal arm rotates in a magnetic field past a stationary conductor, Helmholtz (1875) had to assume that the air between the moving and stationary plates behaves like "*Gleitstellen*," even though it is not a conductor, since otherwise there could not be any change in the electrodynamic potential given the principles of his theory. This simply means that an interaction energy can exist between changing polarization in air and external currents (or magnets).

Align the sliding metallic surfaces in the yz plane, so that the x direction is normal to them. Then if, as a result of the change in potential engendered by the aerial *Gleitstelle*, an electromotive force, represented by E_{ind}, arises, the air between the two metallic surfaces will acquire a polarization P equal to

$$P = \chi(E_{ind} - \partial\phi/\partial x)$$

where ϕ is the electrostatic potential there.

The neighboring surface of the conductor will have upon it a layer of charge that is generated by the air's induced polarization. The sum of the two surface charge densities, polarization (σ_p) and conduction (σ_c), at the interfaces between air and metal will be proportional, Helmholtz continues, to the normal potential gradient generated by their joint action at the interface:

$$-4\pi(\sigma_p + \sigma_c) = -\partial(\phi)/\partial x$$

Together these equations yield

$$-4\pi f\sigma_c = -4\pi\chi E_{ind} + (1 + 4\pi\chi)\partial(\phi)/\partial x$$

where f is $+1$ for the stationary metallic surface and -1 for the moving metallic surface.

If we remain, as Helmholtz insists, with the fundamental structure of his electrodynamics, there can be no electromotive force whatsoever that arises directly as a result of motion. Consequently, the electric potentials at the moving and stationary metallic surfaces must be equal to one another, for if they were unequal, then we would have an electric force *in addition to* E_{ind}, and this would necessarily have had to arise from the motion. The gradient of the scalar potential therefore vanishes, and the conduction charge density reduces to

$f\chi E_{ind}$. The outer surface of the stationary condenser will therefore receive a charge of magnitude χE_{ind}.

Helmholtz decided to probe the result by investigating what would occur if the electromotive force were instead generated directly by motion, and if the air could be polarized by it. In that case the net charges on the two air–conductor interfaces are produced directly by this motional electromotive force (E_{ind}), so that the gradient of the total electrostatic potential will be equal and opposite to it.[1] Under this assumption, then, the conduction charge will be equal to $f(1/4\pi + \chi)E_{ind}$.

Helmholtz used this result in the following way. His and Schiller's experiments require the measured charge to be the same for air separation and for metallic contact, and indeed to be the same as though there were no polarization and the electromotive force acted directly. But suppose that there is polarization. Then the assumption of a direct action by a motional electromotive force produces an *addition* to the conduction charge equal to χE_{ind}. On the assumption of *Gleitstelle* this is the *total* conduction charge. Thus, if χ is extremely large in comparison to $1/4\pi$, the conduction charges would be empirically indistinguishable in the two cases. The existence of polarization can be assumed in either case—it is *necessary* for the *Gleitstelle* theory—under the proviso that the air's susceptibility is so large that it behaves, qua static induction, just as though it were a conductor. Such behavior is the immediate implication of a large susceptibility. In general the ratio between (surface) conduction charge density and polarization charge density is

$$\sigma_p/\sigma_c = -4\pi\chi/(1 + 4\pi\chi)$$

If χ is extremely large in comparison to $1/4\pi$, then σ_p will not of course be infinite, but it will be nearly equal in magnitude to σ_c. What this does is, in effect, to obliterate the air's physical difference from a conductor with respect to electromotive interactions, so that aerial *Gleitstellen* are not distinguishable from metallic *Gleitstellen*. In either case there is no net charge at the interface, whether it is metallic or aerial, but there is conduction charge on the outer metallic surface.

Nevertheless, a difference between the two situations does remain. With metallic *Gleitstelle* the outer charge is ultimately an effect of the continuity equation since the electromotive force at its locus generates a current that terminates at the outer surface. With aerial *Gleitstelle* the effect is electrostatic: the polarization of the *Gleitstelle* engenders statically a conduction charge on the neighboring metallic face. With metals we need only current–current interactions and the continuity equation; with air we need current–polarization interactions as well as polarization–charge interactions.

Already in 1870 Helmholtz had characterized the limiting case that air (and so ether itself) is a dielectric with very high susceptibility as the "Maxwell limit." But this limit has some very un-Maxwellian implications indeed. Take a simple example: a pair of oppositely charged conducting planes that together form a capacitor. Static forces act equally and oppositely on either plate just as though the only charge present were on the other plate. But in Helmholtz's limiting theory polarization charge equal in amount but opposite in sign to the conduction charge occurs at the surface of each plate. Consequently, the conduction and polarization charges located at the surface of the one plate can together exert no force on the other plate. That plate must be pulled in the direction of the other one by the polarization charge at its surface. For such an action to have an effect the polarization charge must be prevented by internal, presumably nonelectric, constraints from following the conducting surface. In other words if the conductor moves, it must successively polarize different regions of an essentially stationary ether, which continuously pull it toward the other plate. Polarization pulls neighboring conduction charge. This does not abolish action at a distance. Rather, it compensates one distance action by another, leaving only a contiguous effect as a net result. Thus, in Helmholtz's version of Maxwell's theory a body is moved by the immediately surrounding ether only because the regions of the ether that bound other conductors have their distance actions annulled. In Maxwellian theory no such compensation is involved, because there are no distance actions at all: everything is determined by the energy gradient at a given locus without regard to what occurs elsewhere at that moment.

∙ ∙

APPENDIX FIVE

Convection in Helmholtz's Electrodynamics

In 1875 the American physicist Henry Rowland, eager to glean knowledge and organizational experience from the great German university system, came to Helmholtz's Berlin laboratory with the express purpose of examining electric convection. Helmholtz himself would certainly not have been profoundly interested in the subject at the time, because it was important for neither of the two major experiments that tested his theory (Schiller's and his own). Little wonder that he took care to emphasize, in reporting Rowland's results, that Rowland, "den Plan für seine Versuche schon gefasst und vollständig überlegt hatte, als er in Berlin ankam, ohne vorausgehende Einwirkung von meiner Seite" (Helmholtz 1876a, pp. 791–92).[1]

Rowland (1879a) performed two considerably different experiments. In the first, radial scratches to prevent circumferential currents were made on a gilded, charged disk set into rotation. Here he detected a positive, albeit purely qualitative, effect since the deflection of sensitive magnetic needles reversed direction with the sign of the disk's charge. In the second set of experiments Rowland scratched the gilded disk along a series of circles concentric to the axis, which now prevented radial currents. These experiments produced no detectable needle deflection at all, which both Rowland and Helmholtz interpreted as evidence that convection currents exert precisely the same effects as ordinary currents (see Buchwald 1985a, p. 76, for details).

This result is not simple to understand in either Helmholtz's physics or in field theory. Since a moving charged conductor is simply a conductor in a charge state that moves, it has no special interaction energy with current-bearing conductors. Consequently, in the pure form of Helmholtz's physics it certainly should not exert electrodynamic effects. Field theory suffers from a similar problem since it abjures sources altogether, but in it a solution lies to hand with displacement currents. Helmholtz's physics can, however, yield convective actions in the same fashion as field theory if we assume that polarization is ubiquitous, that the equation of continuity applies to polarization currents, and that Helmholtz's polarization must be taken in the Maxwell limit. Then the conduction charge on the plates in Rowland's device engenders equal and opposite polarization charge, which will entail a polarization current via continuity.[2] In other words, the electrodynamic action of the moving conductor must be attributed to the changing polarization in the contiguous air. Consequently, Rowland's experiment becomes extremely important as an indirect (but quite powerful) demonstration of the electrodynamic action of polarization currents. Although Helmholtz did not draw that conclusion in reporting the experiment, two years later he included it with his own as consistent with field theory.

> Clerk Maxwell himself has developed his theory only for closed conducting circuits. I have endeavoured during the last few years to investigate the results of this theory also for conductors not forming closed circuits. I can already say that the theory is in harmony with all the observations we have on the phenomena of open circuits: I mean (1) the oscillatory discharge of a condenser through a coil of wire, (2) my own experiments on electromagnetically induced charges of a rotating condenser, and (3) *Mr. Rowland's observations on the electromagnetic effect of a rotatory disc charged with one kind of electricity.*
>
> The deciding assumption which removes the theoretical difficulties is that introduced by Faraday, who assumed that any electric motion in a conducting body which charges its surface with electricity is continued in the surrounding insulating medium as beginning or ending dielectric polarisation with an intensity equal to that of the current. (Helmholtz 1881a, p. 59; emphasis added)

Now in 1881 charge convection was still a difficult point for the Maxwellians, who were just beginning to notice it. For Helmholtz so easily to assume that it is consistent with field theory probably means that he reduces "Maxwell's theory" to the limiting case of Helmholtz's physics with polarization, which he had by then realized requires convection to act electrodynamically.[3]

APPENDIX SIX

Instability in the Fechner-Weber Theory

In 1870 Helmholtz drew a disturbing implication from Fechner-Weber, or rather from Fechner Weber reformulated in terms of Helmholtz's potential function and his general constant k (1870b, sec. 5).[1] If, Helmholtz argued, electricity does have mass, then a correction must be made in Ohm's law since that law must be interpreted mechanically: the electromotive forces and the Ohmic resistance are moving forces that must be equated to the product of electric mass by acceleration. Taking the current C in the Weberean fashion as proportional to electric velocity, and introducing a factor μ to represent the electric mass, Helmholtz could therefore write

Ohm's Law for Massy Currents
$$\mu \partial C/\partial t = -\kappa C - \nabla\phi - A^2 \partial U/\partial t$$

where κ is the resistivity.

Consider with Helmholtz a conducting sphere of radius R over which a quantity m of electricity is distributed; surround it with a concentric conducting shell of radius R covered with electricity M. From the continuity equation, the expression (in Weber's theory) for U, the Poisson equation, and the Ohm law for massy currents, it follows that the static potential ϕ on and within the inner sphere must have the following form:

$$\phi = m/R + M/R + (Ba/\rho)\exp(n_a t)\sin(\pi a\rho/R)$$

where B is a constant, ρ is the distance from the sphere's center, a must be an integer, and n_a must be

$$1/n_a = -(\kappa/8\pi) \pm \sqrt{(\kappa/8\pi)^2 - \mu/4\pi - k(AR/\pi a)^2}$$

Because of the exponential term in ϕ, the static potential of the inner sphere will increase indefinitely with time if the electrodynamic constant k is negative and the mass factor μ satisfies the following inequality:

$$-k(AR/\pi a)^2 > \mu/4\pi$$

Hertz's primary goal in the prize competition (see chap. 5) was to find as small an upper bound for μ as he possibly could and then to place the bound in

356

Helmholtz's inequality to make the instability, as it were, concrete. This could hardly answer Weber's untouchable claim that μ must be unmeasurably small. At least one Weberean, namely, Friedrich Zöllner, implicitly agreed that Helmholtz's critique would be devastating *if* it were correct. He agreed, in other words, that the mass cannot be so small that the instability would *never* occur (particularly in view of the fact that Zöllner wished to base all of physics on Weber's particles). Zöllner instead rejected Helmholtz's analysis itself, arguing that the differential equation of "Kirchhoff" on which it is based simply does not apply to Weberean electrodynamics under all circumstances:

> The labile electric equilibrium to which Helmholtz was led by his derivation shows only that the differential equations developed by Kirchhoff have no general validity. These differential equations are only partially based on Weber's fundamental laws, in other parts on several assumptions, so that Weber's fundamental law cannot in the least be touched by considerations based on those differential equations. (Zöllner 1872, p. LIV)

Helmholtz completely rejected this argument because as far as he was concerned, if Weber's theory was not compatible with Kirchhoff's differential equations for circuits, then that alone was cause to reject it. That position reflects Helmholtz's unbreakable belief that microphysics at best must be based on relations obtained first of all in the laboratory.

APPENDIX SEVEN

Hertz's First Use of the General Helmholtz Equations

In SM 245 Hertz analyzed the case of a conducting cylinder placed within a ooil. He begins with the vector potential for closed currents, which he did not do for a dielectric cylinder, where the currents could hardly be considered closed. That is, Hertz could use a simplified potential A in which the general k terms do not appear: namely, $\int (C/r) d^3r$, where C is the conduction current. Attacked head-on, however, the analysis would be very unwieldy, and so Hertz conceived the extremely fruitful idea of introducing auxiliary quantities that satisfy simpler equations. Here he used two.

Since the apparatus has the same cross section all along the length of the cylinder, Hertz analyzed only a plane section normal to the axis, within which the configuration has cylindrical symmetry. The current C can be reduced to the form $i(-\sin\theta, \cos\theta)$, where θ is the cylindrical angle coordinate. Then Hertz's auxiliary variables I and Π are

$$I \equiv \int_{\rho'}^{\infty} i d\rho' \Rightarrow i = -\frac{\partial I}{\partial \rho'}$$

$$\Pi \equiv \int \frac{I}{r} d^3r$$

The vector potential A in the plane accordingly becomes $e_\theta \partial \Pi/\partial \rho$, where e_θ is the unit vector $(\sin\theta, -\cos\theta)$. From these definitions Hertz had the following basic relations:

$$\nabla^2 \Pi = -4\pi I$$

$$i = (1/4\pi)(\partial/\partial\rho)\nabla^2\Pi$$

$$E_\theta = -\partial A/\partial t = -e_\theta \partial^2\Pi/\partial\rho\partial t$$

where E_θ is the electromotive force in the tangential direction (from symmetry the radial electromotive force must vanish).

Hertz divided the plane into three sections. The first consists solely of the conducting cylinder; the second of the nonconducting space between the cylinder and the coil; the third of the space beyond the coil. The Poisson equation for Π has different source terms in each case:[1]

Case 1: $\nabla^2\Pi = (4\pi A^2/\kappa)\partial\Pi/\partial t + f(t)$

Case 2: $\nabla^2\Pi = \phi(t)$

Case 3: $\nabla^2\Pi = 0$

Here f and ϕ result from the possibility of adding any function solely of time to $\nabla^2\Pi$. Hertz carried the analysis through at great length, eventually considering also what would occur if the conducting cylinder were magnetically polarizable. Since this is the first evidence we have for Hertz's developing knowledge of the wider structure of Helmholtz's electrodynamics it is worth examining briefly.

Referring directly to Helmholtz's equations (19e)–(19f) (1870b) Hertz was able at once to write down the equations for the vector potential in a magnetically polarizable conductor for the case in which the magnetic field vanishes except along the cylinder's e_z axis:

$$-(\kappa/4\pi)\nabla^2 U = -A^2\partial U/\partial t + A\partial^2 N/\partial y\partial t$$

$$-(\kappa/4\pi)\nabla^2 V = -A^2\partial V/\partial t + A\partial^2 N/\partial y\partial t$$

where U and V are the (only) components (A_x and A_y) of the vector potential, and N is the magnetic field. The latter connects to the vector potential and, through it, to Hertz's auxiliary scalar Π in the following way (with θ the magnetic susceptibility):

$$N/\theta = A(\partial U/\partial y - \partial V/\partial x)$$

Consequently, Π remains useful, only now the magnetic force also appears in the diffusion equation for it:

$$-(\kappa/4\pi)\nabla^2\Pi = -A\partial/\partial t(A\Pi - N) + f(t)$$

with

$$N = \theta A\nabla^2\Pi$$

Hertz introduces yet another auxiliary quantity, Ψ, such that

$$\Psi = A\Pi - N$$

thereby obtaining

$$\nabla^2\Psi = (4\pi A^2/\kappa)(1 + 4\pi\theta)\partial\Psi/\partial t + f(t)$$

which has precisely the same form as the equation for Π in case 1 in the nonmagnetic cylinder. All previous conclusions carry over, only for Ψ instead of Π. Note that waves are not at all involved here, because only magnetic, and not also electric, polarization is brought in. These quite sophisticated manipulations indicate that even at this early date Hertz had gone very far in assimilating at least the technical structure of Helmholtz's system.

The numerical example he gave asserts that the currents in the coil and in the cylinder contain an infinite number of terms, all damped, which decay at successively greater rates. The solutions for the currents in the coil and cylinder have the same rates of decay but different amplitudes, and this remains essentially unaltered even if the cylinder is magnetically polarizable, though the decay rates are thereby much decreased (because of increased induction). Since Hertz was concerned only with the overall properties of the solutions over time, he paid no attention at all to the fact that they diffuse.

Hertz on the Induction of Polarization by Motion

Hertz's early analysis of the electrodynamics of moving bodies (SM 245) aimed at specifying the electromotive force that acts on a current element when its distance from the objects with which it can interact changes. Whether the element forms part of a dielectric or of a conductor makes no difference to the analysis. Although Helmholtz had dealt with the problem in 1874a, he had treated it quite generally. Hertz intended to apply the equations to the particular case of a rotating sphere, and that problem had long ago been dealt with by Emil Jochmann, who had used Weber's electrodynamics. Hertz investigated first of all the relationship between Jochmann's and Helmholtz's equations.

Jochmann's analysis (1864) began directly with an expression drawn from the electromotive force exerted by a current C on a conductor located a distance r_d away and moving at a velocity v with respect to it. Assuming that the circuit containing C is closed, and introducing "the values" (in Jochmann's words) $A = \int (C/r_d) d^3r$, Jochmann obtained the following expression for the induced electromotive force E after a great deal of manipulation:

$$E = k[v \times (\nabla \times A)]$$

where k is Weber's electrodynamic constant. This was not, however, quite the same as the expression that Helmholtz obtained in 1874, which included the additional term $k\nabla(v \cdot A)$. The difference between the two sets of equations was important, because Jochmann's term cannot be used to produce any effects that would be significant for dielectrics; only the extra term in Helmholtz's equations works here. This is worthwhile pursuing in order to appreciate the kinds of problems that Hertz faced—problems that required considerable familiarity with, and insight into, expressions that can be quite complicated even using the simplifying apparatus of the modern vector calculus.

First of all Hertz did not give Jochmann's expression for E directly in terms of the vector potential for currents. Because he, like Jochmann in applying his results, intended to consider a magnetic field, Hertz instead introduced a vector potential λ for the magnetization M, and using it he defined an auxiliary, scalar function χ from its divergence:

$$\lambda \equiv \int (M/r_d) d^3r \quad \text{and} \quad \chi \equiv -\nabla \cdot \lambda$$

The corresponding vector potential for the magnetization will, Hertz asserts, then be $-\nabla \times \lambda$, in which case Helmholtz's equations can be written in the following way:[1]

Helmholtz's Equations in Terms of χ and λ

$$E_{\text{HELM}} = A\{-v \times \nabla\chi - \nabla[v \cdot (\nabla \times \lambda)]\}$$

Then, according to Hertz, the term that includes χ corresponds to Jochmann's expression.

An apparent difficulty here is that Hertz's expression $-v \times \nabla\chi$ is not obviously Jochmann's term. In order to express his electromotive force in terms of magnetization, Jochmann had again used "Weber's law" to transform it into the following form:

$$E_{\text{JOCH}} = 2k[v \times \nabla\textstyle\int(\nabla \cdot M/r_d)d^3r]$$

Hertz's form differs from Jochmann's in that Hertz has the divergence outside the integral, whereas Jochmann has it within the integral. Nevertheless, the two forms are equivalent to one another, though to see that this is so requires a high degree of ability to manipulate the complicated forms required by contemporary electrodynamics.[2]

Return now to the expression E_{HELM} for the electromotive force. The Jochmann term in it will vanish, Hertz remarks, if either χ itself vanishes or, in general, if its nth coordinate derivative vanishes for $n \geq 2$. This last condition, Hertz continues, in fact holds for the case of a body at the earth's surface moving in its magnetic field, where χ corresponds to the earth's field.[3] Consequently, if the earth provides the magnetic field, then Jochmann's term will not be significant, and only Helmholtz's term in the curl of λ remains. More important than this however is a point made implicitly in Hertz's analysis.

Although Jochmann's term is no doubt insignificant for reasonable rotations in the earth's field, it is the one that is responsible for the many usual effects of electromagnetic induction due to motion—his term, not Helmholtz's addition, accounts for the generation of currents when objects are in relative motion.[4] A spinning metal globe will, as a result of *this* term, sustain (in general) currents in a magnetic field. But a spinning *dielectric* globe, though no doubt subject to precisely the same electromotive force, cannot sustain any currents at all, because it does not obey Ohm's law. The metal globe, as well as its dielectric replacement, experiences an electromotive force at each point of its surface. This electromotive force is unchanged as long as the globe spins at a constant rate in a given direction. But, though constant in time, the electromotive force varies in general from point to point over the globe's surface. Consequently, the metal globe, which obeys Ohm's law, has currents. The dielectric globe does not obey Ohm's law, and it has no currents because its polarization at a point is proportional to the electromotive force there, and this does not change.

Thus, Jochmann's term is useless for testing the electromagnetic properties of dielectrics.

Helmholtz's term behaves quite differently. Unlike Jochmann's it cannot generate any currents at all, because it involves the gradient of a scalar, and so its integral around any closed path always vanishes—forbidding currents. Helmholtz's term accordingly behaves precisely like an electrostatic force, and as such it can do the one thing that Jochmann's term cannot do: it can produce charge. According to it a spinning globe (metallic or dielectric) in a magnetic field must acquire a surface charge, and *that* might be detectable.

To see the effect of Helmholtz's extra term consider with Hertz an object rotating in the earth's field with an angular velocity ω at latitude b. Taking the z axis along a meridian and the y axis along a parallel of latitude, the electromotive force that results within the object at a distance r from its instantaneous axis of rotation will be

$$E_x = MA\omega_z \cos(b/r^2)$$
$$E_y = 0$$
$$E_z = -MA\omega_x \cos(b/r^2)$$

This is obviously a maximum at the equator. Consequently, Hertz concluded that if the object rotates about the local vertical (ω_z vanishes), the electromotive force must be in the direction of the local meridian. If it rotates about the tangent to the local meridian (ω_x vanishes), the electromotive force will be along the local vertical. This electromotive force will charge the sphere.

APPENDIX NINE

Hertz on Relatively Moving, Charged Conductors

According to Hertz's Helmholtzian point of view, when two charged conductors move with respect to one another, the changing electric distribution that results entails the existence of currents between loci of change. This belief is a fundamental one, and it is symbolized mathematically by the primitive status of the continuity equation in Helmholtzian electrodynamics. Consequently, Hertz had the following four equations with which to attack this problem:

(1) Ohm: $$\kappa C = -\nabla\Phi$$

(2) Continuity + Gauss: $$\partial\nabla^2\Phi/\partial t = 4\pi\nabla \cdot C$$

(3) Continuity at boundary: $$-\partial\sigma/\partial t = C \cdot e_{n_i}$$

(4) Gauss at boundary: $$-4\pi\sigma = \nabla\Phi \cdot (e_{n_i} + e_{n_e})$$

Here the e vectors are inner (subscript i) and outer (subscript e) unit surface normals, C is the current, Φ is the potential, and σ is the charge density.

Hertz considers an object rotating about the z axis with an angular velocity, say λ, and he uses the spherical (he calls them "polar") coordinates ρ, ω, θ. If, Hertz argues, a surface is moving, then all of the time derivatives in his equations must be treated as total derivatives and so replaced by $\partial/\partial t + v \cdot \nabla$, where v is the linear velocity, so that $v \cdot \nabla$ can be replaced in the example by $\lambda\partial/\partial\omega$. If the period of rotation is T, then equations (1) and (3) combine to yield

(5) $$\nabla\Phi \cdot e_{n_i} = \frac{\kappa\partial\sigma}{\partial t} + \frac{(2\pi\kappa/T)\partial\sigma}{\partial\omega}$$

To see how this works, consider with Hertz a spherical shell that rotates uniformly in an external electric potential φ_E. Further, let φ denote the potential of the charge that is induced on the shell. To solve the problem Hertz expanded the external potential in spherical harmonics that are independent of ω (since it is taken to be symmetric about the axis of rotation for this special case). Then Hertz's equations (4) and (5), applied to the steady state, yield the following relation at the shell's surfaces:

$$\frac{\partial\phi_E}{\partial\rho} + (\nabla\varphi) \cdot e_{n_i} = -\frac{\kappa}{2T}\frac{\partial}{\partial\omega}[(\nabla\varphi) \cdot (e_{n_i} - e_{n_e})]$$

This relation make the problem similar in structure to the conductor or dielectric sphere rotating under magnetic influence that he had examined in his inaugural dissertation. There he had drawn the lines of current flow; here he also drew (for the case of a rotating cylindrical shell) the lines of current flow, only they are not permanent flow lines, because over time the charge distribution on a uniformly rotating conductor reaches a steady state if the external potential is independent of the time.

The overall effect depends primarily upon the ratio of the resistivity to the period. In all cases energy is dissipated in Joule heat via the Helmholtzian currents, but in ordinary conductors the magnitude of the dissipation is altogether negligible. In addition, while the currents last, the charge distribution over the conductors alters. If, for example, the ratio is small (if the object is a very good conductor), then the effect is primarily to rotate the charge distribution over the surface through a *very* small angle, to diminish it by an equally "small quantity," and to bring into being a minuscule charge on the inner surface of a spherical shell—in other words, the usual static conditions are marginally altered (during the transient state). However, effects of any kind are hard to find in the laboratory. In ordinary metals the extremely high conductivity completely overwhelms any conceivable effects by nearly at once returning the object to the usual static conditions. Insulators like shellac, on the other hand, have such high resistivities that hardly any "free electricity" at all can form on them, though they do persist in the transient state (where effects occur) for quite some time.[1] However, for certain kinds of objects ("glass . . . mixtures of insulators with conductors in form of powder"; Hertz 1881a) the conductivity would be low but not too low to obliterate the formation of any "free electricity" at all, and effects might be measured. This is why Hertz used "mirror glass" for his experiment. If, he reasoned, he had used a metal plate instead, then the admittedly large free charge induced on it would have redistributed so rapidly in comparison to the motion of the needle that the currents would have lasted for very little time, and so the heating effects would have been altogether negligible. Glass works by increasing the time to reach equilibrium without reducing the free charge to negligible proportions.

APPENDIX TEN

Elastic Bodies Pressed Together

In his papers 1881c and 1882a Hertz set himself the following problem: to find for bodies in contact with one another (1) the surface to which the *region of contact* belongs (the "surface of pressure"); (2) the form and the absolute magnitude of the bounding "curve of pressure" within that surface; (3) the distribution of pressure over that surface; (4) the maximum pressure within that surface; and (5) the mutual approach of bodies under the influence of a given total pressure. Hertz assumed that no tangential stress exists in the "surface of pressure"—that the contact stress is purely normal.

Hertz's first, and most distinctive, step was to set up two independent systems of coordinates, one for each of the two interacting objects. Denote the first system with a subscript 1 and the second with a subscript 2. The z axes of the two systems are collinear and antiparallel, but share only an initially common origin (for reasons we will come to). The bodies are initially in purely geometric contact. The unstressed shapes of the surfaces of the two bodies near this point of geometric contact can be represented approximately by quartics:

$$z_1 = A_1 x^2 + B_1 y^2 + 2H_1 xy$$

$$z_2 = A_2 x^2 + B_2 y^2 + 2H_2 xy$$

Consequently, the distance δ between a pair of points, respectively on each surface, that have the same x, y coordinates is, before the bodies press together,

$$\delta = (A_1 - A_2)x^2 + (B_1 - B_2)y^2 + 2(H_1 - H_2)xy$$

where the signs reflect the convention that the z axes point *into* body 1. Since this distance is necessarily positive (for otherwise the bodies would somewhere interpenetrate),

$$A_1 - A_2 > 0, \quad B_1 - B_2 > 0, \quad \text{and} \quad H_1 = H_2$$

And so we may write:[1]

$$z_1 - z_2 = Ax^2 + By^2$$

When the bodies press together—when the "surface of pressure" is created—the points separated by δ in the unstressed state will be moved in two ways. First, each point is displaced through a distance w_1, w_2 with respect to its proper z axis in the usual elastic manner. Second, since Hertz limits the deformation to a small region near the point of geometric contact, the farther parts of each body, *to which the respective axes are fixed,* displace rigidly; thus, the coordinate systems themselves undergo a change in mutual distance by some amount α. (The origins of either system, that is, are fixed by the rigidly displaced parts of the bodies and not by their point of initial, geometric contact.) As a result the distance δ is *increased* by an amount $w_1 - w_2$ and *decreased* by an amount α, thereby becoming δ':

$$\delta' = Ax^2 + By^2 + w_1 - w_2 - \alpha$$

Over the region of contact, the two points, Hertz asserted, become coincident.[2] Therefore, δ' must vanish there, which provides Hertz's primary condition:

In the region of contact: $w_1 - w_2 = \alpha - Ax^2 - By^2$

This condition must agree with the differential equations within the bodies as well as with the boundary conditions. Seven conditions result from these two requirements:

1. For equilibrium, within each body $\nabla^2 w + (1 + 2\theta)\nabla\sigma = 0$, where $\sigma = \nabla \cdot w$ and θ is a coefficient for elastic compressibility (following Kirchhoff's notation).

2. The displacements vanish at infinity.

3. The tangential stresses vanish over the surface of pressure.

4. On the surface of pressure, but outside the region of contact, the normal stress vanishes; over the region of contact the normal stress is continuous.

5. Within the region of contact, $w_1 - w_2 = \alpha - Ax^2 - By^2 = \alpha - z_1 + z_2$.

6. Within the region of contact the normal stress is greater than zero (since both bodies are compressed), and outside it $w_1 - w_2 > \alpha - Ax^2 - By^2$, since otherwise the bodies would interpenetrate.

7. The integral of the normal stress over the surface of pressure must equal the total pressure.

As was by now usual for him, Hertz solved the problem by introducing a set of auxiliary potential functions, here P and Π, such that

$$\nabla^2 P = 0 \quad \text{within either body}$$

$$\Pi \equiv -\frac{zP}{K} + \frac{1}{K(1 + 2\theta)} \left(\int_z^i P\,dz - J \right)$$

where K is Kirchhoff's elastic coefficient, J is a constant, and i is taken to infinity after the integration; the purpose of J is to make Π finite.[3] From these definitions there follow:

$$\nabla^2 \Pi = -\frac{2}{K}\frac{\partial P}{\partial z}$$

$$w = \nabla\Pi + 2\vartheta P e_z$$

where $\vartheta = 2(1 + \theta)/[J(1 + 2\theta)]$.

Using the relation between displacement and stress for linear elastic bodies, the stresses can be expressed in terms of Hertz's potentials and the elastic constants as follows:

$$X_x = -2K\left[\frac{\partial^2\Pi}{\partial x^2} + \frac{2\theta}{K(1 + 2\theta)}\frac{\partial P}{\partial z}\right] \quad \text{for } x = x, y, z$$

$$X_y = -\frac{2K\partial^2\Pi}{\partial x\partial y}$$

$$X_z = \frac{2z\partial^2 P}{\partial x\partial z}$$

$$Y_z = \frac{2z\partial^2 P}{\partial y\partial z}$$

These expressions are alone sufficient to satisfy all but the last three of Hertz's conditions.

The next step was to assume a particular solution for P. Noting that P satisfies the Laplace equation, Hertz considered it to be due to a fictitious distribution of electricity over the region of contact, which he assumed to be plane and bounded by an ellipse, in such a fashion "that [the electricity] can be considered as a charge which fills an infinitely flattened ellipsoid with uniform density" (1881c). The assumption that the region of contact is a plane ellipse punishingly limited the generality of Hertz's theory, but the solution to this problem was known, and Hertz could show that all remaining conditions were satisfied by the resulting potential. Hence, it constituted a unique solution. The analysis yielded integral expressions that related the axes of the ellipse to the elastic constants and to Hertz's constants A, B, which were in turn functions of the original radii of curvature. It yielded also an integral for the distance α through which the rigidly attached coordinate systems are moved with respect to one another. The integrals could be approximated numerically using Legendre's tables, and Hertz gave a table with sample values.

*   *

APPENDIX ELEVEN

Evaporation's Theoretical Limits

Hertz felt that he had been unable to achieve a "lucid theory" for evaporation, even though his paper (1882c) contains two distinct excursions in that direction, one of which moves almost entirely outside the laboratory. The first, and much the longest of the two, was indubitably written (or at least thoroughly *re*written) after Hertz concluded that he had been mistaken in thinking that his experiments had ever revealed an evaporative limit. These remarks show *why* Hertz's failure had been inevitable; they indicate that none of the devices that he used could possibly have revealed the limit, even if it did in fact exist. Obviously Hertz could not have had anything quite like these considerations in mind when he began experimenting, for if he had then there would have been no point in going further. They were constructed after his failure to transform the experiment into something else—if not into success, at least into something not entirely negative.

Hertz envisioned "two infinite, plane, parallel liquid surfaces kept at constant, but different temperatures." Vapor will move "from the one surface to the other," conserving its heat and establishing, one might say, the requisite interaction between the two surfaces. During this passage, Hertz argued, "it follows from the hydrodynamic equations of motion" that the state of the vapor, including its velocity, remains constant. The vapor, that is, has some particular native pressure and temperature p, T. These values may be different from the temperatures T_1, T_2 of the surfaces and from the pressure P that the vapor exerts on each surface. The latter, Hertz asserts, must be the same for both surfaces, presumably because the force exerted on the vapor at one surface must be the same as the force that the vapor exerts on the other surface.

Hertz's theory—or, better, the reason for the failure of his experiments—hinges on the magnitude of the difference between p, the native vapor pressure, and P, the pressure between vapor and surface. The difference is due entirely to the fact that the vapor streams away from the surface with a certain speed; the difference $P - p$ must therefore be the force that accelerates the vapor to that speed. Suppose the vapor (of density d) has, after acceleration, speed u, and that in unit time a weight m passes from one surface to the next. Then, clearly,

$$m = ud$$

We can use this to establish a link to the pressure P, which is what Hertz measured. In his words:

> For let us suppose the quantity m spread over unit surface, the pressure upon one side of it being P and on the other side p, and its temperature T maintained constant. It will evaporate just as before; after unit time it will be completely converted into vapour, which will occupy the space u and have the velocity u. Hence its kinetic energy is $(1/2)mu^2/g$; this is attained by the force $P - p$ acting upon its centre of mass through the distance $u/2$, so that an amount of work $(P - p)u/2$ is done by the external forces.[1] From this follows the equation $P - p = mu/g$; or, since $m = ud$, $m^2 = gd(P - p)$. (Hertz 1882c, pp. 195–96)

Hertz wanted to use this result to assign limits to the relations between the six quantities that he had at his disposal, namely, T, T_1, T_2, p, P, d, u, and m. That way he could calculate from observational data what the corresponding evaporative limit ought to be. But first he had somehow to establish that there *ought to be* such a limit. This was not easy, but Hertz succeeded in formulating an argument based on two "assertions which, according to general experience, are at any rate exceedingly likely to be correct"—they do not derive in any way at all from theory:

> [1.] If we lower the temperature of one of several liquid surfaces in the same space while the others remain at the original temperature, the mean pressure upon these surfaces can only be diminished, not increased.
> [2.] The vapour arising from an evaporating surface is either saturated or unsaturated, never supersaturated. For it appears perfectly transparent, which could not be the case if it carried with it substance in a liquid state. (1882c, pp. 196–97)

These two rules from "general experience" suffice to establish that the rate of evaporation must have a limit for a given temperature. From the first we know that the surface pressure P must be less than the pressure p_M of saturated vapor at the temperature of the high source (T_1); from the second we know that the actual density d of the vapor is less than the density d_p of *saturated* vapor at the same pressure p. Hertz could therefore write the following inequality:

$$m < \sqrt{gd_p(p_M - p)}$$

Now when the pressure p vanishes, then so, of course, do the other pressures, so that the quantity under the root goes to zero with p. It also goes to zero when p goes to p_M. In between these two points, Hertz noted, it must reach some maximum, which is then the theoretical maximum for the evaporation rate at a given high temperature T_1. No matter how low the temperature of the other reservoir may be, the evaporation rate cannot exceed this value.

Of course, Hertz had not observed any limit at all, and the point of going through this bit of "theory"—based on a pair of decidedly untheoretical rules—was to show why he had not succeeded. To do that he needed to compute numbers, which meant he had to be able to calculate volumes and temperatures from pressures. To do so he assumed, without elaboration, that the Boyle law holds for the vapor; the pressure–temperature relation could be derived from thermodynamics and known experimental values. The calculation gave values for the limit that were vastly greater than the evaporation rates that Hertz had achieved (e.g., at 180° the calculated limit was 20.42 mm/min, whereas Hertz had achieved only 1.67 mm/min), which gave him the confidence to remark of his experiments that "they do not show definitely the existence of the deviation from this rule which probably occurs, and which is of interest from the theoretical point of view."[2]

Hertz's second excursion into theory provided a reason for *why* a limit might exist at all. He based it on the kinetic theory of gases, arguing in effect that the evaporation rate cannot exceed a quantity determined by the "mean molecular velocity of the saturated vapour corresponding to the temperature of the surface." His argument was, however, quite sketchy and undeveloped.

Hertz's Model for Geissler-Tube Discharge

Hertz's argument for the continuity of the discharge current in a Geissler tube had to address three claims that, he asserted, were frequently raised to prove intermittence: first, that "a weak current (*e.g.* such as an induction machine gives) is always discontinuous, and does not become continuous even when the partial discharges succeed each other at the rate of several thousand per second"; second, that the Joule heat in a tube is more nearly proportional to the current than to its square;[1] and, third, that "in accordance with this" second point, the potential across the tube "persists at the value which enables the weakest current to traverse" it even as the current through the tube increases.[2]

In Hertz's figure 83 the tube is represented by the apparatus between the plates A and B. The latter consists of a strongly conducting spring, hung from A, that carries a weight α (also strongly conducting) terminating in a poorly conducting region β of resistance r. A potential p across AB is just large enough to charge β sufficiently to pull the spring down and bring β into contact with the lower plate, which initiates a very weak current (because of β's high resistance).

Now, Hertz continued, if the source of p is capable of maintaining the resulting current p/r, then β can remain in permanent contact with B: because almost the entire potential drop occurs across β, its termini can remain charged strongly enough to overcome the spring tension. If, on the other hand, the source cannot maintain the current, then the charges across β will decrease. At some point the spring will break the contact, and the current will cease—producing an intermittent effect as the process repeats.

To accommodate Hertz's three points (or rather to overcome the first and to embody the second and third) requires adding a very large number of Hertz's spring devices, each of which has a slightly different spring constant, with p being the potential necessary to bring the weakest such device into contact with B. Suppose that the potential across AB increases somewhat. Then the current through the original device—and the heat it dissipates—remain essentially the same as before because the resistance in each device is extremely high, but several more new devices come into contact with B. Each carries more or less the same current, which means that the total current across AB

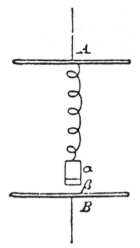

FIGURE 83   Hertz's electromechanical model for Geissler-tube discharge (Hertz 1883c)

simply increases by the number of devices. Since each device dissipates heat at the same rate, the total rate increases directly with the total current and not with its square over the ranges realizable in the contemporary laboratory. The current does change magnitude in a *punctuated* fashion, but it is not formed out of single currents that succeed one another without overlap. On the contrary, once a partial contribution to the current comes into being, it persists as further contributions are added, thereby showing that weak currents can occur without intermittence.

It also follows that the potential will remain nearly constant, though Hertz did not explain his reasoning on either point. As the source potential slightly increases, it brings many more spring-loaded devices into contact with the lower plate. This produces a group of high-resistance circuits in parallel between A and B. If there are n devices touching B, and each has resistance r, then the reciprocal $1/r'$ of the total resistance will be n/r. Assuming again that the current through each device remains pretty much what it was before, the total current i will be directly proportional to the number n of circuits, and so (since the resistances are fixed) to $1/r'$. But in that case Ohm's law requires the potential to be constant. Here, then, we have a way for the current to increase without the potential increasing proportionately.

Hertz's goal was certainly not to provide a working model for tube discharge but rather to demonstrate that aspects of it which had been taken to require its intermittence need not be thought of in that way. The model does produce a permanent current once potential p is achieved, but the current does not grow continuously, since it increases punctually as more of the spring carriers come into action. Each one jumps the current by a finite amount. Nevertheless, this kind of discontinuity is entirely different from the intermittence that Hertz was seeking to exorcize, because in the latter the current magnitude must grow and decrease tremendously in very short periods of time whether or not the action of the source is altered, whereas in Hertz's model the current changes, if at all, comparatively slowly and never decreases unless the action of the source does (e.g., upon removal of elements from the battery).

APPENDIX THIRTEEN

Propagation in Helmholtz's Electrodynamics

13.1. FUNDAMENTALS

The essential structure of high-level polarization theory is extremely simple, amounting to the addition of a single assumption: dielectric polarization can always be generated by electromotive forces as well as by electrostatic ones. This is hardly a small requirement because it is based on the implicit assumption that there is an interaction energy between charges and currents, an energy that Helmholtz never attempted to represent. In order to produce the most general possible set of equations, which will greatly aid our understanding of Hertz's work, we will not follow Helmholtz precisely. His own presentation is prolix and requires careful study to uncover the physical and mathematical principles upon which it was based. Instead, we will examine a set of equations representing the most general possible situation, namely, the case of a polarizable, conducting body.[1]

Begin with the equations that link the electrodynamic potential (U) to the total current (C_{tot}), and the polarization (P) to the static (ϕ_f) and electrodynamic potentials:

Electrodynamic Potentials

$$U(r) = \int \frac{C_{tot}(r')}{r_d} d^3r' + \frac{1}{2}(1 - k)\nabla_r\left\{\int [C_{tot}(r') \cdot \nabla_r' r_d]\right\} d^3r'$$

Electric Polarization

$$P = \chi\left(-\nabla\phi_f - A^2 \frac{\partial U}{\partial t}\right)$$

Note the presence here of $\partial U/\partial t$, which is the explicit effect of the implicit interaction energy between charge and current. Helmholtz tacitly assumes that U can serve mathematically as this potential where the effect of a changing current on a charge is concerned, just as, conversely, he assumes in Ohm's law that ϕ_f can serve mathematically as the potential where the effect of a charge in generating a current is concerned. The Mossotti hypothesis conjoined to the continuity equation requires that $\partial U/\partial t$ be treated as a current in computing electrodynamic interactions via the potential U, so that, with C_{con} representing the conduction current, Helmholtz could write for the total current C_{tot}:

Total Current

$$C_{tot} = C_{con} + \frac{\partial P}{\partial t}$$

Continue with the continuity equation proper, Poisson's equation, and Ohm's law:

Continuity

$$\nabla \cdot C_{tot} + \frac{\partial \rho}{\partial t} = 0$$

Poisson's Equation

$$\nabla^2 \phi_f = -4\pi\rho$$

Ohm's Law

$$C_{con} = \frac{1}{\kappa}\left(A^2 \frac{\partial U}{\partial t} - \nabla\phi_f\right)$$

These six equations are the foundations of Helmholtz's theory. From the expressions for the electrodynamic potential, continuity, and Poisson's equation there result (as we have already seen):

$$\nabla^2 U = (1 - k)\nabla \frac{\partial \phi_f}{\partial t} - 4\pi C_{tot}$$

$$\nabla \cdot U = -k\frac{\partial \phi_f}{\partial t}$$

And from these we have the Helmholtzian version of the Ampère law:[2]

Ampère Law

$$\nabla \times (\nabla \times U) = - \nabla\frac{\partial \phi_f}{\partial t} + 4\pi C_{tot}$$

By retaining all of the available parameters, including both resistivity (κ) and electric polarizability (χ), we can produce a thoroughly general equation for propagation that we can then apply to special cases. To do so combine Ohm's law with the Ampère law and the expressions for $\nabla \cdot U$ and P to obtain

The General Helmholtz Equation for Propagation

$$-\frac{A\partial^2 U}{\kappa\,\partial t^2} + \frac{\nabla(\nabla\cdot U)}{k\kappa} - \chi A^2\frac{\partial^3 U}{\partial t^3} = -\frac{1+4\pi\chi}{4\pi\chi}\nabla\frac{\partial(\nabla\cdot U)}{\partial t} +$$

$$\frac{1}{4\pi}\nabla\times\left(\nabla\times\frac{\partial U}{\partial t}\right)$$

This general equation can be split in two by separating disturbances for which the divergence $(\nabla\cdot U)$ vanishes from those for which the curl $(\nabla\times U)$ does:

$$\nabla\cdot\text{ vanishes}$$

$$\chi\frac{\partial^2 U}{\partial t^2} = \frac{1}{\kappa}\frac{\partial U}{\partial t} + \frac{1}{4\pi A^2}\nabla^2 U$$

$$\nabla\times\text{ vanishes}$$

$$\chi\frac{\partial^3 U}{\partial t^3} = \frac{\nabla^2 U}{k\kappa A^2} - \frac{1}{\kappa}\frac{\partial^2 U}{\partial t^2} + \frac{1+4\pi\kappa}{4\pi k A^2}\nabla^2\frac{\partial U}{\partial t}$$

13.2. PROPAGATION IN POLARIZABLE NONCONDUCTORS

In nonconductors the resistivity, κ, is infinite, and so for the case in which the curl vanishes we have

$$\nabla\times\text{ vanishes: longitudinal waves}$$

$$\frac{\partial^2 U}{\partial t^2} = \frac{1+4\pi\chi}{4\pi k\chi A^2}\nabla^2 U$$

Consequently, the electrodynamic potential can propagate at the speed $\sqrt{(1+4\pi\chi)}/A\sqrt{(4\pi k\chi)}$ in a polarizable nonconductor. Here P can replace U because the basic equations imply that $\nabla^2(\nabla\cdot U)$ is equal to $-4\pi k\nabla\cdot\partial P/\partial t$.[3] Here, then, we have a longitudinal wave (i.e., one wherein the disturbance is normal to the front) for the electrodynamic potential, U, and for the electric polarization, P. This wave is *not* determined simply by the scalar potential ϕ_f but it does lead to a propagation equation for it: simply replace $\nabla\cdot U$ with $-k\partial\phi_f/\partial t$ and drop a time derivative to obtain

$$\frac{\partial^2\phi_f}{\partial t^2} = \frac{1+4\pi\chi}{4\pi k\chi A^2}\nabla^2\phi_f$$

Since this equation was derived under the assumptions that χ is not zero (otherwise infinites occur in deducing it) but that the resistivity is infinite, it applies only to polarizable nonconductors.

For the case in which the divergence vanishes we find

$\nabla \cdot$ vanishes: transverse waves

$$\frac{1}{4\pi\chi A^2} \nabla^2 U = \frac{\partial^2 U}{\partial t^2}$$

Here also P can directly replace U (from the basic expression for P and the fact that the divergence vanishes), so that this equation determines a transverse wave of polarization as well as one of electrodynamic potential, with both propagating at the speed $1/A\sqrt{(4\pi\chi)}$.

These kinds of waves will occur in any nonconductor that can be electrically polarized, but as the theory stands, they will not occur in space. *To force a fieldlike structure Helmholtz assumed that the ether is itself polarizable and that a material body adds its polarization to that of the ether at its location.* This has the effect of requiring a correction to be made for the ether's polarization in order to connect the χ of the propagation equations with measured susceptibilities. If χ_0 is the actual susceptibility of the ether, and χ is the actual susceptibility of a body, then, Helmholtz demonstrated, the body's measured susceptibility $\bar\chi$ is such that

$$1 + 4\pi\bar\chi = \frac{1 + 4\pi\chi}{1 + 4\pi\chi_0}$$

This equation provides a quick, though not particularly revealing, way to see how relationships that characterize Maxwell's theory can be obtained from Helmholtz's as limiting cases of it. In Maxwell's theory the index of refraction for a disturbance passing from a body with a measured dielectric capacity ε_1 to one with a measured capacity ε_2 is given by $\sqrt{(\varepsilon_1/\varepsilon_2)}$. Now the capacity of a body is given by the sum $1 + 4\pi\bar\chi$, where $\bar\chi$ is its measured susceptibility, so that Maxwell's expression can be written in terms of susceptibilities as $\sqrt{[(1 + 4\pi\bar\chi_1)/(1 + 4\pi\bar\chi_2)]}$. However, according to Helmholtz's theory the index of refraction will be given by $\sqrt{\{[(1 + 4\pi\chi_0)(1 + 4\pi\bar\chi_1) - 1]/[(1 + 4\pi\chi_0)(1 + 4\pi\bar\chi_2) - 1]\}}$. Consequently, to reach Maxwell's wave speed from Helmholtz's we must assume that the susceptibility of the ether is effectively infinite.

The existence of a polarizable ether affects all electromagnetic measurements because the quantity A cannot be measured independently of the ether's susceptibility; measurement in fact yields a quantity A' equal to $A\sqrt{(1 + 4\pi\chi_0)}$. In the free ether the speeds of longitudinal and of transverse waves must in consequence be (ignoring magnetic susceptibility)

$$v_{\text{lon}} = \frac{1}{A'} \frac{1 + 4\pi\chi_0}{\sqrt{4\pi k\chi_0}}$$

$$v_{\text{trans}} = \frac{1}{A'} \sqrt{\frac{1 + 4\pi\chi_0}{4\pi\chi_0}}$$

(Helmholtz 1870b, p. 627). This has the important (and, here, obvious) consequences that by setting the ether's susceptibility to infinity *independently* of the value of k, the longitudinal wave can be granted infinite velocity and the Maxwell value ($1/A'$) of the transverse speed can be retrieved. It can in fact be demonstrated generally, as Poincaré did (1890), that the basic Maxwell equations do not, in this limit, depend at all on k.[4]

Nevertheless, almost all commentators assumed that the Maxwell scheme requires both limits (infinite susceptibility and $k = 0$), primarily because Helmholtz had so written, and because little attention was paid to the fact that the constant A cannot be measured by itself given the ether. Indeed, although Helmholtz had himself derived these last two expressions for the wave speeds, nevertheless at the beginning of the same 1870 paper (1870b, p. 557) he had written: "The velocity of longitudinal waves in air is equal to the product of that of the transverse waves multiplied by the factor $1/\sqrt{k}$. In Maxwell's theory the propagatory velocity of the longitudinal electric waves is assumed to be infinite, which corresponds to the value $k = 0$." This statement certainly does not square with the expression just given, which requires the ratio of longitudinal to transverse speed to be $\sqrt{[(1 + 4\pi\chi_0)/k]}$, not just $1/\sqrt{k}$. In other words, Helmholtz's introductory statement does imply that k must be set to zero to obtain Maxwell's equations, whereas his concluding expression does not.

Helmholtz may perhaps not have thought this through carefully, because in 1870 he was more concerned with attacking Weber than clarifying Maxwell. But the apparent contradiction may also reflect a bifurcation that Helmholtz, and perhaps Helmholtzians, drew: between Maxwell's theory as something concerned with open circuits regardless of the ether and Maxwell's theory as grounded exclusively in the ether. Obtaining Maxwell's system in Helmholtz's fashion suggests such a division, since one can say that there is indeed a "Maxwell system" sans ether that corresponds to Helmholtz's equations with zero electric susceptibility and zero k. This system requires different open-circuit effects from those of either Weber's or Neumann's schemes because its electrodynamic potential differs.

13.3. PROPAGATION IN NONPOLARIZABLE CONDUCTORS

In a conducting but nonpolarizable body set χ to zero and obtain for the case of no curl

$$\nabla \times \text{ vanishes}$$

$$\frac{\partial^2 U}{\partial t^2} = \frac{\kappa}{4\pi k A^2} \nabla^2 \frac{\partial U}{\partial t} + \frac{\nabla^2 U}{k A^2}$$

This equation also holds for the scalar potential, and it has the following particular solution S (see Helmholtz 1870b, p. 601, for a broad discussion; what follows is not Helmholtz's solution, because he gives none):

$$S = \frac{\alpha}{r}e^{\gamma r}\cos(k'r - \omega t)$$

$$\text{where} \quad \gamma = \alpha \, \sin\!\left(\frac{\arctan \beta}{2}\right)$$

$$k' = \alpha \, \cos\!\left(\frac{\arctan \beta}{2}\right)$$

$$\alpha = \frac{\omega A \sqrt{k}}{[(1 + \omega^2 k^2)/16\pi^2]^{1/4}}$$

$$\beta = \frac{\omega k}{4\pi}$$

$$\frac{\omega}{k'} = \frac{[(1 + \omega^2 k^2)/16\pi^2]^{1/4}}{A\sqrt{k} \cos(\arctan \beta/2)}$$

Consequently, both the electrostatic potential and the divergenceless electrodynamic potential propagate in nonpolarizable conductors as a *damped wave*. The phase velocity ω/k of the wave is a function of the angular frequency ω, so that the propagation is dispersive. In the limit that κ (the resistivity) vanishes, the velocity becomes $1/A\sqrt{k}$, which would therefore be the phase velocity for these kinds of waves in *perfect* conductors that are completely unpolarizable.

For the case of no divergence we have

$$\nabla \cdot \text{vanishes}$$

$$\frac{A^2 \partial U}{\kappa \, \partial t} = -\frac{1}{4\pi}\nabla^2 U$$

Here we do not have a wave, because the equation is linear in time; instead, we have an equation for the diffusion of a vector disturbance. Thus, in nonpolarizable conductors the electrodynamic potential (and so the electric potential) can propagate as a damped, dispersed longitudinal wave, and it can also diffuse as a transverse disturbance.

13.4. Distinctions between Nonpolarizable Conductors and Polarizable Nonconductors

Helmholtz's equations, then, distinguish in the following ways between the kinds of propagation that can occur in nonpolarizable conductors (NPC) and in polarizable nonconductors (PNC):

Type of Wave	Speed in NPC	Speed in PNC
Transverse U and P	Diffusion	$\dfrac{1}{A\sqrt{4\pi\chi}}$
Longitudinal U and P	$\dfrac{[(1 + \omega^2 k^2)/16\pi^2]^{1/4}}{A\sqrt{k}\cos(\arctan \beta/2)}$	$\dfrac{\sqrt{1 + 4\pi\chi}}{A\sqrt{4\pi k\chi}}$
ϕ_f	$\dfrac{[(1 + \omega^2 k^2)/16\pi^2]^{1/4}}{A\sqrt{k}\cos(\arctan \beta/2)}$	$\dfrac{\sqrt{1 + 4\pi\chi}}{A\sqrt{4\pi k\chi}}$

Only PNCs propagate transverse waves of every kind, but both PNCs and NPCs propagate longitudinal waves, although the latter kind of propagation is both dispersive and damped in NPCs. Furthermore, wherever longitudinal waves occur, there are also waves of static potential that propagate at the same phase velocity.[5] Since any wave that is independent of χ must always exist, in a pure Helmholtzian system the potentials must propagate as longitudinal disturbances in a conducting space whether or not the region is also polarizable.[6]

Given that according to field theory this kind of wave does not exist in any kind of body, whether it be conducting, polarizable, or both, it is important to understand why it must exist in the Helmholtzian system. To see how it comes about, return to Helmholtz's general propagation equation. Longitudinal waves derive from the second term on the left-hand side in NPCs, and from the first term on the right-hand side in PNCs. Both of these terms exist because $\nabla \cdot U$ is equal to $-k\partial\phi_f/\partial t$. That relation in turn arises from combining Poisson's equation with Helmholtz's continuity equation in order to obtain a relationship between the current and the scalar potential, viz., $4\pi\nabla \cdot C_{\text{tot}} = \nabla^2\partial\phi_f/\partial t$. This is the crucial step in the deduction because it introduces time dependence into the variable (C_{tot}) that determines U: combining this expression with the fundamental equation for U yields Helmholtz's propagation equation for $\nabla \cdot U$, and so longitudinal waves.[7]

The ultimate reason for the existence of these kinds of waves in Helmholtz's theory is the form of his continuity equation, which is entirely different from the one that holds in field theory. We can see this by writing both of them in terms of the two possible kinds of E fields, namely, those due to static actions (E_{stat}) and those due to changing currents (E_{ind}):

Maxwell's Continuity Equation

$$\nabla \cdot C_{\text{tot}} = 0 \Rightarrow \nabla \cdot \left[C_{\text{con}} + \frac{\partial P}{\partial t} + \frac{\partial(E_{\text{stat}} + E_{\text{ind}})}{\partial t}\right] = 0$$

using $\quad C_{tot} = C_{con} + \dfrac{\partial D}{\partial t} \quad$ where $\quad D = E_{tot} + P \quad$ and $\quad E_{tot} = E_{stat} + E_{ind}$

Helmholtz's Continuity Equation

$$\nabla \cdot C_{tot} + \frac{\partial \rho_f}{\partial t} = 0 \Rightarrow \nabla \cdot \left(C_{con} + \frac{\partial P}{\partial t} + \frac{\partial E_{stat}}{\partial t} \right) = 0$$

using $\quad C_{tot} = C_{con} + \dfrac{\partial P}{\partial t} \quad$ and $\quad E_{stat} = -\nabla \phi_f \quad$ where $\quad \nabla^2 \phi_f = -\rho_f$

Written in this way we can see that Helmholtz's equation lacks the term in $\partial E_{ind}/\partial t$ that Maxwell's contains. It is absent because only Maxwell considers that a changing electric field *in itself* constitutes a current. Helmholtz by contrast allows exclusively conduction and polarization currents. Thus his term in $\partial E_{stat}/\partial t$ must not be thought of as *in itself* a current but rather as the effect of currents, one that comes from Poisson's equation via continuity. An apparently corresponding term in Maxwell's equation appears only because we have artificially split the Maxwellian E field into static and inductive parts. But here the term does not arise from Poisson's equation in any way at all; indeed, for Maxwell's theory, the continuity equation has nothing directly to do with charge, because it is formulated in terms of conduction and displacement currents. According to it every term contributes on an equal basis to the magnetic field since, for Maxwell, the continuity equation is simply a different form of the Ampère law. But for Helmholtz that equation has little directly to do with magnetic action: its significance lies entirely in specifying the things that can alter charge densities.[8] Longitudinal waves arise because Helmholtz's continuity equation permits the total current to have a divergence that is a function of the rate of change with time of the scalar potential. Maxwell's theory lacks such waves because in it the total current has no divergence.[9]

The proper way thoroughly to grasp the Helmholtzian system requires abandoning attempts to construct parallels between its equations and Maxwell's because Helmholtz does not begin with fields, whereas Maxwell does. In Maxwell's theory, for example, the vector potential does not represent an interaction between objects; it represents an independent state of the field. In Helmholtz's system it specifies the effect of an interaction energy between two things. To deduce an "Ampère law" for the curl of the magnetic force requires artificially separating the action between the two things from the things themselves. Or, better put, magnetic action does not exist unless there are two things— sources—for it to occur between. In Maxwell's theory the magnetic field exists as a state in its own right.

This has an important effect upon the conceptual relationships of the two theories to the continuity equation. In field theory continuity gains its significance from the law that links the magnetic field to currents and to the electric field. In Helmholtz's scheme continuity represents a separate requirement that links currents to charge. Consider, for example, $\nabla \times U$, where U is the vector potential. In field theory this is the magnetic field, and from its relationship to currents and to changing electric fields we may deduce the continuity equation. (And if, as may happen in Maxwellian theory, this relation changes, then so do the terms in the continuity equation.) But in Helmholtz's system we do not know *ab initio* what $\nabla \times U$ signifies. We must deduce its meaning by varying the basic interaction energy. We do, however, have an expression for U together with a separate continuity equation. These two equations are *necessarily* consistent with one another since the first involves only the variables U and C, and the second involves only C and ϕ (or ρ). We may if we like deduce $\nabla \times U$ from the definition of U and term the result an "Ampère law" but it has no deeper significance, and it is analytically consistent with *any link at all* between C and ϕ (or ρ). (See Buchwald 1985a, pp. 9 and 178–81, for details on how to construct a Helmholtzian Ampère law, but keep in mind this caveat.)

Turning next to the transverse waves, we can see that the analytical reason for their existence in PNCs but not in NPCs is quite simple and revealing. In order to generate transverse waves the Ampère law and so the total current C_{con} must contain a term that is proportional to the second derivative with respect to time of U. But in Ohm's law the current is proportional only to the first time derivative of U, so that only diffusion, and not propagation, can occur. Accordingly, propagation occurs in polarizable bodies because in them we replace Ohm's law for the currents C_{con} with a law for the polarization currents C_{pol} that does involve the second time derivative of U:

$$C_{pol} = \chi \frac{\partial}{\partial t} (E_{ind} - \nabla\phi_f + E_{ctm}) \quad \text{where} \quad E_{ind} = -\frac{\partial U}{\partial t}$$

This leads to transverse waves through the Ampère law. But the price that has to be paid to achieve such things in the overall context of Helmholtz's system is quite large because C_{pol}, unlike C_{con}, is not by any means a primitive of the theory. It is a change in something (P, the polarization) that is defined in terms of charge, and only charge is a theoretical primitive. Any relationship between C_{pol} and E_{ind} therefore depends upon the prior link between P and E_{ind}. But there is no more basis for this latter connection than there is for an electromotive action by changing currents upon charges. And there is considerably less apparent warrant for the electrodynamics of Helmholtz's new "current," because one has evidently to postulate that a changing state ($\partial P/\partial t$) can itself constitute a new state that establishes an interaction energy with yet a third state (C_{con}). Assumption must pile upon problematic assumption to produce this new entity.

Here, then, is a (in the end, *the*) major characteristic of Helmholtz's polarization theory. It cannot be eliminated in any simple way, because it derives from the concept of a source upon which the scheme rests: unlike either Weber's or Maxwell's electrodynamics, Helmholtz treats laboratory sources as primitives of the theory. Sources are related to one another through the interaction energies, but every interaction between two kinds of sources requires its own specific representation. Consequently, the unity of Helmholtz's electrodynamics is badly damaged if it is forced to accept a novel interaction without being able also to specify the corresponding interaction energies.[10] Without potentials and the corresponding states there is simply no basis at all for grasping the theoretical meaning of an interaction, which led a puzzled Hertz in the early 1880s to try to circumvent problems of this kind by altering his mentor's concept of source (or state) in a way that would not also abandon the critical notion of Helmholtzian interaction. But his attempt was not likely to succeed, because the scheme cannot easily survive the kind of elimination that Hertz was, no doubt unwittingly, proposing at that time.

13.5. The Problem of Wire Propagation

It is important to grasp the difference between the kind of propagation that we have been considering thus far and the kind that had long been thought to occur in wires. In 1857 Kirchhoff had deduced an equation of propagation for long, cylindrical conducting wires of length l and cross section α in the following fashion (Kirchhoff 1857; Whittaker 1973, vol. 1:230). Suppose that the current is uniform across any section of the wire and that charge accumulates at its surface, where it is surrounded by a nonconducting medium, with a density e per unit length. Calculate by approximation the static potential V that is generated within the wire by the boundary charge. Then calculate the potential w of the current i, again by approximation (and using Weber's formula, though the form of the resulting equation is independent of which of the available formulae are used). Finally, write down Ohm's law (with k the conductivity), including in it both the electrodynamic electromotive force and the electric force generated by the static boundary charges:

$$V = 2e \ln \frac{l}{\alpha}$$

$$w = 2i \ln \frac{l}{\alpha}$$

$$i = -k\alpha^2 \frac{\partial V}{\partial x} + \frac{1}{c^2} \frac{\partial w}{\partial t}$$

As they stand these equations do not imply propagation, because they combine to yield

$$-\frac{2 \ln(l/\alpha)\partial i}{k\alpha^2 \, \partial t} = \frac{\partial^2 e}{\partial t \partial x} + \frac{1}{c^2}\frac{\partial^2 i}{\partial t^2}$$

To convert this into a wave equation requires linking e, the charge density at the surface of the conducting cylinder, with i, the current. This can at once be done with the continuity equation. Since charge accumulates only at the surface, we can use continuity to find that $\partial i/\partial t$ must be just $-\partial e/\partial t$, and so an equation for a damped wave results:

"Kirchhoff's Equation" for Wire Propagation

$$-\frac{1}{2\ln(l/\alpha)\pi k\alpha^2}\frac{\partial i}{\partial t} = -\frac{\partial^2 i}{\partial x^2} + \frac{1}{c^2}\frac{\partial^2 i}{\partial t^2}$$

Clearly waves of current (or, equivalently, charge and potential) propagate in wire cylinders at the speed c according to this result.[11] These kinds of waves can of course only be obtained by allowing the static potential to generate a current at a point.

In 1870 Helmholtz reconsidered the problem of wire propagation from the standpoint of his general electrodynamic potential. This was a different problem from calculating a wave speed for an unbounded medium, because (as Kirchhoff had already shown by taking account of the potential due to surface charge) boundary conditions alter the situation in important ways. Helmholtz treated the wire as an infinitely long, unpolarizable conducting cylinder with a finite radius and brought to bear the general boundary conditions entailed by his equations (but without considering the bounding medium to be polarizable). He was able to retrieve, as he put it, the equation that Kirchhoff had obtained "from Weber's laws" (1870b, p. 611) in the limit that the cylinder's radius goes to zero. That is, he found a (longitudinal) wave form that propagates through the wire at the speed of light in this limit. However, Helmholtz's general result clearly indicated that the velocity in fact depended in a complicated fashion (which he did not reduce) upon his electrodynamic constant k. Only in the case that k vanished altogether did Kirchhoff's result emerge whatever the radius of the wire might be.

In 1876 Oliver Heaviside obtained an equation similar to Kirchhoff's, which was, however, much more general.[12] In his derivation the speed of propagation is expressed as a function of the inductance and capacitance of the wire, so that it will not in general be c. These results in no way at all assume that a changing electric force behaves like a current. That is, they completely ignore any form whatsoever of displacement current. In field theory it is, however, customary to think that propagation occurs only as a result of the presence of that very term. Furthermore, without the term Maxwell's equations are inconsistent with the very existence of varying charge densities. And yet field theory can, as Heaviside demonstrated, yield an equation for wire propagation that is the same in form as Kirchhoff's. Why is this possible?

We are not concerned here with the propagation of a disturbance throughout space but rather with what happens in and on a delimited conducting region—the wire—that is surrounded by a nonconductor. We are not, that is, concerned with the general question of wave propagation in an unbounded conducting medium or, for that matter, even with the propagation of a wave in a dielectric that surrounds a wire conductor. The nonfield theories make use of this limitation to introduce a connection between the current and the charge that is located at the boundary between the conducting and the nonconducting regions (assuming that the current is itself distributed uniformly throughout the conductor). It is this connection, which is obtained through the continuity equation, that converts the term in $\partial^2 e/\partial t \partial x$ into one in $\partial^2 i/\partial t^2$, thereby yielding propagation.

Kirchhoff's waves propagate at the speed c because under his approximations for the cylindrical wire (viz., a uniform distribution of current over its cross section and charge accumulation only near the surface) the static potential V is the same function of e that the electrodynamic potential w is of i. Different approximations, or different wire cross sections, will in general lead to different wave speeds. Heaviside's (1876) equation accommodates this by introducing generalized constants for capacity and self-induction. For a wire of length l, he writes, the resistance is given by kl, the electric capacity by dl, and the coefficient of self-induction by sl, where d and s can be computed from the geometry of the wire (if we ignore magnetic permeability), and k must be obtained from experiment. Heaviside tacitly adopted the same approximations that Kirchhoff had and obtained what later became known as the "telegrapher's equation" (here given, as he did in 1876, in terms of the static potential v):

$$\partial^2 v/\partial x^2 = dk\partial v/\partial t + sd\partial^2 v/\partial x^2$$

From this we at once see that the wave speed depends upon d and s.

For our purpose here the meaning of the telegrapher's equation can best be grasped physically by envisioning the wire as a sequence of capacitors that are connected in parallel, with each one discharging to its neighbor through a coil. Each capacitor represents a cross section of the wire, so that the charge that accumulates at the wire surface corresponds to the charge stored in the capacitor. The coils represent the self-inductance of the wire. Suppose this system to start with a given pattern of charges on the capacitors and of currents in the coils. If each coil–capacitor element in the system formed a closed circuit of its own, the system would be governed by the equation for a damped, harmonic oscillator. But the elements are not isolated from one another; they are connected in parallel. Consequently, each capacitor–coil element in the sequence will be driven by an electromotive force determined by the element preceding it and by the next one after it in the sequence, and so its oscillations will not be in phase with those of either its predecessor or its successor, just as the elements of a compound pendulum do not oscillate in phase with one another. The result is that the current (as well as the voltage and the charge) on the second element will not be the same at any given moment as the current on either of the two elements with which it is directly connected. In the limit that each element in the system becomes infinitesimally small, the resulting pattern obeys the telegrapher's equation. Heaviside put it this way:

> Instead of a line along which capacity, conductivity and self-inductive power are continuously distributed, consider a number of coils, joined in sequence, whose terminals are joined to the earth through condensers. That is, we replace an infinite number of degrees of freedom by a finite number. We have one degree of freedom for every condenser, and one for every coil; say n altogether. Given the state of this system at a given moment, to find its state at any time after, when left to itself without impressed force, requires the solution of a linear ordinary differential equation of the nth degree, in which t, the time, is the independent variable. There are n rates of subsidence, and n normal arrangements of charge and current, any one of which, subsiding at its proper rate, is a possible solution. We require, then, to decompose the initial state of charge and current into these n arrangements. (Heaviside [1892] 1970, vol. 1:149)

From this point of view it is clear why field theory can yield much the same sort of equation for wires as nonfield theories like Kirchhoff's even though the cause of radiative propagation (the displacement current) is not considered. We need only to take the same model in field theory for a wire that we just employed for the nonfield theories.

However, it is critically important to understand that this kind of propagation is quite different from the only kind of wave that, according to field theory, can move in an unbounded and uniform medium that possesses both capacitance and inductance, because only this latter wave depends upon the displacement current. The former kind of propagation, the one that is also implied by the nonfield theories, involves the current of conduction; via approximations, capacitance, and, critically, the continuity equation, it also yields waves for charge and potential. It occurs only in nonuniform media, as can be seen from the capacitor–coil analogy. The essence of the analogy involves distinguishing regions that have capacity and inductance (the wire) from those that do not, or that at least have different inductance and capacity from the wire (the wire's surroundings). The propagation that occurs only in a field theory takes place even in uniform media. If a medium is not uniform, then according to field theory there will in effect be two kinds of propagation: one involving radiation and one involving only the conduction current (via capacitance, charge, and voltage) and resulting from the electrodynamic interactions between neighboring currents of conduction via Ohm's law.

The relationship between these two kinds of waves becomes particularly important in field theory when one tries to construct equations for, say, a finitely conducting cylinder embedded within an infinite dielectric—for, that is, a wire. Here we have radiation within the dielectric, and we also have conduction current. However, at radio frequencies the current in the wire will not be distributed throughout it but will be concentrated into a very thin region near its surface: the higher the frequency and the magnetic permeability of the wire, the thinner the current-bearing region will be. As a result the usual approximations that are made to obtain the telegrapher's equation no longer hold. At radio frequencies wire propagation becomes a special case of radiation, with the wires acting as guides for the waves, which are transmitted only through the dielectric. Here there is no sense in speaking about propagation in the wire, because the current occurs only at its surface as a by-product of the radiation in the surrounding dielectric.

In the near-absence of radiation at low frequencies, when we can assume that the current is distributed throughout the wire, there is a legitimate sense in which we can speak of propagation *through* the wire because the current is not concentrated *on* the wire. Here we are in effect treating the wire as a region that can propagate waves of current, but in doing so we must rely on a highly artificial specification of the current distribution, one that actually violates the boundary conditions that must hold at the wire's surface.[13] Moreover, even here, where we ignore the transmission of energy to large distances by radiation, field theory regards the propagation as adventitious—as a by-product of processes that occur within the surrounding dielectric but that terminate within the substance of the wire. The fact that waves of current can sometimes be fruitfully regarded as existing within the wire merely reflects the variation from point to point along the wire of the surrounding processes under these special, approximate conditions.

From the pure Maxwellian perspective wire propagation does not truly exist as a separate category of propagation: waves, properly speaking, occur only in media that have capacity, so that either the wire has capacity, and so can transmit waves through it, or else it has no capacity, in which case it transmits no waves at all, and everything of consequence goes on around it, not in it. In Maxwellian practice two extreme situations therefore emerged: either radiation was completely ignored, yielding the telegrapher's equation, or else the frequency was taken to be so large that the wire itself could be ignored, concentrating the current at its surface via the skin effect and reducing everything to radiation in the surrounding dielectric.

$$\cdot \; \cdot$$

APPENDIX FOURTEEN

Forces in Hertz's Early Experiments

According to Hertz's understanding, the resonator responded primarily to the forces acting on its parts that were directly opposite its spark gap. The forces consisted of the wire's electrodynamic action and the combined dynamic–static action of the oscillator. In both cases the action was very nearly parallel to its respective source (wire or oscillator),[1] and the net effect on the resonator could be found by projecting the two forces onto the direction of its plane (specified by the angle ϕ in fig. 84). Hertz did not provide quantitative expressions for the effect, either in his laboratory notes or in his printed articles, but he indubitably knew quite well what the following simple mathematics easily shows.

Let W_M and W_A respectively represent the projections of the wire's force and the direct action on the resonator. Then

$$W_M = \cos\phi \cos\left\{\frac{2\pi}{\lambda_M}[v_m t - (z + z_m)]\right\}$$

$$W_A = \sin\phi \cos\left\{\frac{2\pi}{\lambda_A}[v_a t - (z + z_a)]\right\}$$

where λ_A and λ_M are respectively the wavelengths of the air and wire actions, v_A and v_M are respectively the speeds of propagation of the air and wire actions, and z_A and z_M are the respective distances (at $z = 0$) from their origins of the air and wire actions. These combine to form a resultant force S on the resonator. Because the waves have the same frequency, trigonometric decomposition shows that S can be written in the following form:

$$S = W_A + W_M = R\cos\left\{2\pi f[t + \varepsilon(z)]\right\}$$

where $\quad f = \dfrac{v_A}{\lambda_A} = \dfrac{v_M}{\lambda_M}$

and $\quad R^2$

$$= 1 + 2\sin\phi\cos\phi\cos\left[2\pi z\left(\frac{1}{\lambda_M} - \frac{1}{\lambda_A}\right) + 2\pi\left(\frac{z_M}{\lambda_M} - \frac{z_A}{\lambda_A}\right)\right]$$

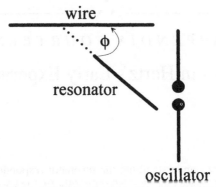

FIGURE 84   Hertz's resonator, wire, and oscillator

R^2 can nicely measure the strength of the resonator sparking at a given position z and angle ϕ since it is the square of the amplitude of the driving force. Hertz's experiments consisted in setting the resonator at ϕ (45°), noting the strength of the sparking, and then rotating the resonator to $-\phi$ and noting the sparking at this new angle. He had, consequently, turned the resonator through an angle 2ϕ and had sought the resultant change in sparking intensity, calling an increase $+$, a decrease $-$, and no change 0. We can represent what Hertz initially thought he had observed as ΔR^2:

$$\Delta R^2 = 2(R_\phi - R_{-\phi}) = \sin(2\phi)\cos\left[2\pi z\left(\frac{1}{\lambda_M} - \frac{1}{\lambda_A}\right) + 2\pi\left(\frac{z_M}{\lambda_M} - \frac{z_A}{\lambda_A}\right)\right]$$

This function accordingly represents in the most general form (i.e., allowing for different wave and air speeds, as well as different origins for both actions) what Hertz originally considered his resonator to have detected. That is, he interpreted his data in accordance with the implications of an expression of this sort, though he certainly did not explicitly provide such an expression. Note that interference vanishes wherever the phase difference between the metal and air actions is an odd multiple of 90°, which would correspond to a quarter-wavelength if the actions propagated at the same rate (and so had the same wavelength). Note also that if the direct action propagated infinitely rapidly (i.e., had an infinite wavelength), the spark-change function would track a standing wave in the wire.

An example shows what, according to Hertz's early understanding, would ideally take place if the ratio of the air to the metal speed were about 1.6 and if the metal wavelength were 5.6 m. In figure 85, the solid line represents the spark curve for the case in which neither action is retarded; the dotted line represents a 1-m retardation for the metal wave; the dashed line represents a 1-m retardation for the direct action. A given spark-intensity locus is shifted toward the origin by retarding the slower (metal) action, whereas it is shifted

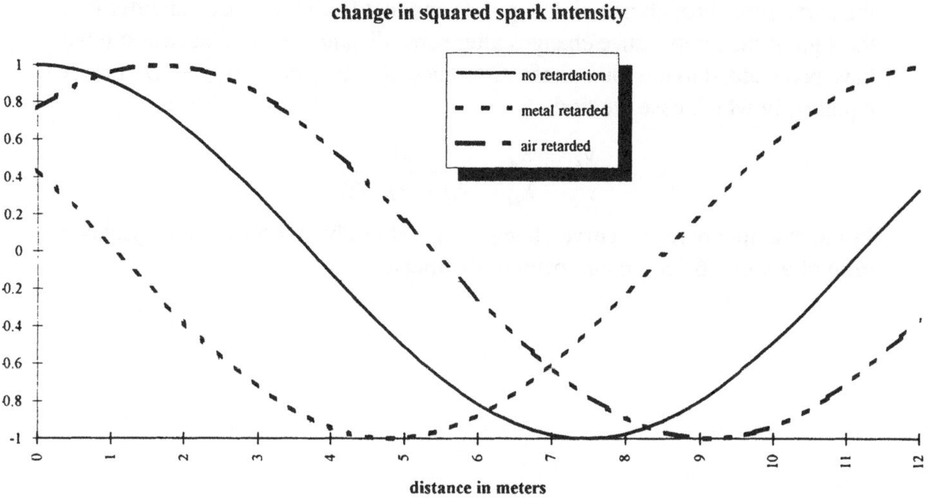

change in squared spark intensity

FIGURE 85   Spark curves for interfering metal and air waves with different retardations

away from the origin by retarding the faster (air) action, which accords with Hertz's claim (1888e, p. 121).

The point can be demonstrated analytically in the following way. Suppose first that the metal wave and the direct action start from the same point. Then, since they have the same frequency, at a distance d from this common origin the phase difference ϕ between them will be (ignoring a common factor of 2π)

$$\phi = d/v_M - d/v_A$$

Suppose next that the metal wave's origin is set back by an amount h. In that case its phase at the original point changes to $(d + h)/v_M$. The resultant of the two waves that had existed at the original point, and that corresponds to a given value of the phase difference, will be shifted through some distance δd to a position $d + \delta d$ from the origin. The total change in the phase for this shifted point must (by definition) vanish:

$$\delta\phi = \frac{\delta d + h}{v_M} - \frac{\delta d}{v_A} = \delta d \left(\frac{1}{v_M} - \frac{1}{v_A}\right) + \frac{h}{v_M} = 0$$

Suppose now that the metal wave has the smaller speed; then $1/v_M - 1/v_A$ must be greater than zero. It follows from the last equation that δd must be less than zero, which means that the spark curve is shifted toward the origin by retarding the slower action; conversely, it is shifted away from the origin by retarding the faster action.

It is also interesting to examine how the curve might be used to infer the ratio of the speed of the direct action to that of the metal wave. Suppose that the curve runs through actual data points and that from these one can infer that the sign of the interference changes after some distance Δz. In that case π must have been added to the total phase difference, that is, $2\pi\Delta z(1/\lambda_M - 1/\lambda_A)$ must equal π, in which case we find

$$\frac{v_A}{v_M} = \frac{\lambda_A}{\lambda_M} = \frac{\Delta z}{\Delta z - (\lambda_M/2)}$$

In the example here the curve changes sign after about 7.5 m, which yields a ratio of about 1.6 for the air speed to the metal speed.

APPENDIX FIFTEEN

Hertz's Quasi Field Theory for Narrow Cylindrical Wires

Hertz 1889a is an attempt to discover what Maxwell's theory implied for wire propagation. This was not an easy task, because it was not a priori certain, in Hertz's opinion, that Maxwell's theory did require wire waves to propagate at the same speed "in" wires that they do in air. This at once marks him as, to this point, incompletely immersed in Maxwellian lore. Maxwellians did know, almost a priori, that the two speeds had to be the same because there was no such thing as a wave "in" a wire (unless the wire was considered to be a finite-sized conducting space with dielectric capacity, in which case it was at once certain from Maxwell's equations that the speeds were equivalent). For Maxwellians a conducting wire, or indeed any conducting surface, acted as a boundary for guiding waves that propagated in the surrounding, nonconducting medium. To Hertz in late 1889 this was not as yet an item of faith. He felt compelled to produce mathematics that, given appropriate Maxwellian conditions, specified the speed "in" wires.

To do so Hertz turned, appropriately (given the cylindrical symmetry of the wire), to Bessel functions and their corresponding differential equations. These functions, which he designated $K(p\rho)$, in which ρ is the distance form the wire as axis, contained the free parameter p. Hertz made use of this freedom to find a general expression for wave speeds on the basis of (his) Maxwell equations, which he then limited to a single choice, in the following way. First, he asserted that the following function Π will satisfy the propagation equation and can therefore be used to build expressions for the fields:

$$\Pi = \frac{2J}{An} \sin(mz - nt) - p^2 K(p\rho)$$

provided that p^2 is equal to $m^2 - A^2 n^2$. The function becomes discontinuous at the z axis, and so, Hertz argued, the corresponding fields represent "an electrical disturbance taking place in a very thin wire stretched along the z-axis." Hertz's Maxwellian "wire," like his dipole, is terra incognita; it marks where analysis stops.

Following out the fields and deducing from them the electric density, and thence (via continuity) the wire current, Hertz found that the disturbance is a sine wave with speed n/m. Since m and n are otherwise constrained only by their relation to p, this meant that, to this point, the wire wave could have any speed less than or equal to $1/A$, that is, less than or equal to the speed of a Maxwellian disturbance in air. In the particular case that p vanishes, the wire speed equals the air speed, and here, Hertz showed, the electric field lines (which are always normal to the wire) become straight instead of curved, pointing directly away from the wire toward infinity. Here, then, the electric field has no component along the wire at all, and this, Hertz wrote, is precisely Maxwell's criterion for a "perfect conductor."

Considerations regarding the Possible Background to Helmholtz's New Physics

When Wilhelm Weber produced his atomistic electrodynamics in the late 1840s, he was extremely careful not to make extravagant claims for it, concentrating instead on demonstrating its power to integrate the several known electromagnetic effects into a single type of action and on pursuing exact electric measurement in the Gaussian tradition. Nevertheless, several historians have argued that the basis of Weber's theory, Fechner's hypothesis, was itself embedded in a much more far-reaching scheme, one that perhaps sought to transform the analytical structure of atoms and forces into something that resonated with metaphysical meaning. According to this view, Fechner, and Weber after him, sought to incorporate the imperatives of German Idealism while avoiding the unacceptable mystification of *Naturphilosophie*.[1] Part of Fechner's goal, in this view, was to *dematerialize* the physics of particles and forces through atomic points: through, that is, unextended centers related to each other by pure, *immaterial* force—force governed by law, but spirit in its essential nature.[2] In his 1855 *Atomenlehre* Fechner put it this way:

> *Geist* [loosely, spirit or mind] steps up and asks, what have I to do with you? And the atoms say: we spread our individualities under your unity; the law [of force] is the commander of our band, but you are the king in whose service he leads.[3]

This dematerialized atomic physics seems to be quite different from Weber's electrodynamics, and Weber himself never drew in print any connections between it and other areas of physics, much less to *Geist*. During the 1850s and 1860s Weber did attempt to use his force law and atoms to provide a universal model for both matter and the ether, a scheme which directly incorporated energy conservation.[4] Nevertheless, *this* work, unlike his vastly more limited electrodynamics, did not fully retrieve the known laws of, for example, even the wave theory of light, and there is little evidence that, outside a very small group, it had much contemporary influence. Furthermore it remains somewhat doubtful that the kinds of metaphysical considerations that seem to have influenced Fechner ever did play a very great role in Weber's thought.[5]

However, one thing does seem reasonably certain, which is that Friedrich Zöllner, perhaps the most ardent proponent of Weberean electrodynamics in the 1870s, understood atoms and electrodynamic force in a different way from that in which French physicists conceived particles and force. Physical connections had little, if any, metaphysical resonance for the French, who, early in the century, had regarded *Naturphilosophie* as ridiculous, and for whom the central question was how to create positive science. Some German electrodynamicists, at least in the 1870s, probably did invest atoms and force with meaning that went beyond positive science to an understanding in which the atoms—necessarily mere points—do not *exert forces* on one another but are rather linked each to another in bipartite relationships whose physical expression *is force*. The world has, as it were, been stripped of space-filling, inert materiality and populated instead with centers of vital interaction.[6]

From the metaphysical point of view Weberean *force* reflects the irreducible and immediate character of bipartite interactions among atoms. Force is not an action by one otherwise dead object on another but a physical symbol of their interconnection. In such a world the ultimate physical objects are in themselves eternally the same, and indeed each is bound to every other ultimate object in a bipartite connection. Consequently, the relational character of Weberean electrodynamics does not extend to the objects with which we have to do in the empirical world, for such things are built up out of a myriad of atomic points, and the relation subsists between the points, not the bodies. Helmholtz's electrodynamics accepts, indeed it is explicitly founded on, the axiom that bodies interact with one another through irreducible bipartite relationships. In the Helmholtzian world, as in the Weberean, things do indeed *interact* in pairs; they do not, properly speaking, *act*. But Helmholtz's electrodynamics goes much further in this vein than Weber's ever did, and with radically different implications.

It is entirely possible to dispense with a relational understanding of force and yet still retain the analytical and physical structure of Weberean theory, including his universal model. Whether point atoms are thought to create and to be guided by independent forces, or whether the forces are symbols of an unbreakable unity between points, does not in any way whatsoever affect the theory's empirically testable consequences. Helmholtzian theory is utterly different. It rejects atoms in Weber's sense; it makes no direct use of forces.[7] Instead of building bodies out of invariant and inaccessible entities (Weber's metaphysically pregnant atoms), it builds them out of small pieces that are entirely similar to the bodies proper—it takes them, that is, essentially as they are in the empirical world. Instead of having forces link atoms (or even energy determined by atom pairs), Helmholtz has the energy of a *system* consisting of a pair of volume elements in given states and at a given separation.[8] A relational understanding of the actions between objects lies at the very core of this enterprise precisely because the relationships are not determined by the invariant natures of the bodies, and this has direct empirical consequences.

The atoms in a Weberean pair are ever joined in the same family. A pair of

Helmholtzian objects can have a potentially infinite number of different relationships with one another depending on their states. Only the laboratory can determine what kinds of states bodies may have and what system energies bodies in given states and at given separations together determine. Here, then, the interactive character of electrodynamics does not need to be inserted into the scheme: it is irretrievably there from the very beginning. And its necessary presence carries a very different message from that of Idealist metaphysics. Since Helmholtz's interactions cannot be known a priori, whereas Weber's are, his relational electrodynamics remains directly tied to the laboratory, to the world of empirical practice. To understand how, and perhaps why, Helmholtz developed his scheme we must begin with his creation of the energy concept itself in 1847.

16.1. INVARIABLE CAUSE

Very early in his career (specifically, when he first formulated energy conservation) Helmholtz had probably held a relational view of force (as Wise 1981, p. 296, remarks), but it was then embedded in an atomistic world. He wrote in 1847:

> Motion is the alteration of the conditions of space. Motion, as a matter of experience, can only appear as a change in the relative position of at least two material bodies. Force, which originates motion, can only be conceived of as referring to the relation of at least two material bodies towards each other; it is therefore to be defined as the endeavour of two masses to alter their positions. *But the forces which two masses exert upon each other must be resolved into those exerted by all their particles upon each other; hence in mechanics we go back to forces exerted by material points.* The relation of one point to another, as regards space, has reference solely to their distance apart: a moving force, therefore, exerted by each upon the other, can only act so as to cause an alteration of their distance, that is, it must be either attractive or repulsive. (1847 [1853, p. 117], emphasis added)

Here we see that Helmholtz begins with force as the originator of motion and insists on its relational (though not necessarily bipartite) character. He then introduces the vector property of force, and from this he leaps to the world of "material points," which enables him to insist on force's uniquely spatial character. The material point, consequently, has two functions: first, to represent the final stage of decomposition and, second, to make force ineluctably spatial. But the point is not *in itself* a foundation of Helmholtz's argument since he presupposes both the vector character of force as well as its purely spatial dependence. The material point provided a uniquely suitable way to embody both characteristics.

For Helmholtz in 1847 the important aspect of a fundamental interaction was its pure dependence on space. The signal importance of spatial relation derived from its prior association with *causal invariability*. "The problem of the sciences," he remarked well before introducing "material points" is

> to evolve the unknown causes of the processes from the visible actions which they present; . . . to comprehend these processes according to the laws of causality. We are justified, and indeed impelled in this proceeding, by the conviction that every change in nature *must* have a sufficient cause. The proximate causes to which we refer phenomena may, in themselves, be either variable or invariable; in the former case the above conviction impels us to seek for causes to account for the change, and thus we proceed until we at length arrive at final causes which are unchangeable, and which therefore must, in all cases where the exterior conditions are the same, produce the same invariable effects. The final aim of the theoretic natural sciences is therefore to discover the ultimate and unchangeable causes of natural phenomena. (1847 [1853, p. 115])

Helmholtz designates cause in general as "force," so that he is here distinguishing between "forces" that are changeable (and therefore not ultimate) and "forces" that are unchangeable (and so final). He develops this in the following way. First, he asserts: "Bodies with unchangeable forces have been named in science (chemistry) elements." He continues: "Let us suppose the universe decomposed into elements possessing unchangeable qualities; the only alteration possible to such a system is an alteration of position, that is motion: *hence* the forces can be only moving forces dependent in their action upon conditions of space" (emphasis added). Helmholtz's unchangeable forces depend by definition only upon distance, and so his conclusion is that the only possible effect of such forces, assuming them to attach only to unchangeable bodies, is to alter the distance—because the qualities of such bodies are themselves fixed. It is essential to remark that the relegation of the effect of an unchangeable force (i.e., an unchangeable cause) to change in position follows from Helmholtz's secondary assertion that unchangeable forces require unchangeable bodies.

In 1847 Helmholtz turned to the material point as an embodiment of this prior requirement because the point is completely invariant except as regards location. Since, at that time, Helmholtz felt that an unchangeable force should be linked to a similarly unchangeable quality, the material point was a natural—indeed, the *only possible*—vehicle for him to employ. But if he had broken the connection and allowed a purely spatial force to affect something other than location, the material point would not have been essential. He would have had to allow forces to produce qualitative change as well as change in position. In 1847 Helmholtz did not even conceive that the connection could be broken; by the end of the 1860s he felt otherwise, but to break the connection he had to alter his understanding of physical cause by separating it from moving force and attaching it instead to the much more powerful notion of "energy."

If we take Helmholtz's introductory remarks seriously, it follows that he did not discover through misapplied technique that Weber's force law must be rejected in view of the energy principle. He knew it before he even developed the principle in technical form because Weberean force necessarily involves time through its dependence on speed and acceleration. It is not, in other words, that time-dependent forces necessarily violate the conservation of *vis viva* (though at the time he certainly thought they did). Rather, they are a priori unacceptable because they are by their very nature variable and so cannot serve as final causes. Helmholtz's rejection of time-dependent forces and his formulation of energy conservation both bore the stamp of his deeply felt conviction that nature can only be comprehended through invariable causes. In 1847 Helmholtz developed a way to remove from force even implicit variability, and this began to effect a very deep change indeed.

He invented what soon evolved (though not initially at his hands) into the concept of energy, or rather what he at first thought of as the quantitative aspect of force. Ultimate force in Helmholtz's sense must depend upon space, not time, and the prime example of such a thing is gravitation. Yet even a force like gravity does change over time (albeit only implicitly) as the distances do. Such a force is invariable as a cause, but in another, less important, way it is not. And here Helmholtz apparently detected a puzzle, because he strongly believed that (as he put it in 1847) "it is impossible by any combination whatever of natural bodies to produce force continually from nothing" (1847 [1853, p. 118]).[9] The conundrum—that even ultimate force does vary, yet that it cannot be produced from nothing—manifests the potency of the concept and the essential ambiguity that accompanies its power, because, of course, there can be no conundrum here *if* we limit force, as the analytically minded French physicists did, to signifying whatever equates to the product of mass by acceleration. Such a thing is not in any sense conserved. Helmholtz's problem emerges from his insistence on conserving something about *force,* not just on conserving.[10]

Helmholtz's analytical resolution of the enigma was simple in appearance because he founded it on ultimate (i.e., purely spatial) actions between "material points." He identified the integral of such a force over distance with a *force quantity,* or "tension" as he called it, and at once arrived at the well-known fact that such a thing equates to the corresponding change in the sum of half the product of mass by the squared velocity. This tightly bound the notion of conservation to points, which meant that it appeared to contain an internal push toward models, though neither at the time nor in the future did Helmholtz himself pursue atomic model making.

During the next two decades an independent concept, that of *energy,* was developed, at first by William Thomson in Scotland and Rudolf Clausius in Germany, and as it congealed, energy was separated not only from force but from models. It gradually became a thing that in certain respects has its own identity and that can always be traced in fixed amount through the operations of nature. Helmholtz after the early 1850s was not intimately involved with these developments.[11] Instead, at Königsberg, Bonn, and then at Heidelberg, he built an impressive and influential career as the master of new fields of research; of physiological acoustics and then optics. He occasionally, and with great power, produced work in mathematical physics (as in his 1857 theory of vortex motion, which arose out of his work in acoustics), but he did not for the most part pursue energy physics. That changed shortly before he left for Berlin to take up Gustav Magnus's chair in physics, in the year that Prussia invaded France and that Imperial Germany was fashioned by Bismarck. Helmholtz began then to develop a way for physics to retain the critical and unique dependence of relations on spatial distance that he considered to be the very essence of a scientific cause, and yet to free physics from the impulse toward atomic modeling that was present even in his own original formulation of conservation and that was soon to be actively indulged by Weber and Zöllner. In the process he very nearly removed "force" altogether.

16.2. THE ABOLITION OF FORCE

Helmholtz announced his new method for physics in 1871 in a lecture to the Prussian Academy of Sciences honoring the memory of his predecessor at Berlin, Gustav Magnus:

> In reference to atoms in molecular physics, Sir. W. Thomson says, with much weight, that their assumption can explain no property of the body which has not previously been attributed to the atoms. Whilst assenting to this opinion, I would in no way express myself against the existence of atoms, but only against the endeavour to deduce the principles of theoretical physics from purely hypothetical assumptions as to the atomic structure of bodies.

Citing Gauss, Franz Neumann, Faraday, Stokes, Thomson, and Maxwell as progenitors of the view he was embracing, Helmholtz continued:

> It is now understood that mathematical physics is a purely experimental science; that it has no other principles to follow than those of experimental physics. In our immediate experience we find bodies variously formed and constituted; only with such can we make our observations and experiments. Their actions are made up of the actions which each of their parts contributes to the sum of the whole; and hence, if we wish to know the simplest and most general law of the action of masses and substances found in nature upon one another, and if we wish to divest these laws of the accidents of form, magnitude and position of the bodies concerned, we must go back to the laws of the smallest particles, *or, as mathematicians designate it, the elementary volume. But these are not, like the atoms, disparate and homogeneous, but continuous and heterogeneous.* (Emphasis added)

Mathematical physics investigates these bodily "elements":

> The characteristic properties of the elementary volumes of different bodies are to be found experimentally. It is thus admitted that mathematical physics

only investigates the laws of action of the elements of a body independently of the accidents of form, in a purely empirical manner, and is therefore just as much under the control of experience as what are called experimental physics. In principle they are not at all different, and the former only continues the function of the latter, in order to arrive at still simpler and still more general laws of phenomena. (Helmholtz 1881b, pp. 17–19)

In declaring the aim and nature of mathematical physics, Helmholtz had subjected it to experience and had excluded Weberean electrodynamics from it. These paragraphs constitute a verbal expression (albeit an incomplete one) of the pattern that Helmholtz in 1870 built into his own electrodynamics, and which he was now extending to all of physics. It amounted to the removal of force from physics as a category because (though the point remains implicit here) the interactions between volume elements cannot be directly characterized by distance forces. Distance forces must only be *deduced* from the more fundamental interaction, which determines a bipartite system energy.[12]

The proximate source of Helmholtz's new physics was most likely W. Thomson and P. G. Tait's *Treatise on Natural Philosophy,* which he and Emil Wertheim translated into German; the first volume was published in 1871. Thomson and Tait's *Treatise* has nothing to do with atoms and forces. Its structure is molded at the deepest level about continuity, energy, and extremum principles for mechanics. Although one cannot be certain, it seems probable that Helmholtz appropriated from the *Treatise* its foundation in the *differential volume element* as the appropriate unit of analysis, and he then transformed its use of energy into a form that better suited his understanding of interaction as a bipartite process. He accordingly invented the concept of *an irreducible system energy density that is determined by the simultaneous states of, and the distance between, a duo of volume elements.* Thomson and Tait showed Helmholtz how to break away from the tyranny of the mass point while nevertheless allowing him to insist on the ineluctably spatial character of the interactions between bodies. To do so he had to replace the mass point with the differential volume element, require the latter to have empirically discoverable "characteristic properties," and replace the a priori assignment of a force between points with the laboratory-based deduction of a system energy that is determined by the qualities of two elements *as well as* by the distance between them.[13]

In 1872, for example, two years after exposing his new approach, Helmholtz linked it in a revealing way to Maxwell's early attempt to model the ether without using atoms and distance forces, using instead rotating elastic cells of continuous matter.

> The idea of such a molecular structure for the space-filling ether might strike the imagination as too artificial, yet it seems to me that Maxwell's hypothesis is very significant, because it provides the proof that there is nothing in apparent electrodynamic behavior that forces us to attribute it to a very deviant kind of natural force, to forces that depend not only upon the posi-

tions of the affected masses but also upon their motion. *Indeed, a complete and mathematically very elegant theory of all apparent electric, magnetic, electrodynamic, and induction behavior can be developed out of Mr. Maxwell's assumption of reciprocal reactions between volume elements of the ether, and this same theory also provides an account of the apparent behavior of light.* (Helmholtz 1872b, p. 639; emphasis added)

Here we see that Helmholtz emphasized two aspects of Maxwell's model: first, that it (unlike Weberean theory) makes no use of anomalous forces and, second, that it relies exclusively on actions between (contiguous) *volume elements* with certain properties. In his own theory of electrodynamics Helmholtz let fall the critical British insistence that only contiguous elements can act directly upon one another, replacing it with the requirement that a duo of elements, however far apart from one another they may be, can establish a system energy.

Neither in his memoir of 1870b nor in this one of 1872 (which was written in large measure to probe Weberean instabilities in reply to critiques launched against his arguments by Weber himself and by Carl Neumann) did Helmholtz explore the novel implications of his new electrodynamics, based on bipartite system energies, for circuit interactions. The next year, 1873, Helmholtz continued his discussion, and this time he did begin to concentrate on his own theory, in major part because critiques of it forced him to lay bare its novel structure. The French physicist Joseph Louis Bertrand, for example, had thoroughly misunderstood Helmholtz's new electrodynamics, claiming it to be "self-contradictory."[14] Helmholtz expended some effort in setting him straight. He continued to elaborate the empirical and theoretical workings of his theory over the next three years, interrupting his analyses only when, in 1876, he became rector of the university. By then his electrodynamics had roused more than the usual scientific interest, for it had elicited the first of many attacks from the ardent Weberean Friedrich Zöllner.

16.3. ATTACK AND DEFENSE

Friedrich Zöllner (b. 1834) was educated at Berlin under Magnus and Dove and became an *Extraordinarius* in astrophysics (he invented and deployed the "astrophotometer") at Leipzig in 1866 and then *Ordinarius* in 1872. In 1872 he published a book entitled *On the Nature of Comets* that began rapidly to mire his career in controversy.[15] In its preface Zöllner attacked Thomson and Tait's remark, preserved in Helmholtz's translation of their *Treatise*, that Weber's atoms and forces are not only useless but even harmful. Zöllner was captured by the notion that the mind, in harmony with nature, could by intuition discover nature's inner workings, and so he insisted that the laboratory could not reveal anything fundamental (Molella 1972, p. 199). Even in 1871 Zöllner was attacking Helmholtz as well as Thomson and Tait. By the end of the decade his philosophical opposition had evolved into a deep-seated hatred of

Helmholtz and his co-conspirator Du Bois-Reymond, with overtones of xeno-
phobia and anti-Semitism, culminating in a posthumous pamphlet that treated
Helmholtz as a dupe of Jewish scientists (Molella 1972, pp. 212–13).[16]

Although the *Volkisch* cast of Zöllner's ideology was not perhaps overt in
his writings in the early 1870s, nevertheless his concern to develop an entire a
priori physics on the basis of Weber's electrodynamics was already apparent in
his 1871 work on comets, appearing full-blown in 1876. During the intervening five years Zöllner had not been inactive; in 1874 he attacked through experiment what he perceptively recognized as a physics thoroughly alien to
Weber's, namely, the one Helmholtz was developing on the basis of system
energies. This stimulated Helmholtz to a direct reply and, in part at least, to a
more thorough exploration of his own scheme. Soon Helmholtz felt it necessary to respond to the underlying thrust of Zöllner's work, which he correctly
perceived to be utterly antithetical to his most deeply held beliefs concerning
science, society, the state, and the university.

Helmholtz took the opportunity to reply to Zöllner in the preface to the second part of his translation of Thomson and Tait. Crum Brown translated the
remarks, which vigorously defend the Scottish scientists, for *Nature* in 1874.
Helmholtz turned nearly at once to the crux of the matter, and it is worth quoting him at some length to catch the anger he clearly felt:

> Judging from what [Zöllner] aims at as his ultimate object, it comes to the
> same thing as Schopenhauer's Metaphysics. The stars are to "love and hate
> one another, feel pleasure and displeasure, and to try to move in a way corre-
> sponding to these feelings." Indeed, in blurred imitation of the principle of
> Least Action, Schopenhauer's Pessimism, which declares this world to be
> indeed the best of possible worlds, but worse than none at all, is formulated
> as an ostensibly generally applicable principle of the smallest amount of dis-
> comfort, and this is proclaimed as the highest law of the world, living as well
> as lifeless.
>
> Now, that a man who mentally treads such paths should recognize in the
> method of Thomson and Tait's book the exact opposite of the right way, or of
> that which he himself considers such, is natural; that he should seek the
> ground of the contradiction, not where it is really to be found, but in all con-
> ceivable personal weaknesses of his opponents, is quite in keeping with the
> intolerant manner in which the adherents of metaphysical articles of faith are
> wont to treat their opponents, in order to conceal from themselves and from
> the world the weakness of their own position. (Helmholtz 1874b, p. 150)

Helmholtz criticized tendencies implicit in Zöllner's vituperative 1871 remarks twice more: first in the summer, then in the fall, of 1877. In a popular
address entitled "On Thought in Medicine" that he delivered on August 2 to
the Institute for the Education of Army Surgeons (which he had been), he
again remarked the lack of politesse among "metaphysicians," noting that as
far as he was concerned "a metaphysical conclusion is either a false conclusion

or a concealed experimental conclusion" (1881b, pp. 234–35). Helmholtz had good reasons for concern, because the increasingly "impolite" remarks of the "metaphysicians"—directed in fact as much at his old friend and colleague Du Bois-Reymond—threatened what Helmholtz strongly felt to be the uniquely German character of the university, its academic freedom. In October he gave his inaugural address as rector at Berlin on that very topic, where he waxed enthusiastic over the great freedom of the German university—and in particular of the "new German Empire"—where "the most extreme consequences of materialistic metaphysics, the boldest speculations upon the basis of Darwin's theory of evolution, may be taught with as little restraint as the most extreme deification of Papal Infallibility." Though "it is forbidden to suspect motives or indulge in abuse of the personal qualities of our opponents, [and] any incitement to such acts as are legally forbidden," nevertheless "there is no obstacle to the discussion of a scientific question in a scientific spirit" (1881b, pp. 255–56). The free teaching of "materialistic metaphysics" (which Helmholtz himself was not engaged in) was precisely what Zöllner and those of like opinion strongly objected to, and Zöllner was, of course, the one who questioned motivations.

Hertz arrived in Berlin one year after Helmholtz had delivered his lecture, and the project that Helmholtz soon set him concerned yet another way of attacking Weberean electrodynamics, or at least of undercutting it. Although Hertz may not at first have been aware of the wider implications of his decision to study in Berlin, nevertheless he could hardly have missed what was surely a topic of discussion among the students. Even if he had somehow remained ignorant of the tremendous social and moral gulf that had opened wide in German physics by the mid-1870s, he could hardly have missed Helmholtz's deep-seated loathing for Weberean electrodynamics, since the first complicated laboratory problem that Hertz undertook was precisely designed to irritate a tender spot in it. In going to Berlin Hertz had implicitly chosen a way of doing physics, a way of thinking about nature, and a specific set of *cultural* attitudes and patterns of behavior that powerfully bound him to Helmholtz's network of similarly minded colleagues.[17] Hertz's particular aptitude was to partially transcend the impress of Berlin by probing with great technical expertise tensions that inhered in Helmholtz's new physics. Here a complex process of rebellion, transformation, adaptation, and assimilation occurred that interwove intellectual pressures with social strains as Hertz sought to differentiate himself within the highly competitive structure of German physics in the 1880s.

APPENDIX 17

Poincaré and Bertrand

One example of misunderstanding nicely reveals how difficult it was to grasp—much less to accept—basic aspects of Helmholtz's new electrodynamics even for non-Webereans. In 1870 Joseph Louis Bertrand was forty-eight years old, professor of analysis at the École Polytechnique, and successor to Biot at the Collège de France. Although primarily a mathematician, Bertrand was at the time engaged in writing a text on thermodynamics, and his earliest publication (at the age of seventeen) concerns Poisson's equation (Struik 1970). He was without doubt quite familiar with Ampère's electrodynamics, and he read Helmholtz's paper that year (1870b) under the impression that one could carry over Ampère's conceptions to it: in particular that the *objects* in Helmholtz's potential law for circuit elements are precisely the same kinds of things that Ampère had deployed. Indeed, Bertrand also thought that the potential must yield Ampère's central force between elements. In 1873, however, Helmholtz explained that his potential entails a torque as well as a force (so that Bertrand's original claim that it is "self-contradictory" fails), and in responding to this, Bertrand revealed how thoroughly he had missed the essence of Helmholtz's new physics.

In 1890 Henri Poincaré discussed the difference between Helmholtz and Bertrand, beautifully capturing its core.

> According to [Bertrand] all these couples, acting on all the elements of a conducting wire traversed by a current and subject to the action of another current or of the earth, should immediately break the wire and reduce it to powder. Helmholtz replied that a magnetic needle does not break under the earth's action, even though a couple acts on each element of its length whose moment is of the order of magnitude of that element. M. Bertrand retorted that no one today any longer believes in the real existence of Coulomb's magnetic fluids and that Helmholtz's response makes no sense; it seems that Helmholtz could have said that one no longer believes in the objective existence of a material current circulating in a conductor. (Poincaré 1890, p. 51)

Poincaré goes on to explain that Bertrand insisted on considering the circuit elements to be irreducible objects, in which case the wire must be "ruptured" by Helmholtz's torques. But, he continues, in fact "Helmholtz supposes that, however far one pushes the division of the matter, each part will always be subject to a couple," so that no final entity is reached which must somehow sustain the full effect of Helmholtz's torque: there are no ultimate entities; there are only differential volume elements and their deformations.[1] In Poincaré's words again, "the couple is only a sort of tendency to turn that properly exists independently of its two components, which may have no determinate point of application. The couple exists whenever the rotation does work."

Although Poincaré gave no evidence that he was aware of the deeper reaches of Helmholtz's electrodynamics (in 1890 he was primarily concerned with using it to reach Maxwell), he well understood that Helmholtz's structure abstracts completely from atomicity. Bertrand not only missed the more technical elements that Poincaré (explicitly) also ignored but remained unwilling to grant Helmholtz his differential volume elements. For Bertrand there had to be fixed entities that exert forces, in Ampère's fashion, on one another—or there had at least to be Weberean particles (though Bertrand does not refer to them).[2] Poincaré's remarks portray a profound impasse indeed, one that divided those who thought in terms of a continuous *field* from those who required continuity to be the averaged result of an underlying discontinuity. That division acquired concrete form in Helmholtz's potential law precisely because his law could not be reduced to actions between discrete, fixed objects.

In presenting Helmholtz's electrodynamics Poincaré himself never made any use of what tied Helmholtz's entire structure together: namely, Helmholtz's unification of electromotive and material forces under the energy principle. Without that unification Helmholtz's theory can be thought of as a formal generalization that simply abstracts from physical hypotheses but that might be integrated with them. In which case the intricate energy problems that powerfully engaged Hertz would never arise. Moreover, if one thinks about Helmholtz's structure in this way, the electrodynamics of his polarizable ether can be thought of rather as one might think of interacting circuits spread throughout space. That, indeed, was precisely how most physicists in the early 1890s *did* think about it (Buchwald 1985a, pp. 183–84): the issues that so stimulated Hertz in the early 1880s, and that Helmholtz explored in part during the 1870s, accordingly never became common property.

APPENDIX EIGHTEEN

Difficulties with Charge and Polarization

Faraday had discovered that many nonconducting bodies can influence the charge on conductors, and so they can affect the force between such things. They are also themselves moved by electric force—by, that is, charged conductors. The prevailing opinion in Germany accepted a form of the hypothesis developed by Ottaviano Mossotti in 1846 to explain the effect of nonconductors on the electric capacities of conductors. Mossotti's hypothesis could also explain their mechanical effects on one another. As understood in Germany, the hypothesis required that nonconductors, or dielectrics in Faraday's terminology, consist of molecules each of which either is already electrically polarized or else becomes polarized when it is subjected to electric force.[1] Then the effects of the insulator on a conductor can be reduced to the effects that would be produced on it by such a collection of microscopic objects.

Since we do not know the structure of these molecules we cannot a priori conclude that changing dielectric polarization does, or does not, act like a current in a conductor to generate electrodynamic and electromotive force. We have assumed that electric force can polarize the molecules in the sense that each molecule generates the same electric force that would be generated by an electric dipole whose moment is proportional to the external electric action. Suppose that the molecules are similar to conducting spheres. Then since, for whatever reason, each has a dipole moment, the charge density over the surface must change magnitude and sign from a negative minimum on one end of a diameter parallel to the moment to a positive maximum at the diameter's other end. If the dipole moment changes, the charge distribution changes, and this necessitates currents because of the continuity equation. Consequently, in this model changing dielectric polarization would implicate currents in precisely the same way that changing charge densities implicate currents in macroscopic conductors: the equation of continuity demands it. But we could certainly not conclude from this alone that electromotive forces due to changing currents can polarize the dielectric. And the reason for this asymmetry, for the ability of (changing) polarization to act electromotively but not necessarily also to be affected electromotively, reduces to the inherent assumption in Helmholtz's system that electromotive forces act only to generate currents.

This involves a subtle but deeply significant question that arises out of

Kirchhoff's assumption, in his circuit theory, that the electrostatic potential must itself be treated like an ohmic tension—that a term appears in Ohm's law which is formally, though not necessarily physically, the same as the electric force that moves charged bodies.[2] In Weberean theory (which Kirchhoff then used) this is not an assumption; it is a direct consequence of the theory's basic premise that currents consist of moving electric particles and that all electromagnetic interactions are due to the forces that act between the particles. But in Helmholtz's system currents and charge densities are linked to one another by necessity only through the continuity equation, so that a force that moves a charged body does not in itself also act to generate a current. Indeed, in Helmholtz's electrodynamics charges strictly speaking never move at all unless the charged body itself moves.

To see the perplexing difficulties that this can raise consider what might seem to be (and in Fechner–Weber is) a simple situation: two neighboring charged conductors are displaced with respect to one another. Since the distribution of charge over the two bodies is, we know from electrostatic principles, different before and after the displacement, the continuity equation demands that during the time that the system is not in, but is approaching, equilibrium, electric currents must exist in both bodies. Now if we assume that the displacement occurs infinitely slowly, we can ignore electromagnetic induction, so that the only forces that can be involved in this process are electrostatic. Here, therefore, is a situation in which currents must evolve although only electrostatic actions occur. But how are we to understand the physics of this almost primordially important process (since, in the end, it explains electrostatic induction) in terms of Helmholtz's principles?

In Helmholtz's unadorned scheme there is no interaction energy between currents and charges, which is why we cannot assume a priori that electric force can function in Ohm's law as though it were an electromotive force. The electric force between charged bodies tends to move them, but Helmholtz's system never actually permits charge *itself* to move as though it were simply a substance. Such an idea is anathema because it suggests the principles of Fechner–Weber, which Helmholtz's scheme avowedly rejects because in the latter, charge is a condition, rather than a distinct entity, that exists on or within a body. Now the old electrostatics and Weberean theory insist that over time the electric potential must eventually become constant over every conducting surface, in the absence of nonelectric actions. But why? Fechner–Weber and traditional theory argue that, were this not true, "charge" would continue to flow on and within conductors, producing electric currents that we know empirically do not occur. But this conclusion is based on the presumption that charge is itself a distinct entity on which electric force acts—and Helmholtz's electrodynamics does not agree, since in it the force is between the charged bodies.

Instead of recurring directly to the absence of currents, suppose we begin with an empirical fact (i.e., something no one doubts)—namely, that eventually a constant potential is indeed reached, but that there is a period of time during which the conducting surface does not have the same potential at every point. From Poisson's equation it follows at once that the charge density changes with time. This is translated through the continuity equation into an actual current, and it is the current that effects the charge redistribution.[3] At every moment an electric force certainly does act between the charged bodies according to the usual Coulomb law, but *this* force has nothing directly to do with the approach to electric equilibrium: because it does not act between the charges proper, it cannot alter charge densities. In order to explain why equilibrium is reached we must therefore introduce an independent hypothesis: namely, the tendency to generate the current at a point is proportional to the force that would tend to move a body with unit charge placed there.

This action is not at all the same as the one that moves charged bodies even though they share the same measure. We might introduce the action as an unexplained hypothesis, in the sense that we do not know why electric force can generate a current at a point, whereas we do know, for example, why charged bodies act to move one another: because there is an interaction energy between them. Since Ohm's law is itself rather mysterious in Helmholtz's system this need not be overly surprising. Or we might say that the action reflects the existence of an appropriate interaction energy between charge density and current. But then we are also forced to accept other consequences of such an energy. Suppose that the energy in question is $\Phi(q, c)$, where q is charge and c is current. By assumption the mere existence of a fixed charge generates an electromotive force that can affect a current's magnitude. Consequently, this action must be obtained by varying the spatial coordinates of Φ while holding the time fixed, which is precisely the reverse of what must be done to obtain the electromotive force exerted by currents on one another (reflecting the difference that current quantity is not conserved whereas charge quantity is).

One might then deploy Helmholtz's energy argument in a fashion similar to that used for other purposes by Hertz to obtain a mechanical force exerted by a current on a charged body. This new force will depend upon the rate of change of the current, and so upon the temporal variation of Φ. In other words, *if* charge can directly generate current via its electric force, then, it seems, Helmholtz's principles require that changing currents must also move charged bodies, which renders our previous table of interactions (table 1) not merely incomplete but perhaps even incapable of being completed with a finite series of terms. In the context of Helmholtz's system this difficult and radical result derives from, and implies, the existence of an otherwise unrecognized charge–current interaction energy, one that has no formal expression in the theory.[4]

The electromotive force that is generated by a changing current behaves in a markedly different fashion from this. It cannot be obtained from the gradient of a potential; instead, its curl is determined by the time rate of change of the magnetic action produced by the generating current. Consequently, if it exists at all, then its integral around a curve does not vanish (unlike that of the electric force), and it can generate reentrant currents if the curve lies entirely within or on a conducting body. If it does not (if, e.g., the curve is interrupted by a nonconducting space), this electromotive force may behave like one that derives from a scalar potential, and it may accordingly produce currents that terminate, leading over time to net charge densities.[5] This brings us back to the difficulty that Helmholtz had with polarization.

To see what is involved, consider a conducting, spherical shell placed near a conducting loop in which the current is changing. Every section of the shell that is parallel to the plane of the loop experiences a circuital electromotive force that produces a current in it. Therefore, the shell as a whole will bear a continuous series of closed, concentric currents each of which is parallel to the loop: it will not be at all polarized electrically. Even if the shell were cut along, say, half of a great circle whose plane is normal to that of the loop—thereby interrupting the circuital paths—charge would *still* not appear, because closed paths exist that do not traverse the slit.

Now suppose that dielectrics consist of microscopic conductors that are separated by a completely nonelectric material of some sort. Place the dielectric near a charged conductor. Then the electric force of the conduction charge will (we suppose) act to generate noncircuital currents in the microscopic conductors that compose the dielectric and so, via the continuity equation, eventually to polarize them. Next bring near a circuit with a changing current. The electromotive force that results from this current will, we also suppose, act to move the dielectric because it acts to move each of the charged objects that compose it (given that we assume an interaction energy between current and charge). But it does not at all follow from this that the electromotive force can itself polarize the dielectric. If the microscopic objects that compose the dielectric are indeed conductors, then the electromotive force will simply create circuital currents on them, in which case the dielectric will show no polarization. If we want to assume that electromotive forces will *always* polarize dielectrics, then we must also assume that they are not composed of microscopic conductors, that their structure is completely different in kind from that of conductors, and that the ability of an external object to move a dielectric when the latter is charged translates directly into that object's ability also to polarize the dielectric through some mysterious process that must be utterly unlike conduction.

Consider in this context the Berlin Academy questions of 1879 that Hertz eventually declined to answer. The first of the two questions does directly concern field theory since it asks explicitly for an experiment to show that the electrodynamic action of changing polarization is "in the intensity as assumed by Maxwell" and not just any intensity. Both Helmholtz and Fechner–Weber can also predict this kind of effect, albeit with different intensities from field theory, since they accept the Mossotti hypothesis for the structure of dielectrics.[6] However, the second question concerns much more than field theory proper. Only Helmholtz's system does not explicitly predict that electromagnetic induction can polarize dielectrics even if changing polarization does exert electrodynamic and electromotive effects, in the first instance because (unlike Fechner–Weber) Helmholtz necessarily has great difficulty envisioning charged objects that are not in some sense conductors,[7] and in the second instance because Helmholtz does not explicitly employ an interaction energy between charges and changing currents. To fit the pattern demanded by Helmholtz's system, polarization current should be treated as a theoretical primitive of the same kind as the conduction current (an addition to it in circumstances where bodies are polarizable as well as conducting) that appears in the interaction energy between currents. But since a current of either kind interacts only with other currents, there need be no direct effect of a changing current on polarization itself.

Now if the Berlin Academy's second question had a negative answer—if the effect did not occur—Helmholtz's scheme would, other things being equal, remain tenable.[8] But if the experiment had a positive result, Helmholtz's system would have to be supplemented by considering the interaction between changing currents and charges, if not through the explicit construction of a new interaction potential, then at least through the phenomenological incorporation of the action into the Mossotti hypothesis. And this latter procedure was precisely what Helmholtz had suggested in the last few sections of his 1870 article, wherein he produced an extremely influential scheme that bore a superficial resemblance to field theory.

Helmholtz had here introduced propagation of a kind, although he ignored the issue of an interaction energy between changing current (or magnetization) and charge.[9] He accomplished this by assuming that electromotive actions can polarize insulators in the same manner that electric actions can, and he additionally assumed that changing polarization constitutes a current (in the sense that it must be incorporated into the electrodynamic potential and into the continuity equation, but not into Ohm's law). These two requirements sit uneasily with one another: given the Mossotti hypothesis and the continuity equation, changing polarization should generate the same forces as a current, since the dielectric consists, in effect, of microscopic conductors. But if the dielectric is so composed, then electromotive forces cannot polarize it in the manner that Helmholtz required (i.e., in such a fashion that the polarization in a small region is proportional to the electromotive force there). Conversely, if electromotive forces can polarize a dielectric in this manner, then it cannot be composed of conductors, in which case it is difficult to understand why changing polarization should behave like a current.

And so in Helmholtz's system (as in all other contemporary forms of electrodynamics) it was simple to predict the first effect queried by the Berlin prize on the Mossotti hypothesis. But the second effect is foreign to the purest form of Helmholtz's electrodynamics, and its existence would force Helmholtz's scheme to incorporate an interaction energy between charge and changing current that lacks any formal expression. But if the second effect occurs, the first effect should, it seems, not occur, because there would be no conductors to provide the currents *it* requires. In other words, the single entirely novel assumption upon which Helmholtz's polarization theory relied was the very one that the Berlin Academy's second question of 1879 proposed to examine. And in the context of Helmholtz's system this was no small assumption.

The words with which Helmholtz introduced the polarization current in 1870 indicate that he considered it to be in itself a difficult conception that is strongly bound to the continuity equation:

> If in the volume element dS the quantity E of positive electricity moves $1/2s$ in the positive x direction, and the quantity of negative [electricity] moves $1/2s$ in the negative x direction, then in that same element an electric moment
>
> $$X = Es$$
>
> will be produced, and at the same time this process corresponds to a current in the element
>
> $$u_0 dt = Es$$
>
> The act of polarization therefore builds a kind of electric motion [*Der Act der Polarisation bildet also eine Art elektrischer Bewegung*]. (Helmholtz 1870b, pp. 615–16)

Helmholtz has chosen his words carefully indeed. He begins with equal but opposite shifts of equal but opposite electric quantities to create an electric moment—to create, that is, an electric, not an electromotive, force. "At the same time," he continues, "this process corresponds to a current." Note the wording—it is not that the generation of an electric moment is in itself a current. Rather, it "corresponds to [*entsprechend*]" a current, so that "the act of polarization therefore builds a kind of electric motion," and this is embodied in Helmholtz's equation $u_0 dt = Es$, which is a direct expression of continuity. Much was at stake here, and so Helmholtz decided to introduce as carefully as he could an assumption that raises great difficulties. Even if continuity may seem plausibly, if rather obscurely, to link the generation of an electric moment to a current, nevertheless the action of an electromotive force in effecting the moment remains as problematic as its ability to move a charged body. Yet the foundation of Helmholtz's analysis requires precisely that assumption.

Helmholtz's theory was, then, an account of what occurs when, given a link between the generation of moments and currents, electromotive actions are permitted to alter polarization charge densities. But should this hypothesis prove to be *empirically* tenable or even necessary, then a disturbing *theoretical* novelty has been introduced, a novelty that does not sit at all well with the requirement that changing polarization must be treated like a current. In the absence of the hypothesis, Helmholtz's system is complete. There are only two kinds of interactions—charge–charge and current–current—and each has a corresponding potential function. There are no problems in understanding or representing the connection between an action and its corresponding source. But permit electromotive forces to act upon charge, and this satisfyingly complete structure fractures. Since there is no energy function to represent the interaction, it must be assumed a priori by incorporating electromotive force into the expression for electric polarization: instead of taking the latter proportional solely to electrostatic force, as the combination of Helmholtz's system with the Mossotti hypothesis would have it, one must also include electromagnetic induction. And then the thing that the electromotive force produces must itself generate an electromotive force if its second derivative with respect to time is nonzero. This is, to say the least, mysterious within the confines of Helmholtz's system even given an interaction potential between charge and current.

This weak spot evidently troubled the young Hertz in the early 1880s. As he sought to comprehend the basic system along with Helmholtz's novel and unnatural addition to it, he came to feel, at first in an inchoate manner, that there was a problem with the very structure of the system itself, with its reliance on a *taxonomy of forces*. This taxonomy, he eventually decided, was in turn based upon a distinction between the nature and the action of a source. Here electric forces are not simply electric forces. They are forces that act upon charges, and they may arise from entirely different kinds of sources. Although it is no doubt true that one cannot tell, at a given moment and a given point, whether the force on a charge derives from a source of type alpha or one of type beta, nevertheless completely different interaction energies are necessarily involved unless alpha and beta are physically identical with one another.

Hertz studied parts of Maxwell's *Treatise* in the early 1880s, and it is clear that he early understood that field theory (which, after all, was what Helmholtz's polarization supposedly emulated) introduced no such distinctions. As a result he found himself in a distinctly uncomfortable intellectual position, which he attempted to resolve in 1884 in a radical fashion by retaining the elementary structure of Helmholtz's system, its basis in energy conservation, but introducing into it a completely incompatible understanding of the relationship between a source and its action. The resulting incoherence plagued him over the next half-decade until, as he rethought the scheme in the light of his experiments on propagation, he began to see that the problem lay at the deepest level of Helmholtz's system, in its foundation on interaction energies. When, therefore, he embraced (or, better, created) a form of field theory, he also completely discarded the potentials. For him they were the vestiges of an outmoded way of thinking which had no place in his new scheme. With them distinctions between actions based on sources were possible; without them the actions could be divorced from the sources and located independently in the field.[10]

NOTES

Chapter 1
1. See, e.g., Kuhn 1978, p. 33. Because the dipole's effects could be analyzed without probing its structure, physicists could deploy it in extremely general circumstances.

Chapter 2
1. See Wiedemann 1885, vols. 2 and 4, for a rational catalogue of current- and charge-generating devices. A contemporary and early colleague of Helmholtz, Wiedemann remained closely tied to laboratory measurement. See Jungnickel and McCormmach 1986, vol. 1: 258–59.

2. This is a force that moves a body rather than one that changes its state in some manner.

3. The magnetic force acts to change a current-bearing circuit's position. Below we will see how, given principle 6, Helmholtz was able to derive 2 from 3 and 3 from 2.

4. This fundamental requirement is nevertheless implicit; see appendix 3 for a discussion of how it is essential to understand Helmholtz's claim that electrodynamic force can be derived from electromotive force and vice versa.

5. In the 1840s, when he was young and attempting to gain a professional foothold, Helmholtz did not explicitly assert that the electrodynamics of the well-known Wilhelm Weber must be rejected on this account. He remarked only that Weber had obtained electromagnetic induction from these kinds of forces, and that no one had "for now" been able to do so using forces that are functions solely of the distance (Helmholtz 1847, p. 61).

6. Perhaps Helmholtz's verbal clarifications of his papers were just as important for Hertz, though Hertz mentions nothing of this in his letters to his parents between 1878 and 1880.

7. If the object moves, then the state of the ambient field changes, which entails an alteration as well in the local energy densities. That, in turn, translates into a stress over the region occupied by the object according to the basic principle of field theory that moving force derives from local energy gradients. See immediately below for a pertinent example.

8. If the latter objects interact locally with the field at some moment, then its changed state will propagate from point to point throughout it, ultimately reaching other objects and affecting the field's state at their positions. Consequently, the interaction between field and object at a given moment depends upon other object–field interactions at previous times.

9. As, e.g., when the moving object is a homogeneous cylinder that rotates about a central axis in a uniform magnetic field that is parallel to the axis.

10. For purposes of clarity I shall present here what is certainly an overly precise specification of the theory, reserving to later chapters the important details that eventually disturbed Hertz.

11. The argument is rather more complicated than this simple remark suggests because it must exclude certain kinds of energy transformations, include other kinds, and the whole must presume the existence of a systemic interaction energy that acts as the sole energy source for system changes. In fact, Helmholtz's several presentations of the energy principle in this form are rather incomplete, even faulty, though their results are unexceptionable. The essential point to grasp about the principle is that the elements that enter it as givens are system energy and system states, with "force" emerging only as an artifact of the necessity to balance energy. See appendix 3 for details.

12. Once the ether is reduced to a qualitatively unalterable seat of fields, rather than being a potential object of investigation in its own right, one can also say that no theoretically fertile *thing* need be supposed to intervene between interacting objects. The ether's previous canonical function is replaced by hidden things that are now taken to form parts of objects. Indeed, Lorentz spent a great deal of effort showing how to reconstruct *macroscopic* field theory on the grounds of retarded interactions between *microphysical* particles.

13. To perceive the distinction I have in mind consider that in Fechner–Weber the force that acts on a given electric atom does not in any way depend upon a *state* in Helmholtz's sense, because the atoms are unalterable in themselves. They do, however, move, and the forces depend directly upon the relationships between their motions and upon the distances between them. In Helmholtz's conception, on the other hand, two bodies moving in certain ways with respect to one another are not thought to have a mutual relationship with one another that depends immediately upon the motion. Rather, they may have electromagnetic states, and these states determine an interaction energy that may change as a result of the motion, but the motion itself has no effect at all.

14. Buchwald (1985a, chap. 27.1–2) examines how Helmholtz's instrumentalist ions permitted him and others, on the basis of an electromagnetic mutation in mechanical equations developed by Helmholtz himself in 1875, to bear down on dispersion and magneto-optics in ways that were unavailable to the Maxwellians and that were of course utterly closed to Webereans. This work is a natural offshoot of the instrumentalist electrodynamics that Helmholtz formulated in 1870. We will return to the point below.

15. Appendix 16 briefly discusses some of the factors that may be involved in Helmholtz's creation of this novel kind of physics.

16. Since Maxwell had not as yet provided a unified structure—since, in particular, his understanding of charge and current remained unclear in 1870—it is hardly surprising that Helmholtz interpreted field theory in a way that would later prove to be a great stumbling block for Continental physicists' (including Hertz's) attempts to understand Maxwell's *Treatise*.

17. The contemporary terminology—closed circuits—can mislead out of context. Consider, for example, a flattened ring with a section cut out of it, leaving a gap between two parts. Such a thing is, in the usual parlance, an open circuit. However, if it is placed in a magnetic field normal to its plane, and the field changes, then currents will be induced in it that are themselves completely closed, although they do not of course span the gap. The important distinction is between currents that are reentrant—that close on themselves—and currents that are not, rather than between closed and open *circuits*.

18. Here, as elsewhere, the positions of the bodies in which C, C' occur are given, respectively, by r, r' and r_d is the distance $r - r'$. ∇_r is taken with respect to r, and $\nabla_{r'}$ with respect to r'.

19. Helmholtz does not indicate at this point why setting k equal to zero leads to Maxwell's expression (which Helmholtz knew about only from Maxwell's 1864–65 papers). However, Maxwell always assumed that the vector potential—his version of U—lacks divergence, in which case $k = 0$ would be necessary for reaching his theory independently of other requirements.

20. Temporal variation does not at all mean that the interaction energy depends in any explicit way on the time. Quite the contrary: as long as the currents C remain the same, so must the interaction. However, if they are allowed to vary, the energy changes as well. Helmholtz's unbreakable insistence on spatial dependence is perfectly embodied in his expressions, only he has now permitted the interacting objects to possess vector properties as well as scalar ones.

21. More precisely, at the terminus of the moving wire a liquid element is, at any given moment, losing its current-carrying state, while at the same time another element is gaining it. Even though the total energy does not change in quantity, nevertheless the objects between which the interaction occurs do change, and this entails a corresponding force.

22. As Helmholtz saw it, the force acts on the constantly changing mercury elements that touch the immersed end of the hanger, and they then carry the hanger along with them.

23. All translations are my own except where the source cited is a translation.

24. That is to say, either a state is an unreducible but quantitatively variable property of an object or else it is the result of the properties (perhaps themselves unreducible) of other things that together constitute the object in question. For Webereans composite objects—which is to say sets of electric particles—have states in the latter sense, but the electric particle itself has no states at all because it has no *properties* that are quantitatively variable. Its position, speed, and acceleration are all characteristics that have physical significance only when taken with respect to another such particle, and so they necessarily characterize the state of a composite object, not the state of the particle itself, which is eternally the same.

25. Helmholtz's argument has been thoroughly discussed by Klein (1972). Bierhalter (1981) provides extensive technical detail.

Chapter 3

1. See the careful analysis in Archibald 1987, pp. 417–31. Archibald discusses the unsuccessful experiments to find new forces, the difficulties raised by convection, and the impetus the experiments gave to Helmholtz's conviction that polarization of a very high order must occur in air.

2. This is a crucial point, because it means that the constant k has no bearing on Helmholtzian experiments as performed during the 1870s or, later, by Hertz. It emerges in the following way. First, the terms under the integrals in Φ_2 can be rewritten (since $ds d\sigma \partial^2 r/\partial s \partial \sigma = (ds \cdot \nabla)(d\sigma \cdot \nabla)r$):

$$\frac{1}{r^3}(r \cdot ds)(r \cdot d\sigma) - \frac{1}{r}(ds \cdot d\sigma)$$

Since ∇r is just r/r, then this becomes

$$\frac{1}{r^2}[(\nabla r \cdot ds)(\nabla r \cdot d\sigma)]$$

But $r^{-2}\nabla r$ is $-\nabla(1/r)$, and so if either circuit is closed, the corresponding integral involves the product of a gradient by a line-length taken around a complete circuit, which vanishes by Stokes's theorem (since the curl of a gradient is identically zero).

3. Working before the explicit enunciation of energy conservation, Neumann nevertheless obtained his potential directly from a conservation principle, in which it appears as a "force function." However, Neumann's force function did not represent the interaction between circuit elements; it represented instead the link between closed circuits as such. This is reflected in Neumann's method of deriving it: he *starts* with the expression implied by Ampère's element–element law for the force between closed circuits; this effectively wipes out the specific character of the Ampère law insofar as the elements are concerned. His function accordingly had significance solely for circuits as complete entities, not for the elements that constitute them. See Archibald 1987, chap. 3.2, for discussion of Neumann.

4. He provides only the result, not the intermediate steps.

5. To obtain the force corresponding to Φ_2 requires treating both circuit elements as isolatable entities with physically meaningful termini. Translated into analytical procedures, this implies that one must always try to partially integrate any integral that contains them in order to separate the terms that apply to the body of the element from those that apply to its endpoints. In the case of Φ_2 this procedure shows that the only nonvanishing terms involve the products of the boundary-value resultants for the two elements. That is, the only forces that result occur between the endpoints of the two elements (which is why they vanish altogether when either circuit is closed, for then the actions of the endpoints of the element that belongs to the closed circuit will be canceled by the actions of its contiguous neighbors). These forces are

$$F_{\Phi_2} = \frac{1-k}{2}A^2 i_{\text{end}} \sum_\sigma j_{\text{end}} \nabla r$$

6. The center–center force results from applying the triple cross product for $ds \times [\nabla (1/r) \times \nabla \sigma]$ in the following expression for the integral of the center–center force:

$$F_{ds} = -A^2 ij \int_\sigma \nabla \frac{1}{r}\left(\frac{ds}{ds} \cdot d\sigma\right) - A^2 ij \int_\sigma \left(\nabla \frac{1}{r} \cdot \frac{ds}{ds}\right) d\sigma$$

7. The *ensemble* of Helmholtz's element–element forces satisfies the principle of action–reaction. This is not obvious from the forces given by the last equations, since the first one (F_{ds}) certainly does not satisfy action–reaction if it is applied *as is* to a pair of elements, i.e., as though they existed independently of circuits. However, a partial integration must also be done over $d\sigma$ for this term, which splits F_{ds} into two parts. One part gives the Ampère element–center to element–center law, and the other part gives new forces between the ends of $d\sigma$ and the body of $ds;$ the first part satisfies action–reaction, and the second part does when taken in conjunction with the k-independent end-forces on ds (see appendix 2). Many of the contemporary element–element force laws violated momentum conservation, either because they could not be deduced at all from a potential or else because absolute velocities appeared in them. See J. J. Thomson 1885 for discussion.

8. Unless something goes on outside them, a point that will be important below.

9. The extension has two advantages. First, the linear theory cannot take account of self-induction since the corresponding coefficient is infinite; second, the extended theory shows explicitly that the novel forces arise only where the volume current diverges (see appendix 2).

10. Which raises the question of the status of Helmholtz's novel actions within the other forms of electrodynamics. On the one hand, the novelties had also to be present in them, because they are embraced by Helmholtz's potential. On the other hand, they simply are *not* there. That is, one cannot begin with Weber's force law and the Fechner hypothesis and obtain the novel actions from them. The reason is simple: the assignation of -1 to k in Helmholtz's potential does not at all yield Weber's theory, as Helmholtz indubitably knew, because for Weber circuit elements are not physical entities. They guide the motions of the electric particles, which exert and are affected by the electrodynamic forces. The potential that Helmholtz wrote for Weber is accordingly the *result,* not the predecessor, of the actual actions. The potential that Weber himself used concerned the particle–particle interaction. One may say that Helmholtz's novel actions do not exist for Weber precisely because Helmholtz's physical objects, the circuit elements, are not active entities for Weber. Any critique of Weber based on Helmholtz's element–element potential would necessarily be suspect in Weberean eyes.

Helmholtz's assertion, then, that the general potential embraces other forms of electrodynamics was itself highly problematic and, in fact, a claim for hegemony. If Helmholtz could persuade that the potential does embrace all forms of electrodynamics, then arguments would have to be carried out in terms of it, i.e., in terms originated and controlled by Helmholtz himself. Weichereans powerfully, vehemently resisted the incorporation, correctly perceiving what was at stake. Maxwellians never took it altogether seriously, because they perceived that Helmholtz's continuity equation, with the many central concepts that orbited about it, was an alien entity (Buchwald 1985a, chap. 21).

11. Another difficulty troubles the history of Helmholtz's earliest experiments. Because they were vitiated by uncontrollable disturbances, Helmholtz never described them in any detail at all, although their general structure can be reconstructed well enough for our purposes from his remarks on what he hoped to do in his report of 1873c, pp. 700–701.

12. Again it is essential to understand that an experimental comparison between the Ampère law and whatever follows from the potential has nothing at all to do with discriminating between different theories of electrodynamics. Since all of these last can supposedly be accommodated by the potential, such experiments can decide only whether the potential itself works to underpin all physical theory.

13. The experiment seeks to measure the aligning of the plane of the capacitor with that of the ring magnet. However, the force (should it exist) that produces the torque will be very small indeed, so that attempting simply to observe the capacitor's stationary position would be nearly impossible. Instead, the most reasonable thing to do is to observe the oscillations of the capacitor and determine their midpoint, much as in a Cavendish experiment to measure the constant of universal gravitation.

14. The "Ampère law" in quotation marks refers to the force that would be exerted on a circuit element by other elements if the element–element interaction is the one specified by the form that Ampère's element–element interaction takes when both circuits are closed.

15. See Archibald 1987, pp. 424–28; Jungnickel and McCormmach 1986, vol. 2:27–28; and Woodruff 1968, p. 308. In light of what we know about Helmholtz's later relationship with Hertz, it is more than likely that Helmholtz in effect told Schiller what to do, or at least suggested the experiment and how to go about it.

16. Figure 9 is derived from Wiedemann 1885, vol. 2:200. He provides an extensive discussion of the device and its relatives.

17. See Wiedemann 1885, vol. 2:228ff. The device was used to generate steady currents via spark gaps as well as to produce large charge densities, the purpose evidently being to investigate, in Wiedemann's words, the "connections between the rotational velocity, the work of the rotation, the current intensity, and the resistance of the conduction" (p. 231). If Helmholtz's novel forces had existed, then these kinds of investigations, insofar as they depended on galvanometric measurements, would at the least have had to be reassessed.

18. A decade later J. J. Thomson nicely described what should occur on Helmholtz's principles: "If we calculate from [Helmholtz's forces] the couple produced by an end on an endless solenoid, or on what is practically the same thing, a ring magnet, we shall find that the couple tending to turn the ring about an axis in its own plane will not vanish, while the couple arising from the forces given by Ampère's law will. Thus if the ring rotates, as it should, according to the potential theory, it must be from the action of the end" (1885, p. 144).

19. Seen in this light Helmholtz's (soon pressing) interest in the electrodynamics of charge convection probably emerged only after the failure of Schiller's first experiment, as a way to modify the system. This interpretation is I think necessary since otherwise there would have been little reason to perform *Schiller's* experiment in the first place—charge convection is much better examined directly, which is precisely what Helmholtz had another visitor to his institute, the American Henry Rowland, do at just the time that he, Helmholtz, discussed the Schiller experiment in print. But when he did publicly commit himself, Helmholtz also left the impression that he had from the beginning thought of the Schiller experiment as tightly meshed with convection, since he remarked in introducing it that it "seemed possible by means of a way I indicated in Borchardt's Journal to obtain current ends with sufficient efficacy by means of electric convection" (Helmholtz 1875, p. 779). In fact in the article he refers to, Helmholtz had not said quite this. Rather, he had there explicitly denied that convection would have an electrodynamic effect, as indeed it should not in Helmholtz's basic system, because there is no interaction energy for it. In his words: "[P]oints through which electricity flows out or in, be it toward the disk of an electrification machine or be it born by the moving air, are to be considered as current ends and would be capable of exhibiting the latter's electrodynamic qualities" (1874a, p. 710). Helmholtz had previously thought that situations in which convection occurs would be good ones to use in looking for the new forces simply because disruptive discharge, which produces convection, is a sign of large, rapidly changing charge densities. Schiller's experiment quickly changed his mind about this.

20. The physics of the situation is a bit subtle. Naturally, even before the conducting element joined the circuit it existed in the same external potential field that it occupies after the link comes into being. However, in order to determine what occurs in potential theory one must suppose that a virtual current exists in the circuit and then determine what happens *to the system energy* as this

same current now flows through the newly added element as well. If the interaction energy of this virtual current with the external sources changes as a result of adding the element to the circuit, then an electromotive force must come into being.

21. A related situation arises when, e.g., a current feeds from its pivot through the radial arm, out a circular conductor touched by the arm, and back to the pivot. According to Fechner–Weber and field theory, the arm rotates. Suppose a powerful electromagnet produces a homogeneous field normal to the plane of rotation. Again, at first thought Helmholtz's system predicts no motion since the configuration of the system is unchanged no matter where the arm sits. However, there is a marked change in the interaction energy between current and field at the end of the arm, where the direction of the current changes from radial to circumferential. The portion of the current in the arm has *no* potential with respect to the currents that generate the magnetic field (since they and it are mutually orthogonal), whereas the circumferential portion has a maximal potential with respect to them. Consequently, the interaction energy changes radically at the sliding point— again, the *Gleitstelle*   so that there, and only there, a mechanical force arises that acts upon the tip of the arm in the direction of the circumference. The arm should therefore rotate, but Helmholtz localizes the force at the end of the arm, whereas Fechner–Weber and field theory distribute it over the arm. (see appendix 2.2.2 for Helmholtz's computation of the *inductive* effects that must also arise here.)

22. However, in the light of what I shall say below, Schiller's experiment could be interpreted as showing that convection currents *do not* have electrodynamic effects. See appendix 5 for details.

23. If one wishes to apply a general principle in describing Helmholtz's argument, one could say that when something does not work out right, the experimenter looks immediately for an explanation in some overlooked pathology of the device or (failing that) in an effect that can be understood without an extreme rupture in understanding what is going on, or in both. Anyone who has spent time attempting to make an experiment work already knows this entirely obvious fact about laboratory behavior.

24. And here we have the major, indeed unbridgeable, difference between Helmholtzian and Maxwellian *theory* even though we are considering the so-called Maxwell limit of Helmholtz's polarization equations. According to Maxwellian theory the absence of force in Schiller's experiment has nothing to do with any kind of cancellation: Maxwell's displacement current is, like Helmholtz's limiting polarization current, equal in magnitude to the bounding conduction current, *but it is also equal to it in sign.* The force does not occur in Maxwell's theory simply because the current that determines the energy of the system *never* has any divergence (being the sum of conduction and displacement currents), so that whether or not the conduction current is inhomogeneous makes no difference at all to electrodynamics. One could hardly ask for a better example of the profound incompatibility between Helmholtzian and Maxwellian conceptions.

But, one may wonder, where in this does convection, in the sense of charge transport, figure? If one has in mind that convected charge *in itself* should (as in Fechner–Weber) be considered a current, then it figures nowhere at all. Helmholtz did not make such an assumption, because he did not consider that a charged body in motion has, qua moving charge, an interaction energy with a current. Nevertheless, he certainly did think, after 1875, that such a body will exert electrodynamic forces, because the polarization of the neighboring air will change as the body moves, producing polarization current, which must act electrodynamically in order to explain the Schiller experiment. (See appendix 5 for details.) And in this respect Helmholtz's attitude was quite similar to the Maxwellian (developed a half decade later), which saw convection as a special case of changing displacement (Buchwald 1985a, appendix 1).

25. More precisely, Helmholtz had essentially two choices. He could (and did), following Mossotti, introduce dielectric polarization into the magnetic model as a shift in position of something that exerts electric force. This requires an interaction energy between conduction charge and polarization and, if the continuity equation is assumed to apply to changes in polarization, also yields the latter's electrodynamic effect. But it leaves completely mysterious the interaction between

changing currents (electromotive forces) and polarization itself. Or the latter could be introduced a priori by assuming that dielectrics are similar to conductors with effectively infinite resistance, but then the interaction between conduction charge and polarization would be mysterious since the latter would only be the integral of a new primitive—the polarization current—over time. For a fuller discussion see appendix 18.

26. In recent years historians and sociologists of science have increasingly used the word "closure" to refer to the ending of experiment in some sort of general agreement. The Berlin experiments in the 1870s never did close except within the distinct ambits of Helmholtzians and non-Helmholtzians. To the former the experiments provided evidence for what they thought of as "Maxwell's theory." But to the latter the experiments could show only that there was no reason to pay any attention to Helmholtz's generalized potential, about which they were both skeptical and confused.

Chapter 4

1. I thank Rudolf H. Hertz for providing me with the family tree. His grandfather was second cousin to Heinrich's father through Schönchen Hertz.

2. This situation was certainly not unusual for a bourgeois family in midcentury Hamburg: see, e.g., Holborn 1964, pp. 493–94, who remarks that often "these *Kulturprotestanten* did not retain anything of the historic faith of the Reformation except an anti-Catholic attitude."

3. Quotations in text in this chapter are all from J. Hertz 1977 unless otherwise noted.

4. Lange, born in 1826 in Krampfen in Prussia, had been teaching in Hamburg since 1848. The town permitted him to set up his own boys' school on December 15, 1850. See Bertheau 1912, p. 1881, for brief details.

5. Albisetti (1983) provides a detailed discussion of German secondary schools during the nineteenth century. He notes (p. 19) that according to von Humboldt mixing practical with classical training produces "neither wholly developed human beings nor fully integrated members of the separate classes."

6. Pyenson (1983, p. 4) quotes George Steiner on the social purpose of schooling in dead languages: "Power relations, first courtly and aristocratic, then bourgeois and bureaucratic, underwrote the syllabus of classic culture and made of its transmission a deliberate process."

7. Albisetti 1983, pp. 62–66. Pyenson (1983, p. 25) discusses the creation in 1859 of two "orders of Realschulen, the first order teaching Latin but not Greek and the second order substituting sciences and modern languages for classics. Graduates of a nine-year course in the Realschulen of the first order—known later as *Realgymnasien*—enjoyed substantially the same privileges in the governmental hierarchy as Abiturienten from the Gymnasien, but Gymnasien in Prussia and in most other German states retained a monopoly on sending students to the university."

8. The director of the *Gewerbeschule,* Jessen, felt that Heins had a superior aptitude for mathematics and should study it intensely. But Heins told his mother, "I should not like that, mathematics is such an abstract science in which one must immerse himself completely, and I should like so much to be involved with people." Hertz was rather gregarious, but his dislike of pure abstraction arose in equal measure from his powerful attachment to physical manipulation and construction.

9. Only twelve out of twenty-three passed (Kelter 1912, p. 187). Hoche was, and was seen to be, a representative of the new Prussian hegemony by Hamburg schoolmasters. They did not like him—"Schliessen Sie die Fenster, Meyer; draussen kommandiert ein Preusse" (Kelter 1912, p. 186)—and he intended to demonstrate that their teaching had been inadequate. The student failures fitted this assessment, and they caused a "scandal" in Hamburg, with Hertz right in the middle as one of the few successful candidates.

10. See Kelter 1912, pp. 183–93. Hoche directed the Johanneum from 1874 to 1888 (Bertheau 1912, p. 82).

11. He had even begun to study Arabic under the tutelage of a Professor Redskob, whose enthu-

siasm for Heinrich was unlimited, stimulated no doubt by the great joy of finding anyone, much less one as intelligent as Heinrich, who showed an interest.

12. Anna Elisabeth wrote, "Their father gladly gave [the children] the freedom of which he had been deprived as a child, and I certainly did not begrudge it, but sometimes I should have liked to have been far away."

13. Not a few professors in the natural sciences were opposed to accepting *Ungebildete* into their precincts. See Pyenson 1983, chap. 4. Until 1870 the *Gymnasium* was the only route to the philosophical faculty at the universities, and hence to a career in natural science. Consequently the division between the *Gebildete* and the *Ungebildete* also embraced, until that date, the division between the scientist and the engineer. Once, however, the first-class, or semiclassical, *Realschulen* could also send students to philosophy, the barriers between the two careers, both formal and social, were no longer quite so firm. It was consequently possible for Heinrich to conceive that starting out in engineering would not preclude his turning to physics, provided only that he obtain an appropriate *Abitur* which he did.

14. Which was also read by Oliver Heaviside early in his career.

15. On June 4 he had written his parents that "I have not yet said goodbye to mathematics, yet I should first like to get a general impression of experimental physics, and a better one than is given in the course on experimental physics."

16. See Cahan 1985 and Jungnickel and McCormmach 1986 for details. Friedrich Kohlrausch—the son of Weber's collaborator, Rudolf—to take but one from a myriad of examples, first used an offer from Zurich in 1870 to obtain more from Göttingen, and then used Göttingen's very nice response to get a better deal from Zurich, where he went, but only for a year, preferring to return to Germany, where he felt so much more at home than among the alien Swiss, who rather objected to the establishment of the Imperial German Reich. See Cahan 1989b for details. Although Kohlrausch was certainly no Zöllner, nevertheless he always felt that "a German is preferable to a Jew" (Cahan 1989b), and he was not overly enchanted with Catholics. After his father's death in 1858 Kohlrausch's education was guided by Weber, the very man whose work embodied for Zöllner the purity and perfection of a truly German science, and by Wilhelm Beetz (who told Hertz two decades later that he could find a laboratory almost anywhere).

17. Cahan (1989a, p. 129) translates the following remark that Kohlrausch made in 1900: "Measuring nature is one of the characteristic activities of our age. Without the [measuring of nature] the progress made during the last century in the natural sciences and technology would not have been possible."

18. Nor was Helmholtz deeply enamored of sensitive instruments designed for practical ends. Visiting Britain in 1884, he wrote his wife that he had "an impression that Sir William [Thomson] might do better than apply his eminent sagacity to industrial undertakings; his instruments appear to me too subtle to be put into the hands of uninstructed workmen and officials, and those invented by Siemens and Hefner v. Alteneck [less sensitive but more robust] seem much better adapted for the purpose" (Koenigsberger 1965, p. 349).

It is hardly that Helmholtz was uninterested in, or uninvolved with, accurate instrumentation. Far from it—he developed new or more accurate devices in several areas, including galvanometry itself. But he was not personally gripped, as William Thomson was, by measurement and industrial economics. By 1887, when the immense Physikalish-Technische Reichsanstalt was taking shape under his direction, Helmholtz had however forged a precision industrial research and testing laboratory (though he was replaced after his death in 1894 by that preeminent apostle of measurement Rudolf Kohlrausch). Nevertheless, Helmholtz's involvement with the Reichsanstalt does not reflect a direct concern with exact measurement but rather involves a complex of motives, including a hoped-for release from teaching, his close friendship with Werner Siemens, and his intense German patriotism, as well as a sense that he was unlikely to produce much more fundamental research. See Cahan 1989b, pp. 67–68 and passim, for a discussion of the Reichsanstalt; Cahan remarks that "Helmholtz's close ties to the court, his military background and connections, his mastery of science, his experience in directing a physical institute, his sympathetic understanding of the importance of science for an industrializing Germany, and his unmatched prestige abroad made him the best and only candidate for the Reichsanstalt presidency." Moreover, Helmholtz treated the Reichsanstalt rather like an inflated Berlin Physics Institute, but with indifferent success since, Cahan writes, despite the research spirit "under [Helmholtz's] administration the [Scientific Section] spent most of its time creating physical standards, instruments, and measuring methods" (1989b, p. 109). The Reichsanstalt was in part at least an attempt to synthesize the measuring goals of the Weberean tradition with the very different aim of physical discovery epitomized by Helmholtz's Physics Institute, a synthesis made pressing and desirable by the rapid development of science-based industry in Germany.

19. He remarked in his autobiography that "it was in Berlin that my scientific horizon widened considerably under the guidance of Hermann von Helmholtz and Gustav Kirchhoff, whose pupils had every opportunity to follow their pioneering activities, known and watched all over the world" (Planck 1949, p. 15).

20. "Helmholtz," he wrote concerning his doctoral dissertation on applying entropy to irreversible processes, "probably did not even read my paper at all. Kirchhoff expressly disapproved of its contents" (Planck 1949, p. 19). Nevertheless, Berlin—which certainly means Helmholtz—would never have appointed him had his work not achieved a considerable measure of respect by the late 1880s. But respect does not always carry with it enthusiastic acceptance, and Helmholtz's early indifference would not likely have turned by itself into the effusive welcome that brought Planck to Berlin.

21. Helmholtz's, Kundt's, and Bezold's support for Planck's membership in the Prussian Academy of Sciences rested on their approval of his ability to derive results in thermodynamics "without having to rely on the hypotheses about molecular motions" (Jungnickel and McCormmach 1986, vol. 2: 53–54).

Chapter 5

1. We will examine in the next section Hertz's revealing efforts to attack the problem; here we will concentrate on the impress that the Berlin experience had upon Hertz as a fledgling physicist.

2. For his experiment, see sec. 5.2 for details.

3. Clearly this kind of experiment could never effect any kind of closure about the issue of electric mass except among those who were already doubtful about its existence.

4. Since neither Hertz nor Helmholtz left clear evidence on the point, we cannot be certain that either of them saw the experiment in this light. However, there would be hardly any point at all in having Hertz perform the experiment if the Weberean claim to invulnerability was entirely persuasive, for then it would make no difference at all how small an upper bound Hertz could establish. For the experiment to have any importance Helmholtz had to believe that a sufficiently small upper bound could be used convincingly.

5. An "extra-current" is the reverse current that is generated in a circuit by electromagnetic induction when the current in it changes. Hertz's experiments first establish a steady current and then open the circuit. Ignoring capacitance, the current would at once fall to zero were it not for self-induction, which has the effect of causing it to decay exponentially instead:

$$ir = Pdi/dt \Rightarrow i = i_0^{-rt/P}$$

6. Since the inductors sit in diagonally opposite branches of the bridge, they will produce induction currents in the *same* direction through the connecting wire that contains the galvanometer.

7. All Hertz had to do was to adjust the resistance in the A branch until no deflection shows on G.

8. Since i is $i_0 e^{-rt/P}$, the flow J is just ri_0/P. Hertz's experiments used different initial currents i_0.

but the measured flows can be adjusted to some standard i_0 simply by multiplication with an appropriate factor (i_s/i_0 for $i_0 < i_s$ or i_0/i_s for $i_0 > i_s$).

9. The picture of the trough in fig. 16 is misleading because it shows the ballistic galvanometer and the battery connected in the same circuit. In practice the trough B in fig. 15 has the battery connected across it as shown, and the bridge is connected across the points where the galvanometer appears in fig. 16. Vice versa, in trough C (fig. 15) the galvanometer is connected as depicted (fig. 16), and where the battery appears, the bridge is in fact connected.

10. The distinction drawn here between an instrumental mutation that corrects *random* errors and one that corrects *systematic* errors, is developed in Galison 1987, pp. 69–72.

11. The alternation is not *in itself* of any electrodynamic significance at all: it merely serves to eliminate background noise.

12. The errors Hertz estimated in calculating and measuring the inductances translate into an error of about 1 part in 25 for the mass upper bound, which is almost exactly the same as the error that arises from the flows alone. See the formula given in the next note, where the fractional errors in the flows are about 1 part in 50.

13. In the new device the ratio of the inductances in the two types of experiments (currents in the same and currents in opposite directions) drops from about 200 to 1 to only about 5 to 1. To appreciate Hertz's error claims, we must consider what the *maximum* value for the expression $P(JP'/J'P - 1)$ can be, since this determimes the upper limit on m. To do so we add positive errors to the numerator and subtract them from the denominator, obtaining

$$m < \frac{JP'(P + \Delta P)}{J'P} \frac{1 + (\Delta J/J)}{1 + (\Delta J'/J')} \frac{1 + (\Delta P'/P')}{1 + (\Delta P/P)}$$

The term $JP'(P + \Delta P)/J'P$ remains essentially the same in all of the experiments, so that the upper limit of m depends for the most part on the second fraction. In it, the terms that contain the flows also remain nearly the same throughout the experiments since Hertz used more or less the same magnitudes of primary currents throughout. Consequently, the differences between the accuracies of the several experiments depend primarily upon the fractional errors in the inductances, and these are vastly reduced in the new arrangement.

14. Hertz easily confirmed the error of about 3% simply by calculating the deviation from the mean flow for the experiments with the large inductance (where mass effects can be ignored in any case) which was very nearly 3.4%.

15. The coefficient m in the induction equations is equal to $l\rho/q\lambda$, where in a unit volume l is the conductor's length, q the cross section, ρ to the actual mass of a unit of electricity, and λ the number of units of electricity. Suppose one can have a pair of conductors whose conductivities are in the ratios of their electric densities, i.e., $\sigma_1/\sigma_2 = \lambda_1/\lambda_2$. Then their mass coefficients m will be reciprocally as their conductivities, in which case $(1/\sigma)di/dt$ will appear in the induction equations. But in Hertz's experiments the conductivities are kept the same, so that if this relationship existed between the conductors in the branches of the bridge, then whatever parts of the flows that are due to mass effects would be the same in all branches and so would cancel one another. Therefore, this relationship must not exist for Hertz's experiments to have any significance at all, which requires that the electric densities in the conductors must all be approximately the same. In August of 1881 Hertz carried out an improved, and very different, experiment that sought to detect the Coriolis force that would exist if the current had mass and the plate bearing it were rotated. This experiment avoided the limitation on conductivities since it did not rely at all on induction. It is quite possible that the arrangement was suggested by Edwin Hall's recent discovery of magnetic deflection (on which see Buchwald 1985a).

Chapter 6

1. These calculations involve the inductance of a doubly wound spiral.

2. This and other Hertz manuscripts are now held at the Science Museum in London. They will

be referred to numerically by item and page, e.g., SM 245, p. 12. The introductory pages of this first manuscript are translated in O'Hara and Pricha 1987. All of Hertz's manuscripts, though not his letters to foreigners, are written in the German script.

3. The introduction was clearly drawn up last because he refers in it to results of the analysis, so that it does not represent what Hertz started with. This is important because Hertz had to learn, probably by attempting to solve the problems he was posing, how Helmholtz's potential functions in combination with dielectric polarization. This was hardly an easy matter, and we shall also see that in certain respects Hertz's solution was highly problematic.

4. O'Hara and Pricha (1987, p. 127) state that the body of the manuscript "contains no calculations relating to point 1 and, in fact, most of the mathematical work adjoined relates to point 3." In fact there are, we shall see, extensive calculations concerning point 1, through they are certainly hidden in a morass of detail. It is, however, correct that about half the manuscript concerns point 3, though we shall also see that Hertz's work here is much more straightforward than for points 1 and 2 since the structure of the problem was considerably clearer (though the conceptual issues are, if anything, more complicated).

5. This is a bit complicated. The potential function in question here depends upon the polarization and the current, which means that there must be other, spatially dependent forces besides the time-dependent one that Hertz chose to examine, just as there must presumably be for the interaction between conduction charge and current.

6. The following passage in Helmholtz 1873c may have been particularly influential: "It is well known that a closed ring magnet or a solenoid of circular currents that corresponds to it does not, according to Ampère's law, act externally. According to the potential law it does not act on closed currents but on the ends of unclosed currents. If one suspends a plane, circular Franklin slab so that it can rotate about its vertical diameter, which coincides with the vertical diameter of the ring, and connects its plates with the ends of the ring's conducting wire, then the charge of the Franklin slab from the ring, which will in this case be oscillatory, will according to the potential law endeavor to make the plane of the slab parallel to that of the ring, whereas according to the Ampère law it would be without effect" (p. 700).

7. Here the primary and secondary currents, inductances, capacitances, and resistances are, respectively, given by (I, p, c, w) and (i, P, C, W).

8. The only difference between Hertz's and Schiller's equations for coupled coils are that Hertz allowed both of the coils to have capacitance, whereas Schiller ignored the capacitance of the inducer. Nevertheless, the structure of the solution for the impulse that will be delivered to a ballistic galvanometer is the same in both cases, and it is extremely likely that Hertz simply generalized Schiller's approach to probe for higher accuracy.

9. Schiller's procedure was a simple exercise in algebra. Generalized by Hertz's equations, it runs as follows. Take Hertz's first equation and multiply it by idt to form a first relation; multiply it by Idt to form a second. Take Hertz's second equation and multiply it by Idt, giving a third equation. Multiply the second Hertz equation by Idt/c, which is a voltage (call it $d\omega$) for a fourth relation; multiply it by idt/C, which is a voltage $d\omega'$, for a fifth. The trick to obtaining the impulses is to recall that all of the variables vanish after infinite time and that only the inducing current is nonzero at the beginning; this permits partial integrations that will reduce the number of different integrals to five, and so the system can be completely solved. This is what Schiller and, after him, Hertz did.

10. Of course, polarization charge occurs only on the outer surfaces of the dielectric, but polarization proper, and so polarization current, exists throughout it.

11. The experimental device would sit within the core of one of Hertz's double-wound spirals. One strand of the spiral would act as primary, the other as secondary, and the force on the element would arise from its interaction with the currents in both coils.

12. The computation does not appear explicitly at this point in the manuscript, but the elements that went into it had already been used by Hertz to calculate the mutual induction between the

primary and the secondary in his double-wound spiral, which he had undertaken for the first prize competition. There, as in Hertz's present experiment, one must compute the effect on a current element of a closed coil. Since, however, in the case of the spiral both circuits are closed, the unique properties of the Helmholtz potential are obliterated by the double integration. Here one has instead a single integration, which leaves these effects intact.

13. One might think that in these experiments Hertz felt that he was testing the Helmholtz potential itself. He might indeed have done just that. But he did not. The point of the experiments was entirely explicit: to examine whether dielectric elements can have the same effects as conducting elements. To that end he intended to look for an action on the former that he gives no indication at all of doubting exists on the latter. In the absence of the specific Helmholtz actions there was no point in examining the highly problematic dielectric element.

14. It might help to consider the following physical analogy. Imagine a ring around whose circumference small bar magnets are arrayed in the plane of the ring and along its radii. Place a large bar magnet along the ring's axis. Nothing happens, because the symmetry cancels the forces. Now place the large magnet off-axis, but still normal to the ring's plane. Those circumferential bar magnets that do not point at the large magnet will be torqued by it, and the torques will vary over the ring because the distances to the large magnet do. The ring itself will twist in some way.

15. The proportionality easily follows from the continuity equation: the changing charge density on the outer plates is proportional to the divergence of i and so varies directly with its magnitude. The corresponding induced currents on the inner apparatus vary directly with the surrounding charge density because they arise from electrostatic induction.

16. A. Grove's cell produced about 2 V. A contemporary Daniell's cell, which Hertz also used, produced about 1 V. See R. Weber 1902, p. 341.

17. Hertz had deduced the following general relationships between the primary and secondary circuits:

$$\int_0^\infty \frac{i^2}{2} = \frac{I_0^2}{2} \frac{\left[W^2(WC + wc) + \frac{c}{C}(wP + Wp)\right]\Pi^2}{Ww\left\{CW\left[pW + Pw + \frac{1}{C}\left(\frac{\Pi^2}{w} + \frac{P^2}{W}\right)\right]\right\} + cw\left[pW + Pw + \frac{1}{c}\left(\frac{\Pi^2}{W} + \frac{p^2}{w}\right)\right] - 2(Pp - \Pi^2)}$$

$$\Pi(wc + WC)\int_0^\infty Iidt = w(CP - cp)\int_0^\infty i^2dt - \frac{\Pi^2cI_0^2}{2}$$

In computing the capacities Hertz considered only the effects between wires in a given strand of the spiral, ignoring the capacitance between the two strands. He was not altogether confident of the result, but he was at this point interested only in finding out whether the effect would be large enough to be detected.

18. This occurs primarily because the ratio of the polarization current to the current in the coil is $\kappa\varepsilon bR_1/R_2$. The polarization current, it seems, depends directly on the resistivity of the coil, which is why it and any effects due to it are so small. Consequently, any attempt in *this* plan directly to detect the electromotive action of the polarization current is essentially hopeless. The only way to improve the situation would be somehow to increase the frequency of the stimulating oscillation by many orders of magnitude. In these experiments, where coils with many windings, and so large inductances, are used, the frequency is comparatively small.

19. Hertz's solution is approximate, based on the requirement that the ratio $k/\sqrt{v}$ is extremely small, where v is $R_1/2\pi\varepsilon boa^2$ and k is $\kappa R_2/2\pi A^2$. Since κ is extremely small and A is on the order of 10^{10}, this is a rather good approximation. It affects, however, only the extra terms—it has nothing to do with the solution for the current in the coil in the absence of any dielectric at all. That solution is purely damped because Hertz has completely ignored the capacitance of the coil in

setting up Ohm's law. Thus, as far as Hertz was concerned, the stimulator's *oscillations* had no direct significance at all for electromotive interactions. This was also substantially true for the electrodynamic experiments, where the existence of oscillations in the primary and secondary coils of the stimulator translated analytically only into a doubling of the possible effect (compare to what would occur if, say, no oscillations at all occurred in the secondary as a result of rapid damping via a spark).

20. He did nothing similar for the dielectric cylinder because there the currents were certainly *not* closed, which meant that Hertz would have had to use the general Helmholtz potential to proceed in this way, vastly increasing the difficulty of the analysis.

21. This work is also significant for the development of Hertz's analytical technique, for here he introduced a method that he frequently deployed in later years: he defined a quantity that behaves like a potential in respect to another quantity in the original equation. In and of itself this new variable, and its descendants a half-decade later, had no direct physical significance, but from it physical implications could be drawn with ease that would otherwise remain hidden behind intricate expressions.

22. The Michelson–Morley experiment, for example, caused exceedingly little stir among Maxwellians. This is not to say that there was *no* interest in such questions. There was, but it remained far from the center of research interest until the end of the decade, when the discipline was reorganized about issues surrounding Larmor's and Lorentz's electron. See Buchwald 1985a and 1988.

23. Since 868 of Hertz's units equal 1/100,000 of a Daniell's cell, a single Hertz unit is about 9.3×10^{-6} V.

24. Hertz's proposal is particularly interesting in historical retrospect because it depends entirely upon the extra term in Helmholtz's expression for the electromotive force produced by motion. This term, which arises from the distinction between total and convective derivatives, can be quite different in form in field theory, depending upon one's assumptions concerning the relative velocity of ether, which is the field site, and matter, which affects the field. It is worthwhile noting a signal characteristic of field theory before the electron here.

In field theory, a field is a particular state of a given element of the ether. The magnetic field certainly represents a different condition from the electric field, but both are states of the ether. Now if either of these states changes for a given element of the ether, then the other state will arise, again for that particular element. Suppose first of all that every ether element remains permanently in contact with the same neighboring elements. Then changes can occur in an element only as a result of processes in situ, though the processes are transmitted to it through its neighbors. But suppose instead that some particular element can move, and thus its neighbors vary depending upon its position, whereas other ether elements cannot move. Suppose further that the magnetic states of the fixed elements remain constant over time but that the states vary from element to element.

As our particular element swims from neighbor to neighbor, it moves into regions with different magnetic states. Its magnetic state depends upon that of its neighbors, and so as it moves, the state at its surface continuously changes. The electric state of the surface must also change in consequence, and both kinds of change occur over time throughout the element as its parts successively encounter the environs of new neighbors. In field theory, then, everything depends completely upon the motion of the elements of the ether with respect to their contiguous neighbors. The moving object is merely an occasion to introduce moving ether. And here, then, is a question for research since it is open to debate whether the velocity coordinates of an ether element that momentarily coincides with the moving object are the same as those of the object with respect to contiguous ether elements.

Chapter 7

1. He did, however, devote a brief section to showing that the Helmholtz term merely generated the same effect as a static electric potential.

2. The "views" that Hertz had in mind were probably Jochmann's himself, since something like this assumption is present in his paper, at least by implication (1864, p. 521). Jochmann remarked that he intended to neglect "the induction that takes place between the several parts of the conductor" as being of second order. He did not, however, mention an infinite series. In his equations the approximation permitted him to neglect entirely the magnetic field of the induced current in computing the inducing electromotive force. "This simplifying assumption," he remarked in a note, "has in fact been tacitly made in all previous investigations in connection with the subject." In the *Treatise* (vol. 2, chap. 12) Maxwell did not consider self-induction, though he did show that (in this approximation) a magnet moving near a conductor produces a current in it whose own magnetic effect can be represented by an infinite trail of moving images of the inducing magnet. This was Hertz's first thorough contact with the *Treatise*, and it is not likely that he would have missed the major implication of Maxwell's method: that one should seek to abstract from the particular interacting objects in order to determine their effects, which stand apart from them. Here the trail of magnetic images entirely replaces an induced current. Neither Helmholtz nor Weber would naturally have thought about interactions in such a manner, and Hertz would probably have found it unusual.

3. Namely, that the spherical shell not have a rotational speed that is greater than a certain quantity that depends on the object's resistivity, radius, and thickness. It is not, however, impossible to exceed the limitation (a copper shell 50 mm in radius and 2 mm thick rotating faster than 87 revolutions per second will do so), and if it fails, "it is no longer allowable to regard the phenomenon as a series of successive inductions, since each one would [in that case] be larger than the preceding one." Here, then, the series representation must not be thought of as a fundamental representation but is rather a convenient tool for calculation under appropriate circumstances. In 1884, when Hertz wished to insist on the general significance of a series that is rather similar in physical characteristics to this one, he was careful to argue for its general convergence.

4. For the degenerate case of the disk (considered to be a flattened sphere) Hertz specified: "The plate is made of copper (thus $\kappa = 227{,}600$) and has a thickness 2 mm. (thus $k = 113{,}500$). The distance of the [magnetic] pole from it is 20 mm. The values of Ψ [the current function] marked give absolute measure when the strength of the pole is 13,700 mm $^{3/2}$mgr $^{1/2}$/sec." (Hertz 1880b, p. 114).

5. Helmholtz had already given ample evidence of this, and the behavior of his dissertation examiners certainly reinforced Hertz's consciousness of having been singled out. Even Kirchhoff, with whom Hertz later had a small run-in (see chap. 8), let him know how exceptional he was. At a large dance party one Saturday—held, it is significant to note, at the home of a cousin of Hertz's father (J. Hertz 1977, p. 123) Kirchhoff "came over to me right away, saying that he was pleased to meet me in another setting, and frequently showed me further kindness during the evening."

6. Though, as was usual with Hertz, both applications of *theory* are closely linked to experiments that he either performed or that could be done.

7. For the technical details of Hertz's approach to the problem through continuity, see appendix 9.

8. Today we would follow Hertz, not Weber, because we do not use the Fechner hypothesis. Since the discovery of the electron, the current in ordinary metallic bodies has been viewed as the flow of one kind of particle, so that flow inhomogeneities do translate directly into current inhomogeneities, which means that the continuity equation suffices for analyzing the problem. Though we would still think physically of interacting particles, our mathematics would bear no mark of that image.

9. One cannot be certain about this because the diary (J. Hertz 1977, p. 141) does not specify the paper he read. However, in late January he read a different paper—the one on elasticity—so

that it seems likely that the paper read in December concerned the relatively moving conductors. It is also probable that Hertz had done the experiments for the paper the previous summer, when he still had time, since his letter of January 25 remarks: "Recently Helmholtz inquired about the results of the work I did last summer; he asked whether I had published anything about it and made me tell him about the experiments. I have also discussed this project with some of my other colleagues who were also interested and who thought that I should by all means publish something on it. Unfortunately I did not write everything down right away, and memory is deceptive, so that I would have to do all the experiments over again, and I have no desire to do that now" (J. Hertz 1977, p. 143). It seems that he decided in the end that it was not necessary "to do all the experiments over again," but that the obvious interest of Helmholtz and his "colleagues' convinced him to go ahead with publication.

Chapter 8

1. In the manuscript (SM 250) the passage in question has only a single marginal comment, but it is boxed by a line. The printed version is considerably different from the manuscript and was probably changed at Kirchhoff's behest.

2. This is again an approximation that is justified by Hertz's assumption that the deformation is not large. It is in fact essentially the same approximation that, e.g., permitted Newton to represent the distance between a plane and a sphere pressed to it by a parabola and requires discarding third- and higher-order derivatives in a Taylor's series expansion for the distance.

3. There is more direct evidence for Kirchhoff's discomfort with Hertz's new system. In the manuscript Hertz had left out the distance α by which the origins move together from his expression for the difference in displacements "in order to avoid being long-winded"; he wished instead to incorporate it as a common element into the displacements proper. Kirchhoff would have none of that and stuck it back in, thereby bringing the (disturbing) novelty of Hertz's coordinate system to the fore.

4. The calculation was made under the approximation that this period is vastly larger than the time taken for waves to propagate across the bodies, so that the objects are always effectively in static equilibrium. (In 1906 Rayleigh extended the theory to the first few moments of a collision in which the limitation fails, concluding that even under these conditions Hertz's theory still works quite well.) Hertz offered a vivid example: "For two steel spheres as large as the earth, impinging with an initial velocity of 10 mm/sec, the duration of contact would be nearly 27 hours" (Hertz 1881c).

5. Hertz's calculations lead to the result that "the dimensions of the surface of pressure increased as the cube root of the pressure," which he found to differ from the experimentally produced mean using a glass lens pressed to a glass plate by 0.022 for a theoretical value of 2.685, i.e., by approximately 1%. The experiment was quite clever. Hertz took a horizontal lever and hung weights from it. Near the fulcrum he fastened one of the two bodies and covered the other one with lampblack; the lever pressed the two together. "If the experiment succeeded," he wrote, "the lampblack was not rubbed away but only squeezed flat; in transmitted light the places of action of the pressure could hardly be detected; but in reflected light they showed as small brilliant circles or ellipses, which could be measured fairly accurately by the microscope."

6. The most recent measurement methods that Hertz cited involved the weight necessary to drive a certain pointed object a given depth into the object to be measured. This, Hertz noted, means that the original state of the object is changed considerably by the measurement process.

7. There is an implicit assumption here, albeit one that evidently remains in modern accounts. Hertz's procedure yields in the end a *continuous* surface each point of which marks an elastic limit for one or more of the principal stresses. It is an assumption that such a surface even exists, because it might conceivably be the case that the principal stresses that define the elastic limits consist of discrete elements. Perhaps, that is, the elastic limit does not change continuously with the principal stresses. Nevertheless, elasticians after Hertz do use such a surface, or rather surfaces (because the

surface may have to be re-created after each new set, supposing that the object returns to linear elasticity). It is known as the "yield surface," and if constructed in Hertz's fashion it is today said to be built in Haigh-Westergaard stress-space. Since yield does not depend on hydrostatic stress, the yield surfaces are generalized cylinders. It is sometimes possible to define a family of yield surfaces by means of a so-called loading function that depends on plastic strain as well as upon stress.

8. There is a *standard* involved in the procedure, a method of calibration, in that the surface of pressure is required to be circular. However, Hertz notes, hardness "could have been defined by assuming for [the surface] any definite ellipticity," as long as the *same* ellipticity is chosen for all materials.

9. Timoshenko (1983, p. 349) remarks that "Hertz's method did not find acceptance, since in ductile materials it is very difficult to find at what load permanent set begins," though one person at least (August Föppl) attempted to use Hertz's method in 1897 (almost certainly because the famed, and now dead, Hertz had developed it).

Chapter 9

1. Hertz's work is perhaps unusual in this respect. Fresnel's investigations in optics, more than a half-century before, show a considerably different pattern, because, unlike Hertz, Fresnel knew that he faced strong resistance to his claims. Hertz had nothing similar to worry about, because the only people who might have reacted with a great deal of antagonism to his work—Webereans—had by the mid-1880s been pushed to the margins.

2. He built a simple device based on a torsion-balanced glass rod that carried tissue paper saturated with calcium chloride.

3. If measurable evaporation occurs, then the system has not yet reached equilibrium, because when it does, the rate of condensation equals the rate of absorption.

4. Suppose, for example, that both surfaces drop to the low temperature T_1. Then, in effect, the high-temperature source simply feeds heat into a single-temperature system: surface 2 uses the heat to form saturated vapor at its (now low) temperature, which then condenses on surface 1 at the same temperature. Alternatively, though Hertz did not initially consider the possibility, the surface temperatures might lie halfway between T_1 and T_2—all else being equal—since then surface 1 would transfer the heat produced by condensation to its reservoir at exactly the rate that surface 2 sucks it in from its source to produce vapor. Such a system differs from one with a single liquid surface in that it has a separate, higher-temperature heat reservoir that can be drawn upon to form vapor. As the vapor forms, surface 2 drops while surface 1 rises. This is a sort of evaporation engine.

5. To see this, suppose that the surfaces initially share a temperature that is intermediate between the source temperatures. If T_2 is large enough, then the rate at which heat flows to surface 2 might be greater than the finite evaporation rate can absorb. And if T_1 is low enough, it may suck heat from its surface faster than heat can be provided to the surface by the condensing vapor generated at surface 2. As a result the temperature of surface 2 will increase toward T_2, and that of surface 1 will decrease toward T_1, until the heat flows balance (by decreasing the fluxes between the surfaces and their respective reservoirs): the system might accordingly be unstable at a common surface temperature. To observe such an effect clearly requires sufficiently high and low temperature sources, but how high and how low cannot be calculated beforehand.

6. He assumed he knew the upper limit on the pressure because, he initially believed, the common surface temperature that would be achieved *if* the evaporation rate were unlimited would be close to that of the low-temperature reservoir. If the rate had a limit, the surface temperatures could differ, but, Hertz evidently thought, the vapor would stay close in temperature to that of the low (condensing) source.

7. The surface pressure must vary with the evaporation rate because it is due predominantly to the "kick" as streaming vapor leaves the surface. The pressure measured in the neck should not

vary, because it is determined by the state of the vapor, which in turn is governed by the low-temperature surface. Significantly, Hertz did not report measuring the pressure on the evaporating surface at several temperatures to see whether it did indeed vary with the rate of evaporation, as it should, but as the pressure in the neck—whose magnitude he did not describe—had *not*. He simply interpreted the existence of the surface pressure as warranting his conclusion that the rapidity of the vapor's egress precluded using the apparatus as he had hoped—not as evidence that the state of the vapor differed from his initial supposition.

8. In the evaporation apparatus Hertz correspondingly measured the temperature of the reservoir that surrounds the manometer, and not the temperature within the mercury itself. Because, he believed, the heat was supplied by convection along the walls of the containing vessel (i.e., at the boundary of the liquid), measurements taken in the mercury could not be reliably used.

9. "I am preparing the paper on evaporation for publication, as much as still remains to be done; I find this work so wearisome that I am almost glad that I need not keep it to its original great length" (J. Hertz 1977, p. 165). The "original great length" no doubt referred to a much more detailed version that Hertz was preparing when his third set of experiments unexpectedly turned negative.

Chapter 10

1. Hertz was very careful not to hide any debts to Goldstein. He acknowledged him in the last paragraph of his first paper on discharge tubes, and the second one (1883c) remarks in its first note: "I was first induced to undertake these experiments by conversations which I had with Dr. E. Goldstein as to the nature of the cathode discharge, which he had so frequently investigated. My best thanks are due to Dr. Goldstein for the ready way in which he placed at my disposal his knowledge of the subject and of its literature while I was carrying out the experiments" (Hertz, 1883c, p. 224). Not everyone might have been so "ready" to share his craft knowledge as Goldstein, and Hertz was well aware of the fact.

2. Other remarks in this article as well as elsewhere make clear that the particles Goldstein referred to formed the residual gas in the tube. They were not, that is, components of the polar ether, which itself transmitted the rays that affected the particles in question.

3. Conduction, on the other hand, is a *state* of an object as a whole, so that it makes no sense to think one part of a conducting object *causes* the state in another part of the same object.

4. See Goldstein 1881b, pp. 266–67. In Britain William Crookes championed the notion that cathode rays involve the motion of electrically charged particles. However, the oft-repeated statement that the British understood cathode rays as moving particles, whereas the Germans took them to involve ether motions, requires careful qualification. With the exception of Crookes, few experimenters were primarily concerned with the rays that emanate directly from the *cathode* and fill the tube only at extremely low pressures. J. J. Thomson was no exception: though he did agree with Crookes that these rays mark the paths of moving, electrified particles, he was much more interested in the discharge phenomena that occur at more moderate pressures, when gas particles are still present in reasonable numbers, and where the direct rays from the cathode are confined to a small, dark space near it. In these circumstances both Thomson and Goldstein agreed that the glowing regions contain essentially *stationary* material particles. Goldstein believed that the particles transform an invisible (longitudinal) into a visible (transverse) ether motion and generate new, invisible rays; if there are no or very few gas particles, then the transformation simply does not occur until the walls of the tube are struck. Thomson's view was more intricate, in a typically Maxwellian way, in that for him the particles determine through their conductivity a discharge path. Moreover, no longitudinal waves can be involved, because such things do not exist in the Maxwellian scheme (see Buchwald 1985a, pp. 52–53, for a bit more detail, and the brief discussion in Heilbron 1964, pp. 59–68).

Schuster later noted that the whole subject of gaseous discharge was not considered to be very interesting in Britain: "a good deal of apathy was shewn, even in this country [Britain], with regard to the theoretical significance of [Goldstein's and Crookes's] experiment, while in Germany the opposition to the corpuscular view was almost universal" (Schuster 1911, pp. 55–56). Schuster attributed the German viewpoint to the notion that "a current of electricity was only a flow of ether." In Helmholtz's electrodynamics this is substantially true, since polarization is not in principle different from *charge* on conductors, and neither of them involves electric particles in the Weberean sense.

5. By 1880 others shared some form of Goldstein's understanding, as one can see from the following remarks of Eilhard Wiedemann, which are much more speculative than Helmholtz would probably have countenanced at the time: "The electricity produced by the machine, which we may imagine as free aether, is accumulated on the surface of the electrodes partly as free electricity, and there is prevented from passing into the surrounding gas by the mutual action between it and the molecules of the metal; and a transference can only take place when its density has reached a sufficient magnitude. At the same time the electricity produces a dielectric polarization in the surrounding medium in such a manner that the aether envelopes of the separate gas molecules become deformed, and, in consequence of the rotation of the molecules on their axes, maintain a definite position. If a discharge takes place, the sudden change of the dielectric polarization thereby produced propagates itself from the electrode through the aether envelopes of the gas molecules, and thereby puts them into vibration. At the same time a transference of free electricity from the electrode may no doubt take place from molecule to molecule" (Wiedemann 1880, p. 419).

6. Hertz nicely pinpointed Goldstein's reasoning on this point: "The position and development of each stria of the glow-light depend upon the preceding stria (in the direction of the cathode): upon this is founded the legitimate view that from the cathode outwards there must be a time-development from one stria to the next. But such a development is not conceivable, if the discharge in all parts persists continuously" (Hertz 1883c, p. 237).

7. Context is critical. To those who did not entirely participate in Hertz's Berlin-centered outlook his experiments had a considerably different character from the one that Hertz intended. To take just one example, in 1911 Arthur Schuster, who had been working with tube processes in the 1880s, recalled that "Heinrich Hertz conducted experiments which were intended to prove that kathode rays produced no magnetic effects, and therefore could not form part of the main process of conduction; but these experiments, as I pointed out at the time, did not support the interpretation which Hertz gave them" (Schuster 1911, p. 56). The experiments that Schuster was referring to were not at all intended to prove "that kathode rays produced no magnetic effects." Nor did Hertz draw that conclusion from them. And yet he did claim to have shown (in distinct experiments), first that cathode rays do not affect magnets through any *nonelectromagnetic* action and, second, that they do not generally follow the current between cathode and anode. These claims are very different from the ones that Schuster and later reinterpreters attributed to him, although Hertz did believe, and felt he could support, the separate claim that cathode rays share no properties with electric currents. (Note that Schuster in 1911 attributed Hertz's claim that the rays "could not form part of the main process of conduction" to a demonstration that they "produced no magnetic effects." We shall see that Hertz never made any such argument, but that he performed an experiment designed specifically to disconnect the rays from "the main process of conduction.")

8. Because the rays begin and end seriatim down the tube, the charge does not, in Goldstein's view, transfer from the end of one ray to the beginning of the next one.

9. An intermittent current is distinguished from oscillations in remaining off for finite lengths of time. But the partial discharges that constitute it may rise at the same rate that they fall, or they may not do so.

10. In fact de la Rue and Müller were well aware of the virtues of a dynamometer here and had tried to construct an appropriate one "but it was found to be far too sluggish for our purpose."

Hertz's rhetorical critique was, like most such efforts, unfair, and he would have known that it was if he had read their paper carefully (which he directly cites).

Despite Hertz's denial, "the accepted theory of induction" does allow for the British scientists' claim provided that the discharge frequency is not extraordinarily large—say, on the order of 100,000 Hz or greater. In 1878, when de la Rue and Müller wrote up their results, Hittorf had not as yet published his battery experiments, which were interpreted as showing that the frequency had to be at least that large. Consequently, they almost certainly had in mind that it might be somewhat larger than could usually be detected audibly or by the mirror, but certainly not so high as, a year later, seemed to be necessary. Moreover, de la Rue himself later acquiesced in this view, and Hertz cited the paper in which he did so. In other words, Hertz criticized the British workers' theoretical competence on grounds that were only later provided. Hertz himself was probably well aware of this, but he built his critique anyway because he thought that they should have used a mirror to cross-check for these comparatively low frequency discharges: "[de la Rue's demonstration of intermittence] could only be carried out under special conditions, and these conditions appeared to be just those under which the rotating mirror would have proved discontinuity." Hertz, that is, would not grant that the frequency could be so high as to be undetectable by the mirror and yet detectable by the galvanometric method.

11. Hertz here used the word "continuous" to mean a current that has a nonzero, though not necessarily constant, value at all times.

12. A dynamometer becomes an ordinary galvanometer when the current through it is steady.

13. Hertz's result can be formally produced by driving the circuit with a harmonic electromotive force, but there is no evidence that Hertz produced the solution on paper. Under these circumstances the apparent resistance of the parallel capacitor–resistor branch will be $\sqrt{(1 + C^2\omega^2)}$, where C is the capacitance and ω the driving frequency.

14. As the only time that Hertz reasoned from a mechanical model it is worth considering in more detail. Appendix 12 provides details.

15. Goldstein's views on the nature of the rays were not altogether clear. He did, however, conceive that the discharge process begins with the production in the ether of "a certain state of labile equilibrium in the ordering of the ether's particles" (1881b, pp. 266–67), with a corresponding tension that reaches maxima and minima over certain loci. Discharge consists in the equilibration of that tension, resulting in the production of "a motion" in the ether at each surface of extreme tension. These motions constitute the cathode rays proper, which on striking whatever material particles remain in the tube excite them to produce a different kind of ether motion, namely, the transverse waves that constitute light. From Goldstein's perspective each of these rays, or motions between extreme tension surfaces, constitutes a species of open current—open because no charge is transferred between the end of one ray and the beginning of the next one. The rays are, as it were, ether currents that shoot out when the ether's tension equilibrates. Hertz sought to break apart Goldstein's understanding of the ray as an ether process from this particular model, and to do so he began by attacking its basis in the assumption that the discharge process must be punctual.

16. One has to take particular care to distinguish between kinds of interactions. To say that the interaction between the ray and the magnet is electromagnetic means that the interaction energy between them has precisely the same functional form as the interaction energy between a current and a magnet. To say that it is nonelectromagnetic means that the interaction function is different in form. Since physical processes are exhaustively described by potential functions combined with whatever restrictions govern the state variables that appear in them, there can be no distinctions between phenomena that have precisely the same functions and state variables under the same constraints.

17. That is, he had shown this supposing only, but essentially, that the rays are not Goldstein's open currents. If they were, Hertz's result would also foreclose an *electromagnetic* action between the termini of the rays as open currents and the magnet. (The detecting magnet will be closer to

some termini than to others since they presumably do not all end at the same point, i.e., at the anode, but just stop where they seem visibly to do so.) Since the electrodynamic potential actually requires such an action to exist, Hertz could have had no such thing in mind. Consequently, in order for Hertz's first experiments to have been convincing, Goldstein's original understanding, that the rays are open currents, had to be thrown out.

18. Here one has evidence beyond the deflection proper to connect the rays to something that is also deflected by magnets, namely, to a moving gas that sustains a discharge.

19. Hertz's argument precludes using an electromagnetic action (Hall) to explain an effect on something (the rays) for which there is no independent positive evidence concerning its electromagnetic character. This is to be contrasted with the point about the moving gaseous discharge, which can be connected to the rays because the latter occur in very similar circumstances.

20. It is once again essential to divest ourselves of views that became common later, namely, that the connection between cathode rays and currents is so tight as to be unbreakable (with the major historical question then reducing to the question of the nature of these currents). As far as Hertz was concerned, the *only* evidence for the rays-as-currents hypothesis was their deflection by magnets, for he had now carefully dismantled every other piece of evidence for the connection.

21. Hon (1987) nicely describes the apparatus and discusses the experiments, though he does not note the critical significance of Hertz's having placed the electrodes so close together, which, we shall see, was precisely what Jean Perrin did not do twelve years later (and why, therefore, Perrin's experiment would not have been unproblematic in 1883).

22. Since, as Hertz noted, the charge in the experimental space does vanish about an hour after the machine is shut off, it follows necessarily that a small current leak must maintain it during the machine's working. If Hertz had been a bit more artful than he was at this time, he might even have made the magnetic deflection into something more positive than troublesome by *requiring* its presence given the persistent electrometric deflection and its decay over time.

23. It is not, however, clear from Perrin's printed article (1895) that he was aware of Hertz's experiment, though he knew that "Goldstein, Hertz, and Lenard" held that the rays were "vibrations in the ether." Perrin wrote that "to my knowledge the electrification [of the rays] has not been shown," which he set out to do, but he nowhere mentioned that Hertz had claimed to have demonstrated precisely the opposite result. The article has no explicit references. Perrin would have tried to neutralize Hertz's assertions had he been aware of them.

24. Perrin himself finally referred to Hertz (albeit still without textual reference) in a 1905 rewriting of the ten-year-old experiment for inclusion in two volumes of essays edited by Abraham and Langevin on the burgeoning subject of "ions, electrons, corpuscles." He there remarked that "Hertz, looking for his part to manifest the electric and magnetic properties of the rays, supposing them to be charged, did not get any results in this sense and adopted the theory of oscillations" (Abraham and Langevin 1902, vol 2:558).

25. E.g., Heilbron (1964, p. 67) remarks that in Perrin's experiments "the electric charge transported by the rays was immediately demonstrated by collecting it in a metal cup inserted in the discharge tube." Heilbron's discussion of Geissler-tube work nevertheless remains admirably concise and clear.

26. J. J. Thomson, in his 1893 *Recent Researches,* was well aware that the current paths must be distinguished from the cathode (or "negative," as he called them) paths, as, we have seen, was George FitzGerald in 1896. Thomson, moreover, had at least perused Hertz's paper, since he refers to it. Yet on the very same two pages in which he refers to the paper and makes the point that the ray and current paths diverge, he also discusses Crookes's 1879 experiment. This showed that two cathode beams diverge from one another, which Thomson interpreted as their having "repelled each other" (Thomson [1893] 1968, pp. 122–23). He does not even mention Hertz's electric experiment, as though it were irrelevant to his own interests, which it was, because at this time Thomson was interested primarily in the discharge process itself, which he connected with molecular dissociation (see Buchwald 1985a, pp. 49–53; Feffer 1989).

27. G. P. Thomson remarks: "Since Hertz's work, fourteen years before, pumps had improved, though they were still incredibly primitive by modern standards. This improvement was largely due to the demands of the electric-lamp industry" (1966, p. 57). Other factors come into play at the low pressures that J. J. Thomson achieved, such as the much higher potential needed to effect discharge, so high that the tube easily shattered. This can be mitigated by increasing the tube's diameter, which Thomson may have realized (on this see Falconer 1985, sec. 3.3.7). In 1883 de la Rue and Müller themselves discovered the existence of a critical pressure below which the potential for discharge increases rapidly.

28. It becomes problematic, in this argument, as to why J. J. Thomson caught charge, because many of his rays were also prevented from reaching the ray catcher. We could argue that the residual gas present in Hertz's tube, but not in Thomson's, itself shielded the charge from detection. This might be a variant of George FitzGerald's explanation for Hertz's not having deflected the rays by electric means (see n. 32 below).

29. Though Hertz does not mention it, there is even an electric analogue of the Faraday effect, namely, the Kerr effect, in which electrically stressed glass becomes doubly refracting.

30. The original reads: "Es war hierbei zweifelhaft, ob die grosse elektrostatische Kraft, welcher das Rohr ausgesetzt war, im Inneren nicht compensiert würde durch eine daselbst eintretende elektrische Verteilung."

31. Discharge would fail because the neutralizing charge would have counteracted the electric action between the plates that rendered the gas conducting.

32. The first such remark on the issue, by FitGerald, did not, however, refer to ions. He argued: "Sufficient account does not seem to have been taken of the shielding action of the conducting gas surrounding the kathode ray" (1896, p. 441). FitzGerald was probably not thinking here of the later claim that Hertz's plates were coated by residual ions that neutralized the field between them. He had in mind, as he wrote, that the gas "surrounding the kathode ray" might itself be conducting and therefore act like a Faraday cage, preventing the ray from being deflected (though the two explanations are compatible with one another if the shielding region in FitzGerald's explanation does not make conducting contact with the plates). Hertz would have rejected this sort of criticism on the grounds that the gas (which he thought of as a mass in contact with the plates) could not have been conducting since no discharge occurred. In any event Maxwellian understanding of gaseous conduction had for some time considered it to involve ionic dissociation (though not necessarily the carriage of the main current by moving ions) in a process similar to, but not identical with, electrolysis. See Buchwald 1985a, chap. 4.3.

33. Helmholtz's wording in the Faraday Lecture (1881a) suggests a conceptual distinction between current and ionic transport; for example, "molecules of the electrolyte, charged with electricity, are carried by the current to the surface of the metal [electrode]" (p. 65). There is, Helmholtz insists, an identity between "electrical and chemical" motion. Now electric motion, in the sense of charge transfer, does translate through the continuity equation into current, but it nevertheless does not follow that the current *is* charge transport, even in electrolytes. It may simply be the case that under those circumstances, the current is accompanied by a chemical process that is de facto associated with charge transport. From this point of view electric resistance is still something that applies to the current, not to the moving ions; however, because the ionic motions are specified by the current, then they are also affected by resistance. Ionic motion, one might say, is slow because the resistance is high—the resistance is not high because the ionic motion is slow. Helmholtz, for example, writes that for certain electrolytes "this resistance [to current] may be very great, *and the motion of the ions very slow,* so slow indeed that we should need to allow it to go on for hundreds of years before we should be able to collect even traces of the products of decomposition" (p. 70; emphasis added).

34. One might argue that Hertz could have found something peculiar had he measured the voltage across the plates in his Daniell's cells experiment, in which discharge did not occur. The plates would have been coated with the residual ions in amounts that would always nearly equal

the charges placed upon them by the battery, and the voltage across them would, one might think, therefore have dropped considerably. J. J. Thomson was convinced that the coating explained the absence of the effect in Hertz's device by neutralizing the field. However, one must distinguish between the field neutralization and the voltage across the plates, because the voltage is maintained at a constant value by the battery. It would drop only if the region between the plates behaved like a conductor in the fullest sense, that is, if it carried a continuous current, which in Thomson's explanation it does not. His account makes the region between the plates (in Hertz's device) behave as though it were a conducting mass that is not in contact with them (since no current flows across), and such a thing will not affect the potential difference across the plates as long as the latter is maintained by the battery, though it will certainly shield the region from the field. The situation is akin to a conducting cube placed between, but not in contact with, the plates of a capacitor connected across a battery. The voltage between the plates remains the same whether the cube is there or not.

35. The dating here is a bit difficult because the evidence is scanty. From the published article we know that Hertz did not use his battery for the electric experiments. The battery was, however, working well as late as the end of November 1882. There are no further pertinent remarks until January 20, 1883, when he returned to work on "the Geissler tubes." On February 11 he was writing up the autumn's experiments and complaining about the corrosion the battery had acquired "during the months that I did not use it." By February 17 he had finished the paper. Since he would have used the battery if he could have, it is reasonably certain that Hertz had not performed the electric experiments by the end of November. And since the battery corroded during "months" of non-use, then January 20, when we know he was back working on the tubes, is the earliest date for his having begun the electric experiments. They were accordingly completed to his satisfaction in, at the very most, twenty days, and more likely within a week.

36. Normally a *Privatdozent* had no regular salary and had to survive entirely on student fees.

37. Jungnickel and McCormmach (1986, vol. 2:43–46) provide details of the ministerial consultations.

Chapter 11

1. He sought unsuccessfully for an interaction between magnetization and the electromotive potential of copper in water.

2. The paper has been most recently and informatively discussed by S. D'Agostino (1974). Although D'Agostino does not comment on the peculiar characteristics of Helmholtz's electrodynamics that inform Hertz's analysis and render it problematic, he does note that the account was limited to closed circuits. Hertz began from Helmholtz's theory, in which the vanishing of the vector potential's divergence (a necessary condition for Hertz) requires the vanishing of the current's divergence and hence its closure if the constant k in the interaction energy does not vanish. The vanishing of k was however widely thought to be one of the requirements for obtaining "Maxwell's" theory from Helmholtz's, so that Hertz's analysis could have applied to unclosed circuits if he had adopted this "Maxwell" limit, but he preferred not to do so in order to remain completely general.

3. Principle I does not assert that currents can be produced in closed paths by electrostatic actions, as they can be by inductive actions, since that claim would obviously be false. Rather, Hertz was arguing that there is no difference in kind between inductive and static forces as there is, say, between magnetic forces and electric forces. There may, and in fact must, be a difference between the distributions of the forces of a given kind that are produced by different kinds of sources, say, by varying currents as opposed to static charges. But according to the principle one cannot place a charge at a given point and at a given moment and tell whether, at that point and time, the force on it derives from other charges or from changing currents or magnetizations. Moving such a test charge from one region to another, however, to probe the distribution of the force would enable detection of the nature of the source.

4. In the situation that Hertz considered, which involves closed currents, these forces are equal and opposite, and Hertz explicitly mentioned that he would use the principle of "action and reaction as applied to systems of closed currents" (1884, p. 274). Despite this, Hertz actually directly used only an appropriate potential, one that represents the interaction of magnetic currents. His principles, taken together, guarantee its existence.

Helmholtz remarked (1873c, pp. 691–92, referring to the four forces that involve the ends of a pair of elements—two of which also involve the body of the other element—and that combine with an element-center to element-center interaction) that the general electrodynamic potential for current elements must in fact satisfy action–reaction (i.e., the net force on one element is equal and opposite to the net force on the element with which it interacts; see appendix 2), because the potential is a function of the distance between the centers of the elements. In that case the derivative of the potential with respect to the x_i coordinate of the center of the one element must be equal and opposite to the derivative with respect to the corresponding x'_i coordinate of the center of the other element. Consequently, the Helmholtz potential guarantees the conservation of momentum for every pair of interacting entities. This held in his theory *even when the ether was included in it:* currents in ether interact with other currents precisely in the same way that currents in conductors interact with one another. On the other hand, if the ether does act like a dielectric with very high polarizability, then all currents will be effectively closed, and therefore the non-Maxwellian end-forces that, in Helmholtz's electrodynamics, guarantee momentum conservation between material circuit elements will necessarily be undetectable. As a result the principle of action–reaction will be experimentally violated for material circuit elements, leading implicitly to questions about the motion of current-bearing ether elements (which will have to sustain the reaction force, and so transmit the missing momentum). Current elements in ether and in matter may still be said to satisfy the action–reaction principle if the Helmholtz potential continues to govern the processes, except that now the ether elements cannot be isolated from the material elements (so that an independent test of the principle becomes impossible).

One may say that in Helmholtz's ether theory momentum conservation actually required that ether elements bearing currents must be moved locally just like material elements carrying currents. From a thoroughly different perspective Maxwellians had always assumed that the ether could serve as a momentum carrier. J. J. Thomson (1885, p. 110) remarked, for example, of Clausius's force law, which violates action–reaction between elements, that (if the theory were correct) the violation could be understood "by supposing that the ether possesses a finite density, and that the momentum lost or gained by the bodies is added to or taken from the surrounding ether." He did not at the time develop the concept for the Maxwellian field; that required, at the least, assimilating the implications of the Poynting energy-transfer theory, which Thomson had just learned about. Eight years later (1893) he provided a highly developed account of momentum transport by tubes of displacement moving through the ether.

5. Hertz's argument is rather longer than this because he introduces a fictitious electric dipole spread over the apertures of the toroids to generate electric forces equivalent to those that are produced by the changing magnetizations. This nicely embodies the unimportance of the source since electric dipoles qua physical objects are not at all the same things as changing magnetization. It also represents physically the presence of an interaction term in the system energy as a product of electric dipole moments. I will return to the specific terms of Hertz's argument below in examining reactions to it.

6. Usual electrodynamics, he remarks in a note, is one that can be formulated in terms of Franz Neumann's potential U_d, according to which the electrodynamic forces that move bodies are given by $\nabla \times U_d$ and the electromotive forces by $-dU_d/dt$. Hertz's claim is almost obvious for Helmholtz's system because the only forces that can move bodies derive from $\nabla \times U_d$ or from the gradient of the electrostatic potential (∇U_s), and neither of these involves the time.

7. In Weber's theory, however, it was not immediately obvious (to someone not thoroughly familiar with the scheme) that such a force is absent, because, unlike Helmholtz's, the theory is

not based directly on a potential and, again unlike Helmholtz's, it requires the Fechner hypothesis for the current. Nearly every contemporary demonstration that the Weber law integrated around a closed circuit yields the same result as the Ampère law under these conditions also presumed that the current is constant, so that, contra Hertz, these results do not obviously exclude the Hertz interaction between ring magnets. See, e.g., Riemann 1880, pp. 333–36. Nevertheless, the term in the Weber law in the acceleration of the distance between the particles is simply $(dv/dt - dv'/dt) \cdot r$, and this vanishes under the Fechner hypotheses, which means that the forces between the conducting elements proper are indeed Amperean. However, non-Webereans were often not familiar with the worked implications of the law. I thank Olivier Darrigol for discussions on this point.

Hertz himself was not clear about what Fechner–Weber does and does not assume. It does not satisfy even his principle I, though he thought that it did. There are situations, e.g., in which Fechner–Weber predicts that the effect of a closed circuit in which the current changes differs from the effect on the same body of an electric dipole layer that is spread over the region surrounded by the circuit in such a fashion as to produce the same electromotive force as the changing current: the effect depends on what kinds of processes are going on in the affected body. This violates both of his principles. I will return to the point below.

8. It can be granted an artificial meaning only if the action between objects depends solely upon their mutual distance. Then, and only then, the value of the force can be assigned an unambiguous meaning in relation to space.

9. This argument cannot be made once ether and matter are completely disentangled from each other, because it conflates actions on ether with actions on the material structures that are embedded in the ether, which was the only way that ponderomotive forces were assumed to arise until the positing of the Lorentz force. The essential assumption, which was more a practical maxim than a rigorous principle, can be stated in the following way: if the energy density of the ether depends upon the locus within it of a material structure, then the corresponding energy gradient translates into a force on that structure, as well as into a force on the ether itself. Although the latter no doubt produces negligible displacement of the ether proper, it must be taken into account when balancing energy and momentum. Many non-Maxwellian accounts of field theory took it for granted that the ring magnets would move one another; e.g., Boltzmann (see chap. 13 below), Drude 1894a, p. 321, and Poincaré 1890, p. 123.

10. In the light of an electron-based field theory the principle is simply incorrect. Electron theory requires that ring magnets do not exert net ponderomotive forces on one another, for the simple reason that such a force requires the presence of a net quantity of charge, which neither ring magnet has.

The field interaction analysis given immediately above as an instance of why Hertz's principle looks reasonable in field theory (an analysis that typifies many, though not all, Maxwellian analyses of field processes in the 1880s) does raise the problem that charge–charge interactions are not energetically parallel to current–current interactions, because only the latter draw upon energy from external sources. It raises in particular the issue of what possible potential energy might be involved here. We have an interaction between charge and changing current, or between changing currents. Hertz's assertion evidently requires that these must not differ energetically from charge–charge interactions, which would seem to require that one can determine the moving force simply by taking the negative derivative with respect to the material coordinates of this new contribution to the *electric* interaction energy. Although currents are involved here, which might seem to vitiate this requirement (since current–current interactions require the positive coordinate derivative of their *magnetic* interaction energy as a result of electromagnetic induction), in fact the interaction involves a novel state, one of *changing current.*

Nevertheless, it is very difficult to see how one might construct a Lagrangean representation for the medium proper that could accommodate this requirement—that could yield an appropriate *potential* interaction from which moving forces could be deduced as gradients—because one of the two field energy densities (electric or magnetic) must certainly be treated as kinetic. If the magnetic energy is kinetic, the Ampère law, for example, yields electric energy as a function of the curl of the medium's physical displacement. The Hertz interaction would then seemingly require that the acceleration of the displacement coordinate should appear where it does not—in the potential function, which involves only the displacement proper. If we remain with the Lagrangean for the medium itself, and we do not consider material coordinates at all, then nothing like the Hertz interaction appears, because there is nothing for it to occur between: the medium is the only thing involved. It is only when we introduce material coordinates that the issue arises, because we are then forced to consider how *sources* interact through the fields they produce. This is a tremendous irritant in a full fledged field theory because material and medium coordinates are uneasy bedfellows. In more prosaic terms, the intractable problem of the ether–matter link once again raises its head.

Despite these apparent difficulties at the highest theoretical level, some Maxwellians at least admitted the Hertz interaction, almost certainly on the simple grounds that the magnetic current (viz., changing electric current) contributes to the electric field. In his review of Hertz's *Miscellaneous Papers* FitzGerald, for example, remarked: "Hertz applies [an energy] argument to the case of a ring magnet changing in strength and producing magnetic force [*sic*] on another ring magnet in its neighbourhood, and doing work there, *and shows thereby* that there should be a magnetic force due to a changing electric field exactly corresponding to the electric force due to a changing magnetic field" (1896, p. 439; emphasis added). Hertz's force was electric, not magnetic, but this was probably a simple slip on FitzGerald's part given that a constant magnetic current produces a constant electric field (and given the remainder of the sentence). In any case FitzGerald clearly did admit a work-producing interaction between ring magnets, from which, he noted, Hertz had obtained the magnetic effect of a changing electric field via an energy argument. The only questionable thing here for FitzGerald was the energy argument itself—not the interaction on which it was based. It therefore seems likely that the interaction seemed reasonable to many, though not to all, field scientists, who seem generally to have assumed that objects which generate fields of type *x* interact with one another. Oliver Heaviside, who never undertook the elaborate, field-creating exercises that J. J. Thomson, FitzGerald, Glazebrook, and others avidly pursued, did avoid the interaction (at the cost of establishing a different—and in retrospect equally odd—one between a displacement field and a magnetic current) by creating an energy-flow analysis that contains a velocity *v* which represents simultaneously the velocity of matter and the (large-scale) velocity of ether within it (Buchwald 1985b). Heaviside never played with the ether's energy structure, and he never multiplied fields in what was, by the mid-1880s, standard Maxwellian practice, and it is therefore not surprising that his analysis was, to say the least, hardly clear to many of his British contemporaries (and was in any case subject to doubt because of its reliance on a particular assumption concerning the ether–matter link, as well as because of its idiosyncratic "equation of activity"). I thank Olivier Darrigol for illuminating discussions about these difficult issues.

Chapter 12

1. Of course, $-\nabla\phi$ cannot generate a current in a closed path; it can only act at a point to produce a current there.

2. This depends only on the empirical fact that the rate at which Joule heat is generated is $\kappa C^2 d^3 r$.

3. It is precisely this assumption of such a *system energy* that cannot be sustained from the standpoint of field theory. Without it, Helmholtz's argument fails; with it, the argument succeeds.

4. The variation is not simple to do in a rigorous manner because the bodies containing the currents cannot be supposed rigid, but Helmholtz provided a general deduction. See appendix 2.

5. This condition, we shall see, meant to Hertz that his analysis was limited to reentrant currents, a conclusion that reflects the foundation of his work in Helmholtz's system. Zatskis (1965) claims to show that the new vector must lack divergence. His proof fails, however, because it

presumes the result. In particular (translating his into our notation), Zatskis first derives the result that A_i acts as a potential for $\nabla \times \partial A_C / \partial t$ "provided we can prove that" $\nabla \cdot A_i = 0$. He solves this equation for A_i and then proves in an appendix that the curl can be taken out from under the integral sign, in which case A_i of course has no divergence. But the fact that A_i acts as a potential results from a derivation which already requires that it have no divergence. Zatskis has therefore demonstrated only that potentials which satisfy a vector Poisson equation in which the source is a curl must lack divergence. The point was, however, to reach the appropriate Poisson equation in the first place, and that already required the very assumption that Zatskis seeks to derive.

6. Hertz himself turned directly to this system of currents I and showed how to obtain an expression for it in terms of $\partial A_C / \partial t$. This at first rather obscures his analysis. However, his purpose was to exhibit clearly how the I system parallels the C system, and therefore how the energy could be used.

7. Hertz, as Zatskis (1965) remarks, did not demonstrate that the curl, which operates under the integral sign only on the vector source, could be taken out from under it. However, anyone schooled in Helmholtz's electrodynamics would at once have understood from the many partial integrations the theory requires that such an operation was legitimate provided, as physical reality demands, the vectors are localized. See, e.g., Buchwald 1985a, appendix 10, for some examples involving divergence operators.

8. Hertz writes almost nothing at this point, remarking only that the A_i must determine "magnetic" forces unless they are constant (Hertz 1884, p. 282).

9. An equivalent expression to what follows is given in Havas 1966 and Zatskis 1965. It defines A_{comp} directly rather than recursively by introducing the integral operator pot such that pot(A) is $\int (A/r) d^3 r$:

$$A = \text{pot} \sum_{n=0}^{\infty} \left(-\frac{1}{4\pi} \right)^n \frac{\partial^{2n}}{\partial t^{2n}} [(\text{pot})^n C]$$

10. It is not immediately obvious that Hertz's series will converge. However, he attempts to show that it will by examining the contribution dA_{comp} due to $C d^3 r$ in the particular case that C varies harmonically with the time as $\sin(nt)$. If the product $n^2 r^2$ is less than 1, then, Hertz shows, the series has a limit. Zatskis (1965) and Havas (1966) disagree concerning the validity of Hertz's series for the total vector potential, but in any case Hertz himself was concerned solely with illustrating the likelihood of convergence.

11. This relationship is rather hidden by the detour through the associated vector potentials, or as Hertz put it, the problem is that the "vector-potentials of electric and magnetic currents have hitherto occurred as quite separate, and from them the electric and magnetic forces were deduced in an unsymmetric manner." He was referring here to his own deduction, in which the vector potential for the magnetic current (i.e., the first-order correction to A) was obtained *from* the vector potential for electric currents (i.e., from the zero-order term in A) via his argument. If instead we concentrate on the hierarchy of interaction energies, it is very clear that there is no essential distinction between the derivations of the electric and the magnetic forces since each electric interaction entails a succeeding magnetic interaction and vice versa. The only difference is that the series begins with the electric currents and uses principle II to deduce interactions between magnetic currents.

12. Notice, however, that the focus of Hertz's account was upon the direct connection of the forces proper: the E and the H. Maxwell instead concentrated upon the generation of fluxes from forces and elaborated a scheme in which each element—electric, magnetic, charge, or current— had its place as one of these two kinds of things. The generative link between electric and magnetic forces, which gripped Hertz, only indirectly interested Maxwell. In Maxwell's theory the electric current C and the displacement current $\partial D/\partial t$ generate the curl ($\nabla \times H$) of magnetic force. The time change of the magnetic flux $\partial B/\partial t$ generates the negative curl ($-\nabla \times E$) of electric force. Consequently, the relationship between E and H—between the electric and magnetic forces—is

given only through their respective relationships to changing magnetic and electric fluxes. Since the distinction between flux and force is at the core level of Maxwell's theory, it cannot be ignored without provoking a complicated series of changes that affect the concepts of charge and current (Buchwald 1985a). This is, however, not the major point at which Hertz's 1884 analysis fails to make contact with field theory, as we shall now see.

It is important to remark at once (and again below) that the recursive procedure described immediately above would certainly have been unacceptable to a Maxwellian, because in field theory one cannot consider *sequential* contributions as each separately satisfying the basic field equations when those equations apply to the entire field itself: if a field has n time derivatives, this does not mean that each of them separately satisfies the cognate Maxwell equation and that they can therefore be summed together to produce a total field in Hertz's fashion. Rather, the field proper satisfies the Maxwell equations.

13. This defect was first discovered by Peter Havas in 1966. Havas and W. Pauli (in Pauli 1973, p. 148) note that the series for the retarded vector potential with a source C must be

$$A = \sum_{n=0}^{\infty} \frac{1}{n!} (-A)^n \frac{\partial^n}{\partial t^n} \int |r - r'| C_{r'} d^3 r'$$

Hertz would hardly have been likely to know this a priori because it first of all requires a thorough assimilation of the meaning and mathematics of retardation, which at the time was an exceedingly obscure area, both mathematically and physically (on which see Archibald 1987, pp. 312–37). Only then would he have known that the *sole* difference between the time-independent and the time-dependent solutions for A is that the source under the integral sign must, in the latter case, be a function of $t - r/_c$ (where r is the distance between the point at which the value of A is desired and the appropriate position of the source at a previous time).

14. Havas (1966) argues that Hertz's series can only represent "standing waves" because (at least for point sources) his series results from superposing the advanced and the retarded solutions.

15. Rayleigh in Britain was a prominent exception since he treated such equations to some extent in his *Theory of Sound* (vol. 2, sec. 275). This fact is important in understanding George FitzGerald's discussion of radiation in the early 1880s.

16. Care must be taken, however, not to misinterpret Hertz. He probably did not think that each term in his series generates its successor in a physically meaningful sense. Rather, he had discovered through his principles that the interaction energy between current-bearing bodies must contain an infinite number of terms. Each of these contributes its own separate, distinguishable portion to the energy as a whole, so that every term in his series for the vector potential has its seat in some identifiable part of the total energy. In this sense Hertz's series does indeed consist of physically distinct components. This is the effect of the missing displacement current: Hertz has to mimic its contribution through the infinite sequence of terms in the interaction energy.

17. The delayed propagation of an effect at once violates Helmholtz's electrodynamics since the latter presumes that systemic energy is determined by the simultaneous conditions of a pair of objects in given states and at given distances from one another.

18. Interaction energy, in the sense used here, means a part of the complete energy that is a joint function of the coordinates of the two objects in question and that can behave like a potential. Such a part might also be a function of the coordinates of other objects, but if it is not, the behavior of each object in a pair will apparently depend solely on the behavior of its sibling. The interaction will accordingly be "bipartite." In general, bipartite interactions occur in field theory if the energy is proportional to the square of the magnitude of a vector that is the sum of vectors that each depend solely on a single object. In electromagnetism, for example, field theory yields a bipartite interaction between charged objects because the electric field energy varies as $E \cdot E$, and E is the vector sum of the partial fields E_i, each of which is a function solely of a single charged object. Therefore, the interaction energy between the ith and the $(i + 1)^{th}$ object is $E_i \cdot E_{i+1}$, which does not involve the coordinates of any other charged object. But if the field energy were, say, $(E \cdot E)^4$,

things would become vastly more complex, because this expression cannot be separated into purely bipartite components. And the fundamental principles of field theory cannot guarantee that such a situation will never arise. For example, the field energy that Maxwellian understanding of the Hall effect did demand was only *approximately* bipartite: if the magnetic fields of the currents that are being affected are themselves included in the Hall action, the energy cannot be separated into a sum of self-energies and bipartite terms; see Buchwald 1985a, p. 115, n. 3.

19. Note, however, that Hertz did not question the foundation of Helmholtz's deduction, namely, the existence of a system energy. He raised only the point that use of the principle presumes that all the relevant states (i.e., all the relevant determinants or dissipaters of the energy) have been taken into account. Helmholtz system energy, in any form, is itself incompatible with field theory, for unlike a field energy density, it is not local. There are other, less-precise incompatibilities. For one, Helmholtz's scheme sought to construct the system energy from a partial specification of effects and an exhaustive specification of contributing and dissipating states; other effects involving these states can then be obtained by appropriate variation. Maxwellian practice in the 1880s usually *conjectured* an energy density and deduced effects from it, all of which had to involve the unitary fields (electric and magnetic). One could not pragmatically work backward from, say, the effects through Lagrange's equations to an appropriate Lagrangean (see Buchwald 1985a, chap. 14, for an example of Maxwellian procedure in this regard), whereas the route from effect (given the states) to system energy was precisely what Helmholtz's scheme followed.

20. I have found no trace of any responses to his 1885 paper, but we shall see that the "Fundamental Equations" (1884) paper generated a great deal of reaction in Germany.

21. Hertz qualifies this by writing "for purposes of calculation," but he only means by this that they are of course not physically the same thing: the dipoles are electric objects; the magnetic currents are not. However, in respect to the forces they exert they are not to be distinguished *at all* from one another.

Chapter 13

1. Boltzmann clearly regarded this work as extremely important, remarking in his evaluation of Aulinger's dissertation that it concerned "a problem of the most fundamental significance for all of theoretical electricity." See Jungnickel and McCormmach 1986, vol. 1:68.

2. Specifically, he remarked: "Before I give other problems, however, I must mention an obscurity in Hertz's interpretation of his principle, to which Hr. Boltzmann drew my attention in conversation" (Aulinger 1886, p. 120).

3. Since we know that the ring magnet does exert an electromotive force, the AB principle requires that it must exert an electric force on the dipole layer, because electric forces are unitary and are completely specified by the action on a stationary unit electric charge.

4. Aulinger also demonstrated that Weberean theory fails to yield the ponderomotive forces between changing ring magnets that are implied by the AB principle. To do so he first proved that the force does not occur between a pair of arbitrarily oriented closed circuits, and then he simply built the ring magnets out of sets of these pairs.

5. Boltzmann had already undertaken experimental investigations on this subject at Helmholtz's Berlin laboratory. See Jungnickel and McCormmach 1986, vol. 1:212. They note that Boltzmann's position as *Ordinarius* in mathematics prevented him from pursuing experiment as avidly as he wished, though his "good relations with Stefan," the director of the Vienna physics institute, did not preclude it.

6. "For the first suggestion of the idea to make use of alternating charge, which later proved to be so fruitful, I thank a written communication from Herrn Geheimrates Helmholtz, before the autumn of 1872" (Boltzmann 1873, p. 480).

7. So, e.g., Boltzmann considered indifferently that changing magnetic polarization (and thus changing currents) can both move charge and effect dielectric polarization. These are all cases of distance forces which exist in the same sense that, e.g., distance forces exist between Weberean

particles. In other words, for him (at least when attempting to link Maxwell's field to Helmholtz's ether) "charge" retained an essentially substantive character: it was a thing more than it was a state, albeit a decidedly enigmatic thing.

8. Here we catch a glimpse of the gulf that would soon divide Boltzmann from his somewhat younger contemporaries as they pursued the physics of interacting states while he pursued the physics of molecular collectives.

Chapter 14

1. Documentary evidence is unfortunately scanty, consisting of the published texts (together with, for some, the submitted manuscripts) as well as letters. However, as we have already seen in his work on evaporation, Hertz had not yet learned to cover his tracks thoroughly in writing public accounts. Many clues to the twisted course of his work remain in the printed texts: sequential remarks that seem to have little to do with one another, experimental byways hinted at that seem far from the main point, and so on. By the early 1890s Hertz had learned much better how to hide these kinds of things, and his later (experimental as well as theoretical) publications have a much more coherent, polished structure than these early efforts. He himself noted, e.g., that the published account of the work we will examine in this chapter "gives, generally in the actual order of time, the course of the investigation as far as it was carried out up to the end of the year 1886 and the beginning of 1887" (1962, p. 2).

2. E.g., the spark gap might, even during discharge, have a high enough resistance to overdamp the current sufficiently rapidly to produce detectable induction.

3. Figure 58 is a modification of the one published by Hertz (1887a), which does not contain the wire connecting knob 2 to discharge point D. Hertz, however, first discusses a configuration like this one, albeit with a long shunt instead of a loop.

4. The published account first describes the diverting ability of the micrometer gap and then links it to the efficacy of such gaps as lightning protectors. Following this Hertz introduces the possibility that a low-resistance shunt might make the effect go away but reports that it does not. This likely does not represent the actual course of his work, because the intellectual jump from the Riess spirals to the shunted micrometer in a gap-bridged linear discharge circuit is simply too vast.

5. In his published account Hertz presents this result as a *confirmation* of the prior assertion that the voltage drop accompanying large current cannot explain the sparking. But he would initially have had no reason at all for suspecting such a thing. His raising of the possibility indicates that he probably did at first assume that the odd behavior resulted from a very large current (with its attendant large potential drop in the micrometer wire). Had this suspicion proved out, it would have had much to say about the unusual behavior of the spark gap in the discharge circuit: namely, that the gap did not break down until a tremendous quantity of charge had been stored on the halves of the circuit that it divided (thereby producing, after breakdown, an extremely large current, provided that the gap's resistance went down at least to that of the metal parts of the discharge circuit).

6. Except, of course, at the terminating surfaces of capacitors, like Leyden jars, where the charge had to accumulate.

7. The change in charge value had been important since the 1850s for understanding signal behavior in telegraphy, but even there charge was not treated as propagating, but rather as diffusing, through the circuit. See Smith and Wise 1989, chaps. 13 and 19, for a discussion of this in the British context.

8. Early breakdown would mean a very weak effect since the current strength would not be large. Slow breakdown, even if late, would mean that the potential would equalize along the circuit long before the current built up to a significant level.

9. Hertz certainly knew this from Kirchhoff rather than from Helmholtz, since Helmholtz's analysis of propagation in conductors concerns something entirely different that, e.g., would not

even occur at the (Maxwellian) value of 0 for the disposable constant k in his electrodynamic potential (see appendix 13).

10. Hertz presents this experiment as indicating that the single-pulse explanation, which he gives first, is incomplete. This very likely reflects his having discovered the incompleteness through manipulation. There would have been no reason for him to have suspected it initially.

11. Hertz here remarks that as a result he "felt convinced that the problem of the Berlin Academy was now capable of solution." This seems unlikely, at least as a precise record of events, since at this point he probably did not suspect that the discharge circuit itself oscillates. It is doubtful that the discovery of oscillations in the *Nebenkreis* would have been enough to suggest that a related process might be going on in the discharge circuit.

12. See the brief but illuminating discussion of Hertz's apparatus in Bryant 1988, from which I have adopted the informative phrase "spark-switched oscillator" for the evolved form of Hertz's device.

13. "These" refers here to the induction effects.

14. As Hertz wrote in his introduction to *Electric Waves* (p. 2), the device might have behaved in a "turbulent and irregular" fashion.

15. Hertz refers in his account to "certain [unspecified] accessory phenomena" as having suggested regular oscillations to him as an explanation for the discharge circuit's inducing strength. Whatever they might have been, his reference to them further indicates that he did not at all start out with this assumption. The original impetus to the experiments, recall, was to probe the odd behavior of the spark gap, not to demonstrate some preordained belief.

16. He made no remarks on this point, which only emphasizes his signal concentration on the discharge circuit proper as a novel device. Later understanding revealed that the self-induction of the coil apparatus had to be vastly greater than that of the linear discharge circuit. Because the discharge current changes rapidly with time, it will be effectively excluded from returning to the coil by the much greater back–electromotive force that it generates there.

17. Though in this case electrostatic effects superposed on electrodynamic ones because the affected circuit was now open. To show this Hertz pulled the oscillator's knobs so far apart that sparking in it stopped. *Presumably* the separate halves no longer oscillate, the necessary actuation of the fast-switching gap having been removed, and yet sparks still occur in the neighboring open circuit. Hertz attributed these to electrostatic induction, since C and C' will still be charged by the induction coil. To make this effect go away Hertz connected the micrometer gap in the affected circuit with a "bad conductor," which was, however, not so bad that the charging of the micrometer's knobs could not follow that of the inducing coil in the oscillator (on C and C')—or, better put, Hertz tried out "bad conductors" until he found one that made the micrometer sparking stop. Starting up the oscillator by narrowing its gap then restarts the sparking in the affected circuit, which is now presumably due solely to electrodynamic action.

18. Hertz did not specify what he had in mind here, but it was certainly not the kinds of interactions that had for the most part fascinated Helmholtz during the 1870s, for these had been ponderomotive, not electromotive. Hertz's device, as a stimulator of interactions between "rectilinear open circuits" might perhaps address the term in Helmholtz's expression for electromotive force that involves the second derivative with respect to time of the charge density at a circuit terminus: this should affect the electromotive force that produces current in the other circuit. Indeed, *only* Hertz's device stood any chance of doing such a thing, because only in it might the rate of change of the charge density at a terminus be rapid enough to produce any kind of detectable effect.

19. In *Electric Waves,* the English translation of the German "*resonanzartige*" is "symphonic," which, as Aitken (1985, p. 76) remarks, does not capture the original. I will translate "*resonanzartige*" simply as "resonant" rather than Aitken's "syntonic."

20. In the absence of resonant action, the increase of capacity, which simply raises the natural period of the *Nebenkreis,* no doubt makes it less responsive to aperiodic impulsive actions emanat-

ing from the discharge circuit, just as a slow-swinging pendulum will hardly be budged by rapid shocks.

21. In breaking the connection between knobs *2* and *4* Hertz created in effect a supplementary circuit, namely, *2bdca13eghf4*, in which knobs *2* and *4* behave like knobs *1* and *2* in a detached circuit *1acdb2*, so that sparking can occur between *2* and *4*, requiring a node in the wire connecting *1* and *3*. This does not, however, disturb the previous configuration, which means that the earlier nodes persist.

22. Poincaré provided them in 1890 in *Électricité et optique*, chap. 8. In his expression, however, Hertz's 0.75 is replaced by 1, reflecting Poincaré's disagreement with Hertz's assumption of a uniform current density within the conductor.

23. Hertz retrospectively discusses the error in *Electric Waves*, n. 6.

Chapter 15

1. Doncel (1991, p. 9, n. 20) has remarked the apparent contradiction and dates the experiments to December 8–21, 1886.

2. I thank Manuel Doncel for his remarks to me on these points.

3. It seems probable that Hertz's confidence in the spark-producing inefficacy of electrostatic forces, and his consequent belief that the resonator is effectively closed, were not due solely, or perhaps even mainly, to the potential argument. Recall that he had worried enough in the original experiment that the large conductor *C* (fig. 62) might indeed spark the secondary conductor electrostatically. He had eliminated the possibility in his usual way: by moving *C* to point *g* and showing that the sparking which occurs when it is connected to *h* no longer takes place. His confidence was based as much on early experience with the device as on theoretical properties, though here the two reinforced one another.

Hertz had even more reason to ignore static effects. According to his remarks in the introduction to *Electric Waves* (p. 4), he had first thought to see if a dielectric could be used to upset the "neutral point of a side-circuit"—that is, to displace the position of the no-sparking point in the secondary. But this position, he was quite certain, had to depend very strongly upon static actions: even though the presence or absence of current in the secondary did not depend upon these kinds of actions, they would certainly affect the sparking loci quite directly by static induction on the secondary's termini. His decision not even to try this kind of experiment was therefore based on a conscious determination to produce a configuration that was explicitly designed *not* to respond to static force.

4. "Above all," he continued to his father, "my present mood is such that I would rather return to the work that was interrupted by this observation; and it is disagreeable to me to have others on my trail only inasmuch as I am now forced to give priority to this project, to the extent that I want to develop it myself."

5. Doncel (1991) presents a careful and frequently insightful discussion of the dating of events during the fall and early winter of 1887, based on Hertz's *Versuchsprotokolle*, or laboratory notes, for this period—the only ones currently known to exist. For these events we have direct evidence from a single page (1d) in the laboratory notes, three pages of diary entries, and the only publication that dates directly from this time, namely, the account of his dielectric experiments that Hertz sent to Helmholtz for insertion in the *Sitzungsberichte* on November 5 (1887c) (the last is denoted document 6$_{AK}$ by Doncel since it is the original version of the one that appeared in the *Annalen* during 1888 (1888d) and was printed as document 6 in *Electric Waves*). There is no *Sitzungsberichte* version of Hertz's *Annalen* account of the force-distinguishing experiments Hertz conducted in the fall and winter of 1887 (see sec. 15.2), and using a letter to Wiedemann Doncel demonstrates conclusively that the *Annalen* account of these experiments (1888c) was actually written by Hertz in February 1888 as a logical prelude to two other (now rewritten) papers (1888d, 1988e), which were appropriately modified as well. Hence we cannot use this paper (1988c) to conclude very

much about Hertz's thoughts on propagation before early November, since when he wrote it, his research program was already in radical transformation. Nevertheless, the single apposite page of the laboratory notes does indicate that Hertz had undertaken the experiments that I denote below by series A and B during October of 1887. Further, the original dielectric report (6_{AK}) both describes a device that required Hertz to have already performed many of the manipulations he later discusses and also reaches conclusions that similarly depend upon a reasonably clear knowledge at least of how to distinguish closed from unclosed contributions to the force that drives the resonator.

6. Doncel (1991) points out the complete absence of the "theory" for the resonator in the laboratory notes, although this is not necessarily significant since these notes are very clearly missing quite a few things that Hertz did during these fall weeks.

7. These results were reached by September 7 since they are explicitly mentioned in the laboratory notes.

8. These results may not have been obtained until some time later, because in the laboratory note entry for October 10 Hertz remarks only that here "there is a greater variety of phenomena" than in series A.

9. The notes for September 7 and October 10 do not discuss these points explicitly, although Hertz does remark here that the maxima in A are the directions of the "electric force" and that the positions of zero spark length are normal to it. The "electric force" is distinguished from the "magnetic force," which "is everywhere perpendicular" to a plane containing the oscillator. The former is therefore the open-circuit, and the latter the closed-circuit, action. By November 5, when Hertz sent his paper on the dielectric effect (1887c) off to Helmholtz, the distinction was well and explicitly drawn. Further, by that time Hertz was also distinguishing between electrostatic and electrodynamic contributions to the open-circuit force, if only to note circumstances in which the latter, but not the former, can be changed by shifting the resonator's center downward relative to the oscillator.

10. The conclusions detailed immediately below are not discussed in the laboratory notes and therefore appear only in Hertz (1888c), which was crafted after his work on insulators and on finite velocity (1888d, 1888e) had been completed. He may not have thoroughly developed these conclusions until sometime near the middle of December.

11. There seems to me to be little doubt that quite early in these experiments Hertz was thinking in terms of mapping the direction of the total force at any point in space. This has nothing at all to do with forces as fields; it concerns just what affect the total force acting at each point of the resonator will have upon the resonator.

12. The sparking reveals only the magnitude, not the sign, of the sum $\alpha + \beta\sin\theta$, and so fig. 68 shows the sum's absolute value.

13. Suppose that the projection of the electromotive force onto the plane of the resonator is E. This force yields by projection the force Σ that acts at a given point s of the resonator's circumference. If s is an angular distance of θ from E, then the force that acts upon it will be $E\sin(2\pi s/S - \theta)$, projecting E along the tangent to the circumference. This splits into a term that depends on $\sin(2\pi s/S)$, and which therefore does not activate the resonator, and a term $-E\sin\theta\cos(2\pi s/S)$, which does activate it. Consequently, Hertz's rule of thumb emerges from this "theory" for Σ, with β equal to $-E$.

14. This transformation that Hertz effected is more complicated than it might seem, because the 1887 configuration differs in two essential respects from the 1884 arrangement. In 1887, we shall see, the very existence of object b (the dielectric) in an appropriate state is due to an unhypothetical interaction between it and object a (the oscillator), viz., statically induced polarization. Consequently, a and b interact with one another on an entirely different basis from their interactions with object c (the resonator). Moreover, to parallel the 1884 configuration would seem to require considering the resonator and the oscillator to be the two objects in the same state, with the dielectric then the object with which each of them interacts; that is, the interaction pairs would be dielectric–resonator and dielectric–oscillator. The 1887 configuration, however, requires the

significant pairs to be resonator–dielectric and resonator–oscillator. Hertz had accordingly to switch the roles of dielectric and resonator.

15. Doncel (1991, p. 15), evidently on the basis of the diary entries, dates Hertz's investigation of interference effects between a conductor and a dielectric, in conception at least, to September 17–23. This is certainly possible. The diary entries refer to "experiments on the relative positions of the circuits" (September 19) and to his looking "for large asphalt block" (September 21). The entry for the twenty-third is "Sketched and made preparations for new experiments." The entry for the nineteenth most plausibly refers to interference experiments between conductors, which would have preceded ones with dielectrics in order to stabilize a known effect before probing a presumably similar unknown one.

16. Fig. 70 is adapted from Hertz's figure 24 in 1888d; the latter is reproduced in full in fig. 71.

17. Hence the apparatus oscillated at about 50 MHz; see Bryant 1988, p. 29.

18. Hertz never went into such detail, but he had indubitably thought this through because the paper on dielectrics which he communicated to Helmholtz on November 5 contains assertions that depend upon it. On October 12, according to his diary, he had made "Experiments on bringing conductors towards each other (phase differences)." These were not likely the first such experiments, but they may have been the critical, careful ones that he referred to in his papers.

19. Hertz may not have pursued in detail these phase-effect experiments before he tried dielectrics. He may, that is, have done just enough to see what C's effect was before he tried the same thing with a dielectric in its place.

20. Another possible objection—though Hertz could not see how to verify it—would be that the effect is due to static, rather than dynamic, action, since the dielectric is, of course, charged by static induction. Hertz decided to follow his usual procedure in such circumstances: to remove the objection by experiment. Arguing that the dielectric could not have any substantial external static effect if its surface is bounded by (static) lines of force, he moved it into an appropriate position for this to hold, where however the effect still appeared. Note that Hertz was also taking it for granted that the electrodynamic inducing force emanating from AA' would not alter his conclusions. Or, rather, he probably *knew* that it would not do so, for the following reason (which he did not provide). The dynamic inducing force from AA' is always opposed to the static force between its terminal plates. The oscillator is in the first instance statically imaged in the dielectric block. In addition, the dynamic induction from AA' will also polarize the block. But since the latter opposes the (static) force between the oscillator's plates, then, given the configuration of Hertz's device (in which AA' sits very close to the block's surface), the dynamic induction will actually *increase* the induced polarization; that is, it will strengthen the image.

21. This was certainly not true. Röntgen had touched it quite firmly, and Hertz was unlikely not to have known about it. Perhaps by "untouched" he meant unanswered in an appropriately simple, revealing manner.

Chapter 16

1. It is no part of my purpose to explain in modern terms what went on in Hertz's experiments, particularly these early ones, though it is comparatively simple to do so since Hertz himself produced an explanation (1889a) after he developed radiation theory on the basis of Maxwell's equations in 1888. It is nevertheless helpful to keep the following points concerning the operation of the device in mind. First, the oscillator is driven by an induction coil whose primary connection to a battery is interrupted a number of times per second. Second, the spark gap breaks down rapidly and sufficiently completely to reduce the resistance across it to metallic levels. Third, the gap also reforms as an insulator quite rapidly. Fourth, the oscillator damps very rapidly as it radiates. Fifth (which is actually an implication of 2 and 3) during radiation the oscillations are essentially harmonic, or, in Hertz's original terms, "regular." This last result was precisely what Hertz had produced in his early laboratory work, although he had at first thought the oscillations would be quite irregular, i.e., aharmonic, in which case the concept of resonance would not have applied.

2. This is strikingly illustrated by Hertz's dropping of the source term in considering the propagation equations "in empty space." The resulting "Maxwell equations" accordingly also lack the source, which means that they do not even raise the issue of continuity at all, whereas continuity was itself an essential feature of Helmholtz's deduction.

3. We have seen, however, that Hertz's theory was not completely successful, because his second principle imports a problematic (in Helmholtz's electrodynamics) concept of the source. Hertz remained only vaguely conscious of this problem until, at the earliest, the spring of 1888.

4. To be precise, a's efficacy will of course weaken with distance from the oscillator, whereas we assume for the moment that w only oscillates with distance. In that case the effect of a on w will be more marked closer to the oscillator, but the periodicity of the interference will belong entirely to w.

5. Hertz in this way easily avoided the problem of "experimenter's regress" discussed by H. M. Collins (1992), which is particularly interesting since Hertz was trying to produce and to detect something that, in contemporary context, was just as problematic as high-flux gravity waves were in the late 1960s (Weber's attempts to detect the latter constituting Collins's prime example of regress).

6. From fig. 72 we see that Hertz did not at first let the wire end freely; he terminated it in another plate. This complicates the effect since the wave termini here lie along the plate edges rather than at the pointlike wire end.

7. The diagram in his laboratory notes (fig. 72) has "300" between the three marked nodes, which yields a wavelength of 600 (presumably cm), or just about what Hertz did later use. The "wavelength" in the November 10 notes no doubt refers to the half-length.

8. According to the diary the date was November 10, but there is no mention of this in the laboratory notes for that date. Indeed, on the evening of the tenth Hertz was still busy examining the wire waves. Doncel (1991, p. 16) suggests that the diary entry should in fact be November 11, which accords with the laboratory notes entry for "11–12 November."

9. Doncel (1991, p. 17) believes this is unambiguous evidence for Hertz's having set out to measure the propagation speed. Hertz's remark must, however, be taken in context, because his experiment could in fact actually *measure* the speed only if it were either infinite or else very nearly equal to that of the wave in the wire. These experiments, like most undertaken in the spirit of Helmholtz's scheme, were intensely qualitative—not in the sense that they provide no numbers but that the numbers they do provide lie within very broad ranges, the goal being to assign the result to a range rather than to find a precise value for it.

10. The available evidence (Hertz's remarks in *Electric Waves,* the published papers, the diary, and the laboratory notes) indicates that Hertz tried to detect "interference" between the direct action and the wire wave in two different ways. The first one that he actually described (in the laboratory notes entry for November 18) was not the one that he used a month later. Since, however, this first-described experiment follows by a week the "futile" ones of November 11–12, it seems likely that these earliest attempts utilized a different method. That method, no doubt in a comparatively primitive form (see below), was probably the one that he did ultimately use extensively since there is no indication of a third technique. This was in any case the one that made use of both wire and oscillator.

11. Hertz's diagram shows both a circular and a rectangular resonator. He initially worked with the circular form, but his experiments in December used both. The circular resonator is particularly convenient for experiments in which the resonator plane contains the wire.

12. This is graphically illustrated by the pushing of the force lines away from the oscillator, as compared to what their distribution would be for an electrostatic dipole of the same size.

13. We will see below that the device will not work properly if it is set first at, say, L_1 and then turned all the way to L_2. It must be recalibrated before every deviation.

14. In this case (supposing some sparking to exist and supposing also that the direct action cannot have a variable speed of propagation) either (1) the wire-wave induction or the direct action

vanishes altogether at this point, (2) the wire-wave action is so much greater than that of the direct wave that no effect of the latter can be detected with Hertz's device, or else (3) the total direct action at a given point along the wire no longer reverses direction between L_1 and L_2 (even if it still varies in magnitude, reaching a minimum between the two loci, in position 1). Hertz never explicitly distinguished these cases.

15. The laboratory notes date the new experiments to November 18. The diary entry for November 17 refers to "further experiments" with "equally negative results"; the entry for November 18 reads in part "Repeated the experiments very carefully."

16. This distance corresponds loosely to a quarter of the wire wavelength in his previous experiments. The wire wavelength would be the same as the direct-action wavelength if the latter propagated at the same rate as the wire wave, in which case a 1.5-m remove would have generated another quarter-wave delay in these experiments, yielding a total 180° phase difference and hence cancellation.

17. This result holds with reasonable certainty, however, only for a cylindrical wire.

18. This calculation does not involve polarization at all. It is, rather, Helmholtz's own theory for wire propagation using the general electrodynamic potential.

19. Under those circumstances the force exerted at a given moment by the direct action would reach effectively instantly all parts of the wire that Hertz could examine.

20. The diary records this entry for December 16, but the laboratory notes describe experiments on December 15.

21. To make the statement fully general Hertz could have referred to a 90° phase difference, since even if the direct action propagates infinitely rapidly, the phase of the spark-length curve at such a point would be 90° (see appendix 14).

22. It is difficult to be certain from Hertz's discussion and diagram, but it seems as though he measured a wire-reversal rather than an oscillator-reversal, because he refers to rotating the normal to the resonator from plate P and toward plate A'. However, one cannot be certain about this: in the case of a wire-reversal the head (say) of the resonator normal would turn from P to A'; in the case of an oscillator-reversal the head turns from P until it points directly away from A', so that its tail now points toward A'. The resonator normal lies along the same line in both cases but points in opposite directions. One might loosely say, as perhaps Hertz did, that even in the case of an oscillator-reversal the normal is directed at A' (i.e., intersects it).

23. Hertz writes nothing about this, except to say that "small rotation" produces weakening but the sparking eventually comes back with further deviation, "so that" the wire has to be disconnected each time.

24. He could stably reverse the effects of the deviations by putting the resonator's gap at the bottom rather than at the top; nothing depended on a 180° rotation of the resonator's plane. If the driving plate to which the wire is connected is shifted from one oscillator plate to the other, the effects again reverse (in Hertz's words, they remain the same "relative to the [driving] plate"). Sometimes the wire length had to be changed to make effects stand out. The null position, in which the resonator's spark gap here lies in a horizontal plane that contains both the wire and the oscillator, is displaced up or down depending on whether the wire is, respectively, above or below the resonator. With the resonator lying horizontally, the spark strength depends both on whether the gap is near or opposite the wire and on which of the oscillator's plates drives the wire.

25. In fig. 73 this corresponds to moving the wire to the far side of oscillator plate A and putting P directly opposite oscillator plate A'.

26. In retrospect we might say that Hertz's first critical experiments were done at about 55 MHz. Using his previous method, which equated the oscillator with a simple inductance–capacitance device, Hertz obtained 70 MHz for the frequency. This value incorporated the error due to incorrect estimation of capacitance, but in any case Hertz was quite certain by the time of publication that "it is doubtful whether the ordinary theory of electric oscillations gives correct

results here." This means that Hertz knew that he could not use his device to measure the speed of electrodynamic propagation directly; he could find the speed only in relation to the speed in wires.

27. See Doncel 1991, p. 18, for remarks and translation.

28. Doncel (1991, pp. 17–21) argues that in view of other results he obtained on December 23, the unhappy letter was likely written in the morning. The results he discusses in the letter were in any event obtained before Hertz turned sometime that day to a different orientation for the resonator.

29. They probably do not represent distances from + to − or from − to +, as a quick glance at the table immediately reveals.

30. As suggested by Doncel 1991, p. 21.

31. The diary records for the twenty-third that he "obtained an indication of the finite velocity of propagation of the induction effect in the afternoon."

32. In this configuration Hertz's open-circuit term, $\beta \sin\theta$, vanishes because here the open-circuit force parallels the resonator. In this configuration the resonator sparking effected by the oscillator alone would be midway in strength between maximum sparking with the resonator's gap parallel and near the oscillator's gap and minimum sparking with the resonator gap rotated 180°. In all positions except the one Hertz has here chosen for the gap (and its diametric opposite) both static and dynamic open-circuit forces from the oscillator come into play.

33. In the printed versions of the table Hertz switched the significance of the + and − designations and suppressed the uncertain measurements at 5 m (see Doncel 1991, p. 22, n. 61).

34. More precisely, between 3 and 6 m the interference does not change in as marked a fashion as it does between 0 and 3 m, which is indeed clear from Hertz's earlier results. Whereas in every case below 3 m there are at least two types of interference change, there is at most one such change between 3 and 6 m.

35. The figure that accompanies the following remarks (fig. 76) appears in the laboratory notes for December 29 (Doncel 1991, p. 22).

36. Both results would have been particularly interesting to Helmholtz, and Hertz did refer to "the electrodynamics of open circuits" in his November 5 communication to Helmholtz of the paper on dielectric effects that was printed in the *Sitzungsberichte* and that was redone for the *Annalen* as 1888d. This may—though not with certainty—indicate that Hertz had these results by early November. Until he became thoroughly convinced that he had manufactured, and could manipulate, air waves he continued to emphasize something else that his experiments were good for, independently of propagation: namely, the standing questions concerning open-circuit electrodynamics.

37. The dynamic force might be substantially modified if the constant k were not equal to 1, since then the following extra term appears in the electrodynamic potential:

$$\varphi_2 = -A^2 \frac{1-k}{2} ij \iint \frac{1}{r^2} [(\nabla r \cdot ds)(\nabla r \cdot d\sigma)]$$

This term vanishes if either of the interacting circuits is closed, whatever the value of k may be, but Hertz already thought that both resonator and oscillator must be treated as open. To accommodate Hertz's observations this extra term would have to strengthen the force parallel to the oscillator alone the normal line through its center and weaken the parallel force along the line of the oscillator proper. That is, it would have to supplement the Neumann action in the normal direction and counter it in the parallel direction.

Chapter 17

1. See Doncel 1991, pp. 2–6, for a thorough discussion of the origins of the trilogy and of the differences between the trilogy papers on dielectrics and on propagation and the versions of them that were printed in the *Sitzungsberichte*.

2. Doncel (1991, p. 9) makes this point.

3. The discussion immediately following owes much to Doncel (1991), who makes the point that during most of December Hertz was not thinking in any definite fashion about electric waves but that he was instead concentrating solely on witnessing propagation itself.

4. Hertz first referred to the "induction wave" and the "air wave" on December 27. See Doncel 1991, p. 22.

5. Had Hertz been asked at the time, he would certainly have agreed that experimental success requires that the direct action must present the same succession of phases past a given point as does the wire wave since otherwise his device would simply not respond at all.

6. Either the strong force still wins but is much weakened in effect by acting at the gap, or else the weak force wins, but since it is less powerful than the strong force, it cannot produce as much sparking under the same conditions.

7. Hertz's quantity Π is a scalar because of the dipole symmetry. However, his analysis can be generalized (and was in subsequent years) to transform the quantity into a vector, in which case the E and H fields can be represented in terms of it as follows:

$$E = \nabla \times (\nabla \times \Pi) \quad \text{and} \quad H = \partial(\nabla \times \Pi)/\partial t$$

Hertz's auxiliary R is just $\nabla \times \Pi$, and Q is $(1/2\pi)\!\int\! E \cdot dS$, integrating over a surface. See Doncel and Roqué 1990, vol. 2, appendix 4, for an illuminating generalization of Hertz's analysis.

8. This can easily be seen in the following way. Hertz's E field is $Re_\rho + Ze_z$, where e_ρ and e_z are unit vectors. Further, ∇Q is just $\partial Q/\partial \rho e_\rho + \partial Q/\partial z e_z$. It follows at once from the defining conditions for Q that $E \cdot \nabla Q$ vanishes, implying that E lies in the level surfaces of Q.

9. Simply replace r in Hertz's expression for Π with its equivalent, $\sqrt{(\rho^2 + z^2)}$, and differentiate, noting that $\sin\theta$ is ρ/r.

10. "I based my first interpretation of these experiments upon the older views, seeking partly to explain the phenomena as resulting from the co-operation of electrostatic and electromagnetic forces. To Maxwell's theory in its pure development such a distinction is foreign. Hence I now wish to show that the phenomena can be explained in terms of Maxwell's theory without introducing this distinction" (Hertz 1889a, p. 137).

11. The tangent of the inclination is just $2\pi/\lambda$, and the faster wave has the greater wavelength.

12. Hertz no doubt felt entirely confident in this assumption, because the resonator does not spark when its plane is orthogonal to the oscillator.

13. In the words of the discovery paper, "the retardation of phase proceeds more rapidly in the neighbourhood of the origin than at a distance from it" (Hertz 1888e, p. 118).

14. In the introduction to *Electric Waves* Hertz remarked that an "interpretation of the experiments [from the *Gesamtkraft*] point of view could certainly not be incorrect, but it might perhaps be unnecessarily complicated" (Hertz [1893] 1962, p. 15). Because the introduction was going to be read by Helmholtz, Hertz would certainly have taken care not to imply that the master's scheme was in any sense "incorrect" particularly since the *Gesamtkraft* nicely fitted the interference experiments. The question, as he posed it here, was to see whether the "special limiting case of Helmholtz's theory," namely, the Maxwell equations, in which the distinct identity of the forces altogether vanishes, can still fit the experiments.

15. Which raises an interesting, but entirely hypothetical, question. Suppose that Hertz's experiments had in fact revealed what his Maxwell theory later required, namely, that the interference changes character once below about 3 m. Would he have still concluded, on the basis of *Gesamtkraft* reasoning, that the air wave has a finite speed? He probably would have, because an infinite speed requires two character changes, not one, the second becoming clearly apparent by 3 m (which is just what had taken place in his first series of experiments). The single change would have increased the speed he computed for the air wave, but it would also have puzzled him, because the speed calculated from these observations would then have been considerably higher than the speed he calculated from observations made farther out. And here he would not have been able to rely on an infinitely rapid electrostatic force, so he would have been squarely faced with a powerful

anomaly. One might say that Hertz was lucky his discovery experiments did not nicely fit the demands of Maxwell's equations, else he might have felt seriously stymied.

16. The reduction of the forces to singularity at the axis reflects Hertz's definition of the thin conducting wire as a region that does not support electric processes. In one respect, then—in respect to his mathematical techniques—the wire was simply a line, and here the notion of a wave "in" it could have no meaning. In the second respect as well the wave could not be "in" the wire, because Hertz's definition of a "good conductor" prohibited anything at all from going on much below the surface. He later came to grips with this issue directly when he used very thin wires to fabricate a device that simulated a finitely thick wire in order to show that the relevant (high-frequency) processes took place at the object's surface and not within it.

17. For this to make sense the longitudinal wave should appear only in the wire and not also in air. In fact, unless one pays very careful attention to the implications of Helmholtz's equations, it does look as though longitudinal waves that do not require ether polarizability at all can be produced in wires if k is nonzero (see appendix 13.3).

Chapter 18

1. There is a difference, however, between the Hertz before and the Hertz after mid-1888, because after mid-1888 Hertz in fact produced the very foundations of radiation theory that are nowadays quite familiar.

2. Of course, to elucidate these effects I have quite deliberately shown how they emerge from a way of considering the electrodynamic potential—namely, as representing the system energy of directly interacting circuit elements—that modern theory forbids.

Appendix 1

1. Thomson's two differential equations (second order for the dielectric, first order for the conductors) can be solved in series using Bessel functions since the coaxial cable has cylindrical symmetry. The mathematics is intricate though entirely straightforward, and Thomson proceeds immediately to approximations. He finds that for very large wavelengths in relation to the diameter of the inner conductor the phase velocity depends upon the frequency, as does the damping coefficient. For example, if the conductors are copper and have diameters of 4 mm and 10 mm, then the phase velocity for waves with periods of a hundredth of a second is about 80,000 km/sec, and the wave will drop in amplitude to $1/\varepsilon$ of its initial value after 128 km. At much higher frequencies, where the wire's diameter now becomes considerably greater than the wavelength (though the latter must still, in Thomson's approximation, be vastly larger than optical lengths), the dispersion becomes very small, the wave's decay is practically governed by the conductivity of the outer conductor, and the currents in the wire are confined to a very narrow region near its surface (yielding the "skin effect").

2. Buried deep within a complex technical article written by the Cambridge mathematician and Maxwellian Horace Lamb in 1883 was the near-approach. Lamb decided to investigate, using Maxwell's equations, the behavior of a conducting sphere either left free after starting in disequilibrium (i.e., with a potential that varies from point to point over its surface) or else while forcibly driven. Through most of the article Lamb ignores radiation entirely by assuming that the speed of propagation is infinite in order to obtain results of "practical value." In section 12 he abandons the approximation and allows radiation, but then he at once introduces a second approximation that obliterates radiation resistance, so that in the end he deduces only that in a free sphere the initial "non-uniform electrification" will decay exponentially with a constant $-\rho\Theta/4\pi R^2$, where R is the sphere's radius, ρ is its resistivity, and Θ is a number of order π determined by zonal harmonics.

3. Note Lodge misses the fact that Lamb and Niven ignored precisely the radiative effects that damped the oscillations independently of resistance. This oversight is not surprising. Lodge was no mathematician, either by training or by inclination: he could certainly appreciate the overall

structure of the analysis, but he could not easily penetrate deeper to uncover its hidden assumptions.

4. Lodge remarked in the August 30 issue: "In most commercial circuits the loss by radiation is probably so small a fraction of the whole dissipation of energy as to be practically negligible; but one is, of course, not limited to the consideration of commercial circuits or to alternating machines as at present invented and used. It may be possible to devise some less direct method—some chemical method, perhaps—for supplying energy to an oscillating circuit, and so converting what would be a mere discharge or flash into a continuous source of radiation" (1888, p. 417). Lodge's intuition was correct about the rapidity of the damping on a cylinder, but not about the reason for it: the rapid damping reflects the fact that radiation is not sent from one part of the cylinder to another part of it, as occurs in wave guidance, but is let loose into the dielectric. J. J. Thomson developed a pictorial understanding of the difference between the processes (radiation and wave guidance) using tubes of force in his *Recent Researches*, which was written in the light of Hertz's experiments.

5. Aitken (1985, pp. 94–95) emphasizes Lodge's lack of interest in confirming Maxwellian electrodynamics through free-space propagation. This is, no doubt, correct: Lodge, like other Maxwellians, felt no very pressing need to support Maxwellian theory in this manner. Nevertheless, even if he had felt pressed to do so, he would not have known how to go about it since he remained trapped by the apparent opposition between the radiative "flash" of a cylinder and the persistence of wave guidance.

6. In 1884 FitzGerald had written J. J. Thomson concerning some "experimental test" that Thomson proposed, remarking: "I hope you will succeed in getting your experimental test of Maxwell's theory tried. The great difficulty is something to feel these rapidly alternating currents with. Would Langley's bolometer do? I was working at a receiver whose period of oscillation should be the same as that of the current and which would consequently 'resound' to the vibration and integrate the energy of a large number of vibrations" (Rayleigh 1969, p. 22). FitzGerald naturally had in mind the persistent oscillations in a nearly closed radiator, or waveguide, of just the kind that Lodge later deployed. Such things, he knew, do not radiate very much, and so one needed a very sensitive detector. He proposed two. One, the Langley bolometer, would respond to electric heating effects and so would not be affected by the fact that the electric field alternates in direction. The other foreshadows Lodge's resonating circuits, but not Hertz's detector, which did not have to accumulate individually small effects since the amount of radiation was quite large.

7. These developments are very nicely examined by Aitken (1985), who delineates the opposition between spark-switched radiators that, like Hertz's, vibrate at a frequency determined by their own capacitance and inductance and the desire for a precise "syntonic," or tuned, relation between the transmitter and the receiver in wireless signaling. As long as the radiator had to be kicked into natural oscillation, the two desiderata remained incompatible, though Lodge tried to argue otherwise (see esp. Aitken 1985, pp. 142–58).

Appendix 2

1. Helmholtz remarked in 1873c, p. 691, that the four end–end forces satisfy action–reaction because the potential is a function solely of the relative distance between ds and $d\sigma$. This can be extended to the net force on the element: the force along say x_s, where the coordinate applies to the center of ds, is given by the derivative of the potential with respect to it; since the potential is a function of the relative distance, the derivative with respect to the corresponding coordinate x_σ for the center of $d\sigma$ must be equal and opposite to the previous one, thereby guaranteeing the equality of action–reaction.

2. Specifically, it is $-A^2(ij/r^2) [2ds \cdot d\sigma - 3(r \cdot ds)(r \cdot d\sigma)]$.

3. Here the gradient is always taken with respect to r. It is essential to understand that the variation does not involve C directly but only the body in which it exists. What happens to C

during the displacement is governed by the constraint that the absolute current intensities within the body must remain constant.

4. That is, Helmholtz required the flux of current $C \cdot dS$ to be independent of direction and to be maintained constant during virtual displacement.

5. The result is that both the displacement and the conduction currents must appear in the equation. In other words, field theory avoids the term in the changing charge density by altering the definition of the current. Here, then, is a major difference between field theory and Helmholtzianism but it can only be examined in situations where the C are not reentrant, that is, in experiments with open circuits.

6. This is a particular instance of the general fact that in Helmholtz's electrodynamics all of the effects that result from motion are due to precisely those actions that are functions of current inhomogeneities and that are *completely lacking* from both field theory and from Weber's electrodynamics, as well as from the original Ampère law. In this sense, therefore, Helmholtzian endeffects are not at all marginal: from his point of view they are amply and essentially present in every dynamo or electric motor. Indeed, they are the reasons such things work at all. His laboratory goal in the 1870s was to seek out situations that uniquely revealed their working.

Appendix 3

1. They probably read it in Helmholtz 1874a, sec. 19, on which the analysis in this appendix is based. In order to concentrate upon the system energy, or what Helmholtz termed the "work equivalent of the electric currents," I have, however, drawn out assumptions Helmholtz left implicit. Winters 1985 provides an excellent analysis of Helmholtz's route to the general concept of a system energy, which he indubitably did *not* possess in 1847, and which he was still developing during the early 1880s.

2. As Hertz in effect noted in 1884, Helmholtz's method requires an exhaustive specification of the dissipating factors.

3. From a rigorous point of view this and indeed nearly all versions of the electromagnetic energy balance equations are incorrect because they usually omit material kinetic energy. This is not, perhaps, a significant oversight, because the material velocity, or the mass, is taken to be extremely small, so that the energy transformed into motion is always vanishingly small compared to the electromagnetic system energy. The essential point is the *assumption* that such a systemic energy exists. That assumption was absent, or at least completely undeveloped, in Helmholtz's 1847 argument linking the Ampère force to electromagnetic induction. See Winters 1985 for a careful, detailed analysis.

4. This itself assumes that the resistivity κ does not itself depend on v, since if it did, we could not separate the terms in this way. This is more obvious in a variational computation based on a virtual displacement, because there we would have had to take account of a change in the Joule heating due to $\delta\kappa$.

Appendix 4

1. By contrast, in our previous case, where the motional electromotive force did not exist and E_{ind} was due to *Gleitstellen*, this gradient vanished. The distinction between motional electromotive force (E_{mot}) and electromotive force due to *Gleitstellen* (E_{glt}) is critical but confusing. The difference between the two is that the former, but not the latter, has an independent physical existence in, but only in, the moving conductor, whereas the latter occurs only at the interface between air and the moving conductor. E_{mot} produces its effect by generating a temporary current in the moving conductor that terminates in a charge at its end which is maintained by E_{mot}. It behaves, on the one hand, like an electromotive force due to a changing current and, on the other, like an electrostatic force, which is why the gradient of the static potential must be equal and opposite to it. This is not to Helmholtz's liking.

E_{glt} produces its effect in a much more congenial manner (congenial, that is, to Helmholtz's physics). The air–metal interface behaves like *Gleitstellen*. Consequently, there, and only there, polarization arises, as will an associated conduction charge by static induction. At the other interface a similar, but opposite, effect takes place. But there can be no force in the region between the interfaces, because there is nothing to produce it. All that happens is that the *Gleitstelle* effect acts in the end to produce a conduction charge at the terminus of the stationary conductor, which is what also occurs when the *Gleitstelle* is metallic rather than aerial. But the effect depends in the aerial case on the susceptibility of the air, whereas in the metallic case it is independent of any property of the conductor other than its mere ability to carry a current.

Appendix 5

1. Rowland's experiment soon became widely known, despite this explicit disavowal of credit, as "Helmholtz's experiment." This forced Rowland rather angrily to write the *Philosophical Magazine* to insist on his priority in the matter. He also remarked: "Unfortunately for me, Helmholtz had already experimented on the subject with negative results; and I found, in travelling through Germany, that others had done the same. The idea occurred in nearly the same form to me eleven years ago; but as I recognized that the experiment would be an extremely delicate one, I did not attempt it until I could have every facility, which Helmholtz kindly gave me" (Rowland 1879b). The negative experiment that Rowland here refers to could only be Schiller's, because Helmholtz's rotating capacitor measures electrostatic and not electrodynamic force, and because Helmholtz nowhere refers to an earlier convection experiment. The Schiller experiment of course did give a negative result in the sense of no deflection, just as one of Rowland's did. This could, however, be interpreted as proving the electrodynamic action of convection (in a pure Helmholtzian theory), but it could also be interpreted in terms of polarization (in the modified Helmholtzian account). Rowland's recollection may have been a bit confused, or else he may have had in mind that assuming polarization currents do cancel the effects of the conductor's endpoints, then the absence of any deflection would apparently indicate that the convection currents streaming from the conductor's ends do not have electrodynamic effects. But in itself the experiment is inconclusive because it is also consistent with the absence of any polarization effects and the positive action of convection. But in this last case the convection would have to stream away at precisely the right speed to cancel the action of the endpoints, which seems unlikely indeed.

2. See Buchwald, 1985a, appendix 1, for how this arises in field theory. The analysis is in this case similar for Helmholtz's theory in the Maxwell limit, with the signal difference that the continuity equation must be extended to include polarization currents in Helmholtz's scheme, whereas it necessarily includes displacement currents in field theory.

3. Whether he realized this via the continuity equation can be questioned since it took the Maxwellians until the late 1880s to do so (Buchwald 1985a, appendix 1). I think it likely that the point would have been easier for Helmholtz than for them. The Maxwellians' attention was ever focused on what occurs in the field, so that they at first sought the effect of convection in the displacement currents engendered by it throughout space. Helmholtz tended to focus instead on the boundary between conductors and air, where, e.g., *Gleitstellen* arise, which naturally raises the question of how to consider the contiguous polarization currents in relation to the conductor's charge. This is yet another instance of the fact that Maxwellians generally ignore sources, whereas partisans of both Fechner–Weber and Helmholtzian physics concentrate on them, albeit in thoroughly different ways.

Appendix 6

1. Helmholtz's analysis employs the potential function to obtain the terms that appear in Ohm's law according to Fechner–Weber, but the result does not depend essentially upon the potential. The same result can be reached by using Fechner's hypothesis and the Weber force law *if* we are permitted to use differential equations in a Weberean context (which Zöllner for one refused to allow). The argument (A) presented in this appendix develops an empirically realizable situation,

unlike an otherwise similar critique that Helmholtz had developed and that *begins with* extremely small distances or high speeds between electric particles. In a related argument (B) Helmholtz criticized Fechner–Weber for the fact that it requires an apparent redefinition of kinetic energy in the conservation law. Argument B is subject to a considerable amount of doubt because (as Helmholtz was well aware by 1870) Fechner–Weber certainly does not fail *qua* conservation: it is just that the terms which enter the conservation equation behave in strange ways under certain extreme conditions. Weber himself replied to a form of A (albeit in a highly restricted fashion, recurring almost at once to atomistics while ignoring Helmholtz's laboratory examples). Carl Neumann directed his attention to B and criticized Helmholtz's restrictions on the form that energy conservation laws may take. Weber admitted that the instabilities which Helmholtz derived in A do exist analytically (given only the force law), but he insisted that they do not occur physically, because the mass associated with the electric current is extremely small, because the speeds of the particles never exceed the speed of light, or else because the particles can never come closer than a certain minimum distance (all of which take the argument directly into Weber's atomistics, which he was extensively developing during the 1870s). Helmholtz eventually became quite indignant and replied to the Webereans at length in 1872b and 1873b. See Archibald 1987, pp. 383–416, for details of argument B and for the examples of argument A that Weber directly addressed. Though Helmholtz gave the example discussed here, only Hertz directly used it.

Appendix 7
1. Case 1 derives from Ohm's law since i represents the tangential currents and the function f depends only on t.

Appendix 8
1. The transformation requires the assumption that $\nabla^2\lambda$ vanishes. Helmholtz obtained his general expression simply by calculating what changes in the vector potential occur at a given object as it moves with respect to another one. He calculated, in terms that later became common, the so-called convective derivative of the vector potential. This was all that he had to do because by *Helmholtzian* principles nothing else could possibly have any effect.

2. Here is what Hertz had to perceive, given homogeneity:

$$\nabla_r \cdot \int M'/r_d d^3r = \int M' \cdot \nabla_r (1/r_d) d^3r' = -\int M' \cdot \nabla_{r'}(1/r_d) d^3r' \text{ since } r_d = r - r'$$

$$\int M' \cdot \nabla_{r'}(1/r_d) d^3r' = \int \nabla_{r'} \cdot (M'/r_d) d^3r' - \int [(\nabla_{r'} \cdot M')/r_d] d^3r$$

The first term on the right in the last equation vanishes on using Green's theorem and integrating with a surface at infinity, and so the required equivalence does hold.

3. This is to because the earth's field has a very weak gradient.

4. And, to make an ahistorical point, Jochmann's term is the one that corresponds formally to the magnetic part of the Lorentz force.

Appendix 9
1. Since the rotational effects are transient, any experiment that seeks to bring them out must balance between increasing the duration of the transient state and keeping the charges involved sufficiently large. Metals satisfy this last condition but not the former. Good insulators satisfy the former condition but not the latter.

Appendix 10
1. The constants A and B are functions of the principal radii of curvature of the surfaces at the point of geometric contact since these are the reciprocals of $2A_{1,2}$ and $2B_{1,2}$. Specifically, $2(A + B)$ is equal to the sum of the reciprocals of the radii of curvature.

2. This is itself an approximation because it assumes that points which have initially the same

x, y coordinates will always do so. It is, however, an approximation of essentially the same order as representing the original surfaces by quartics.

3. Hertz had originally omitted *J* and set *i* to infinity, but Kirchhoff enforced greater rigor.

Appendix 11

1. Hertz does not justify the factor of 1/2 here, nor is it a simple matter to see what the factor should be. The force *P* presumably acts only very near the surface since it is by assumption the force exerted by the surface on the vapor. The force *p* comes into action as soon as the vapor leaves the surface. We must somehow imagine that the vapor is shot from the surface by *P* but is slowed down by *p*. After the action ceases—when the ejected vapor experiences *p* on either side—it has speed *u*. Hertz convinces himself that the *distance* over which a net force acts on the vapor "centre of mass" is actually *u*/2, at the end of which the "centre of mass" has speed *u*. Hertz may have reasoned that, at the surface proper, the speed is zero, whereas a distance *u* away from the surface the speed is *u*, so that the mean speed of the vapor mass that fills the space *u* is *u*/2, and consequently that after unit time the "centre of mass" of the vapor will be *u*/2 away from the surface.

2. Note the peculiar phrasing here: "they do not show *definitely*" (in the original, "*sie zeigen aber nicht mit Bestimmtheit*"). The experiments in fact show *nothing at all* with regard to such a limit.

Appendix 12

1. This, it was thought, speaks for a sequence of discrete discharges. Only the time-averaged current can be measured, so that, say, in a time *t* the measured current would be *nt*, where *n* is the number of discharges that take place during that period. The heat generated will be proportional to the number of such discharges as well, and so it will also be proportional to the measured current itself.

2. Since the rate at which an electromotive force *p* feeds energy into a current *i* must still be *pi*, the electromotive force remains constant.

Appendix 13

1. In so doing I will, however, ignore magnetic polarization because it merely adds symbolic confusion without conveying anything of importance for the situations that we are interested in, namely, wave propagation. However, since Hertz did later use the version of the equations that included magnetic polarization it is worthwhile making the following remarks.

Helmholtz (1870b, pp. 119–21) first obtained the vector potential U_M as $\int M \times \nabla (1/r_d)d^3r'$. He then introduced a secondary vector equal to $\int M/r_d d^3r'$ and asserted that U_M could be written $-\nabla \times \int M/r_d d^3r'$. This equivalence, however, holds only if $\nabla \times M$ vanishes. Since the term that Helmholtz retained can itself be transformed into a surface integral, this means that he had discarded volume contributions and retained only surface ones, though he may perhaps not have developed the equations far enough to see this.

2. But remember that Helmholtz generates the Ampère law from a primitive expression for the electrodynamic potential, whereas in field theory the law is given ab initio.

3. The procedure is simple. Operate with $\nabla \cdot$ on the equation for longitudinal waves and then make the replacement on the right-hand side. Drop a time derivative, operate with ∇^2, and make the replacement on the left-hand side. Drop the divergence (which is legitimate since here $\nabla \times$ vanishes) and obtain longitudinal waves for *P* as well as for *U*.

4. A general process of renormalization shows that Maxwell's equations (except with respect to certain sorts of effects, like Hall's, that fascinated British Maxwellians and that alter the field equations in a nonlinear fashion) result from Helmholtz's equations solely by making the ether's susceptibility infinite (Darrigol 1993, pp. 236–39). We shall, however, see in chapter 17 that Hertz, who evidently realized this fact sometime in 1888, nevertheless continued to think that longitudi-

nal wave forms based on a nonzero value for k were yet possible. Maxwell's equations are wrong or else these wave forms simply are not taken into account in Maxwell's equations and might exist independently of them in some fashion. This is because the Maxwell equations are not as a group equivalent to the Helmholtz equations except for infinite ether polarizability; the latter remain more general, incorporating an action that is absent from the former.

5. It is important to note that the differences between Maxwell and Helmholtz over the role of the static potential do not derive from whether or not it can be considered separately from the electromotive force due to induction. In both theories it can (and in Helmholtz's theory it must) be. Rather, the difference originates in the continuity equation since in the Helmholtzian system the propagation of static potential follows from this equation. Among many Maxwellians the static potential was thought not to propagate since the vector potential was usually taken to have no divergence.

6. This kind of propagation should not be conflated with the transmission of waves by means of wires, a point I will return to in the next section.

7. In PNCs all forms of waves also require the presence of $\delta U/\delta t$ in the fundamental expression for P. In NPCs the waves require the presence of $\nabla\phi_f$ in Ohm's law. We will, however, see in the next section that current waves in wires of the kind that had been analyzed by Kirchhoff fifteen years before Helmholtz's polarization theory appeared do *not* require this term in Ohm's law, and this reflects the fact that they are entirely different in kind from the longitudinal waves that we are considering here (though we shall also see how easy it is to conflate the two).

8. The most striking effect of this difference between Maxwell's and Helmholtz's continuity equations concerns their compatibility with the Maxwell and Helmholtz Ampère laws. It has often been asserted that the Ampère law in the form $\nabla \times H = C$, with C the conduction current, requires the conduction current to have no divergence and so to be incompatible with the production of charge by current. Something is therefore missing. This is certainly correct for this form of the Ampère law, but then it is hardly the only form possible, for Helmholtz's system without polarization requires it to be $\nabla \times H = -A\partial\nabla\phi/\partial t + 4\pi C$. Not only is this consistent with the Helmholtz continuity equation, it is implied by it. In Maxwell's system, one may say, the appropriate terms in a continuity equation are implied by whatever form the Ampère law has, whereas in Helmholtz's system the Ampère law derives from the combination of the continuity equation with the electrodynamic potential.

9. More precisely, in order to obtain Helmholtz's longitudinal waves—in NPCs or PNCs—the current that appears in the fundamental expression for the electrodynamic potential U must have nonvanishing divergence.

10. Here, e.g., one cannot formulate interaction energies between polarization and charge, on the one hand, and changing polarization or conduction currents, on the other hand. One can formulate an interaction potential between changing polarization and conduction current, but only at the considerable cost of specifying that a changing state can itself *be* a new state.

11. I have followed Whittaker in using quotation marks ("Kirchhoff's equation") (Whittaker 1973, vol. 1: 231) because Kirchhoff did not give the equation in this pure form. Rather he at once separated the variables into the product of two factors, one of which depends only on time, while the other depends only on distance (Kirchhoff 1857, p. 204). In this way he avoided having to deal directly with partial derivatives since the separation of variables produces two ordinary differential equations. This is not perhaps the simplest way to solve the equation, although Heaviside took essentially the same approach in 1876.

12. Three years before Kirchhoff, William Thomson derived the equation without the term that depends upon induction; that is, he found the diffusion equation for low-frequency disturbances in wires. He was well aware that he had neglected induction but considered that in the problem he was concerned with (heavily insulated cables), the capacitance of the system vastly outweighs any inductive effects. This had the unfortunate result in later years of turning into a practical dogma for telegraph engineers, who, until the 1890s, considered induction to be a pernicious factor in

signal transmission but one that could be ignored where the wires proper are concerned. Although reasonable for the low-frequency signals involved in telegraphy, Thomson's electrostatic approximation begins to fail markedly at telephonic frequencies, particularly if iron wires are used. Heaviside's early work, and his reputation, revolved about correcting wire transmission theories for the intricate effects of induction. See Jordan 1982 for a thorough history of the issue.

13. If the current is spread throughout the conductor, the E field within it and tangent to its surface cannot vanish. Since E_{tan} must be continuous across the boundary, it follows that such a field also exists in the surrounding dielectric. But if this is so, then, according to the Poynting theorem, there must be a flow of energy into the wire from outside. This can be accomplished only by radiation, so that the condition that permits us to use the telegrapher's equation—that we completely ignore radiation—is strictly speaking inconsistent with the necessary continuity of E_{tan}.

Appendix 14

1. The static force parallels the oscillator because Hertz placed the resonator's center along the baseline, which intersects the center of the oscillator's spark gap.

Appendix 16

1. The two best sources in recent years for this argument are the unpublished 1972 dissertation by Arthur Molella and a 1981 article by Norton Wise. I have extensively relied on both in this section. Wise and Molella do not, however, agree on all points, and the differences are important here. We will discuss them at the appropriate point.

2. Molella remarks in this vein: "In dynamic atomism, Fechner found a physical model reconciling aspects of *Naturphilosophie*, an early faith which he apparently never abandoned, with accepted concepts in physical science. Combining the point-atom with the dynamics of immaterial forces allowed him to avoid kinetic atomism and the material continuum, both equally alien to his fundamental natural philosophy. It is no surprise that some contemporary Idealists considered Fechner an ally in spite of his assaults on the Naturphilosophen. In his famous historical diatribe against materialist philosophy, the eminent Idealist historian Friedrich Lange saw Fechner's atoms as extensionless force-centers and proper vehicles for ridding science of its materialistic taint" (1972, p. 68).

3. Translated by Wise (1981, p. 285), who (p. 284) claims the murky, relational monadology of Johann Friedrich Herbart was the proximate source for Fechner's attempt to accommodate Idealist strivings. However, Herbart was only one of several Idealist philosophers who debated atomistics and who were discussed by Fechner. Moreover, Molella points out (1972, p. 81) that Fechner "carefully disavowed any debts to Herbartian atomism, a doctrine which he insisted contradicted physical atomism." Molella instead considers Rudolf Hermann Lotze, whom Fechner had taught at Leipzig, to be closest to Fechner's atomism. We need not concern ourselves here with the intimate details of Idealist metaphysics, however, since our aim is solely to point out its probable influence on Fechner and its prevalence among some Weberean physicists, particularly in the 1870s.

4. Wise (1981, pp. 276–83) discusses Weber's universal model in some detail.

5. To my knowledge there are only two paragraphs in all of Weber's work, published and unpublished, that explicitly invoke metaphysics. The first is from a letter to Fechner that Fechner published in his *Atomenlehre* (see Molella 1972, p. 136). The second is an originally unpublished remark included in his *Werke* (translated in Wise 1981, p. 283). Wise notes Weber's close collaboration and friendship with people who were concerned with contemporary metaphysics, but the documentary evidence is simply too weak to settle the point unequivocally. Wise's claim (1981, p. 283) that "Weber wrote with the traditions of German Idealism in mind" remains undemonstrated and perhaps undemonstrable. Nevertheless, there can be no doubt that Zöllner, who was closely

associated with Weber in the critical 1870s, certainly did have metaphysics on his mind and that he invested Weber's electrodynamics as well as his universal model with an Idealist aura.

6. Wise (1981, pp. 279–83) goes further to argue that Weber's ultimate scheme is "coherent" *only* on the supposition that, for him, "force is an [irreducible] relation of two atoms." Wise supports the claim by remarking that Weber's universal model posits that points of "like" kind interact with a different intensity from points of "unlike" kind, so that the "character of both" points is an essential factor. Yet one could say precisely the same thing of *any* force law that contains factors which depend upon both objects, such as the simple Coulomb force. In it, the force between + and − is opposite in direction to the force between + and +, or between − and −; the "character of both" interacting objects is, it seems, necessarily implicated. If the intensity is also different, then this merely adds a change in magnitude to one in direction. But French deployers of the scheme certainly did not invest points with Idealist connections. Wise may be correct to argue that Weber, and certainly Zöllner, did, but nothing in the technical structure even of Weber's universal model *requires* such a notion for "coherence"—unless the concept is, and has always been, necessary.

Wise additionally places much weight (1981, p. 279) on Weber's rather simple, and purely analytical, discovery that his electrodynamic force can be obtained from the gradient of a potential that contains only relative distance and speed, whereas the force law itself also includes relative acceleration. Wise argues that relative acceleration to Weber could have meant the irreducible involvement of a third object "through Newton's second law of motion," whereas relative distance and speed are purely bipartite. This may be so—Weber does consider the possibility—but it puts much weight on an analytical trick: the absence of acceleration from the potential function follows from the relation $\partial v/\partial t = (\partial v/\partial r)(\partial r/\partial t) = v(\partial v/\partial r)$. Relying on a remark Weber made in 1871, Wise further argues that energy, as symbolized in the acceleration-free potential function, had by then replaced force for him as a physical (and so, presumably, as a metaphysical) primitive. This may also be true, but taking potential instead of force as a primitive when there is only the one interaction between points seems hardly to make much difference. In Helmholtz's physics, by contrast, the difference is profound because it has far-ranging theoretical and empirical consequences.

7. Interactions always emerge through variation of the system energy, so that "force" appears merely as a term in an energetic process.

8. The full development of Helmholtzian electrodynamics later in the 1870s produced a well-defined system energy *density*. Such a thing specifies the energy for a pair of *volume elements* in given states and at a given distance from one another. Since the elements are differentials, they lack extension when considered in themselves, so that a pair of them determines a unique distance. However, unlike Weber's points, Helmholtz's volume elements can interact in a myriad of different ways.

9. Thus, vital forces, forces that go beyond the mechanical, are inadmissible because they would in the end have effects that violate the purely mechanical conservation principle.

10. Wise (1981, p. 296) makes a rather different point from this one, remarking that "Helmholtz had continually to move back and forth between concepts of force as moving force and as *vis viva* and to reconcile the two ideas." It was not, I think, this aspect of the problem per se that stimulated him but rather the apparent failure of even an invariable force (in Helmholtz's causal sense) to have something about it that remained forever constant. In other words, the conundrum derived from the coupling of the ambiguous, powerful image he held of force as a causal principle with the "assumption," as he called it, of force conservation. The latter, it has been argued, was itself an essential aspect of his rejection of vital actions, as Lenoir (1982, p. 209) discusses, because it guaranteed that the various processes of "force conversion" that were so widely deployed in Germany during this period would not lead to actions that surpassed what was physically available at any one moment. Lenoir's analysis has, however, recently been vigorously challenged by Caneva

(1990), who rejects almost its entire structure, including Lenoir's emphasis on the antiteleological impetus behind Helmholtz's work.

11. Although a civilized controversy with Clausius forced him to refine his conception of conservation, and he also extended and applied the concept of a potential to the distribution of currents, with the goal in mind of producing results that would be useful in Du Bois-Reymond's physiological investigations. See Archibald 1987, pp. 280–300.

12. Wise (1981, p. 299) quotes Helmholtz's remark that "his law" (Wise's words; referring to the electrodynamic potential that Helmholtz introduced in 1870 as the first expression of his new physics) was "no elementary expression of the ultimate acting forces," from which Wise apparently concludes that Helmholtz remained committed to finding a reduction to distance forces. This, I believe, misrepresents his meaning, in part because Wise translates Helmholtz's *"letzden wirkenden Kräfte"* as "ultimate acting forces" instead of as "final acting forces," with "ultimate" carrying metaphysical weight. Helmholtz's following remarks clarify his meaning: "These expressions for the potential of each pair of current elements are obviously not elementary expressions of the final acting forces; they lead, namely, to at least two forces for each current element considered as a rigid body, or to a force and a force pair. Induction phenomena are only indirectly derived from the electrodynamic potential through the mediation of the law of conservation of energy" (1872b, pp. 637–38). He is asserting only that the actual, measurable forces which come into play to produce induction must be obtained indirectly, by deduction from the potential law via energy conservation. Since the potential law does not give them directly, *it* is not an "elementary expression of the final acting forces," to wit, of the force and force pair it implies for a rigid element. Here *"letzden"* carries a mathematical sense as the end of analysis, not a metaphysical sense as an invocation of the "ultimate forces" in nature, which no longer exist for Helmholtz.

13. Smith and Wise (1989) erect an elaborate structure to demonstrate that Thomson's energy-based dynamics and his geometrical kinematics directly reflect wide-ranging trends in British mathematics, science, and society of the 1840s and 1850s, as well (they claim) as a particular insistence on practice. I do not think that their argument, in its furthest reaches, is ultimately persuasive. Nevertheless, they have certainly shown that the British insistence on developing physical effects through transmitted actions in macroscopic theory is probably related to a complex of other developments that are to be found only in Britain—so that Helmholtzianism, a uniquely German design, was not at all likely to have been founded on the kind of macroscopic structures insisted upon by the British. For a critique of Smith and Wise, see Buchwald 1990.

15. For the details and substance of Zöllner's career I have relied heavily on Molella's (1972) account, which discusses Zöllner's increasingly angry beliefs.

16. German society as a whole became more overtly anti-Semitic after about 1873, when economic collapse produced a twenty-year depression during which, to quote one recent account, the "evils of capitalism and Jewish banking interests soon became one in the minds of many Germans" (Rowe 1986, p. 430). Rowe also notes the increasing presence of hitherto repressed Jews at the universities. Helmholtz's biographer, the mathematician Leo Koenigsberger, was one of the few Jews to penetrate the system during the 1860s.

17. Although I will not here pursue Helmholtz's own work beyond his electrodynamics in the 1870s, the new, energy-based physics that he began to develop permeated almost every area that he subsequently touched, albeit in different ways. For example, in optics he produced in 1875 an account of anomalous dispersion based on a set of twin equations, one for matter, the other for ether. In doing so he relied primarily on the principle of action and reaction to determine the ether–matter connection, and he did not provide any microphysical model at all. Instead, he almost immediately generated differential equations (see Buchwald 1985a, chap. 27, for details). This procedure reflects his new physics by recurring immediately to a differential, energy-based relation (action–reaction) that completely bypasses hypothetical forces between particles. And then he rapidly developed the consequences for the laboratory.

Even, perhaps *especially*, in his electrochemistry Helmholtz avoided atomistics, remarking in 1881: "We shall try to imitate Faraday as well as we can by keeping carefully within the domain of phenomena, and, therefore, need not speculate about the real nature of that which we call a quantity of positive or negative electricity" (1881a, p. 60). Instead of detailed atomic structures and forces, Helmholtz employed generalized ions and molecules, which he treated as though they were small bodies, and he founded his analysis on the energy principle.

Appendix 17

1. Helmholtz always referred to *real* bodies, so that for him perfectly rigid objects cannot exist, in which case Bertrand's objection fails. The torque must be compensated in each differential element by the mechanical reaction to the element's elastic deformation, or as Helmholtz explicitly remarked, Bertrand's reduction "is only admissible for piece *Ds*, which is to be considered as an absolutely rigid body, whereas the application involved here is tied to the supposition that the parts of the conductor are on the contrary flexible" (Helmholtz 1874a, p. 720; see also Helmholtz 1873c, pp. 699–700). This interpretation requires Helmholtz's torque to have a marked effect if the circuit contains plastic regions that can react to it. Helmholtz himself did not look for the torque, but in 1879 Hertz designed an experiment that, though not aimed at producing evidence for the torque per se, nevertheless relied upon its existence to obtain other effects. See chap. 6.

2. By the end of the 1870s some Webereans at least had realized that Helmholtz's electrodynamics requires its objects to be deformable, and they were consequently able to follow, and even to re-create, a technical analysis using it. But, at the same time, they did not understand the most fundamental characteristic of Helmholtz's physics: its assumption that the system energy is an irreducible expression of object relations. One particularly interesting example of this is Eduard Riecke, who took over Weber's laboratory after Weber retired. Riecke was a convinced Weberean, but by 1880 he had grasped how Helmholtz's potential works as a tool. In an article that contains nothing at all that Helmholtz had not himself already pointed out (an article that consequently represents one Weberean's attempt to review the two theories publicly), Riecke remarked: "Even if the potential law finds itself in agreement with the experimental facts, one should nevertheless set out upon the broader challenge of dissolving the complicated total action, whose expression it shapes, into the true fundamental forces exerted between the particles engaged in the galvanic current. Only through such a tracing back would the formal connection that the potential law produces between the different kinds of electrodynamic actions be replaced by an inner connection grounded in the nature of electric particles and in the different states of motion in which they are found. *The problem still to solve for the potential law is to discover a fundamental law for electric action–reaction from which the potential can be derived in the same way as the Ampère law is from Weber's fundamental laws*" (1880, p. 297; emphasis added). The last sentence asks for precisely what Helmholtzianism thoroughly rejected. Yet Riecke remained completely unaware of this fundamental presumption. He was too thoroughly embedded in Weberean reductionism to see that the quarrel did not concern the form of the atomic force law but whether there should be *any* such law at all.

Appendix 18

1. Mossotti himself conceived of each molecule as surrounded by an electric ether whose density changes from one side of the molecule to the other: "In the hypothesis we proposed concerning the constitution of bodies, within which, following Franklin's assumption of a single kind of ether, each molecule is surrounded by an ethereal atmosphere, polarization consists in the condensation of the atmosphere on one side and rarefaction on the other, giving rise in the condensed part to a repulsion of external fluid and permitting in the rarefied part the disclosure of the molecule's attraction for that same fluid" (Mossotti 1846, p. 51).

2. The assumption appears in the table of interactions in chapter 2 as es = emo.

3. Lest this issue seem too abstract and divorced from the actual practice of contemporary

physics, I hasten to add that Hertz himself considered precisely this problem in 1881 and in exactly the way I describe it here. See chap. 7.2.

4. Interesting results follow from this line of thought. A changing current can, we have now argued, move a charged body. If we assume that any action that can move a charged body can also generate current (though nothing in Helmholtz's electrodynamics compels us to do so), a change in this last action will affect the original current's intensity, which will have a cognate action on the charged body, and so on ad infinitum. Hertz reached this conclusion at some point in 1884 and extracted a radical moral from it.

5. An electric (electrostatic) force has no curl because it can be expressed as the gradient of a potential. Such a thing cannot generate reentrant currents, because its integral around a closed path vanishes. It can, however, produce currents between two distinct points on a conductor and thereby alter the charge density on it. Since the currents it generates cannot be closed, they must have nonzero divergence, and so from the continuity equation they result in changing charge densities.

6. However, it is possible, given a variant of the Mossotti hypothesis that was discussed by Boltzmann (see chap. 13.2), for changing polarization not to exert electrodynamic force. If instead of assuming the microscopic objects to be conducting spheres, we take them to be orientable electric dipoles, then changing polarization reduces to a motion of the dipoles, and this need not exert electrodynamic force unless charge convection does so—and that is itself a contentious issue in Helmholtz's system. Here we might have a situation in which an electromotive force could move a charged body, but the moving charge could not itself generate either an electrodynamic or an electromotive force—even though such an interaction energy, as we saw above, does imply that the charge's electric force can act to generate non-reentrant currents.

7. That is, in order for electromotive forces to polarize dielectrics under all circumstances, they must not be composed of microscopic conductors, as we have just seen. But then a dielectric must in some manner already be polarized at the microscopic level, and an electromotive force must act to move the charged objects that compose it.

8. Though by 1879 even Helmholtz required electrodynamically active polarization in order to deal with the results of experiments that had been performed in his Berlin laboratory. Both field theory and Fechner–Weber were bound to the requirement that electrodynamic induction must polarize dielectrics because electrostatic action does.

9. In principle, this interaction energy is the reason that propagation can occur at all in a polarizable medium.

10. In Hertz's earliest recorded words (written in 1889) on this point: "The results of the experiments on rapid electric oscillations which I have carried out appear to me to confer upon Maxwell's theory a position of superiority to all others. Nevertheless, I based my first interpretation of these experiments upon the older views, seeking partly to explain the phenomena as resulting from the cooperation of electrostatic and electromagnetic forces. To Maxwell's theory in its pure development such a distinction is foreign. Hence I now wish to show that the phenomena can be explained in terms of Maxwell's theory without introducing this distinction" ([1893] 1962, p. 137). Only Helmholtz's electrodynamics maintains a distinction between "electrostatic" and "electromagnetic" ("electromotive" in the sense I have been using the term) forces, a distinction based on the differences between the sources giving rise to the forces and represented analytically by the several interaction energies.

BIBLIOGRAPHY

Archival Sources

Hertz MSS, Science Museum, London, including:
244 Theorie des Magnetismus. 1878.
245 Nachweis elektr. Wirkung in Dielectricitat. 1879.
246 Obere Grenze für die kinetische Energie des bewegten Elektricität. 1881.
247 Über eine die elektrische Entladung begleitende Erscheinung. 1883.
248 Über die verdunstung der Flüssigkeiten. 1882.
249 Über das Gleichgewicht schwimmender elastichen Platten. 1884.
250 Über die Berührung fester elastichen Körper und über die Harte. 1882.
251 Graphische Methode zur Bestimmung der adiabatischen Zustandsänderungen leuchten Luft. 1884.
252 Über einen einfluss des ultraviolotten Lichtes auf die elektrische Entladung. 1887.
253 Über sehr schnelle Elektr. Schwingungen. 1887.
254 Über elektrodynamische Wellen in Luftraum und deren Reflexion. 1888.
255 Über die Einwirkung einer geradlinigen Schwingung auf eine benachbarte Strombahn. 1888.

Hertz *Versuchsprotokolle.* Courtesy of Prof. Dr. H. G. Hertz, Karlsruhe, and Prof. M. Doncel, Barcelona.
Letters to and from Heinrich Hertz. Deutsches Museum, Munich.

Other Sources

Abraham, H., and Langevin, P.
1905 *Ions, électrons, corpuscules: Mémoires réunis et publiés.* 2 vols. Paris: Gauthiers-Villars.
Aitken, H. G. J.
1985 *Syntony and spark: The origins of radio.* Princeton: Princeton University Press.
Albisetti, J. C.
1983 *Secondary school reform in Imperial Germany.* Princeton: Princeton University Press.
Archibald, T.
1987 "Eine sinnreiche Hypothese": Aspects of action-at-a-distance electromagnetic theory, 1820–1880. Ph.D. diss., University of Toronto.
1988 Tension and potential from Ohm to Kirchhoff. *Centaurus,* 31:141–63.

1989   Physics as a constraint on mathematical research: The case of potential theory and electrodynamics. In *The history of modern mathematics*, ed. D. E. Rowe and J. McCleary, vol. 2:29–75. New York: Academic Press.

Aulinger, E.
1886   Über das Verhaltniss der Weber'schen Theorie der Elektrodynamik zu dem von Hertz aufgestellten Princip der Einheit der electrischen Kräfte. *Ann. Phys. Chem.*, 27:119–32.

Bertheau, F. R.
1912   *Chronologie zur Geschichte der geistigen Bildung und des Unterrichtswesens in Hamburg von 1831 bis 1912.* Hamburg: Lucas Gräfe.

Bertrand, J.
1871   Note sur la théorie mathématique de l'electricité dynamique. *Comptes Rendus*, 73:965–70.

Bierhalter, G.
1981   Zu Hermann von Helmholtzens merchanischer Grundelegung der Wärmelehre aus dem Jahre 1884. *Arch. Hist. Ex. Sci.*, 25:71–84.

Blackmore, J. T.
1972   *Ernst Mach: His work, life, and influence.* Berkeley and Los Angeles: University of California Press.

Boltzmann, Ludwig
1873   Experimentaluntersuchung über die elektrostatische Fernwirkung dielektrischer Körper. *Wien. Ber.*, 68:81–155. In *Wiss. Abh.*, vol. 1:472–536.

1874a   Experimentelle Bestimmung der Dielektrizitätskonstante einiger Gase. *Wien. Ber.*, 69:795–813; *Ann. Phys. Chem.*, 155:403. In *Wiss. Abh.*, vol. 1:536–55.

1874b   Über einige an meinen Versuchen über die elektrostatische Fernwirkung dielektrischer Körper anzubringende Korrektionen. *Wien. Ber.*, 70:307–41. In *Wiss. Abh.*, vol. 1:556–86.

1874c   Über die Verschiedenheit der Dielektrizitätskonstante des kristallisierten Schwefels nach verschiedenen Richtungen. *Wien. Ber.*, 70:342–66. In *Wiss. Abh.*, vol. 1:587–606.

1886   Bemerkung zu dem Aufsatze des Hrn. Lorberg über einen Gegenstand der Elektrodynamik. *Ann. Phys. Chem.*, 29:598–603.

1887   Einige kleine Nachträge und Berichtungen. *Ann. Phys. Chem.*, 31:139–40.

1889   Über Deformationsströme; insbesondere die Frage, ob dieselben aus magnetischen Eigenschaften erklärbar sind. *Ann. Phys. Chem.*, 37:107–27.

1890   Bemerkung über Deformationsströme. *Ann. Phys. Chem.*, 39:159–60.

1893   On the methods of theoretical physics. *Proc. Phys. Soc. London*, 12:336–45.

1968   *Wissenschaftliche Abhandlungen.* 3 vols. New York: Chelsea Publishing Co.

1982   *Vorlesungen über Maxwells Theorie der Elektricität und des Lichtes.* Introduction by Walter Kaiser. Reprint of the 1891–93 original. Wiesbaden: Akademische Druck.

Broda, Engelbert
19??   *Ludwig Boltzmann: Mensch Physiker Philosoph.* Berlin: Deutscher Verlag der Wissenschaften.

Bryant, John H.
1988   *Heinrich Hertz: The beginning of microwaves.* New York: Institute of Electrical and Electronics Engineers.
Buchheim, G.
1968   *Zur Geschichte der Elektrodynamik: Briefe Ludwig Boltzmanns an Hermann von Helmholtz. NTM,* 5:1125–31.
1971   Hermann von Helmholtz und die klassische Elektrodynamik. *NTM,* 8:23–36.
Buchwald, Jed Z.
1985a   *From Maxwell to microphysics.* Chicago: University of Chicago Press.
1985b   Oliver Heaviside, Maxwell's apostle and Maxwellian apostate. *Centaurus,* 28:288–330.
1988   The Michelson experiment in the light of electromagnetic theory before 1900. In *The Michelson era in American science, 1870–1930,* ed. S. Goldberg and R. H. Stuewer. AIP Conference Proceedings, vol. 179:55–70.
1989   *The rise of the wave theory of light.* Chicago: University of Chicago Press.
1990   The background to Heinrich Hertz's experiments in electrodynamics. In *Nature, experiment, and the sciences,* ed. T. H. Levere and W. Shea, pp. 275–306. Dordrecht: Kluwer Academic Publishers.
1992a   Helmholtzianism in context: Object states, laboratory practice, and anti-Idealism. Forthcoming.
1992b   Scientific kinds and the wave theory of light. *Stud. Hist. Phil. Sci.,* 23:39–74.
1993   Design for experimenting. In *World Changes: Thomas Kuhn and the nature of science,* ed. Paul Horwich, pp. 169–206. Cambridge: MIT Press.
Cahan, D.
1985   The institutional revolution in German physics, 1865–1914. *Hist. Stud. Phys. Sci.,* 15:1–65.
1989a   *An institute for an empire: The Physikalish-Technische Reichsanstalt, 1871–1918.* Cambridge: Cambridge University Press.
1989b   Instruments, institutes, and scientific innovation: Friedrich Kohlrausch's route to a law of electrolytic conductivity. *Osiris,* 5.
1990   From dust figures to the kinetic theory of gases: August Kundt and the changing nature of experimental physics in the 1860s and 1870s. *Ann. Sci.,* 47:151–72.
Caneva, K.
1978   From galvanism to electrodynamics: The transformation of German physics and its social context. *Hist. Stud. Phys. Sci.,* 9:63–159.
1990   Review of Lenoir 1982. *Ann. Sci.,* 47:291–300.
Cantor, G. N., Hodge, M. J. S. (eds.)
1981   *Conceptions of ether.* Cambridge: Cambridge University Press.
Cazenobe, J.
1980   Comment Hertz a-t-il eu l'idée des ondes hertzienns? *Revue de Synthèse,* 101:345–82.
1983   *La vie et l'obstacle: Études et documents sur la "préhistoire de l'onde hertzienne."* Cahiers d'histoire et de philosophie des sciences, vol. 5. Paris: Centre de Documentation Sciences Humaines, Sociéte Française d'Histoire des Sciences et des Techniques.

468   BIBLIOGRAPHY

Collins, H. M.
1992   *Changing Order.* Chicago: University of Chicago Press.
D'Agostino, S.
1971   Hertz and Helmholtz on electromagnetic waves. *Scientia,* 106:637–48.
1974   Hertz's researches on electromagnetic waves. *Hist. Stud. Phys. Sci.,* 6:260–323.
1986   Scienza e cultura nella Germania dell'800: Le basi filosofiche della fisica-matematica nell'opera di Helmholtz. *Cultura e Scuola,* 25:265–82.
Darrigol, O.
1993   The electrodynamic revolution in Germany as documented by early German expositions of "Maxwell's theory." *Arch. Hist. Ex. Sci.,* 45:189–280.
Doncel, M.
1991   On the process of Hertz's conversion to Hertzian waves. *Arch. Hist. Ex. Sci.,* 43:1–27.
Doncel, M., and Hertz, H. G.
1994   Heinrich Hertz's laboratory notes of 1887. *Arch. Hist. Ex. Sci.* Forthcoming.
Doncel, M., and Roqué, X.
1990   *Heinrich Hertz: Las ondas electromagneticas.* 2 vols. Bellaterra: Universidad Autónoma de Barcelona.
Drude, P.
1894a   *Physik des Aethers auf elektromagnetische Grundlage.* Stuttgart: E. Enke.
1894b   Zum Studium des elektrischen Resonators. *Ann. Phys. Chem.,* 53:721–68.
Elkana, Y.
1974   *The Discovery of the conservation of energy.* Cambridge: Harvard University Press.
von Ettinghausen, A.
1880   Bestimmung der absoluten Geschwindigkeit fliessender Elektrizität aus dem Hall'schen Phänomen. *Sitzungsber. Akad. Wiss. (Wien),* 81:441–52.
von Ettinghausen, A., and Nernst, W.
1886   Über das Hall'sche Phänomen. *Sitzungsber. Akad. Wiss. (Wien),* 94:560–610.
Falconer, I.
1985   Theory and experiment in J. J. Thomson's work on gaseous discharge. Ph.D. diss., Bath University.
Fechner, G. I.
1855   *Über die physikalische und philosophische Atomenlehre.* Leipzig: H. Mendelssohn.
Feffer, Stuart M.
1989   Arthur Schuster, J. J. Thomson, and the discovery of the electron. *Hist. Stud. Phys. Sci.,* 20:1–61.
FitzGerald, George
1896   Hertz's miscellaneous papers. In *Writings,* pp. 433–42.
1902   *The scientific writings of the late George Francis FitzGerald.* Ed. J. Larmor. Dublin: Hodges, Figgis, and Co.
Friedman, R. M.
1989   *Appropriating the weather: Vilhelm Bjerknes and the construction of a modern meteorology.* Ithaca: Cornell University Press.

Galison, P.
1987   *How experiments end.* Chicago: University of Chicago Press.
Giere, R.
1988   *Explaining science: a cognitive approach.* Chicago: University of Chicago Press.
Goldstein, E.
1880   On the electric discharge in rarefied gases, part I. *Phil. Mag.*, 10:173–90.
1881a   Ueber elektrische Lichterscheinungen in Gasen. *Ann. Phys. Chem.*, 12:90–109.
1881b   Ueber die Entladung der Electrizität in verdünnten Gasen. *Ann. Phys. Chem.*, 12:249–79.
Gutting, G. (ed.)
1980   *Paradigms and revolutions.* Notre Dame: University of Notre Dame Press.
Harman, P. M.
1982   *Metaphysics and natural philosophy.* Sussex: Harvester Press.
Havas, P.
1966   A note on Hertz's "derivation" of Maxwell's equations. *Amer. J. Phys.*, 34:667–69.
Heaviside, O.
1950   *Electromagnetic theory.* 3 vols. in 1. Reprinted from the first editions of 1893, 1899, and 1912. New York: Dover Publications.
1970   *Electrical papers.* 2 vols. Reprinted from the first edition of 1892. New York: Chelsea.
Heilbron, J. L.
1964   A history of the problem of atomic structure from the discovery of the electron to the beginning of quantum mechanics. Ph.D. diss., University of California, Berkeley.
Helmholtz, Hermann von
1847   Über die Erhaltung der Kraft. *Sitzungsber. Phys. Gesell. Berlin,* 1:12–75. Translated in 1853 as "On the conservation of force: A physical memoir" by John Tyndall, in J. Tyndall and W. Francis, *Scientific memoirs, natural philosophy.* London: Taylor and Francis.
1851   Über die Dauer und den Verlauf der durch Stromesschwankungen inducirten elektrischen Ströme. *Ann. Phys. Chem.*, 1:505–40. In *Wiss. Abh.,* vol. 1:429–62.
1853   Über einige Gesetze der Vertheilung elektrischer Ströme in körperlichen Leitern mit Anwendung auf die thierisch-elektrischen Versuche. *Ann. Phys. Chem.*, 89:211–33 and 353–77. In *Wiss. Abh.,* vol. 1:475–519.
1861   Über eine allgemeine Transformationsmethode der Probleme über elektrische Vertheilung. *Verh. d. Naturhist.-Med. Ver. z. Heidel.,* 2:185–88 and 217. In *Wiss. Abh.,* vol. 1:520–25.
1869   Über elektrische Oscillationen. *Verh d. Naturhist.-Med. Ver. z. Heidel.,* 5:27–31. In *Wiss. Abh.,* vol. 1:531–36.
1870a   The axioms of geometry. *The Academy,* 1:128–31.
1870b   Über die Bewegungsgleichungen der Elektrizität für ruhende leitende Körper. *J. Reine und Angwndte. Math.,* 72:57–129. In *Wiss. Abh.,* vol. 1:545–628.

1870c   Über die Gesetze der inconstanten elektrischen Ströme in körperlichen ausgedehnten Leitern. *Verh. d. Naturhist.-Med. Ver. z. Heidel.*, 5:84–89. In *Wiss. Abh.*, vol. 1:537–44.

1871   Über die Fortpflanzungsgeschwindigleit der elektrodynamischen Wirkungen. *Monatsber. d. Berl. Akad.*, pp. 292–98. In *Wiss. Abh.*, vol. 1:629–35.

1872a   The axioms of geometry. *The Academy*, 3:52–53.

1872b   Über die Theorie der Elektrodynamik. *Monatsber. d. Berl. Akad.*, pp. 247–56. In *Wiss. Abh.*, vol. 1:636–46.

1873a   *Popular lectures on scientific subjects.* Trans. E. Atkinson. With an introduction by Professor Tyndall. New York: D. Appleton and Co.

1873b   Über die Theorie der Elektrodynamik. *Borchardt's J.f.d. Reine und Angwndte. Math.*, 75:35–66. In *Wiss. Abh.*, vol. 1:647:87.

1873c   Vergleich des Ampère'schen und Neumann'schen Gesetzes für die elektrodynamischen Kräfte. *Monatsber. d. Berl. Akad.*, pp. 91–104. In *Wiss. Abh.*, vol. 1:688–701.

1874a   Die elektrodynamischen Kräfte in bewegten Leitern. *Borchardt's J.f.d. Reine und Angwndte. Math.*, 78:273–324. In *Wiss. Abh.*, vol. 1: 702–62.

1874b   Helmholtz on the use and abuse of the deductive method in physical science. Trans. Crum Brown. *Nature*, 11:149–51 and 211–12.

1874c   Kritisches zur Elektrodynamik. *Ann. Phys. Chem.*, 93:545–56. In *Wiss. Abh.*, vol. 1:763–73.

1875   Versuche über die im ungeschlossenen Kreise durch Bewegung inducirten elektromotorischen Kräfte. *Ann. Phys. Chem.*, 98:87–105. In *Wiss. Abh.*, vol. 1:774–90.

1876a   Bericht betreffend Versuche über die elektromagnetische Wirkung elektrischer Convection. *Ann. Phys. Chem.*, 98:487–93. In *Wiss. Abh.*, vol. 1:791–97.

1876b   The origin and meaning of geometrical axioms. *Mind*, 1:301–21.

1878   The origin and meaning of geometrical axioms (II). *Mind*, 3:212–25.

1881a   On the modern development of Faraday's conception of electricity: The Faraday Lecture. *J. Chem. Soc.*, 39:227–304. In *Wiss. Abh.*, vol. 3:52–87.

1881b   *Popular lectures on scientific subjects.* 2d ser. Trans. E. Atkinson. New York: D. Appleton and Co.

1881c   Über die auf das Innere magnetisch oder dielektrisch polarisirter Körper wirkenden Kräfte. *Ann. Phys. Chem.*, 13:385–406. In *Wiss. Abh.*, vol. 1:798–820.

1882   *Wissenschaftliche Abhandlungen.* 3 vols. Leipzig: J. A. Barth.

1884   Principien des Statik monocyklischer Systeme. *Jl. Reine und Angwndte. Math.*, 97:111–40 and 317–36.

1889   Wohlvorschlag für Heinrich Hertz. In Kirsten and Körber 1975, 114–15.

Hermann, D. B.

1982   *Karl Friedrich Zöllner.* Biographien hervorragender Naturwissenschaftler, Techniker, und Mediziner, vol. 57. Leipzig: B. G. Teubner.

Hertz, H. G.

1988a   Festvortrag: Die Entdeckung von Heinrich Hertz—Persönlicher und historischer Hintergrund. In *Festakt zum Heinrich-Hertz-Jubiläum "100 Jahre Ra-*

diowellen" und Rechenschaftsbericht, 1986–1987, ed. Rektorat der Universität Karlsruhe, pp. 23–55. Karlsruhe: University of Karlsruhe.

1988b   Heinrich Hertz: Persönlicher und historischer Hintergründe der Entdeckung. *Fridericiana, Zeitschrift der Universität Karlsruhe,* 41:3–37.

Hertz, H.

1880a   *Ann. Phys. Chem.,* 10:414–48. Translated as "Experiments to determine an upper limit to the kinetic energy of an electric current," in *Papers,* pp. 1–34.

1880b   Inaugural diss. In *Werke,* vol. 1:37–134. Translated as "On induction in rotating spheres," in *Papers,* pp. 35–126.

1881a   *Ann. Phys. Chem.,* 13:266–75. In *Werke,* vol. 1:135–44. Translated as "On the distribution of electricity over the surface of moving conductors," in *Papers,* pp. 127–45.

1881b   *Ann. Phys. Chem.,* 14:581–90. In *Werke,* vol. 1:145–54. Translated as "Upper limit for the kinetic energy of electricity in motion," in *Papers,* pp. 137–45.

1881c   *J. Reine und Angwndte. Math.,* 92:156–71. Translated as "On the contact of elastic solids," in *Papers,* pp. 146–62.

1882a   *Verh. d. Vereins z. Bef. d. Gewerb.* Translated as "On the contact of rigid elastic solids and on hardness," in *Papers,* pp. 163–83.

1882b   *Verh. d. Vereins z. Bef. d. Gewerb.* Translated as "On a new hygrometer," in *Papers,* pp. 184–85.

1882c   *Verh. d. Vereins z. Bef. d. Gewerb.* Translated as "On the evaporation of liquids, and especially of mercury, in vacuo," in *Papers,* pp. 186–99.

1882d   *Ann. Phys. Chem.,* 17:193–200. Translated as "On the pressure of saturated mercury-vapour," in *Papers,* pp. 200–206.

1883a   Dynamometrische Vorrichtung von geringem Widerstande und verschwindender Selbstinduktion. *Zeit. Instrumentkunde,* 3:17–19. In *Werke,* vol. 1: 227–32.

1883b   *Ann. Phys. Chem.,* 19:78–86. In *Werke,* vol. 1:233–41. Translated as "A phenomenon which accompanies the electric discharge," in *Papers,* pp. 216–23.

1883c   *Ann. Phys. Chem.,* 19:782–816. In *Werke,* vol. 1:242–76. Translated as "Experiments on the cathode discharge," in *Papers,* pp. 224–54.

1883d   Über das Verhalten des Benzins als Isolator und als Rückstandsbildner. *Ann. Phys. Chem.,* 20:279–84. In *Werke,* vol. 1:277–82.

1884   *Ann. Phys. Chem.,* 23:84–103. In *Werke,* vol. 1:295–314. Translated as "On the relations between Maxwell's fundamental electromagnetic equations and the fundamental equations of the opposing electromagnetics," in *Papers,* pp. 273–90.

1885   *Ann. Phys. Chem.,* 24:114–18. In *Werke,* vol. 1:315–19. Translated as "On the dimensions of a magnetic pole in different systems of units," in *Papers,* pp. 291–95.

1887a   On very rapid electric oscillations. In *Electric waves,* pp. 29–53.

1887b   *Sitzungsber. d. Kön. Preuss. Akad. d. Wissen.,* pp. 487–90. Translated as "On an effect of ultra-violet light upon the electric discharge," in *Electric waves,* pp. 63–79.

1887c   Über Inductionserscheinungen hervorgerufen durch die elektrischen Vorgänge in Isolatoren. *Sitzungsber. d. Kön. Preuss. Akad. d. Wissen.,* pp. 885–96.

1888a   Über die Ausbreitungsgeschwindigkeit der elektrodynamischen Wirkungen. *Sitzungsber. d. Kön. Preuss. Akad. d. Wissen.*, pp. 197–210.

1888b   Über Strahlen elektrischer Kraft. *Sitzungsber. d. Kön. Preuss. Akad. d. Wissen.*, pp. 1297–1307.

1888c   On the action of a rectilinear electric oscillation upon a neighbouring circuit. In *Electric Waves*, pp. 80–94.

1888d   On electromagnetic effects produced by electrical disturbances in insulators. In *Electric Waves*, pp. 95–106.

1888e   On the finite velocity of propagation of electromagnetic actions. In *Electric Waves*, pp. 107–23.

1888f   On electromagnetic waves in air and their reflection. In *Electric Waves*, pp. 124–36.

1888g   On electric radiation. In *Electric Waves*, pp. 172–85.

1889a   The forces of electric oscillations, treated according to Maxwell's theory. In *Electric Waves*, pp. 137–59.

1889b   On the propagation of electric waves by means of wires. In *Electric Waves*, pp. 160–71.

1890a   On the fundamental equations of electromagnetics for bodies at rest. In *Electric Waves*, pp. 195–240.

1890b   On the fundamental equations of electromagnetics for bodies in motion. In *Electric Waves*, pp. 241–68.

1891   On the mechanical action of electric waves in wires. In *Electric Waves*, pp. 241–68.

1895   *Gesammelte Werke.* 3 vols. Leipzig: J. A. Barth.

1896   *Miscellaneous papers.* Trans. D. E. Jones and G. A. Schott. Leipzig: J. A. Barth.

1892   *Untersuchungen über die Ausbreitung der elektrischen Kraft.* Leipzig: J. A. Barth.

1962   *Electric Waves.* Trans. D. E. Jones. Reprint of the 1893 edition. New York: Dover Publications.

Hertz, J.
1977   *Heinrich Hertz: Memoirs, letters, diaries.* 2d enl. ed., prepared by M. Hertz and C. Susskind. Trans. L. Brinner, M. Hertz, and C. Susskind. San Francisco: San Francisco Press.

Herwig, H.
1874   Über eine Modification des elektromagnetischen Dreversuches. *Ann. Phys. Chem.*, 153:262–67.

Hiebert, E.
1970   The genesis of Mach's early views on atomism. *Boston Studies in the Philosophy of Science*, 6:79–106.

1980   Boltzmann's conception of theory construction: The promotion of pluralism, provisionalism, and pragmatic realism. In *Pisa Conference Proceedings*, ed. J. Hintikka, D. Gruender, and E. Agazzi, vol. 2:175–98.

Holborn, H.
1964   *A history of modern Germany, 1648–1840.* New York: Alfred A. Knopf.

Hon, G.
1987   H. Hertz: "The electrostatic and electromagnetic properties of the cathode rays are either *nil* or very feeble" (1883), a case-study of an experimental error. *Stud. Hist. Phil. Soc.,* 18:267–82.

Hörz, H.
1982   Helmholtz und Boltzmann. In *Ludwig Boltzmann: Ausgewählte Abhandlungen, Internationale Tagung anlässlich des 75. Jahrestages seines Todes,* ed. R. Sexl and J. Blackmore, pp. 191–205. Braunschweig: Friedr. Vieweg and Sohn.

Hunt, B. J.
1983   "Practice vs. theory": The British electrical debate, 1888–1891. *Isis,* 74:341–55.
1991   *The Maxwellians.* Ithaca: Cornell University Press.

Indorato, L., and Masotto, G.
1989   Poincaré's role in the Crémieu-Pender controversy over electric convection. *Ann. Sci.,* 46:117–63.

Jochmann, E.
1864   On the electric currents induced by a magnet in a rotating conductor. *Phil. Mag.,* 27:506–28 and 28:347–49. Published first in 1863 in the *Annalen der Physik* and in the *Journal für die Reine und Angewandte Mathematik.*

Jordan, D. W.
1982   D. E. Hughes, self-induction, and the skin-effect. *Centaurus,* 26:123–53.

Jungnickel, C., and McCormmach, R.
1986   *Intellectual mastery of nature.* 2 vols. Chicago: University of Chicago Press.

Kelter, E.
1912   *Hamburg und sein Johanneum im Wandel der Jarhhunderte 1529–1912.* Hamburg: Lütcke and Wulff.

Kirchhoff, Gustav
1857   Über die Bewegung der Elektrizität in Drähten. *Ann. Phys. Chem.,* 176:193–216.

Kirsten, C., and Körber, H.-G.
1975   *Physiker über Physiker.* 2 vols. Berlin: Akademie-Verlag.

Klein, M.
1972   Mechanical explanation at the end of the nineteenth century. *Centaurus,* 17:58–81.
1974   Boltzmann, monocycles, and mechanical explanation. In *Philosophical foundations of science,* ed. R. J. Seeger and R. S. Cohen, pp. 155–75. Boston: Reidel.
1985   *Paul Ehrenfest: The making of a theoretical physicist.* 3rd ed. Amsterdam: North-Holland.

Koenigsberger, L.
1896   *Hermann von Helmholtz's Untersuchungen über die Grundlagen der Mathematik und Mechanik.* Leipzig: B. G. Teubner.
1965   *Hermann von Helmholtz.* Trans. F. A. Welby. Reprint of the original English translation published in 1906. New York: Dover Publications.

Kuhn, T. S.
1970   *The structure of scientific revolutions.* Chicago: University of Chicago Press.
1978   *Black-body theory and the quantum discontinuity, 1894–1912.* New York: Oxford University Press.
1982   Commensurability, comparability, communicability. *Philosophy of Science Association,* vol. 2.

Kurylo, F., and Susskind, C.
1981   *Ferdinand Braun: A life of the Nobel prizewinner and inventor of the cathode-ray oscilloscope.* Cambridge: MIT Press.

Lakatos, I., and Musgrave, A.
1970   *Criticism and the growth of knowledge.* Cambridge: Cambridge University Press.

Lemmerich, J.
1988   Die Hertzsche Entdeckung im Briefswechsel zwischen Hermann von Helmholtz, Emil du Bois-Reymond und Karl Runge. *Physikalische Blätter,* 44:218–20.

Lenoir, T.
1982   *The strategy of life: Teleology and mechanics in nineteenth century German biology.* Dordrecht: Reidel.

Lodge, O.
1887a   Modern views of electricity, part I. *Nature,* pp. 532–36 and 558–61.
1887b   Modern views of electricity, part II. *Nature,* pp. 582–84.
1888–89 Modern views of electricity, part IV. *Nature,* pp. 388–93, 416–19, 590–92, 10–13, and 319–22.
1894   *The work of Hertz and some of his successors.* London: "The Electrician" Printing and Publishing Co.

Lorberg, H.
1886   Bemerkung zu zwei Aufsätzen von Hertz und Aulinger über einen Gegenstand der Elektrodynamik. *Ann. Phys. Chem.,* 29:666–72.
1887   Erwiderung auf die Bemerkungen des Herrn Boltzmann zu meiner Kritik zweier Aufsätze von Hertz und Aulinger. *Ann. Phys. Chem.,* 31:131–37.

Love, A. E. H.
1944   *A Treatise on the mathematical theory of elasticity.* Reprint of the 4th ed., 1927. New York: Dover Publications.

McKendrick, J. G.
1899   *Hermann Ludwig Ferdinand von Helmholtz.* London: T. Fisher Unwin.

Molella, A. P.
1972   Philosophy and nineteenth-century electrodynamics: The problem of atomic action at a distance. Ph.D. diss., Cornell University.

Mossotti, O. F.
1846   Discussione analitica sull'influenza che l'azione di un mezzo dielettrico ha sulla distribuzione dell'elettricità alla superfice di più corpi elettrici disseminati in esso. *Mem. di mat. e di fis. della Soc. Ital. delle Sci.,* 24:49–74.

Mulligan, J.
1987   The influence of Hermann von Helmholtz on Heinrich Hertz's contribution to physics. *Amer. J. Phys.,* 55:711–19.

1989a   Heinrich Hertz and the development of physics. *Physics Today,* 42:50–57.
1989b   Hermann von Helmholtz and his students. *Amer. J. Phys.,* 57:68–74.
Neumann, Franz
1845   Die mathematischen Gesetze der inducirten elektrischen Ströme. Reprinted
         in 1889 in Ostwald's *Klassiker der Exakten Wissenchaften,* vol. 10. Leipzig:
         Wilhelm Engelmann.
O'Hara, J. G., and Pricha, W.
1987   *Hertz and the Maxwellians.* London: Peter Pergrinus, in association with the
         Science Museum.
Pauli, W.
1973   *Pauli lectures on physics.* Vol. 1: *Electrodynamics.* Cambridge: MIT Press.
Perrin, Jean
1895   Nouvelles propriétés des rayons cathodiques. *Comptes Rendus,* 121:1130. In
         Oeuvres, pp. 3–5.
1950   *Oeuvres scientifiques.* Paris: CNRS.
Planck, M.
1949   *Scientific autobiography and other papers.* Trans. F. Gaynor. London: Wil-
         liams and Norgate.
Poincaré, H.
1890   *Électricité et optique: Les théories de Maxwell et la théorie électromag-
         nétique de la lumière.* 2 vols. in 1. Paris: Deslis Frères.
1891a   Sur le calcul de période des excitateurs Hertziennes. *Arch. des Sci. Phys. Nat.,*
         25:5–25. In *Oeuvres,* pp. 6–19.
1891b   Sur la résonance multiple des oscillations Hertziennes. *Arch. Des Sci. Phys.
         Nat.,* 25:609–27. In *Oeuvres,* pp. 20–32.
1891c   Sur la théorie des oscillations Hertziennes. *Comptes Rendus,* 113:515–19. In
         Oeuvres, pp. 33–37.
1894   Les oscillations électriques. Paris: Gauthiers-Villars.
1954   *Oeuvres.* Vol. 10: *Physique mathématique.* Paris: Gauthier-Villars.
Price, D. de Solla
1983   Sealing wax and string: A philosophy of the experimenter's craft and its role
         in the genesis of high technology. AAAS Sarton lecture. Unpublished type-
         script.
Pupin, M. I.
1926   *From imigrant to inventor.* New York: Scribner's.
Pyenson, L.
1983   *Neohumanism and the persistence of pure mathematics in Wilhelmian Ger-
         many.* Philadelphia: American Philosophical Society.
Ramsauer, C.
1953   *Grundversuche der Physik in historischer Darstellung.* Berlin: Springer-
         Verlag.
Rayleigh, Lord (J. W. Strutt)
1886a   The reaction upon the driving-point of a system executing forced harmonic
         oscillations of various periods, with applications to electricity. In *Works,* vol.
         2, sec. 134.
1886b   On the self-induction and resistance of straight conductors. In *Works,* vol. 2,
         sec. 135.

1906   On the production of vibrations by forces of relatively long duration, with application to the theory of collisions. In *Works*, vol. 5, sec. 314.

1917   Cutting and chipping of glass. In *Works*, vol. 6, sec. 417.

1964   *Collected works*. 6 vols. in 3. Reprinted from the first edition of 1899–1920. New York: Dover Publications.

1969   *The life of Sir J. J. Thomson*. Reprinted from the first edition of 1942. London: Dawson's.

Richards, J. L.

1980   The art and the science of British algebra: A study in the perception of mathematical truth. *Historia Mathematica*, 7:343–65.

1988   *Mathematical visions: The pursuit of geometry in Victorian England*. Boston: Academic Press.

Riecke, E.

1880   Über die electrischen Elementargesetze. *Ann. Phys. Chem.*, 11:278–315.

Riemann, B.

1880   *Schwere, Elektrizität und Magnetismus*. Hannover: Carl Rümpler.

Rompe, R., and Treder, H. J.

1984   Zur Grundlegung der theoretischen Physik: Beiträge von H. von Helmholtz und H. Hertz. Berlin: Akademie-Verlag.

Röntgen, W. C.

1885   Versuche über die elektromagnetische Wirkung der diëlektrischen Polarisation. *Sitzungsber. d. Kön. Preuss. Akad. d. Wissen.*, pp. 195–98.

1888   Über die durch Bewegung eines im homogenen elektrischen Felde befindlichen Dielektricums hervorgerufene elektrodynamische Kraft. *Sitzungsber. d. Kön. Preuss. Akad. d. Wissen.*, pp. 23–28.

Rowe, D. E.

1986   "Jewish mathematics" at Göttingen in the era of Felix Klein. *Isis*, 77:422–49.

Rowland, H. A.

1879a   On the magnetic effect of electric convection. *Amer. J. Sci.*, 15:30–38. In *Physical papers*, pp. 128–37.

1879b   Note on the magnetic effect of electric convection. *Phil. Mag.*, 7:442–43. In *Physical papers*, p. 138.

1889   On the electromagnetic effect of convection-currents. *Phil. Mag.*, 27:445–60. In *Physical papers*, pp. 251–65. Co-authored with C. T. Hutchinson.

1902   *Physical papers*. Baltimore: Johns Hopkins University Press.

de la Rue, W., and Müller, H. W.

1879   Experimental researches on the electric discharge with the chloride of silver battery, part II. *Phil. Trans.*, 169:155–233.

1883   Experimental researches on the electric discharge with the chloride of silver battery, part IV. *Roy. Soc. Proc.*, 35:292.

Schreier, W.

1973   Über einige Beziehungen zwischen den Arbeiten von Helmholtz und Maxwells Aufbau der elektromagnetischen Feldtheorie. *Wissenschaftliche Zeitschrift der Humboldt-Universität zu Berlin, Mathematisch-Naturwissenschaftliche Reihe*, 23:345–47.

Schiller, N.
1874   Einige experimentelle Untersuchungen über elektrische Schewingungen. *Ann. Phys. Chem.,* 152:535–65.

Schuster, A.
1911   *The progress of physics during 33 years (1875–1908).* Cambridge: Cambridge University Press.

Smith, C., and Wise, M. N.
1989   *Energy and empire: A biographical study of Lord Kelvin.* Cambridge: Cambridge University Press.

Struik, D. J.
1970   Bertrand, Joseph Louis François. In *DSB,* vol. 2:87–89.

Swoboda, W. W.
1982   Physics, physiology and psychophysics: The origins of Ernst Mach's empirio-criticism. *Rivista di Filosofia,* 73:234–74.

Thomson, G. P.
1966   *J. J. Thomson: Discoverer of the electron.* New York: Doubleday.

Thomson, J. J.
1885   Report on electrical theories. *Brit. Assoc. Reports,* pp. 97–155.
1888   *Applications of dynamics to physics and chemistry.* London: Macmillan and Co. Reprinted in 1968 by Dawson's of Pall Mall, London.
1893   *Notes on recent researches in electricity and magnetism.* Oxford: Clarendon Press, Reprinted in 1968 by Dawson's of Pall Mall, London.
1937   *Recollections and reflections.* New York: Macmillan and Co.

Thomson, W. (Lord Kelvin)
1904   *Baltimore lectures on molecular dynamics and the wave theory of light.* London: C. J. Clay and Sons.

Timoshenko, S. P.
1983   *History of strength of materials.* New York: Dover Publications.

Todhunter, I.
1960   *A history of the theory of elasticity.* 3 vols. New York: Dover Publications. Originally published in 1886 (vol. 1) and 1893 (vol. 2, pts. 1 and 2).

Treder, H.-J.
1973   Helmholtz' Elektrodynamik und die Beziehungen von Kinematik, Dynamik und Feldtheorie. *Wissenschaftliche Zeitschrift der Humboldt-Universität zu Berlin, Mathematisch-Naturwissenschaftliche Reihe,* 23:327–30.

Weber, R.
1902   *Problems in electricity.* London: E. and F. N. Spon.

Weber, W.
1893   *Werke.* 6 vols. Berlin: Springer. (Vols. 3 and 4 cover galvanism and electro-dynamics.)

Whittaker, Sir Edmund
1973   *A history of the theories of aether and electricity.* 2 vols. New York: Humanities Press.

Wiedemann, E.
1880   On the thermic and optic behaviour of gases under the influence of the electric discharge. *Phil. Mag.,* 10:419.

Wiedemann, G. H.
1882–85 *Die Lehre von der Elektrizität.* 5 vols. Braunschweig: Friedrich Vieweg und Sohn.

Wien, W.
1930   *Aus dem Leben einer Physikers.* Ed. K. Wien. Leipzig: J. A. Barth.

Winters, Stephen M.
1985   Hermann von Helmholtz's discovery of force conservation. Ph.D. diss., Johns Hopkins University.

Wise, M. N.
1981   German concepts of force, energy, and the electromagnetic ether: 1845–1880. In Cantor and Hodge 1981, pp. 269–307.
1988   Mediating machines. *Science in Context,* 2:77–113.

Woodruff, A. E.
1962   Action at a distance in nineteenth century electrodynamics. *Isis,* 53:439–59.
1968   The contributions of Hermann von Helmholtz to electrodynamics. *Isis,* 59:300–311.

Yavetz, Ido
1988   From obscurity to enigma: The work of Oliver Heaviside, 1872–1889. PhD. diss., Tel Aviv University.

Zatskis, H.
1965   Hertz's derivation of Maxwell's equations. *Amer. J. Phys.,* 33:898–904.

Zehnder, L.
1938   *Persönliche Erinnerungen an Röntgen und über die Entwicklung der Röntgenstrahlen.* Basel: Raschen.

Zenneck, J.
1946   Zum 90. Geburtstag von Heinrich Hertz. *Die Naturwissenschaften,* 8:225–30.

Zöllner, F.
1872   *Über die Natur der Cometen.* Leipzig: Engelmann.
1874   Über einen elektrodynamischen Versuch. *Ann. Phys. Chem.,* 153:138–43.
1876a   *Principien einer elektrodynamischen Theorie der Materie.* Leipzig: Engelmann.
1876b   Zur Widerlegung der elementaren Potentialgesetzes von Helmholtz durch elektrodynamische Versuche mit geschlossenen Strömen. *Ann. Phys. Chem.,* 158:106–20.
1877   Über die Einwendungen von Clausius gegen das Weber'schen Gesetz. *Ann. Phys. Chem.,* 160:514–37.

INDEX

Printed and bound by CPI Group (UK) Ltd, Croydon, CR0 4YY

27/10/2024

14580401-0004